T0343342

ASCARIS: THE NEGLECTED PARASITE

ASCARIS: THE NEGLECTED PARASITE

Edited by

CELIA HOLLAND
Trinity College, University of Dublin, Dublin

AMSTERDAM • BOSTON • HEIDELBERG • LONDON
NEW YORK • OXFORD • PARIS • SAN DIEGO
SAN FRANCISCO • SINGAPORE • SYDNEY • TOKYO

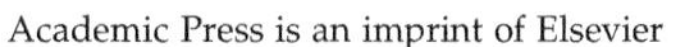
Academic Press is an imprint of Elsevier

Academic Press is an imprint of Elsevier
32 Jamestown Road, London NW1 7BY, UK
225 Wyman Street, Waltham, MA 02451, USA
525 B Street, Suite 1800, San Diego, CA 92101-4495, USA

Copyright © 2013 Elsevier Inc. All rights reserved.

No part of this publication may be reproduced, stored in a retrieval system or transmitted in any form or by any means electronic, mechanical, photocopying, recording or otherwise without the prior written permission of the publisher.

Permissions may be sought directly from Elsevier's Science & Technology Rights Department in Oxford, UK: phone (+ 44) (0) 1865 843830; fax (+ 44) (0) 1865 853333; email: permissions@elsevier.com. Alternatively, visit the Science and Technology Books website at www.elsevierdirect.com/rights for further information.

Notice

No responsibility is assumed by the publisher for any injury and/or damage to persons or property as a matter of products liability, negligence or otherwise, or from any use or operation of any methods, products, instructions or ideas contained in the material herein. Because of rapid advances in the medical sciences, in particular, independent verification of diagnoses and drug dosages should be made.

British Library Cataloguing-in-Publication Data
A catalogue record for this book is available from the British Library

Library of Congress Cataloging-in-Publication Data
A catalog record for this book is available from the Library of Congress

ISBN: 978-0-12-396978-1

For information on all Academic Press publications
visit our website at www.store.elsevier.com

Typeset by TNQ Books and Journals

Dedication

This book is dedicated to the memory of Huw Vaughan Smith whose dedication to parasitology and generous spirit will never be forgotten.

Contents

3. *Ascaris* – Antigens, Allergens, Immunogenetics, Protein Structures

MALCOLM W. KENNEDY

4. Implications of *Ascaris* Co-infection

FRANCISCA ABANYIE AND TRACEY J. LAMB

II
MODEL SYSTEMS

5. Larval Ascariasis: Impact, Significance, and Model Organisms

CELIA V. HOLLAND, JERZY M. BEHNKE, AND CHRISTINA DOLD

IV

HOST AND PARASITE GENETICS

12. Genetics of Human Host Susceptibility to Ascariasis

SARAH WILLIAMS-BLANGERO, MONA H. FENSTAD, SATISH KUMAR, AND JOHN BLANGERO

V
CLINICAL ASPECTS AND PUBLIC HEALTH

13. *Ascaris lumbricoides* and Ascariasis: Estimating Numbers Infected and Burden of Disease

SIMON J. BROOKER AND RACHEL L. PULLAN

14. Impact of *Ascaris suum* in Livestock

STIG MILAN THAMSBORG, PETER NEJSUM, AND HELENA MEJER

Some figures in this book are available online in color at:
http://booksite.elsevier.com/9780123884251

Foreword

I spend considerable time educating students and health professionals about the neglected tropical diseases (the NTDs), frequently referring to them as the most common infections of the world's poor and also as "the most important diseases you have never heard of." In many respects, human ascariasis is the ultimate NTD, both because of its devastating impact on child health and the general lack of awareness of its true global importance. When I present helminth infections and NTDs to new audiences I am quick to point out that ascariasis may not only represent the most common NTD but quite possibly the most prevalent infection of humankind. Many estimates indicate that between 800 million and 1.2 billion people are infected with *Ascaris lumbricoides* at any given time, with a disproportionate impact on the growth and development of children.

Despite the fact that *Ascaris* infections are so ubiquitous we actually know very little about the pathogenesis of this disease, its true public health impact, and how we can best control it in resource poor settings.

For instance, we know that ascariasis disproportionately affects children because on average children harbor substantially a greater number of worms than adults, yet no one really knows the biological basis of this observation. Ascariasis is the most common infection of the world's children whose families live on less than $1–2 per day, but no one really knows why children are the most affected. Similarly, why is it that a child harboring a large number of ascariasis worms experiences stunted growth? Could ascariasis, perhaps together with hookworm infection, represent the world's most common endocrinopathy? And what about the intellectual and cognitive effects resulting from *Ascaris* worms in the gut? Although it is still considered controversial by some investigators, there are suggestions that ascariasis is responsible for a significant burden of the world's childhood neuropsychiatric disturbances. Intuitively we might suspect this observation has some nutritional underpinnings, but the truth is we really have no clue.

I want to congratulate Dr. Celia Holland and her colleagues for bringing together the world's experts in ascariasis for this important new volume. The fact that she has produced a new book specifically devoted to ascariasis could further help to focus global attention on this important disease.

In addition to addressing some of the key scientific issues highlighted above, this new book will also provide important information on the molecular biology and pathogenesis of ascariasis. For several decades molecular and cellular studies conducted on the free-living nematode *Caenorhabditis elegans* have led to a revolution in our basic understanding of the life processes for multicellular organisms. Such efforts have led to the awarding of multiple Nobel Prizes. We now need to begin applying these path-breaking discoveries to unravel the complexities of *Ascaris* biology and its adverse effects on human and animal health, and this volume will summarize much of what we know to date.

Finally, *Ascaris – The Neglected Parasite* will further contribute to our understanding of global efforts to control or even eliminate ascariasis through mass drug administration. It will provide insights on whether mass treatments with existing benzimidazoles will be adequate to effect global elimination – as some would suggest – or whether we require new and improved biotechnologies, as will be required for hookworm and other helminth elimination strategies.

This multi-authored effort represents a timely and important contribution to the biomedical literature. It will be of great use for undergraduates, graduate students, and postgraduate students in the health and biological sciences. It is also one which I hope might serve as a standard for other books that may follow.

Peter Hotez MD, PhD, FAAP, FASTMH
Dean, National School of Tropical Medicine,
Baylor College of Medicine
Texas Children's Hospital Endowed Chair in
Tropical Pediatrics
President and Director, Sabin Vaccine Institute &
Texas Children's Center for Vaccine Development
Baker Institute Fellow in Disease and Poverty,
Rice University

Contributors

Francisca Abanyie Emory University School of Medicine, Emory Children's Center, Atlanta, GA, USA

Roy M. Anderson Imperial College London, London, UK

María-Gloria Basáñez Imperial College London, London, UK

Jerzy M. Behnke University of Nottingham, Nottingham, UK

Martha Betson The Royal Veterinary College, London, UK

John Blangero Texas Biomedical Research Institute, San Antonio, TX, USA

Simon J. Brooker London School of Hygiene and Tropical Medicine, London, UK

Luis Caraballo University of Cartagena, Colombia

Philip J. Cooper Liverpool School of Tropical Medicine and Hygiene, Liverpool, UK and Pontificia Universidad Católica del Ecuador, Quito, Ecuador

Charles D. Criscione Texas A&M University, College Station, TX, USA

Christina Dold University of Oxford, Oxford, UK

Mona H. Fenstad Norwegian University of Science and Technology, Trondheim, Norway

Camila A. Figuieredo Universidade Federal da Bahia, Salvador, Brazil

Albis Francesco Gabrielli World Health Organization, Geneva, Switzerland

Robin B. Gasser The University of Melbourne, Parkville, Victoria, Australia

Peter Geldhof Ghent University, Belgium

Andrew Hall University of Westminster, London, UK

Ross S. Hall The University of Melbourne, Parkville, Victoria, Australia

Celia V. Holland Trinity College, Dublin, Ireland

T. Déirdre Hollingsworth University of Warwick, Coventry, UK and Liverpool School of Tropical Medicine, Liverpool, UK

Aaron R. Jex The University of Melbourne, Parkville, Victoria, Australia

Malcolm W. Kennedy University of Glasgow, Glasgow, Scotland, UK

Satish Kumar Texas Biomedical Research Institute, San Antonio, TX, USA

Tracey J. Lamb Emory University School of Medicine, Emory Children's Center, Atlanta, GA, USA

Bo Li BGI-Shenzhen, Shenzhen, PR China

Yingrui Li BGI-Shenzhen, Shenzhen, PR China

Shiping Liu BGI-Shenzhen, Shenzhen, PR China

Aaron G. Maule Queen's University Belfast, Belfast, UK

Helena Mejer University of Copenhagen, Denmark

Antonio Montresor World Health Organization, Geneva, Switzerland

Peter Nejsum University of Copenhagen, Denmark

Rachel L. Pullan London School of Hygiene and Tropical Medicine, London, UK

Lorenzo Savioli World Health Organization, Geneva, Switzerland

Paul W. Sternberg California Institute of Technology, Pasadena, CA, USA

J. Russell Stothard Liverpool School of Tropical Medicine, Liverpool, UK

Antony O.W. Stretton University of Wisconsin-Madison, Madison, WI, USA

Stig Milan Thamsborg University of Copenhagen, Denmark

James E. Truscott Imperial College London, London, UK

Johnny Vlaminck Ghent University, Belgium

Martin Walker Imperial College London, London, UK

Jun Wang BGI-Shenzhen, Shenzhen, PR China

Sarah Williams-Blangero, PhD Texas Biomedical Research Institute, San Antonio, TX, USA

Huanming Yang BGI-Shenzhen, Shenzhen, PR China

Neil D. Young The University of Melbourne, Parkville, Victoria, Australia

Introduction

Ascaris lumbricoides remains the most prevalent of the geohelminths or soil-transmitted nematodes. *Ascaris* is a remarkably infectious and persistent parasite and its large size, migratory pathway through the tissues, and allergenicity further enhance its public health significance. Chronic ascariasis is known to contribute to insidious morbidity, including growth retardation and effects on cognitive development in children. Larval migration causes both liver and lung inflammation and disease. Pulmonary symptoms, described as Loeffler's syndrome, range from debilitating to lethal, particularly in children. Acute ascariasis, associated with heavy burdens of the large adult worms can result in serious complications including intestinal obstruction, and, in extreme cases, death. *Ascaris suum*, the ascarid of pigs, is also very prevalent and its contribution to morbidity – with, for example, consequent liver condemnation – has significant economic impact.

Why, then, is *Ascaris* so neglected? Some of the reasons are common among the geohelminths, such as chronicity of morbidity and the challenge of parasite control in regions dominated by inadequate sanitation and poverty. However, the lack of a rodent animal model, in which the parasite completes its life-cycle in a manner similar to humans, is an undoubted disadvantage, as is the declining number of investigators that make *Ascaris* their focus. The *A. suum* pig model is a very useful system for understanding human ascariasis, but does lack some of the versatility of defined inbred strains of rodents and their associated immunological reagents. The recent publication of the pig genome is likely to begin to address these lacunae.

One of the defining parameters that contribute to our understanding of *Ascaris* epidemiology is the intensity of infection or worm burden in individual hosts. Worm burden is the most important measurable epidemiological variable with respect to parasite transmission, population dynamics, and the degree of individual and community morbidity. Recent work, utilizing new and more sophisticated statistical approaches seeks to explain the contribution of infection intensity, aggregation, and predisposition to the observed epidemiology of ascariasis.

Our knowledge of the immunology of ascariasis lags behind that of other geohelminths and helminths in general. Immunity to *Ascaris* is characterized by humoral and cellular responses directed at the larval stages, with the most consistent observation being elevated-total and *Ascaris*-specific IgE antibody. Recent work has highlighted the complexity of the relationship between human cytokine responses and *Ascaris* infection, suggesting potential differential impacts of de-worming programs. Furthermore, *Ascaris* as an important allergen has been explored extensively, in an experimental context, in order to understand the allergic response in humans. More broadly speaking, the impact of *Ascaris* and other geohelminth infection on the host immune system is now known to have consequences for important microparasite infections such as malaria, HIV, and TB. This greatly enhances the public health significance of ascariasis, but to an unknown degree.

The practical implications of our understanding of the epidemiology of ascariasis and other geohelminth infections have led to improved strategies for control. Initially, the so-called selective approach was viewed as a potentially attractive option, whereby the most heavily infected individuals were treated. Subsequently, the World Health Organization placed the treatment of school-aged children at the heart of its commitment to the control of geohelminths, with an initial target of at least 75% of school-aged children by 2010. Estimates vary, but it is apparent that this ambitious target remains challenging. More recently, a focus on preschool children has highlighted relatively high levels of *Ascaris* infection in their relatively small bodies with consequent morbid effects. Despite the undoubted benefits of rolling out large-scale chemotherapeutic control efforts, the impact of such interventions – in terms of epidemiological pattern and process, the potential for anthelmintic resistance, parasite genetic heterogeneity and the impact on other infections – is largely unknown. The involvement of mathematical biologists in the development of new models to ascertain the most effective strategies for anthelmintic treatment is crucial to our understanding as large-scale control programs continue to expand worldwide.

Pioneering work on the contribution of host genetics to the observed variation in *Ascaris* infection has been honed using pedigree analysis of a unique human population, the Jirel people of Nepal. The key reason for analyzing host genetics in the context of a parasite like *Ascaris* is to identify the actual genes underlying human variation in disease risk. A number of biologically convincing candidate genes have been identified and will continue to be evaluated. Genetic variation among *Ascaris* worms is another source of variation operating at a number of levels. The debate concerning the genetic relationship between *A. lumbricoides* and *A. suum* continues unabated. New approaches are being utilized in genetic epidemiology, including the use of landscape genetics as a means of identifying transmission foci and the measurement of effective population size, in the context of parasite populations.

The use of *Ascaris* as a model organism for the study of neurobiology of nematodes began in the late 19th century. Recent work, combining the advantages of studying electrophysiology and biochemistry of single neurons with the opportunities to silence selected gene transcripts using

RNAi, provides a powerful platform for parasite gene function studies at the single-cell level. Furthermore, the central role of neuromuscular function in successful chemotherapeutic agents means that nematode neurobiology research resonates beyond discovery biology to parasite control. The recent publication of the *Ascaris suum* draft genome is a very important milestone in helminth biology. The characterization of key genes and biological pathways linked to *Ascaris* migration and its immunobiology, reproduction, and development provides unprecedented prospects for fundamental and applied research. New work on the molecular structure of unique proteins produced by *Ascaris*, such as the ABA-1 allergen, further enhances the applicability of *Ascaris* as a model organism for fundamental molecular biological studies.

It may be a surprise to readers that a book focused solely upon *Ascaris* has not been published for over 20 years. This book presents chapters written by a diverse range of experts (field-based epidemiologists, mathematical biologists, veterinarians, members of the world health community, immunologists, neurobiologists, and molecular biologists) who have chosen to focus upon particular topics that are novel and interesting and that I hope will appeal to modern biologists. The book also includes chapters on both *A. lumbricoides* and *A. suum*. The recent Neglected Infectious Diseases strategy published by the Bill and Melinda Gates Foundation in January 2012 referred to *Ascaris* as "one of six neglected diseases where there may be opportunities to reduce the disease burden but where there are knowledge gaps." It is my passionate hope that this book will engender increased interest in ascariasis, both as a disease entity and as a model organism, not only among helminth biologists but also among the broader scientific community.

Acknowledgments

I extend very warm thanks to Mary Preap, Linda Veersteg and Julia Haynes of Elsevier for their encouragement of the project and all their practical help and speedy responses throughout the process. Grateful thanks are also extended to Malcolm Kennedy, Christina Dold, Fiona Moloney, Mary Foody, Peter Stafford, and Paula Murphy for their ongoing moral support. Most of all, I thank Rory, Kate, and Róisín for all their love and care.

Celia Holland
Dublin, December 2012

PART I

BIOLOGY OF *ASCARIS*

CHAPTER

1

Immunology of *Ascaris* and Immunomodulation

Professor Philip J. Cooper[*], *Professor Camila A. Figuieredo*[†]

[*] Liverpool School of Tropical Medicine and Hygiene, Liverpool, UK; Pontificia Universidad Católica del Ecuador, Quito, Ecuador

[†] Universidade Federal da Bahia, Salvador, Brazil

OUTLINE

Ascaris: The Neglected Parasite
http://dx.doi.org/10.1016/B978-0-12-396978-1.00001-X

Copyright © 2013 Elsevier Inc. All rights reserved.

INTRODUCTION

The large roundworms, *Ascaris lumbricoides* and *A. suum*, are ubiquitous pathogens of humans and pigs respectively with worldwide distributions. *A. suum* and *A. lumbricoides* are highly related genetically and are able to cross-infect pigs and humans. Although it has been suggested that they represent the same species,[1] this remains an area of controversy[2] (see Chapter 10).

Both *A. suum* in pigs and *A. lumbricoides* in humans have similar life-cycles in which the mammalian host is infected orally with infective ova. Larvae hatch in the small intestine and migrate to the caecum and proximal colon where they penetrate the mucosa,[3] and migrate to the liver through the portal system. The larvae reach the lungs via the systemic circulation where they embolize in the pulmonary capillaries, penetrate the alveoli, and are coughed up and then swallowed reaching the small intestine about 2 weeks following infection.[4] Roundworms reach sexual maturity in the small intestine where female and male adults mate. Fecund females release hundreds of thousands of eggs daily into the host feces.

In pigs, the majority of *A. suum* larvae are expelled between 2 and 3 weeks after infection[5] and among those that survive and develop into adults, most are expelled from the pig intestine within 6 months.[6] In the case of *A. lumbricoides* in humans there are no published data to our knowledge on the survival of larvae and adults although adult *A. lumbricoides* may survive 1 to 2 years in the human gut.[7]

Host susceptibility and genetics play an important role in the distribution of parasites within human and pig populations characterized by an aggregated distribution of parasites in which few individuals harbor the majority of parasites while most have few or no parasites[6,8] (see Chapter 7). Resistance to infection and the acquisition of resistance upon exposure are determined largely by the host innate and adaptive immune response. Recent research, particularly of *A. suum* in the pig, has identified specific immune parameters associated with the development of protective immunity that may be relevant also to our understanding of immunity to *A. lumbricoides* in humans. The long-term survival of parasites in the host depends on the modulation of the host's immune response by the parasite. Understanding the mechanisms by which chronic infections modulate the immune response will allow us to understand better immune modulation and homeostasis in mucosal and other tissues.[9,10] It has been suggested that the relative absence of chronic parasitic infections such as *A. lumbricoides* and their immune modulatory effects may have a role in the emerging epidemic of inflammatory diseases such as autoimmune and allergic diseases in populations living in industrialized countries[11–14] and also urbanizing populations in non-industrialized countries.

This chapter will review what is known of the immune response to *Ascaris* in pigs and humans and how infections may modify the immune and inflammatory response of the host. We will also discuss the potential clinical effects of this immune modulation.

IMMUNOLOGY OF ASCARIASIS

Immunology of *Ascaris suum* in Pigs

Great progress has been made over the past 20 years in our understanding of the immune response to *A. suum* in pigs. Earlier studies showed that the inoculation of pigs with infective eggs was associated with the development of specific systemic IgG1, IgA, and IgM antibody responses,[5,15] but that the predominant antibody secreting cells (ASC) in the mucosa of the proximal jejunum,[5] duodenum,[16] and bronchi[16] were of the IgA isotype.[5] That the immune response to *A. suum* in the pig is predominantly Th2 is reflected by an elevated peripheral blood eosinophilia,[17,18] elevated frequencies of parasite-specific monocytes secreting IL-4 in peripheral blood and intestinal lymph nodes,[18] and the increased expression of IL-4 in plasma and Th2 cytokines (e.g. IL-4 and IL-13) in the intestine[19,20] and liver.[21] An increased expression of IL-10 has been observed in the intestine and other tissues during infection[19] where it may have a role in the regulation of inflammation.[22]

Resistance to *A. suum* in pigs has been observed following multiple oral infections with *A. suum* eggs,[17,23,24] with radiation-attenuated *A. suum* eggs,[25,26] and following immunization with parasite-derived products.[27] Partial protection has also been achieved by giving colostrum from immunized sows[28] and by passive transfer of hyperimmune sera.[28] Protective immunity to *A. suum* in swine may occur in the liver, lungs, and in the intestine. Different infection or immunization strategies may have stronger effects on immunity generated in a particular tissue. A strong liver white-spot reaction, reflecting the host inflammatory response to the presence of the larvae in the liver, is a typical consequence of a secondary challenge exposure to infective eggs,[29] although this reaction diminishes over time with trickle infections.[24]

Chronic natural and multiple inoculations with eggs has been associated with the development of pre-hepatic intestinal immunity[29,30] while immunization with irradiated eggs or with *A. suum* antigens[27,31] is associated with post-hepatic protective immunity. Relatively little is known of the specific immunological mechanisms by which protection is achieved. The intestinal expulsion of larvae that have migrated through the lungs has been associated with an increase in the frequency of intestinal mucosal ASCs producing parasite-specific IgA[5] and with mast cell-mediated immediate hypersensitivity responses to parasite antigens.[4,32] Non-specific immunological mechanisms may also have a role: pigs previously infected with transmissible gastroenteritis (TGE) virus were resistant to infection.[33] Intestinal immunity to *A. suum* eggs does not appear to be directly related to the number of adult worms in the host because removal of adults by anthelmintic treatment or the direct transplantation into the intestine did not affect protective immunity following challenge with *A. suum* eggs.[17]

Immunology of Ascariasis in Humans

Chronic infections with *A. lumbricoides* in humans are associated with the production of high levels of specific and non-specific antibodies of all isotypes and IgG subclasses[34,35] and a cytokine response characterized by the production of Th2 cytokines (i.e. IL-4, IL-13, and IL-5) by peripheral blood monocytes (PBMCs) and leukocytes (PBLs in whole blood cultures).[36–38] Other Th2 effector responses are also prominent during infection, reflected by elevated numbers of peripheral blood eosinophils[39] and increased expression of eosinophil degranulation products.[10] The production of IFN-γ by PBMCs/PBLs stimulated with *Ascaris* antigens is not prominent in ascariasis[37,38,40,41] but IL-10 production may be increased in infected individuals.[37,42] An increased production of IL-10 has been observed also to occur spontaneously (i.e. in the absence of antigen stimulation) by PBMCs/PBLs of individuals with chronic infections,[10,38] an observation previously made for other chronic helminth infections.[43–45] Albendazole treatment of individuals co-infected with HIV and *A. lumbricoides* was associated with a decline in plasma IL-10 compared to co-infected individuals receiving placebo,[46] providing further evidence that IL-10 may be upregulated non-specifically in infected individuals.

IL-10 is considered to be a key cytokine mediating immune regulation during chronic helminth infections[10,38,43–45,47] and the combination of elevated Th2 cytokines with IL-10 has been referred to as a modified or regulated Th2 response.[48,49] Another cytokine that has been associated with immune regulation during chronic helminth infections is TGF-β[43]: evidence for the increased expression of this cytokine during ascariasis is inconsistent with some studies suggesting increased production or expression of TGF-β1[34,47] but other authors have observed no such effect.[10,41] It should be remembered that individuals living in endemic areas are often infected with more than one soil-transmitted helminth (STH) parasite and many children with *A. lumbricoides*, particularly those with high parasite burdens, have a high probability of recent or current STH co-infections (see Chapter 4). This makes it more difficult to attribute specific immunologic effects to ascariasis alone even when blood cells are stimulated *in vitro* with *Ascaris* antigens because of the high degree of immunological cross-reactivity between STH antigens.[50]

The existence of protective immunity to ascariasis in humans remains controversial but is suggested by the observation that, under conditions of high levels of transmission, the prevalence and intensity of *A. lumbricoides* declines with age.[51] Non-immunological factors, such as changes in behavior with age, are reasonable alternative explanations for this observation. Even so, a number of investigators have tried to identify immunological parameters associated with protection.

A study of reinfection following treatment was unable to attribute a protective role of specific antibodies to adult and larval stages of *A. lumbricoides* against current infection or reinfection,[35] while other studies have suggested that specific IgE might have a protective role.[52,53] A randomized clinical trial giving anti-IgE for the treatment for atopic diseases, which reduced circulating levels of IgE to negligible levels, did not show a significant effect of anti-IgE treatment on susceptibility to infection with STH parasites, although a trend of increased risk of infection (mainly ascariasis) was seen among asthmatic patients receiving anti-IgE compared to placebo.[54]

It seems reasonable to suggest, therefore, that IgE, particularly specific IgE, has a role in protective immunity to *A. lumbricoides* infection either in the initiation of allergic-type inflammatory responses against the parasite or in the amplification of other Th2-mediated mechanisms. Individual Th2 effector mechanisms could be important in mediating protection against different life-cycle stages of *A. lumbricoides*: IgE-mediated hypersensitivity in the intestine may be important in expulsion of juvenile and adult parasites while other Th2 effector mechanisms, which may or may not include IgE, could be important in the killing of parasite larvae in the liver and lungs.[55]

A study conducted in a hyperendemic community for ascariasis showed that the age-dependent decline in prevalence of infection was mirrored by an increase in Th2 cytokine production (IL-4, IL-5, IL-9, IL-10, and IL-13) by antigen-stimulated PBLs from older individuals,[56] and attributed the age-dependent decline in infection to an increase in Th2 cytokines with age. An infection–reinfection study of individuals infected with *A. lumbricoides* and *T. trichiura* showed that the elevated production of Th2 cytokines (IL-5 and IL-13) by PBLs stimulated with STH antigens was associated with resistance to reinfection.[57] An alternative approach has been to identify individuals that are resistant to infection (i.e. that are free of infection despite residence in a highly endemic community), so-called putative immunes (PIs), and compare immune parameters with infected individuals from the same community. Such a study in Nigeria observed that resistance to infection was associated with increased levels of innate inflammatory markers such as C-reactive protein in peripheral blood and elevated IgE to the *Ascaris* allergen ABA-1 among those with ABA-1-specific IgG.[52]

Single anthelmintic treatments given to infected children appear to have little effect on the host cytokine response to *Ascaris* antigens[58] but giving periodic anthelmintic treatments over time to suppress infections has been associated with increased Th2 cytokine production to parasite antigens.[59,60] Thus, although Th2 responses are prominent during active infections, such responses and perhaps also Th1 responses,[60] are suppressed during chronic infections, an effect that may or may not be mediated by IL-10.[40,58,60] Neutralization of IL-10 in PBMC cultures had no effect on frequencies of cells expressing Th1 or Th2 cytokines.[40] Such modulation of immunity in highly endemic communities may start in early life: a study of cord blood from newborns of infected and uninfected mothers showed that the frequencies of antigen-stimulated CD4+ cells expressing IL-4 and IFN-γ were increased among newborns of infected compared to uninfected mothers suggesting prenatal immune sensitization.[61] There are several birth cohort studies in progress that are addressing the role of exposure to *A. lumbricoides* in early life in molding the immune response to ascariasis.[62,63]

IMMUNE MODULATION IN ANIMAL MODELS

Parasite persistence in the host may require the modulation of the host immune response through a complex host–parasite interaction.[64] The effects of such immune modulation may have important non-specific effects on the regulation of inflammation in the host. Several research groups have explored the effects of antigen extracts derived from *A. suum* parasites on inflammation in murine models of inflammatory disease. The findings show that different antigen extracts of *A. suum* have distinct effects on inflammatory responses with some having strong immunosuppressive effects on immune responses and therapeutic effects in experimental models of asthma[65] and arthritis,[66] while other studies show that such extracts may also induce inflammation, particularly allergic inflammation.[67] This latter observation is hardly surprising given that *A. suum* infections and/or *A. suum* antigen preparations are widely used for the induction of bronchial hyper-reactivity in experimental models of asthma in primates and other experimental animals.[68,69]

Co-immunization of mice with ovalbumin (OVA) and *A. suum* antigen fractions have been shown to suppress immune responses to OVA including the production of specific antibodies of all antibody isotypes including IgE[70] and anaphylactic IgG1,[71,72] and delayed type hypersensitivity (DTH) to ovalbumin (OVA).[73] High molecular weight fractions of *A. suum* body fluid seemed to account for these effects and co-immunization of mice with OVA and a high molecular weight fraction inhibited immediate hypersensitivity and DTH to OVA, the proliferation of lymph node cells to OVA, and Th1/Th2 cytokine and antibody production.[74] The active component of the high molecular weight fraction was identified as a 200-kDa protein called PAS-1 that suppressed immune responses in a dose-dependent fashion.[75]

PAS-1 appears to mediate its suppressive effects on cell-mediated responses and Th1 cytokine responses to heterologous antigens such as OVA through the production of large amounts of IL-4 and IL-10[72] while effects on Th2 responses are associated with the elevated production of IL-10.[76] PAS-1 has been shown to downregulate Th2 responses associated with the development of asthma in mouse models of *A. suum*-induced[67] and OVA-induced asthma.[77] PAS-1 can also suppress pro-inflammatory cytokine responses to LPS-induced inflammation.[72] The suppressive effect of PAS-1 in OVA-induced asthma has been attributed to the induction of CD4+CD25+ regulatory T cells because adoptive transfer of these cells from PAS-1 immunized mice suppressed allergic responses in the airways of OVA-immunized and challenged mice.[76]

Phosphorylcholine-containing glycosphingolipids from *A. suum* suppressed IL-12 production by LPS/IFN-γ-stimulated peritoneal macrophages and reduced B cell proliferation.[78] Pseudocoelomic body fluid from adult *A. suum* worms (ABF) suppressed DTH responses to OVA in OVA-sensitized mice, an effect on the induction rather than the effector phase of the immune response, and which was mediated by IL-4 and IL-10.[79] The potent immunosuppressive effects of *A. suum* extract on the host immune response may be related partly to the downregulation of the antigen-presenting ability of dendritic cells via an IL-10-mediated mechanism.[80]

Overall, these data from experimental animals indicate that some components of *A. suum* are potent inducers of inflammatory and allergic responses while other components have a diverse range of effects in the modulation of the host immune response that are mediated at least partly through IL-10-dependent mechanisms.

EVIDENCE FOR CLINICALLY RELEVANT IMMUNE MODULATION BY ASCARIASIS DURING NATURAL INFECTION

As already discussed, there is evidence that *A. lumbricoides* infection modulates the human immune response and, as observed in experimental animal models, such effects are associated with an increased production of IL-10 during chronic infections,[10,38] although in humans there are limited data to support a mediating role for this cytokine. There may be differences between acute and chronic helminth infections with respect to the effects on inflammation – acute infections appear to be strong inducers of pro-inflammatory and allergic effects while allergic-type reactions are rare and may be suppressed during chronic infections.[39] There is some epidemiological data in humans[81] and experimental data in mice[82] to support the relevance of this paradigm to ascariasis. A number of clinically relevant effects have been attributed to infections with ascariasis including possible impairment of vaccine immunity and effects on the risk of allergic sensitization (i.e. atopy) and allergic diseases. STH infections (i.e. *Trichuris suis* and the hookworm *Necator americanus*) have been used experimentally for the treatment of respiratory allergy with negligible therapeutic effects[83–85] and for inflammatory bowel diseases with at best modest clinical benefit.[86–88] There is no evidence to date to suggest that *Ascaris* infections are likely to be of therapeutic benefit and experimental infections with this parasite might be expected to induce strong inflammatory responses in the respiratory tract of individuals living in areas where there is little or no transmission of infection. More promising, perhaps, will be the development of *Ascaris*-derived molecules with anti-inflammatory effects for potential therapeutic use.

Vaccine Immune Responses

Live attenuated oral vaccines are less immunogenic in poor populations in developing countries compared to those living in developed countries requiring an increase in the dose or number of doses administered to achieve adequate vaccine immune responses.[89] It has been suggested that concurrent STH infections might interfere with immune responses to oral vaccines through effects on mucosal immune responses.[40,90] A trial in Ecuador that randomized children with ascariasis to receive anthelmintic treatment or placebo prior to vaccination with a single dose of a live-attenuated oral cholera vaccine showed that prior treatment of STH infections enhanced titers of vibriocidal antibodies and rates of seroconversion.[36] Further, Th1 cytokine responses to cholera toxin B-subunit, a component of the vaccine used, were elevated after vaccination among children receiving albendazole.[58]

Another study, using a similar design, in Ethiopia showed that giving anthelmintic treatment to individuals with ascariasis before parenteral BCG vaccination was associated with enhanced IFN-γ responses to PPD.[91] However, *A. lumbricoides* and other STH infections alone are unlikely to explain impaired immunity to oral vaccines. A recent study investigating the impact of *A. lumbricoides* infection on responses to oral BCG Moreau failed to demonstrate post-vaccination increases in the frequencies of tuberculin-stimulated PBMCs expressing IFN-γ among children with either active infections or those who had received either short or long courses of anthelmintics before vaccination.[92] The same vaccine and vaccine dose showed strong boosting of post-vaccination IFN-γ responses in wealthier populations,[93,94] indicating the presence of a mucosal barrier to oral vaccination among children living in the rural tropics that, under some circumstances may be accentuated by *Ascaris* co-infections, but which are primarily determined by other factors that interfere between an oral vaccine and the mucosal immune response such as environmental enteropathy,[89] small intestinal bacterial overgrowth,[95] and intestinal microbiota.

Atopy

Chronic STH infections including ascariasis are associated with the suppression of allergic inflammatory responses directed against the parasite[9] although it is less clear if such infections can modulate allergic inflammation directed against non-parasite allergens such as aero-allergens that have been most commonly associated with allergic inflammatory processes (see Chapter 2). Allergic sensitization or atopy in humans can be determined by measurement of allergen-specific IgE in serum or skin test reactivity to allergen extracts. Chronic STH infections are strongly inversely associated with skin prick test reactivity to aero-allergens (SPT) and protective effects have been shown for *A. lumbricoides*, *T. trichiura*, and hookworm.[96] However, the protective effects of *A. lumbricoides* were not independent of those of *T. trichiura* in studies of co-infected children in Brazil[97] and Ecuador.[98] If geohelminths were to be actively suppressing skin test responses, anthelmintic treatment might be expected to reverse this effect.

Intervention studies have provided evidence that anthelmintic treatment of children may increase SPT,[99–101] although a randomized trial in Ecuador was unable to replicate these findings.[102] The latter trial administered 400 mg of albendazole every 2 months for a year, and such treatment was much more effective against *A. lumbricoides* than *T. trichiura*, the most prevalent STH infections in the study population.[102] Similarly, a study of long-term repeated anthelmintic treatments with ivermectin over a period of up to 15 years showed that such treatment was associated with an increase in SPT and such effects were primarily attributable to a lower prevalence of *T. trichiura*.[98] Thus, it is not entirely clear whether ascariasis, *per se*, modulates allergic effector responses against aero-allergens (i.e. SPT) independent of the presence of other STH infections.

Asthma

Seasonal exposures to ascariasis have been associated with asthma symptoms for many years.[103,104] The syndrome, known as Loeffler's syndrome, eosinophilic pneumonitis, or *Ascaris* pneumonia, is a potentially severe but self-limiting illness associated with the pulmonary migration of *Ascaris* larvae. It has been suggested that human infections with *A. suum* are more likely to cause a larva migrans syndrome with pulmonary symptoms.[7,105] A meta-analysis of cross-sectional analyses showed a positive association between the detection of *A. lumbricoides* infections in stool samples and the prevalence of asthma symptoms.[106] Several more recent studies have reported positive associations between the anti-*Ascaris* IgE[107–109] or active *A. lumbricoides* infection[110–112] and asthma symptoms or bronchial hyper-responsiveness (BHR) in low prevalence populations.

A study in rural Ecuador attributed almost 50% of recent wheeze cases to the presence of ani-*Ascaris* IgE but no association was seen with active infections[113]. Repeated anthelmintic treatments for ascariasis have been associated with an improvement in asthma symptoms and a decreased use of asthma medications among asthmatics living in an area of low prevalence for ascariasis but studies conducted in high prevalence populations have showed no effect of anthelmintic treatment on wheeze symptoms.[98,101,102] Asthma symptoms associated with ascariasis could be caused by the migration of *Ascaris* larvae through the lungs, particularly among those with a tendency to mount allergic responses to the parasite (i.e. those with anti-*Ascaris* IgE) or the potentiation of inflammatory responses to other stimuli such as aeroallergens or endotoxin.

CONCLUSION

The immune response during natural infections with ascariasis in humans and pigs is associated with the production of Th2 cytokines and, in some but not all cases, with the elevated production of IL-10 that is presumed to have an immune modulatory function. Th2 cytokine response and associated effector mechanisms are considered to mediate protective immunity to ascariasis in pigs and humans. Some antigen extracts of *A. suum* appear to have potent immune modulatory effects in experimental murine models that may be mediated through IL-10 or related immune regulatory pathways while other antigens or extracts are potent initiators of inflammation. There is some, although inconsistent, evidence that human ascariasis may suppress protective immune responses to vaccines and immediate hypersensitivity reactions to aeroallergens in the skin. More consistent are the observations of potent pro-inflammatory effects of ascariasis in human populations, particularly among those where the prevalence of infection is high, that cause asthma-like symptoms through direct or indirect effects. Specific molecules or antigen fractions derived from *Ascaris* parasites with anti-inflammatory effects could be developed as novel therapies for inflammatory diseases but there is no indication for the use of natural infections in the treatment of allergic or autoimmune diseases.

A better understanding of the host–parasite interaction in humans and how this interaction affects immune modulation during ascariasis and the clinical effects of such modulation will be provided by observational and intervention studies that are currently in progress and that have followed and sampled cohorts of children living in endemic communities from the time of pregnancy. In many regions of the world, the prevalence of human ascariasis is progressively declining with economic development and greater access to sanitation and clean water. However, it could be argued that while the clinical effects of immune modulation associated with chronic infections will inevitably diminish in such populations, the inflammatory consequences of acute and intermittent infections will grow considerably (e.g. asthma), particularly among the urbanizing poor whose conditions of life are unlikely to improve sufficiently to eliminate the infection but rather will ensure its continued survival.

References

1. Leles D, Gardner SL, Reinhard K, Iñiguez A, Araujo A. Are *Ascaris lumbricoides* and *Ascaris suum* a single species? *Parasit Vectors* 2012;**5**:42.
2. Nejsum P, Betson M, Bendall RP, Thamsborg SM, Stothard JR. Assessing the zoonotic potential of *Ascaris suum* and *Trichuris suis*: looking to the future from an analysis of the past. *J Helminthol* 2012 Jun;**86**(2):148–55. Epub 2012 Mar 19.
3. Murrell KD, Eriksen L, Nansen P, Slotved HC, Rasmussen T. *Ascaris suum*: a revision of its early migratory path and implications for human ascariasis. *J Parasitol* 1997 Apr;**83**(2):255–60.
4. Roepstorff A, Eriksen L, Slotved HC, Nansen P. Experimental *Ascaris suum* infection in the pig: worm population kinetics following single inoculations with three doses of infective eggs. *Parasitology* 1997 Oct;**115**(Pt 4):443–52.
5. Miquel N, Roepstorff A, Bailey M, Eriksen L. Host immune reactions and worm kinetics during the expulsion of *Ascaris suum* in pigs. *Parasite Immunol* 2005 Mar;**27**(3): 79–88.
6. Roepstorff A, Mejer H, Nejsum P, Thamsborg SM. Helminth parasites in pigs: new challenges in pig production and current research highlights. *Vet Parasitol* 2011 Aug;**180**(1–2):72–81.
7. Pawłowski ZS. Ascariasis. *Clin Gastroenterol* 1978 Jan;**7**(1):157–78.
8. Hall A, Anwar KS, Tomkins A, Rahman L. The distribution of *Ascaris lumbricoides* in human hosts: a study of 1765 people in Bangladesh. *Trans R Soc Trop Med Hyg* 1999 Sep–Oct;**93**(5):503–10.
9. Cooper PJ. The interactions of parasites with allergy. *Curr Opin Allergy Clin Immunol* 2009;**9**:29–37.
10. Reina MO, Schreiber F, Benitez S, Broncano N, Chico ME, Dougan G, et al. Gene and microRNA expression and cytokine production associated with chronic ascariasis in children in the rural tropics. *PLoS Neglected Tropical Diseases* 2011;**5**:e1157.

11. Elliott DE, Summers RW, Weinstock JV. Helminths as governors of immune-mediated inflammation. *Int J Parasitol* 2007 Apr;**37**(5):457–64.
12. Zaccone P, Burton OT, Cooke A. Interplay of parasite-driven immune responses and autoimmunity. *Trends Parasitol* 2008 Jan;**24**(1):35–42.
13. Cooper PJ. Mucosal immunology of geohelminth infections in humans. *Mucosal Immunol* 2009;**2**:288–99.
14. Fleming JO. Helminths and multiple sclerosis: will old friends give us new treatments for MS? *J Neuroimmunol* 2011 Apr;**233**(1–2):3–5.
15. Lind P, Eriksen L, Nansen P, Nilsson O, Roepstorff A. Response to repeated inoculations with *Ascaris suum* eggs in pigs during the fattening period. II. Specific IgA, IgG, and IgM antibodies determined by enzyme-linked immunosorbent assay. *Parasitol Res* 1993;**79**(3):240–4.
16. Frontera E, Roepstorff A, Serrano FJ, Gázquez A, Reina D, Navarrete I. Presence of immunoglobulins and antigens in serum, lung and small intestine in *Ascaris suum* infected and immunised pigs. *Vet Parasitol* 2004 Jan 5;**119**(1):59–71.
17. Jungersen G, Eriksen L, Roepstorff A, Lind P, Meeusen EN, Rasmussen T, et al. Experimental *Ascaris suum* infection in the pig: protective memory response after three immunizations and effect of intestinal adult worm population. *Parasite Immunol* 1999 Dec;**21**(12):619–30.
18. Steenhard NR, Kringel H, Roepstorff A, Thamsborg SM, Jungersen G. Parasite-specific IL-4 responses in *Ascaris suum* and *Trichuris suis*-infected pigs evaluated by ELISPOT. *Parasite Immunol* 2007 Oct;**29**(10):535–8.
19. Dawson HD, Beshah E, Nishi S, Solano-Aguilar G, Morimoto M, Zhao A, et al. Localized multigene expression patterns support an evolving Th1/Th2-like paradigm in response to infections with *Toxoplasma gondii* and *Ascaris suum*. *Infect Immun* 2005 Feb;**73**(2):1116–28.
20. Steenhard NR, Jungersen G, Kokotovic B, Beshah E, Dawson HD, Urban JF, et al. *Ascaris suum* infection negatively affects the response to a *Mycoplasma* hyopneumoniae vaccination and subsequent challenge infection in pigs. *Vaccine* 2009 Aug;**27**(37): 5161–9.
21. Dawson H, Solano-Aguilar G, Beal M, Beshah E, Vangimalla V, Jones E, et al. Localized Th1-, Th2-, T regulatory cell-, and inflammation-associated hepatic and pulmonary immune responses in *Ascaris suum*-infected swine are increased by retinoic acid. *Infect Immun* 2009 Jun;**77**(6):2576–87. Epub 2009 Mar 30.
22. Schopf LR, Hoffmann KF, Cheever AW, Urban Jr JF, Wynn TA. IL-10 is critical for host resistance and survival during gastrointestinal helminth infection. *J Immunol* 2002 Mar 1;**168**(5):2383–92.
23. Kelley GW, Nayak DP. Acquired immunity to migrating larvae of ascaris suum induced in pigs by repeated oral inoculations in infective eggs. *J Parasitol* 1964 Aug;**50**: 499–503.
24. Nejsum P, Thamsborg SM, Petersen HH, Kringel H, Fredholm M, Roepstorff A. Population dynamics of *Ascaris suum* in trickle-infected pigs. *J Parasitol* 2009 Oct;**95**(5): 1048–53. Epub 2009 Aug 12.
25. Tromba FG. Immunization of pigs against experimental *Ascaris suum* infection by feeding ultraviolet-attenuated eggs. *J Parasitol* 1978 Aug;**64**(4):651–6.
26. Urban JF, Tromba FG. An ultraviolet-attenuated egg vaccine for swine ascariasis: parameters affecting the development of protective immunity. *Am J Vet Res* 1984 Oct;**45**(10):2104–8.
27. Urban JF, Romanowski RD. *Ascaris suum:* protective immunity in pigs immunized with products from eggs and larvae. *Exp Parasitol* 1985 Oct;**60**(2):245–54.
28. Kelley GW, Nayak DP. Passive immunity to *Ascaris suum* transferred in colostrum from sows to their offspring. *Am J Vet Res* 1965 Jul;**26**(113):948–50.

29. Urban JF, Alizadeh H, Romanowski RD. *Ascaris suum*: development of intestinal immunity to infective second-stage larvae in swine. *Exp Parasitol* 1988 Jun;**66**(1):66–77.
30. Eriksen L, Nansen P, Roepstorff A, Lind P, Nilsson O. Response to repeated inoculations with *Ascaris suum* eggs in pigs during the fattening period. I. Studies on worm population kinetics. *Parasitol Res.* 1992;**78**(3):241–6.
31. Hill DE, Fetterer RH, Romanowski RD, Urban Jr JF. The effect of immunization of pigs with *Ascaris suum* cuticle components on the development of resistance to parenteral migration during a challenge infection. *Vet Immunol Immunopathol* 1994 Aug;**42**(2): 161–9.
32. Ashraf M, Urban Jr JF, Lee TD, Lee CM. Characterization of isolated porcine intestinal mucosal mast cells following infection with *Ascaris suum*. *Vet Parasitol* 1988 Sep;**29**(2–3):143–58.
33. Gaafar SM, Dugas S, Symensma R. Resistance of pigs recovered from transmissible gastroenteritis against infection with *Ascaris suum*. *Am J Vet Res* 1973 Jun;**34**(6):793–5.
34. Cooper PJ. Immunity in humans – ascaris. In: Kennedy MW, Holland CV, editors. *Ascaris – World Class Parasites*. Kluwer Academic Press; 2002. p. 89–104.
35. King EM, Kim HT, Dang NT, Michael E, Drake L, Needham C, et al. Immuno-epidemiology of *Ascaris lumbricoides* infection in a high transmission community: antibody responses and their impact on current and future infection intensity. *Parasite Immunol* 2005 Mar;**27**(3):89–96.
36. Cooper PJ, Chico M, Losonsky G, Espinel I, Sandoval C, Aguilar M, et al. Albendazole treatment of children with ascariasis enhances the vibriocidal antibody response to the live attenuated oral cholera vaccine CVD 103-HgR. *J Infect Dis* 2000;**182**:1199–206.
37. Geiger SM, Massara CL, Bethony J, Soboslay PT, Carvalho OS, Corrêa-Oliveira R. Cellular responses and cytokine profiles in *Ascaris lumbricoides* and *Trichuris trichiura* infected patients. *Parasite Immunol* 2002 Nov-Dec;**24**(11–12):499–509.
38. Figueiredo CA, Barreto ML, Rodrigues LC, Cooper PJ, Silva NB, Amorim LD, et al. Chronic intestinal helminth infections are associated with immune hyporesponsiveness and induction of a regulatory network. *Infect Immun* 2010;**78**:3160–7.
39. Cooper PJ, Nutman TB. IgE and its role in parasitic helminth infection: implications for anti-IgE based therapies. In: Fick RB, Jardieu P, editors. *IgE and Anti-IgE Therapy in Asthma and Allergic Disease. Lung Biology in Health and Disease,* **164**. New York: Marcel Dekker; 2002. p. 409–25.
40. Cooper PJ, Chico M, Sandoval C, Espinel I, Guevara A, Kennedy MW, et al. Human infection with *Ascaris lumbricoides* is associated with a polarized cytokine phenotype. *J Infect Dis* 2000;**182**:1207–13.
41. Matera G, Giancotti A, Scalise S, Pulicari MC, Maselli R, Piizzi C, et al. Risk factors for atopic and non-atopic asthma in a rural area of Ecuador. *Thorax* 2010 May;**65**(5): 409–16.
42. Hagel I, Cabrera M, Puccio F, Santaella C, Buvat E, Infante B, et al. Co-infection with *Ascaris lumbricoides* modulates protective immune responses against *Giardia duodenalis* in school Venezuelan rural children. *Acta Trop* 2011 Mar;**117**(3):189–95.
43. King CL, Mahanty S, Kumaraswami V, Abrams JS, Regunathan J, Jayaraman K, et al. Cytokine control of parasite-specific anergy in human lymphatic filariasis. Preferential induction of a regulatory T helper type 2 lymphocyte subset. *J Clin Invest* 1993 Oct;**92**(4):1667–73.
44. Mahanty S. Nutman TB. Immunoregulation in human lymphatic filariasis: the role of interleukin 10. *Parasite Immunol* 1995 Aug;**17**(8):385–92.
45. Sasisekhar B, Aparna M, Augustin DJ, Kaliraj P, Kar SK, Nutman TB, et al. Diminished monocyte function in microfilaremic patients with lymphatic filariasis and its relationship to altered lymphoproliferative responses. *Infect Immun* 2005 Jun;**73**(6): 3385–93.

46. Blish CA, Sangaré L, Herrin BR, Richardson BA, John-Stewart G, Walson JL. Changes in plasma cytokines after treatment of ascaris lumbricoides infection in individuals with HIV-1 infection. *J Infect Dis* 2010 Jun 15;**201**(12):1816–21.
47. Turner JD, Jackson JA, Faulkner H, Behnke J, Else KJ, Kamgno J, et al. Intensity of intestinal infection with multiple worm species is related to regulatory cytokine output and immune hyporesponsiveness. *J Infect Dis* 2008 Apr 15;**197**(8):1204–12.
48. Platts-Mills T, Vaughan J, Squillace S, Woodfolk J, Sporik R. Sensitisation, asthma, and a modified Th2 response in children exposed to cat allergen: a population-based cross-sectional study. *Lancet* 2001 Mar 10;**357**(9258):752–6.
49. Fallon PG, Mangan NE. Suppression of TH2-type allergic reactions by helminth infection. *Nat Rev Immunol* 2007 Mar;**7**(3):220–30.
50. Kennedy MW, Qureshi F, Fraser EM, Haswell-Elkins MR, Elkins DB, Smith HV. Antigenic relationships between the surface-exposed, secreted and somatic materials of the nematode parasites *Ascaris lumbricoides*, *Ascaris suum*, and *Toxocara canis*. *Clin Exp Immunol* 1989 Mar;**75**(3):493–500.
51. Bundy DA, Medley GF. Immuno-epidemiology of human geohelminthiasis: ecological and immunological determinants of worm burden. *Parasitology* 1992;**104**(Suppl). S105–19.
52. McSharry C, Xia Y, Holland CV, Kennedy MW. Natural immunity to *Ascaris lumbricoides* associated with immunoglobulin E antibody to ABA-1 allergen and inflammation indicators in children. *Infect Immun* 1999 Feb;**67**(2):484–9.
53. Turner JD, Faulkner H, Kamgno J, Kennedy MW, Behnke J, Boussinesq M, et al. Allergen-specific IgE and IgG4 are markers of resistance and susceptibility in a human intestinal nematode infection. *Microbes Infect* 2005 Jun;**7**(7–8):990–6.
54. Cruz AA, Lima F, Sarinho E, Ayre G, Martin C, Fox H, et al. Safety of anti-immunoglobulin E therapy with omalizumab in allergic patients at risk of geohelminth infection. *Clin Exp Allergy* 2007 Feb;**37**(2):197–207.
55. Cooper PJ, Ayre G, Martin C, Rizzo JA, Ponte EV, Cruz AA. Geohelminth infections: a review of the role of IgE and assessment of potential risks of anti-IgE treatment. *Allergy* 2008 Apr;**63**(4):409–17.
56. Turner JD, Faulkner H, Kamgno J, Cormont F, Van Snick J, Else KJ, et al. Th2 cytokines are associated with reduced worm burdens in a human intestinal helminth infection. *J Infect Dis* 2003 Dec 1;**188**(11):1768–75. Epub 2003 Nov 14.
57. Jackson JA, Turner JD, Rentoul L, Faulkner H, Behnke JM, Hoyle M, et al. T helper cell type 2 responsiveness predicts future susceptibility to gastrointestinal nematodes in humans. *J Infect Dis* 2004 Nov 15;**190**(10):1804–11.
58. Cooper PJ, Chico M, Espinel I, Sandoval C, Guevara A, Levine M, et al. Human infection with *Ascaris lumbricoides* is associated with suppression of the IL-2 response to recombinant cholera toxin B-subunit following vaccination with the live oral cholera vaccine CVD 103 HgR. *Infect Immun* 2001;**69**:1574–80.
59. Cooper PJ, Moncayo AL, Guadalupe I, Benitez S, Vaca M, Chico ME, et al. Repeated albendazole treatments enhance Th2 responses to *Ascaris lumbricoides* but not aeroallergens in children from rural communities in the Tropics. *J Infect Dis* 2008;**198**: 1237–42.
60. Wright VJ, Ame SM, Haji HS, Weir RE, Goodman D, Pritchard DI, et al. Early exposure of infants to GI nematodes induces Th2 dominant immune responses which are unaffected by periodic anthelminthic treatment. *PLoS Negl Trop Dis* 2009;**3**(5). e433. Epub 2009 May 19.
61. Guadalupe I, Mitre E, Benitez S, Chico ME, Cordova X, Rodriguez J, et al. Evidence of intrauterine sensitization to *Ascaris lumbricoides* infection in newborns of infected mothers. *J Infect Dis* 2009;**199**:1846–50.
62. Elliott AM, Kizza M, Quigley MA, Ndibazza J, Nampijja M, Muhangi L, et al. The impact of helminths on the response to immunization and on the incidence of infection

and disease in childhood in Uganda: design of a randomized, double-blind, placebo-controlled, factorial trial of deworming interventions delivered in pregnancy and early childhood [ISRCTN32849447]. *Clin Trials* 2007;**4**(1):42–57.
63. Hamid F, Wiria AE, Wammes LJ, Kaisar MM, Lell B, Ariawan I, et al. A longitudinal study of allergy and intestinal helminth infections in semi urban and rural areas of Flores, Indonesia (ImmunoSPIN Study). *BMC Infect Dis* 2011 Apr 1;**11**:83.
64. van Riet E, Hartgers FC, Yazdanbakhsh M. Chronic helminth infections induce immunomodulation: consequences and mechanisms. *Immunobiology* 2007;**212**(6): 475–90.
65. Araújo CA, Perini A, Martins MA, Macedo MS, Macedo-Soares MF. PAS-1, a protein from *Ascaris suum*, modulates allergic inflammation via IL-10 and IFN-gamma, but not IL-12. *Cytokine* 2008 Dec;**44**(3):335–41.
66. Rocha FA, Leite AK, Pompeu MM, Cunha TM, Verri WA, Soares FM, et al. Protective effect of an extract from *Ascaris suum* in experimental arthritis models. *Infect Immun* 2008 Jun;**76**(6):2736–45.
67. Itami DM, Oshiro TM, Araujo CA, Perini A, Martins MA, Macedo MS, et al. Modulation of murine experimental asthma by *Ascaris suum* components. *Clin Exp Allergy* 2005 Jul;**35**(7):873–9.
68. Pritchard DI, Eady RP, Harper ST, Jackson DM, Orr TS, Richards IM, et al. Laboratory infection of primates with *Ascaris suum* to provide a model of allergic bronchoconstriction. *Clin Exp Immunol* 1983 Nov;**54**(2):469–76.
69. Patterson R, Harris KE. IgE-mediated rhesus monkey asthma: natural history and individual animal variation. *Int Arch Allergy Immunol* 1992;**97**(2):154–9.
70. Soares MFM, Mota I, Macedo MS. Isolation of components of *Ascaris suum* components which suppress IgE antibody responses. *Int Arch Allergy Immunol* 1992;**97**:37–43.
71. Faqim-Mauro EL, Coffmann RL, Abrahamsohn IA, Macedo MS. Mouse IgG1 antibodies comprise two functionally distinct types that are differentially regulated by IL-4 and IL-12. *J Immunol* 1999;**163**:3572–6.
72. Oshiro TM, Enobe CS, Araújo CA, Macedo MS, Macedo-Soares MF. PAS-1, a protein affinity purified from *Ascaris suum* worms, maintains the ability to modulate the immune response to a bystander antigen. *Immunol Cell Biol* 2006 Apr;**84**(2):138–44.
73. Macedo MS, Barbuto JAM. Murine delayed type hypersensitivity is suppressed by *Ascaris suum* extract. *Braz J Med Biol Res* 1988;**21**:523–5.
74. Faqim-Mauro EL, Macedo MS. The immunosuppressive activity of *Ascaris suum* is due to high molecular weight components. *Clin Exp Immunol* 1998;**114**:245–51.
75. Ferreira AP, Faquim ES, Abrahamsohn IA, Macedo MS. Immunization with *Ascaris suum* extract impairs T cell functions in mice. *Cell Immunol* 1995 May;**162**(2):202–10.
76. de Macedo Soares MF, de Macedo MS. Modulation of anaphylaxis by helminth-derived products in animal models. *Curr Allergy Asthma Rep* 2007 Apr;**7**(1):56–61.
77. Lima C, Perini A, Garcia ML, Martins MA, Teixeira MM, Macedo MS. Eosinophilic inflammation and airway hyper-responsiveness are profoundly inhibited by a helminth *(Ascaris suum)* extract in a murine model of asthma. *Clin Exp Allergy* 2002 Nov;**32**(11):1659–66.
78. Deehan MR, Goodridge HS, Blair D, Lochnit G, Dennis RD, Geyer R, et al. Immunomodulatory properties of *Ascaris suum* glycosphingolipids – phosphorylcholine and non-phosphorylcholine-dependent effects. *Parasite Immunol* 2002 Sep–Oct;**24**(9–10): 463–9.
79. Paterson JC, Garside P, Kennedy MW, Lawrence CE. Modulation of a heterologous immune response by the products of Ascaris suum. *Infect Immun* 2002 Nov;**70**(11): 6058–67.
80. Silva SR, Jacysyn JF, Macedo MS, Faquim-Mauro EL. Immunosuppressive components of *Ascaris suum* down-regulate expression of costimulatory molecules and function of

antigen-presenting cells via an IL-10-mediated mechanism. *Eur J Immunol* 2006 Dec;**36**(12):3227–37.

81. Spillmann RK. Pulmonary ascariasis in tropical communities. *Am J Trop Med Hyg* 1975 Sep;**24**(5):791–800.
82. Schopf L, Luccioli S, Bundoc V, Justice P, Chan CC, Wetzel BJ, et al. Differential modulation of allergic eye disease by chronic and acute ascaris infection. *Invest Ophthalmol Vis Sci* 2005 Aug;**46**(8):2772–80.
83. Feary J, Venn A, Brown A, Hooi D, Falcone FH, Mortimer K, et al. Safety of hookworm infection in individuals with measurable airway responsiveness: a randomized placebo-controlled feasibility study. *Clin Exp Allergy* 2009 Jul;**39**(7):1060–8.
84. Feary JR, Venn AJ, Mortimer K, Brown AP, Hooi D, Falcone FH, et al. Experimental hookworm infection: a randomized placebo-controlled trial in asthma. *Clin Exp Allergy* 2010 Feb;**40**(2):299–306.
85. Bager P, Arnved J, Rønborg S, Wohlfahrt J, Poulsen LK, Westergaard T, et al. *Trichuris suis* ova therapy for allergic rhinitis: a randomized, double-blind, placebo-controlled clinical trial. *J Allergy Clin Immunol* 2010 Jan;**125**(1):123–30.
86. Summers RW, Elliott DE, Urban Jr JF, Thompson RA, Weinstock JV. *Trichuris suis* therapy for active ulcerative colitis: a randomized controlled trial. *Gastroenterology* 2005 Apr;**128**(4):825–32.
87. Summers RW, Elliott DE, Urban Jr JF, Thompson R, Weinstock JV. *Trichuris suis* therapy in Crohn's disease. *Gut* 2005 Jan;**54**(1):87–90.
88. Daveson AJ, Jones DM, Gaze S, McSorley H, Clouston A, Pascoe A, et al. Effect of hookworm infection on wheat challenge in celiac disease – a randomised double-blinded placebo controlled trial. *PLoS One* 2011 Mar 8;**6**(3):e17366.
89. Levine MM. Immunogenicity and efficacy of oral vaccines in developing countries: lessons from a live cholera vaccine. *BMC Biol* 2010 Oct 4;**8**:129.
90. Kapulu MC, Simuyandi M, Sianongo S, Mutale M, Katubulushi M, Kelly P. Differential changes in expression of intestinal antimicrobial peptide genes during *Ascaris lumbricoides* infection in Zambian adults do not respond to helminth eradication. *J Infect Dis* 2011 May 15;**203**(10):1464–73.
91. Elias D, Wolday D, Akuffo H, Petros B, Bronner U, Britton S. Effect of deworming on human T cell responses to mycobacterial antigens in helminth-exposed individuals before and after bacille Calmette-Guérin (BCG) vaccination. *Clin Exp Immunol* 2001 Feb;**123**(2):219–25.
92. Vaca M, Moncayo AL, Cosgrove CA, Chico ME, Castello-Branco LR, Lewis DJ, et al. A single dose of oral BCG Moreau fails to boost systemic IFN-γ responses to tuberculin in children in the rural Tropics: evidence for a barrier to mucosal immunization. *J Trop Med* 2012;**2012**:132583.
93. Monteiro-Maia R, Ortigão-de-Sampaio MB, Pinho RT, Castello-Branco LR. Modulation of humoral immune response to oral BCG vaccination by *Mycobacterium bovis* BCG Moreau Rio de Janeiro (RDJ) in healthy adults. *J Immune Based Ther Vaccines* 2006 Sep 6;**4**:4.
94. Cosgrove CA, Castello-Branco LR, Hussell T, Sexton A, Giemza R, Phillips R, et al. Boosting of cellular immunity against *Mycobacterium* tuberculosis and modulation of skin cytokine responses in healthy human volunteers by *Mycobacterium bovis* BCG substrain Moreau Rio de Janeiro oral vaccine. *Infect Immun* 2006 Apr;**74**(4):2449–52.
95. Gracey M. The contaminated small bowel syndrome: pathogenesis, diagnosis, and treatment. *Am J Clin Nutr* 1979;**32**:234–43.
96. Feary J, Britton J, Leonardi-Bee J. Atopy and current intestinal parasite infection: a systematic review and meta-analysis. *Allergy* 2011 Apr;**66**(4):569–78.
97. Rodrigues LC, Newcombe PJ, Cunha SS, Alcantara-Neves NM, Genser B, Cruz AA, et al. Social change, asthma and allergy in Latin America. Early infection with

Trichuris trichiura and allergen skin test reactivity in later childhood. *Clin Exp Allergy* 2008 Nov;**38**(11):1769–77.

98. Endara P, Vaca M, Chico ME, Erazo S, Oviedo G, Quinzo I, et al. Long-term periodic anthelmintic treatments are associated with increased allergen skin reactivity. *Clin Exp Allergy* 2010 Nov;**40**(11):1669–77.
99. Lynch NR, et al. Effect of anthelmintic treatment on the allergic reactivity of children in a tropical slum. *J Allergy Clin Immunol* 1993;**92**:404–11.
100. van den Biggelaar AH, Rodrigues LC, van Ree R, van der Zee JS, Hoeksma-Kruize YC, Souverijn JH, et al. Long-term treatment of intestinal helminths increases mite skin-test reactivity in Gabonese schoolchildren. *J Infect Dis* 2004 Mar 1;**189**(5):892–900. Epub 2004 Feb 18.
101. Flohr C, Tuyen LN, Quinnell RJ, Lewis S, Minh TT, Campbell J, et al. Reduced helminth burden increases allergen skin sensitization but not clinical allergy: a randomized, double-blind, placebo-controlled trial in Vietnam. *Clin Exp Allergy* 2010 Jan;**40**(1):131–42.
102. Cooper PJ, Chico ME, Vaca MG, Moncayo AL, Bland JM, Mafla E, et al. Effect of albendazole treatments on the prevalence of atopy in children living in communities endemic for geohelminth parasites: a cluster-randomised trial. *Lancet* 2006 May 13;**367**(9522):1598–603.
103. Loeffler W. *Zur Frage der fluchtigen Lungeninfiltrate mit Bluteosinophilie Acta allergol* 1953;**6**(Supp. 3):186.
104. Gelpi AP, Mustafa A. *Ascaris* pneumonia. *Am J Med* 1968 Mar;**44**(3):377–89.
105. Maruyama H, Nawa Y, Noda S, Mimori T, Choi WY. An outbreak of visceral larva migrans due to *Ascaris suum* in Kyushu, Japan. *Lancet* 1996 Jun 22;**347**(9017):1766–7.
106. Leonardi-Bee J, Pritchard D, Britton J. Asthma and current intestinal parasite infection: systematic review and meta-analysis. *Am J Respir Crit Care Med* 2006 Sep 1;**174**(5): 514–23. Epub 2006 Jun 15.
107. Hagel I, Cabrera M, Hurtado MA, Sanchez P, Puccio F, Di Prisco MC, et al. Infection by *Ascaris lumbricoides* and bronchial hyper reactivity: an outstanding association in Venezuelan school children from endemic areas. *Acta Trop* 2007 Sep;**103**(3):231–41.
108. Hunninghake GM, Soto-Quiros ME, Avila L, Ly NP, Liang C, Sylvia JS, et al. Sensitization to *Ascaris lumbricoides* and severity of childhood asthma in Costa Rica. *J Allergy Clin Immunol* 2007 Mar;**119**(3):654–61.
109. Takeuchi H, Zaman K, Takahashi J, Yunus M, Chowdhury HR, Arifeen SE, et al. High titre of anti-Ascaris immunoglobulin E associated with bronchial asthma symptoms in 5-year-old rural Bangladeshi children. *Clin Exp Allergy* 2008 Feb;**38**(2):276–82. Epub 2007 Dec 7.
110. Camara AA, Silva JM, Ferriani VP, Tobias KR, Macedo IS, Padovani MA, et al. Risk factors for wheezing in a subtropical environment: role of respiratory viruses and allergen sensitization. *J Allergy Clin Immunol* 2004 Mar;**113**(3):551–7.
111. Pereira MU, Sly PD, Pitrez PM, Jones MH, Escouto D, Dias AC, et al. Nonatopic asthma is associated with helminth infections and bronchiolitis in poor children. *Eur Respir J* 2007 Jun;**29**(6):1154–60. Epub 2007 Mar 1.
112. da Silva ER, Sly PD, de Pereira MU, Pinto LA, Jones MH, Pitrez PM, et al. Intestinal helminth infestation is associated with increased bronchial responsiveness in children. *Pediatr Pulmonol* 2008 Jul;**43**(7):662–5.
113. Moncayo AM, Vaca M, Oviedo G, Workman LJ, Chico ME, Platts-Mills TAE, et al. Effects of geohelminth infection and age on the associations between allergen-specific IgE, skin test reactivity and wheeze: a case-control study. *Clin Exp Allergy* 2013;**43**: 60–72.

CHAPTER

2

Ascaris and Allergy

Luis Caraballo

University of Cartagena, Colombia

OUTLINE

Ascaris: The Neglected Parasite
http://dx.doi.org/10.1016/B978-0-12-396978-1.00002-1

Copyright © 2013 Elsevier Inc. All rights reserved.

INTRODUCTION

The relationships between parasitic and allergic diseases were perceived before the discovery of IgE. The subject has been of interest to several disciplines and progressively has been focused on particular parasites and specific allergies such as asthma. The search for explanations to the worldwide increase in the prevalence of immune-mediated diseases has led to increased attention that research groups in experimental allergy are giving to this problem. Currently there are several strategies for studying the interactions between *Ascaris* and allergy, focusing mainly on clinical cohorts for analyzing the impact of ascariasis on asthma pathogenesis, diagnosis and severity; genetic epidemiology surveys to explore the genetic and evolutionary basis of the Th2 responses; and animal models to study asthma pathogenesis at experimental level (Figure 2.1).

Ascariasis is a geohelminth disease where adult parasites live in the intestine while larvae migrate through tissues including the lungs. Although similar to the allergic reaction, the immune response to this nematode is modulated to variable degrees by parasite-induced immunosuppression. Atopy is the genetic predisposition to produce high levels of serum total IgE or specific IgE, while allergy (or allergic disease) is atopy plus the clinical symptoms. Ascariasis has been controlled in

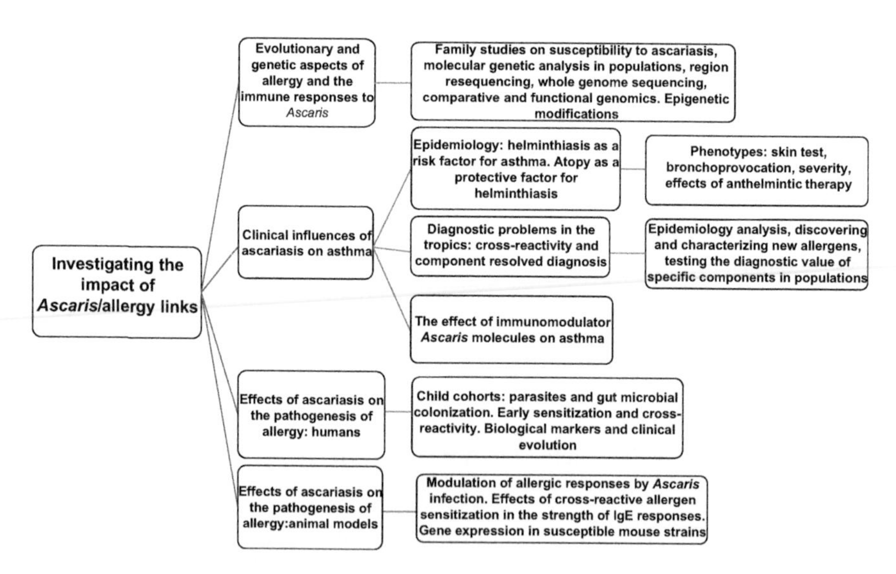

FIGURE 2.1 **Several aspects of the *Ascaris*-allergy association are currently under investigation.** Results from most of these studies are discussed in this chapter. Component resolved diagnosis employs purified allergens instead of the whole extracts for determining the immune reactivity (mainly specific IgE) of patients.

a number of countries but is still present in tropical zones; therefore, two general types of associations with allergy may currently be found: in parasite-free zones, the lack of helminth-induced immunosuppression may have contributed to the increase of allergic diseases. Besides, in countries where helminthiases are still prevalent, ascariasis is a risk factor or a protective factor for allergy; alternatives seem to be related to the intensity of the infection, which in turn depend not only on the level of exposure but also on the genetic background of the population. In addition, the existence of immunological cross-reactivity among molecules from *Ascaris* and domestic mites, which are very frequent in tropical zones, has special implications for the pathogenesis and diagnosis of allergy. Therefore, since ascariasis, whether being present or absent, is, in some way, related to the allergic responses, its impact is general and not limited to underdeveloped countries in the tropics.

BASICS CONCEPTS OF THE ALLERGIC RESPONSE: HOW SIMILAR IS IT TO THE IMMUNE RESPONSE TO *ASCARIS*?

The allergic response is a complex trait influenced by several genetic variants and characterized by a Th2-type hyper-responsiveness to environmental allergens that may lead to severe inflammation in various tissues, for example the lung of asthmatic patients. It includes a pro-Th2 immune activation of different cells through a variety of innate receptors and signaling pathways;[1] the production of a number of cytokines, among them IL-3, IL-4, IL-5, IL-9, and IL-13 from the cluster of chromosome 5q31-33, and others such as IL-33, IL-25, and thymic stromal lymphopoietin (TSLP);[2] also the hyper-production of total and specific IgE antibodies[3] and the parallel activation of cellular B and T regulatory networks.[4,5] Epithelial and dendritic cells are involved from the very beginning of this process and important in the definition of the Th2 character of the response.[6] The participation of the Th17[7] and Th9 responses in this inflammatory process has been also documented[2] and that of Th1 is mainly related to severe asthma.[8] In addition, some cytokines, such as tumor necrosis factor alpha (TNFA-α) and IL-13, which are considered important defense mechanisms against helminths,[9,10] are also activated during the allergic response and act directly on the bronchial epithelium and smooth muscle, increasing the effects of the IgE-mast cell-mediated inflammation in the lung.[11–13] This property has been also described for IL-17A.[14]

The participation of innate type 2 cells such as natural helper cells, nuocytes, and multipotent progenitor type 2 cells, originally described in association with experimental helminth infections, is under intense investigation and has been detected in experimental allergic asthma.[15,16] Basophils are cells traditionally related to the effector phase of the allergic response and immunity to helminths. Recently their role during the early Th2 response to allergens, supplying IL-4 or acting as antigen-presenting cells, has also raised considerable interest[17] and controversy.[18] The characteristic infiltrate of eosinophils, mononuclear cells, and mast cells, as well as the hyper-production of mucus by goblet cells, are consequences of the activation of diverse genes and cytokines; these cells, in turn, will help to increase the severity of the reaction if the allergenic stimulus persists.

The allergen-specific IgE bound to receptors on mast cells and basophils is central to initiate and maintain the chronic allergic inflammation and in some acute allergic manifestations, such as anaphylaxis, is the main trigger of the cascade leading to the release of vasoactive and inflammatory molecules that cause symptoms.[19] The allergic response, as detected by specific IgE to allergens, occurs in around 30% of the general population (considered atopic) and in almost 100% of subjects with allergic diseases. The fact that only a small percentage of atopic individuals develop allergic diseases suggests that clinical allergies such as asthma, rhinitis, and atopic eczema have additional genetic predispositions. The immunology of *Ascaris* and the immunomodulation it induces is described in Chapter 1 of this book. Most of the mechanisms of the allergic reaction are shared by the immune response to helminths and particularly to *Ascaris*; however, there are important differences.

First, the degree of immunomodulation is higher in helminthiases and increases with parasite load and disease severity. It is parasite-induced (several products are already known and others predicted by genome sequencing data[20]) and mediated in several ways, including the action of immunoregulatory cells (such as Treg, Breg, and alternatively activated macrophages) as well as cytokines such as IL-10 and transforming growth factor beta (TGF-β).[21,22] In addition, total IgE production during helminthiases is associated with immunomodulation, although the mechanisms are not defined. As has been long suspected and supported recently,[23–26] the polyclonal production of total IgE during severe helminth infections may inhibit the biologic pro-inflammatory effects of specific IgE, diminishing the expected allergic symptoms potentially induced by the infection itself in a normal population or symptoms in the allergic population. In contrast, the immunomodulation is weak during the allergic responses, which is characterized by an uncontrolled Th2 hyper-responsiveness,[27] also present during the chronic inflammatory process of asthma.[28] This deficiency makes the identification of the parasite-driven immunosuppressor effects of heavy ascariasis on allergic inflammation easier. Whether the high total IgE level characteristics of asthma have a causal effect upon this disease is unknown, but they are usually associated to disease severity[29] and there is no evidence of downregulation in the allergic response in non-parasited populations.

Second, the effector role of specific IgE in ascariasis is less relevant than in the allergic response. Although specific IgE is involved in protection against *Ascaris*,[30–32] recent discoveries suggest that innate and adaptive cellular and cytokine networks may be more important for immunity to helminths.[33–36] In contrast, specific IgE to allergens is a central phenotype of asthma and other allergies, some of them depending on the biological activity of these antibodies upon histamine-releasing cells, as occurs during allergic reactions to foods, drugs or other allergen sources. In summary, although involving common cells and inflammatory mediators, the immunopathogenesis of helminth infections and allergic diseases has important qualitative and quantitative differences. Traditionally the two processes have been considered as very similar; however, recognizing the differences will help to better understand the *Ascaris* parasitic relationships as well as the evolutionary roots of the allergic response.

ALLERGIC DISEASES IN TROPICAL REGIONS

Although it is believed that allergic diseases are infrequent in underdeveloped countries, the fact is that they are very prevalent, especially in urbanized middle-income areas.[37] The prevalence of asthma and rhinitis has been evaluated demonstrating that these problems are as frequent as in industrialized societies[38–41] and in some countries they have increased during the last decade.[42] One particular characteristic of allergic diseases in the tropics is that sensitizer allergens come mainly from domestic mites, with pollens, cockroaches, pets, and mold allergens being less clinically important.[43,44] In addition, the progression of allergic symptoms during infancy that frequently starts with respiratory symptoms seems to be different to the "allergic march" observed in European countries, where atopic dermatitis usually precedes the manifestation of asthma and rhinits. In a survey covering the general population of Colombia, among children from 1 to 4 years old, the prevalence of wheeze was 40.1% with 7.6% requiring hospitalization in the last 12 months while the prevalence of atopic dermatitis was 4.9%.[39] More recently, this pattern was confirmed in a birth cohort designed to study the allergy symptoms and risk factors for allergy in the tropics.[45]

It is probable that these particularities of allergies in the tropics are derived from the adverse (and hopefully transitory) circumstance of permanent exposure to domestic mites and *Ascaris*, two of the most allergenic sources known. However, the extent of the influence of helminth infections and mite co-exposure on the prevalence, age of inception, severity, and frequency of allergic episodes is unknown and other more complex questions persist. For example, urbanization is associated with higher prevalence of asthma and allergy in developing countries,[46] but the risk factors leading to this finding remain unclear. Differences in the level of parasite exposure and intensity of helminth infections in rural and urban settings could be an explanation. In addition, to what extent could ascariasis underlie the apparent contradictory fact of high allergy prevalence in places where, according to the hygiene hypothesis, the prevalence should be low? These and other questions will be addressed in the following comments.

ASCARIASIS, ALLERGIC SENSITIZATION, AND ALLERGY SYMPTOMS

There is a great deal of epidemiological and experimental work, both in humans and other animals, supporting the idea that ascariasis modifies the allergic response and the pathogenesis of asthma and other allergic diseases (reviewed in[47,48]). These investigations have detected that, depending upon several factors related to the type of parasites, the host, timing of exposure, infection intensity, and the environment, nematode infections may induce either severe immunosuppression or enhancement of the Th2 responses. Ascariasis brings together, but in different proportions, immunomodulation and IgE hyper-responsiveness; the latter is also a feature of the allergic responses and strongly depends on the genetic background of the host.[49]

The possibility that helminth allergens have an important role in allergies has been suspected for a long time because there are helminth infections associated with allergic, IgE-mediated symptoms. The most typical is anisakiasis that induces an asthma-like syndrome, urticaria and anaphylaxis.[50,51] In this case, the relationship with allergic symptoms is so evident that some authors consider it more an allergy condition than an infection[50] and the list of allergens of *Anisakis simplex* officially accepted by the WHO/IUIS Nomenclature Committee is the largest for any parasite. But before anisakiasis was discovered, clinicians from distinct disciplines had been dealing with hydatidosis, also known as echinococcosis. The rupture of hydatid cysts that may be located in different places of the body is a well-known cause of anaphylaxis, bronchospasm, and urticaria.[52,53] In addition, allergy symptoms associated with the migration of *Strongyloides* spp. and *Toxocara* ssp. are frequently observed in endemic areas[54–57] and adverse allergy symptoms have been described after vaccination with a recombinant *Ancylostoma* secreted-protein, *Na*-ASP-2.[58]

Ascariasis is also a recognized cause of allergy symptoms, including Loeffler's syndrome.[59–61] In addition, there are several human and animal experimental models showing the capacity of *Ascaris* antigens to induce parasite-specific IgE response and allergic symptoms.[61–64] However, since the relationship with allergy is not as evident as in anisakiasis, and some reports have demonstrated the immunosuppressive effect of chronic, heavy load infections in rural populations,[65] there is the erroneous belief that ascariasis only induces immune downregulation.

ASCARIASIS INFLUENCES THE PREVALENCE OF ASTHMA

Ascariasis has been associated with significantly enhanced likelihood of asthma[66] but this topic is controversial because studies have shown conflicting results. In some population surveys, the infection is a predisposing factor for IgE sensitization and asthma,[66–74] while in others it is protective.[75–78] There are several reasons for these apparently contradictory findings; the most important is that, in fact, ascariasis induces both suppression and enhancement of allergy symptoms according to the intensity of the infection.[79–82] Most of the studies concluding that this parasitic infection is associated with a low prevalence of allergy (or negative skin tests to common allergens) have been undertaken in heavily infected populations[79] and vice versa; those associated with higher prevalence of asthma have been performed in urban, lightly infected communities. Why the infection results in these two contrasting effects is unknown. Whether it reflects just a higher exposure to the parasite or is the result of more complex gene–environment interactions remains to be solved.

When analyzing the effect of ascariasis on the prevalence of asthma, it is necessary to define how the diagnosis was made. Some surveys use the specific IgE to *Ascaris* extract as a marker for ascariasis, a sensitive way to evaluate exposure to this parasite,[83] but disregard the potential bias from cross-reactivity between mites and *Ascaris* on the specificity of the test.[84] There is evidence suggesting that mite–*Ascaris* cross-reactivity influences the results of epidemiological surveys.[49] Not all authors adjust for this confounding variable, but in a study showing that specific IgE against *A. lumbricoides* extract was a risk factor for the number of positive skin tests and bronchial hyper-reactivity, the logistic regression analysis showed that the significance disappeared when correcting for specific IgE to mites and cockroaches.[69]

We will further analyze the mite–*Ascaris* cross-reactivity in this chapter. Here it is important to say that in spite of these and other limitations related to the study design and the heterogeneity of the populations surveyed, the number of publications supporting the association between ascariasis and a high prevalence of asthma or asthma-associated phenotypes is increasing,[85–87] probably because the surveys have been expanded to include urban zones where parasitism is less endemic but still present.

CAN HELMINTH INFECTIONS INCREASE THE ALLERGIC RESPONSE?

In addition to the downregulation of allergic responses detected during some nematode infections (more evident and better studied in schistosomiasis[88]), a strong IgE response dominates in human *Ascaris* infections, a phenotype that, for a long time, has been interpreted as potentially pro-allergenic and probably related to the complex life-cycle and the antigenic composition of this nematode. Also, high total IgE levels are typical of helminthiases, which seem to be the result of polyclonal B cell stimulation by parasite products.[89,90] Table 2.1 shows a list of issues supporting the hypothesis that, in some conditions, ascariasis enhances IgE responses to environmental allergens and allergies.

After penetration of the intestinal mucosa, *Ascaris* larvae migrate to the liver, inducing the formation of granulomas, extensive inflammation, and tissue injury.[91] Surviving larvae reach the lungs and generate an inflammatory infiltrate in the airways dominated by severe peri-alveolar eosinophilia.[92,93] Larvae induce antibody production; high levels of polyclonal and specific IgE are a hallmark of the infection and, in humans and pigs, immunity includes the generation of parasite-specific IgE antibodies against larvae and adult worms.[30,31] Experiments show that in

TABLE 2.1 Experimental and epidemiological findings supporting that ascariasis may enhance the IgE response

Findings	References
Natural infection is associated with a polarized Th2 cytokine response and high levels of total and anti-*Ascaris* IgE	61, 70, 89, 95, 171, 177–181
In some individuals, natural infection induces IgE-mediated allergic respiratory and cutaneous symptoms	59, 182, 183
In experimental human and other animal models, bronchial challenges with *Ascaris* extract induce asthma symptoms	60, 62, 63
Experimental ascariasis in animals enhances IgE response to bystander antigens	96–99
Several epidemiological surveys have found that ascariasis is a risk factor for asthma and atopy	66–72
IgE responses to *Ascaris* allergens is more frequent and stronger in mite-sensitized asthmatic patients	73, 110, 162, 171

humans and infected animals *Ascaris* brings sensitization and IgE-mediated asthma, including immediate-type cutaneous reactivity and airway responses after aerosol challenge with parasite extract.[59,60,62,63,70] For example, Hagel et al. found that specific IgE levels to *Ascaris* and positivity of skin test with the nematode extract were associated with bronchial hyper-reactivity in children from a rural area of Venezuela. Also, the percentage of forced expiratory volume in 1 second (FEV1) predictive values correlated inversely with anti-*Ascaris* IgE levels. In contrast, in urban children the same associations were with specific IgE to *Dermatophagoides pteronyssinus*.[70]

It has therefore been proposed that the systemic enhancement of the Th2 response during helminth infection induces an allergic polarization to bystander antigens, such as aeroallergens.[61,94,95] There is some experimental evidence supporting this contention. Earlier studies described that antigens of *A. suum* potentiate IgE response to ovalbumin in guinea-pigs.[96,97] Also, *Ascaris* pseudocelomic body fluid and the purified allergen ABA-1 prolonged the response to ovoalbumin as third-party allergen but they did not enhance the IgE levels to OVA.[98] In another investigation, co-administration of hen egg lysozyme with the excretory/secretory products of *Nippostrongylus brasiliensis* results in generation of egg lysozyme-specific lymphocyte proliferation, IL-4 release, and IgG_1 antibody responses, supporting the role of some nematode products as adjuvants for third-party antigens.[99] Furthermore, it has been shown that unidentified components in the body fluid of *Ascaris* promote a Th2 response and are adjuvants for specific IgE synthesis to some parasitic allergens like ABA-1.[100] Since, in addition to this allergen, *Ascaris* extract has at least 11 human-IgE binding components, the potentiating effect may be more generalized,[84] and because of co-exposure, this could happen for cross-reactive and non-cross-reactive mite allergens.

Based on their findings from early epidemiological studies, Lynch et al.[82] suggested that the prevalence of allergies may be lower in individuals with high parasite burdens of geohelminths compared to those with low burdens. This idea is now widely accepted and has been related to the acute and chronic clinical phenotypes observed in helminth-infected humans.[101] In addition, intermittent universal de-worming programs in pre-school and school-aged children reduce parasite burdens and boost the immune response to the soil-transmitted helminthes, because reinfections may elicit immune responses different in nature from the original primary infections. Therefore, it is theoretically possible that, in the presence of intermittent infections with low worm burdens, exposure to *Ascaris* promotes allergic sensitization and asthmatic symptoms by increasing the synthesis of parasite-specific, mite-specific, and mite–parasite cross-reacting IgE antibodies. The clinical impact may be particularly important in urban zones of underdeveloped countries, because in rural areas the infections are usually more intense and associated with higher degrees of immunosuppression. Also, differences in mite fauna and levels of mite-allergen exposure may influence the type of sensitization and, in consequence, the relevance of cross-reactivity.

THE ALLERGENS OF A. *LUMBRICOIDES*

As already mentioned, domestic mites are the most important risk factor for asthma in the tropics and their allergens have been extensively characterized. To study the effects of the immune responses to *Ascaris* on allergy, molecules that induce allergy symptoms, those that generate a protective IgE immune response, and those that promote both effects should be identified. Some antigenic components of *Ascaris* (e.g. As14, As16, As24, As37, PAS-1) have been analyzed,[102–106] but information about the allergenic composition and its clinical impact in humans is still very limited. Less is known about the influence of the IgE/IgG immune responses to other nematode molecules, such as glycoproteins and glycolipids[107,108] on the evolution of allergy. Also, the role of chitin as adjuvant of the Th2 responses[109] is particularly interesting since it is an important component of *Ascaris* cuticle and could partially explain the prominent allergenic properties of this parasite. A systematic search of antigen identification and allergens of *Ascaris* inducing immune responses in humans is currently ongoing in our laboratory.

Asc s 1, also known as ABA-1, and Asc l 3, *Ascaris* tropomyosin, are the officially accepted (WHO/IUIS) allergens of this nematode. In addition, several IgE binding components have been detected using one-dimensional (1D) and two-dimensional polyacrylamide-gel electrophoreses (2D-PAGE). Among the 12 allergenic components we have found,[84] there are seven that are cross-reactive with mite allergens and five that are *Ascaris* specific. This explains the high degree of cross-reactivity between mite and *Ascaris* extracts and supports the idea that ascariasis may lead to an enhanced IgE response to several mite allergens. Tropomyosin and glutathione-transferase (GST) are two of these cross-reactive allergens.

Recently, the immunochemical properties of *Ascaris* tropomyosin (Asc l 3) were revealed. Very high allergenic cross-reactivity between the natural Asc l 3 and *Blomia tropicalis* tropomyosin, Blo t 10, was found using sera from asthmatic patients.[84] These results were confirmed using a recombinant *Ascaris* tropomyosin expressed in a bacterial system.[110] IgE antibodies to rAsc l 3 represented a high proportion (~50%) of the total IgE response to an unfractionated parasite extract and there was allergenic equivalence between rAsc l 3 and the native counterpart in the *Ascaris* extract. Moreover, anti-tropomyosin IgE antibodies from sensitized subjects reacted against Asc l 3 and induced mediator release in effector cells, both *in vivo* and *in vitro*. ABA-1, although present in other nematodes, is not cross-reactive with any of the *B. tropicalis* or *D. pteronyssinus* allergens, which means that it can be useful for component-resolved diagnosis of allergic diseases in the tropics.

Possible Effects of *Ascaris* Allergens on the Clinical Evolution of Asthma

One potential mechanism to enhance the IgE responses to allergens in asthmatic patients living in the tropics is cross-reactivity (reviewed in[111]). It can act at several stages of ascariasis or asthma, and the tropical environment provides appropriate conditions for this relationship.

First, there is the possibility of early life co-exposure to allergens from mites and *Ascaris*. The complex interactions elicited by allergenic molecules from different sources acting together on the innate and adaptive immune responses are not yet clearly defined, but one possible outcome is the enhancement of the allergic responses.[47] Early IgE responses to mites and *Ascaris* have been observed in children from tropical regions and some studies have found clinical relevance.[70,112]

Second, parasited children receive anthelminthic therapy during intermittent universal de-worming programs aimed at pre-school and school age. Since the socioeconomic causes of the infections are not eliminated, children become reinfected several times and modified secondary immune responses may boost IgE reactivity against cross-reactive allergens from other sources (e.g. mites). Repeated anthelmintic treatments significantly increase the production of Th2 cytokines, IL-5, and IL-13 and decrease the production of IL-10 by peripheral blood leukocytes after stimulation with *Ascaris* antigens, although no changes were observed when stimulating with *D. pteronyssinus* and cockroach[113]; it has also been found that long-term periodic treatments in a community with various helminthiases, including ascariasis, is associated with increase of allergen skin reactivity.[114]

In schistosomiasis, there is evidence that anthelminthic treatment influences the evolution of several mechanisms of immunity, including increasing the proportion of effector T cells and the switch to protective antibody isotypes such as IgE,[115,116] probably because of higher loads of antigens from dead parasites,[115,117] and the removal of immunosuppressive parasite products.[118] Reinfections can stimulate memory cells,[119] and the mentioned effects of anthelmintics may be also present during the treatment of other helminthiases such as ascariasis.

Third, as it is well known, mite-allergens exposure is perennial and very intense in the tropics; therefore, in the *Ascaris*-infected population (current or past) susceptible to asthma, this may be another cause of increasing IgE responses to cross-reactive allergens. It can be speculated that patients predisposed to asthma, with a strong pro-Th2 genetic background, infected at an early age with reinfections and permanently exposed to mite allergens probably have a stronger IgE response to allergens and more severe clinical symptoms. It is also important to consider that cross-reactivity between *Ascaris* and other nematodes, such as *Ancylostoma duodenale, Strongyloides stercoralis, Trichuris trichiura, Necator americanus*, and *A. simplex* has been described.[120–124] In addition to confounding the serologic diagnosis of helminthiases, cross-reactivity could play a role in the pathogenesis of allergic diseases in the tropics, especially during helminth–helminth co-infections.

The Role of ABA-1 (Asc s 1) and Tropomyosin (Asc l 3) in the Ascariasis/Allergy Relationship

Although both allergens are able to induce IgE response, there are data suggesting that ABA-1 and Asc l 3 have different capacities to induce protective antibody immunity or allergy sensitization. In humans, the IgE and IgG responses to ABA-1 have been more related to protection[32] than to allergic symptoms; in contrast, tropomyosin is a well-recognized invertebrate pan-allergen and Asc l 3 has a high degree of homology and cross-reactivity with mite tropomyosins. However, the possibility that ABA-1 may induce allergic symptoms has not been ruled out. In addition, IgE response to tropomyosin has been associated with resistance to helminths.[125] Further studies are therefore necessary to define the particular impact of each of these allergens. In a case–control survey, evaluating the relationship between IgE, *Ascaris* extract, and asthma, we found a statistically significant association that disappeared when adjusting for mite-specific IgE, which confirms the importance of mites as triggers of asthma in the tropics. This weak association could be due to mite–*Ascaris* cross-reactivity because no association was found with the IgE antibodies to ABA-1, which has no cross-reactivity with mite allergens and seems to have low allergy-inducer potential.

Among the mite–*Ascaris* cross-reactive allergens, *Ascaris* tropomyosin is a good candidate. In the same study, the frequency of sensitization to rAsc l 3 was greater in asthmatics than in controls. Although adjusting for covariates such as specific IgE to mites lowered the significance to borderline, when analyzed as a continuous variable, specific IgE levels to rAsc l 3 were significantly higher in asthmatic patients than in controls.[49] These findings, although not defining an independent association of Asc l 3 with asthma, suggest that it may be a risk factor for asthma symptoms in the tropics, especially to individuals with the predisposition to recognize Asc l 3 cross-reactive epitopes. The C-terminal end of this allergen has high sequence similarity with IgE binding epitopes found in tropomyosins from mite, cockroach and other invertebrates. The actual role of the IgE responses to cross-reactive and non-cross-reactive epitopes on the allergic response and resistance to *Ascaris* needs to be evaluated empirically.

ASCARIASIS INFLUENCING THE DIAGNOSIS OF ALLERGY

High total IgE levels are strongly associated with asthma;[126,127] and this biological marker has been used to assess the allergy component in asthma diagnosis, even though a causal effect has not been demonstrated and there are other conditions associated with elevated total IgE. In tropical regions, where helminthiases are still public health problems and are frequent inducers of high serum total IgE,[128] this test is not useful for diagnosing asthma or other allergic diseases in individual subjects. Probably because of its unspecific nature,[129,130] total IgE is being replaced by specific IgE to allergens as a good marker of both atopy and allergic diseases. But in the tropics there is an additional problem: the impact of cross-reactivity in the diagnosis of mite allergy, the most important risk factor for asthma. *Ascaris* cross-reactive allergens, such as tropomyosin and GST, may influence mite allergy diagnosis when using the whole mite extracts, as is routinely done *in vitro* and for skin testing. Therefore, in the tropics the use of complete mite extracts for diagnosis could lead to false positive results. Component-resolved diagnosis, including mite and *Ascaris* purified allergens, could help to resolve this problem.

Another limitation of serologic allergy diagnostics in patients with ascariasis and other helminthiases is the presence of the carbohydrate epitope galactose-α1,3-galactose (α-Gal), expressed in nonprimate mammalian proteins, such as the cat allergen Fel d 5 (cat IgA). It has been reported that the serologic diagnosis of cat allergy may be impaired among parasited patients.[131] However, since patients of this study had several helminthiases, it is difficult to define the particular importance in *Ascaris*-infected subjects. There are other reports suggesting that intestinal helminths have α-Gal,[132] but immunochemical studies on *Ascaris* extract show that deglycosilation treatment does not significantly affect the specific IgE antibody binding to the extract.[84] Further research is needed to confirm the presence of this cross-reactive carbohydrate in *Ascaris*.

It has been reported that skin test reactivity to clinically relevant allergens is low among children with chronic severe helminthiases, including ascariasis,[65] though the IgE response to allergens can be demonstrated in those patients using serological methods. Several mechanisms, among them the effects of total unspecific IgE competing with specific IgE for FcER1 binding, and the anti-inflammatory action of cytokines, mainly IL-10, may underlie this phenomenon. Also, parasite-derived immunomodulators, such as ES-62 from filarial nematodes, could contribute by directly inhibiting mast cell release of allergy mediators.[133] This type of molecule may be present in *Ascaris* because sequence alignment shows that it has an orthologue product. In addition, reduction in the basophil responsiveness has been reported recently in mice chronically infected with filarial nematodes[134] and humans infected with *Ascaris* and other helminths.[135] Although the mechanisms are not clear, the fact is that allergy diagnosis by skin testing is impaired in parasited populations in the tropics. Since the immunosuppression may also diminish allergy symptoms, this epidemiologic finding may be not visible at the individual patient–physician level. However, if sensitization to allergens is evaluated in these communities using skin tests, false negative results will be common, which may lead to the opinion that not only allergy symptoms but also allergen sensitization are infrequent in parasited populations. Whether this is what actually occurs is a matter of current investigation in our birth cohort.[45]

ASCARIASIS AND ASTHMA SEVERITY

There are few studies evaluating the role of ascariasis on the severity of asthma. This may be because of the inherent general difficulties in the investigation of risk factors for asthma severity, but also because the studies should be done in places where most of the population does not receive the appropriate asthma treatment. One indirect analysis of the problem is to observe the effect of anthelmintic treatment on clinical symptoms. In one study the treatment ameliorated the frequency asthmatic crisis or other markers of severity, suggesting that ascariasis increases asthma severity in endemic zones.[136] In a more direct way, Hunninhake et al. studied the problem in a cross-sectional survey in asthmatic children from Costa Rica[69] and found that sensitization to *Ascaris* was associated with increased severity and morbidity of asthma. Although in this study the role of co-sensitization with mites, one of the main causes of asthma, was not totally excluded, the side effects of the IgE response to *Ascaris* cannot be ruled out. As has been analyzed in this chapter, most of the evidence that supports an increase of the Th2 inflammatory response in the lungs induced by ascariasis is related to the immune responses to the nematode.

The direct effect of pro-inflammatory cytokines on the airways is another possibility when analyzing the relationship between ascariasis and asthma severity. This is especially problematic because it is expected that such a mechanism may occur during the most severe infections by *Ascaris*, which are supposed to be the majority inducers of immunoregulation. However, it should be mentioned that cytokines such as TNF-alpha, IL-13, and IL-17, which are central mediators of allergic pulmonary inflammation in humans[137–139] and asthma severity markers, are one of the most important mechanisms of helminth elimination in experimental animal models.[9,140] It can be speculated that the highly *Ascaris*-resistant population (whether atopic or not) may also overreact with this cytokine-mediated defense mechanism, causing tissue damage and increasing the severity of pulmonary symptoms. In this regard, it has been reported that during an experimental helminth infection there was first a strong IL-17 response that caused lung damage, which was resolved by a parasite-induced Th2 responses that included IL-10 and M2 macrophages.[141] However, the actual role of these cytokines on the severity of asthma deserves more investigation. The current possibility of detecting very low levels of proteins in the circulation makes it possible to simultaneously evaluate the immunomodulator molecules and these cytokines during human ascariasis.

PARASITE INFECTIONS, ALLERGY, AND THE HYGIENE HYPOTHESIS

The hygiene hypothesis, formulated after the epidemiologic finding of lower prevalence of hay fever associated with poor hygiene conditions and other protecting factors,[142] predicts that allergic diseases are more frequent where the improvement of hygiene conditions has been really successful, making microbial infections infrequent in early childhood. Several mechanisms explain how the hygiene hypothesis works. Immune deviation to a predominant Th1 response due to bacterial and viral infections was, at the beginning, the most obvious explanation; but there is recent evidence that one of the most potent Th1 inducers, tuberculosis, instead of protecting is positively associated with allergy.[143] An important point is that, according to the hypothesis, it is expected that allergic diseases have a low prevalence where hygiene conditions are poor, and this actually occurs at least in some geographical locations, but this is likely to be because of chronic helminth infections.[65,82,144] More interestingly, in several moderate- to low-income countries of the tropical zone, the prevalence of asthma and other allergic diseases is high and concurs with early exposure to bacterial and viral infections.[38–41,46,145–150] Therefore, helminthiases may explain not only why poor hygiene conditions are associated with low allergy frequency,[151,152] but also why the increasing trend in the prevalence of allergies is more general and not restricted to affluent countries with good hygiene conditions.[153]

Typically, soil-transmitted helminth infections are susceptible to change in frequency when hygiene conditions improve. During the last decade, the immunosuppressive effects of chronic, heavy helminth infections have been described in both humans and other animals, being stronger than any described immunomodulator phenomenon accompanying bacterial or viral infections excepting human immunodeficiency virus (HIV) infection. This has reinforced the idea that parasite infections have played a major role in controlling the allergic responses; and the lack of this control, because of an improvement in hygiene conditions, may be a relevant factor in allergy prevalence.[154] Therefore, the high prevalence of asthma that is observed in some urbanized zones of the tropics, where helminth infections, such as ascariasis, are still present but with less intensity than in the past, may be explained, among other factors, as a consequence of the particular historical moment of the ancient and complex relationships between parasites and the immune system: a point where, because of several reasons, the immunostimulator effects of helminths on the Th2/IgE responses predominate. As mentioned, the type and distribution of parasites, as well as the frequency, intensity, and immunomodulator effects of helminth infections, are not the same throughout the world. In the tropics, both the immunosuppressive and the Th2-immunopotentiating effects can be detected, the latter being more frequent at the population level. Therefore, in this changing world, three distinct relationships between helminths and the human immune system can be recognized: one with chronic, heavy intensity infections and mainly immunosuppressive; others of intermittent low parasite-load infections, predominantly IgE-enhancer and associated with urbanization; and a third, with absence of infections, where there is no parasite-derived immunoregulation.

The possible consequences of each of these conditions on the development of allergic diseases have been already analyzed, but three additional comments are pertinent. First, allergic diseases are highly dependent on the effects of other environmental factors; for example, the potential effects of changes associated with urbanization, differences in diet and lifestyle, physical activity and housing[155] should also be considered. Second, a comprehensive study of the human–helminth relationships should include the genetic and evolutionary components, which may provide relevant information about the current host–parasite interactions.[156–158] Third, since the pathogenesis of ascariasis and other helminth infections, as well as that of asthma, are highly influenced by genetic factors, these will limit the proportions of individuals that establish any of the proposed relationships in a given population.[49]

ALLERGY AS A PROTECTIVE FACTOR FOR ASCARIASIS: THE OTHER FACE OF CO-EVOLUTION

Although this chapter is mainly about the effects of ascariasis on allergy, some words about the contrary are necessary as they have to do with the evolutionary history of both processes. Because of the scarce empirical research in this particular field, most analyses are supported by hypothesis and interpretations of the theory of evolution through natural selection. In addition, during the last decade, theoretical models of population genetics, genomic scans for selection and bioinformatics have been used to explore the selective pressure of helminths and other infectious organisms on immunity-related genes.[159–161] The following is a brief and personal view about this interesting topic.

From an evolutionary perspective, it is theoretically plausible that strong Th2 immune response that overcomes immunosuppression protects against ascariasis and other helminthiases, and, in fact, there is evidence that atopic individuals have lower intensities of helminth infections.[162–164] A number of factors and organisms, including parasites, have influenced the evolution of defense mechanisms of humans but, because of the multiple phenotype similarities, the evolutionary process of the allergic response and the immune response to helminths are supposed to be the same, or at least very close.[21,165,166] However, a fundamental unresolved question is if current genotypes predisposing to allergy were selected by helminth infections or, to the contrary, those genotypes (shaped by factors different to parasites), generating resistance to helminths, avoided the potential lethal effects of these infections and exerted selective pressure on them.

One of the possible consequences of the long co-evolution of *Ascaris* and different hosts is the selection, among both species, of those organisms with the best genotypes to establish and maintain a parasitic relationship. Considering the complexity of these relationships, this implies a large number of genotypes covering a variety of physiologic processes, but, based on the latest advances in our understanding of the nature of human–*Ascaris* interactions, examples of these organisms may be, for the parasite, those able to induce immunomodulation and for the host, those susceptible to such immunomodulation. This condition does not presume that immunosuppressor properties made successful parasitic helminths totally harmless to their hosts,[167,168] but suggests that the selective pressure exerted by helminths could positively select polymorphisms determining host resistance to immunomodulation and in turn this could be an important mechanism underlying the different degrees of "resistance phenotypes" in humans.

Among the human genotypes resulting from this bilateral adaptive process and involving the "resistance" phenotype, two can be hypothesized as described in Figure 2.2. Based on this model, several subpopulations and outcomes are theoretically possible in regard to resistance to *Ascaris* and the effects of both the infection and parasite eradication on the prevalence of symptoms. It could be that a group of highly allergic subjects are insensitive to immunomodulation of any type, including that

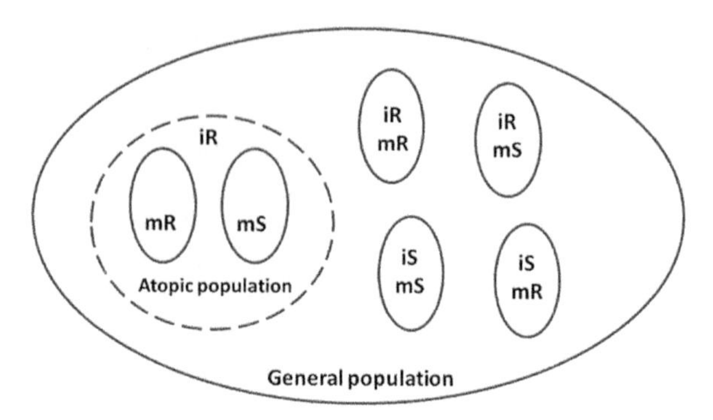

FIGURE 2.2 **Hypothetical distribution of human genotypes involved in the host–*Ascaris* relationships.** Two groups of genes are considered: immune-related genes (i) and genes influencing the susceptibility to parasite-induced immunomodulation (m). In the atopic population immune-mediated resistance (iR) is supposed to be present, therefore, variations in the genotypes depend on the resistance or susceptibility to immunomodulation (mR and mS, respectively). In the non-atopic population, four genotypes may be predicted: iR/mR (underlying a high-resistant phenotype); iR/mS, iS/mR, and iS/mS, the latter being present in the most susceptible subjects. The model assumes that the high-resistance phenotype is not exclusive of the atopic population. Since several variants are expected to occur in the genes of each genotype, in the general population the degree of resistance/susceptibility will appear as a gradient.[49] Note that the separation of "i" and "m" genes is done to get the simplest presentation of the hypothesis; however, this classification is very artificial due to the difficulty in delimiting "immune-related genes"[161] and because the immune response is just one component of the complex host–parasite interactions during infection. Furthermore, "m" and "i" genes may have similar effect in terms of defense.

from *Ascaris*, and have the lower levels of infection and better immunity. Provided with the necessary environmental factors, they will have allergy symptoms whether they are infected or not with *Ascaris*, and the infection may increase symptoms. Other groups of allergy-prone individuals are susceptible to immunomodulation; they will have more symptoms if parasitic infections were removed. In addition, most of the population is not predisposed to allergy, have variable immune-related resistance to *Ascaris*, and variable susceptibility to helminth-induced immunomodulation. Here the most typical forms of parasitic relationships will be present; *Ascaris* control will have no effect on the prevalence of allergy in this subpopulation.

The protective effect of allergy from ascariasis may have no great individual clinical relevance, but it has potential epidemiological impact and contributes to the analysis of the genetics of both allergy and *Ascaris* resistance. For a long time it has been supposed that allergy is a legacy of the evolutionary adaptive process of the protective response to helminths and, as a consequence, genotypes associated with helminth resistance are expected to be the same as those predisposing to allergy. Research conducted under this premise has found some supporting data.[169,170] However, looking at the problem from another perspective, whether genes associated with clinical allergy or the immune responses to allergens are associated with resistance to *Ascaris* should be explored empirically. In this regard, a case–control study evaluating for the first time, in the same population and under the same environmental exposures, polymorphisms possibly associated with the IgE responses to *Ascaris* antigens, domestic mite allergens and asthma, found variants associated only with the IgE responses to *Ascaris*.[171] Since this study was focused on particular genes (*TNFSF13B*, *LIG4* and *IRS2*) located in a region previously found to be linked to ascariasis resistance,[172,173] further research on other chromosomes of this population is ongoing.

In spite of the great progress on allergy, immunoparasitology, and genomics during the last decade, important theoretical aspects of the evolutionary relationships between ascariasis and allergy remain unsolved and testing more hypotheses is guaranteed. For example, the existence of a population strongly resistant to *Ascaris* and non-atopic should be kept in mind. Further studies on this subgroup will help to elucidate the resistance mechanisms of human helminthiases and, besides, the genetics of allergic diseases, which involves more genes than those of type 2 immunity. In addition, polymorphisms implicated in the resistance to ascariasis have been searched for among immune-related genes, but exploring genetic variations in the pathways of parasite-induced immunomodulation may provide interesting information about the genetics of human–*Ascaris* relationships. Empirical genomic research in populations affected with both helminthiases and allergy, as well as bioinformatics analyses oriented by evolutionary genetic concepts, will provide significant signals deserving confirmation by functional studies.

THE POSSIBLE THERAPEUTIC IMPACT OF *ASCARIS* IMMUNOMODULATORY PRODUCTS ON ALLERGY

It is now widely recognized and well documented that helminth infections, including ascariasis, are associated with variable degrees of immune response downregulation. However, the parasite-derived products that dampen the immune system are not fully recognized. It is known that helminths produce immunomodulator substances (downregulators being the best studied) that can modify phenotypes of the immune responses. The possibility of using them for controlling the adverse effects of the inflammation associated with human immune disorders, such as autoimmunity or allergy, is currently under investigation. Detailed aspects of the immunomodulation induced by *Ascaris* can be found in Chapter 1. Here, some general comments about the possibility of using parasite-derived immunomodulator components to ameliorate allergic inflammation are presented.

The immunomodulation is evident in chronic and heavily parasited individuals; this could mean that high doses are needed to achieve an appropriate downregulation of the allergic responses, but also may reflect, as already mentioned, that a particular subpopulation is highly sensitive to parasite-derived immunomodulation (Figure 2.2). Since, at the population level, this degree of sensitivity may appear as a gradient, it is likely that dosage of a particular potential therapeutic product should be adapted to this circumstance. In addition, if the hypothesis of a genetic predisposition to helminth-induced immunosuppression were true, immunosuppressor products will not exert their effects in all allergic patients.

Since immunosuppressant as well as immunostimulant effects have been observed during helminthiases, therapeutic immunomodulation should be induced with helminth-purified products and not with live helminth parasites. The use of "non-pathogen" or natural-infecting helminths (e.g. *Ancylostoma* spp.) does not avoid the risk of sensitization to cross-reactive allergens or the unspecific boosting the allergic responses. In addition, helminths are not harmless and in some cases serious health problems may arise. Using low doses of parasites and short-term infections, in addition to providing contradictory outcomes, has not significantly ameliorated asthma or rhinitis symptoms[174,175] and may not be a good strategy because, as has been described in this chapter, some epidemiological studies have shown that mild helminth infections are associated with more allergy symptoms. As in any other health problem, ethical issues are paramount in this field of therapeutic research, especially when using live parasites.

In addition to PAS-1, which has immunosuppressor activity on both Th1 and Th2 responses,[176] the potential *Ascaris*-derived immunomodulator molecules comprise a large number of the secretome[20] but none of them has been fully characterized. Given the different pathogenic mechanisms of helminthiases, especially the larval migratory phases and the final localization of mature forms, it is possible that not all helminths produce the same immunomodulators; the search for those downregulating the allergic responses without adverse effects upon the protective immune response will provide the most useful results in this field.

CONCLUSION

Ascariasis influences allergy, namely asthma, in several ways. Other intestinal helminthiases may potentiate its effects on allergy, but it is hoped that appropriate socioeconomic progress coupled with control efforts will lead a significant reduction in soil-transmitted helminthiases.

In the meantime, the *Ascaris*-allergy relationship will continue to have an impact upon the origin and management of allergy in the tropics and may be used as a model to better understand the allergic response in humans. It is possible that, to the extent helminthiases decrease, allergies will increase. However, genetic predisposition to allergic diseases is restricted to a percentage of individuals; therefore, the increase will be limited by the genetic background of populations. In addition, since ascariasis is not the only factor that can affect the prevalence and severity of asthma or other immune-mediated diseases, the magnitude of the effects of its eradication have to be evaluated in the context of multiple environmental factors. For the large population currently affected with ascariasis, control campaigns should be reinforced, especially with programs based on social improvements. For the atopy-prone population, the benefits of immunomodulatory therapy using helminth-derived pharmaceutical products will be available in the future.

References

1. Pulendran B, Artis D. New paradigms in type 2 immunity. *Science* 2012;**337**:431–5.
2. Holgate ST. Innate and adaptive immune responses in asthma. *Nat Med* 2012;**18**: 673–83.
3. Galli SJ, Tsai M. IgE and mast cells in allergic disease. *Nat Med* 2012;**18**:693–704.
4. Noh G, Lee JH. Regulatory B cells and allergic diseases. *Allergy Asthma Immunol Res* 2011;**3**:168–77.
5. Palomares O, Yaman G, Azkur AK, Akkoc T, Akdis M, Akdis CA. Role of Treg in immune regulation of allergic diseases. *Eur J Immunol* 2010;**40**:1232–40.
6. Lambrecht BN, Hammad H. The airway epithelium in asthma. *Nat Med* 2012;**18**: 684–92.
7. Souwer Y, Szegedi K, Kapsenberg ML, de Jong EC. IL-17 and IL-22 in atopic allergic disease. *Curr Opin Immunol* 2010;**22**:821–6.
8. Poon AH, Eidelman DH, Martin JG, Laprise C, Hamid Q. Pathogenesis of severe asthma. *Clin Exp Allergy* 2012;**42**:625–37.
9. Bancroft AJ, McKenzie AN, Grencis RK. A critical role for IL-13 in resistance to intestinal nematode infection. *J Immunol* 1998;**160**:3453–61.
10. Hayes KS, Bancroft AJ, Grencis RK. The role of TNF-alpha in *Trichuris muris* infection II: global enhancement of ongoing Th1 or Th2 responses. *Parasite Immunol* 2007;**29**:583–94.
11. Grunig G, Warnock M, Wakil AE, et al. Requirement for IL-13 independently of IL-4 in experimental asthma. *Science* 1998;**282**:2261–3.
12. Nakae S, Ho LH, Yu M, et al. Mast cell-derived TNF contributes to airway hyperreactivity, inflammation, and TH2 cytokine production in an asthma model in mice. *J Allergy Clin Immunol* 2007;**120**:48–55.
13. Wills-Karp M, Luyimbazi J, Xu X, et al. Interleukin-13: central mediator of allergic asthma. *Science* 1998;**282**:2258–61.
14. Kudo M, Melton AC, Chen C, et al. IL-17A produced by alphabeta T cells drives airway hyper-responsiveness in mice and enhances mouse and human airway smooth muscle contraction. *Nat Med* 2012;**18**:547–54.
15. Hams E, Fallon PG. Innate type 2 cells and asthma. *Curr Opin Pharmacol* 2012;**12**(4): 503–9.

16. Barlow JL, Bellosi A, Hardman CS, et al. Innate IL-13-producing nuocytes arise during allergic lung inflammation and contribute to airways hyperreactivity. *J Allergy Clin Immunol* 2012;**129**:191–8. e1–4.
17. Sokol CL, Chu NQ, Yu S, Nish SA, Laufer TM, Medzhitov R. Basophils function as antigen-presenting cells for an allergen-induced T helper type 2 response. *Nat Immunol* 2009;**10**:713–20.
18. Eckl-Dorna J, Ellinger A, Blatt K, et al. Basophils are not the key antigen-presenting cells in allergic patients. *Allergy* 2012;**67**:601–8.
19. Gould HJ, Sutton BJ. IgE in allergy and asthma today. *Nat Rev Immunol* 2008;**8**:205–17.
20. Jex AR, Liu S, Li B, et al. *Ascaris suum* draft genome. *Nature* 2011;**479**:529–33.
21. Allen JE, Maizels RM. Diversity and dialogue in immunity to helminths. *Nat Rev Immunol* 2011;**11**:375–88.
22. Fallon PG, Mangan NE. Suppression of TH2-type allergic reactions by helminth infection. *Nat Rev Immunol* 2007;**7**:220–30.
23. Godfrey RC, Gradidge CF. Allergic sensitisation of human lung fragments prevented by saturation of IgE binding sites. *Nature* 1976;**259**:484–6.
24. Jarrett E, Mackenzie S, Bennich H. Parasite-induced "nonspecific" IgE does not protect against allergic reactions. *Nature* 1980;**283**:302–4.
25. Mitre E, Norwood S, Nutman TB. Saturation of immunoglobulin E (IgE) binding sites by polyclonal IgE does not explain the protective effect of helminth infections against atopy. *Infect Immun* 2005;**73**:4106–11.
26. Xiong H, Dolpady J, Wabl M, Curotto de Lafaille MA, Lafaille JJ. Sequential class switching is required for the generation of high affinity IgE antibodies. *J Exp Med* 2012;**209**:353–64.
27. Akdis M, Verhagen J, Taylor A, et al. Immune responses in healthy and allergic individuals are characterized by a fine balance between allergen-specific T regulatory 1 and T helper 2 cells. *J Exp Med* 2004;**199**:1567–75.
28. Robinson DS. The role of the T cell in asthma. *J Allergy Clin Immunol* 2010;**126**:1081–91. quiz 92–3.
29. Naqvi M, Choudhry S, Tsai HJ, et al. Association between IgE levels and asthma severity among African American, Mexican, and Puerto Rican patients with asthma. *J Allergy Clin Immunol* 2007;**120**:137–43.
30. Hagel I, Cabrera M, Buvat E, et al. Antibody responses and resistance against *Ascaris lumbricoides* infection among Venezuelan rural children: the influence of ethnicity. *J Trop Pediatr* 2008;**54**:354–6.
31. Turner JD, Faulkner H, Kamgno J, et al. Allergen-specific IgE and IgG4 are markers of resistance and susceptibility in a human intestinal nematode infection. *Microbes Infect* 2005;**7**:990–6.
32. McSharry C, Xia Y, Holland CV, Kennedy MW. Natural immunity to *Ascaris lumbricoides* associated with immunoglobulin E antibody to ABA-1 allergen and inflammation indicators in children. *Infect Immun* 1999;**67**:484–9.
33. Jackson JA, Turner JD, Rentoul L, et al. T helper cell type 2 responsiveness predicts future susceptibility to gastrointestinal nematodes in humans. *J Infect Dis* 2004;**190**: 1804–11.
34. Moro K, Yamada T, Tanabe M, et al. Innate production of T(H)2 cytokines by adipose tissue-associated c-Kit(+)Sca-1(+) lymphoid cells. *Nature* 2010;**463**:540–4.
35. Neill DR, Wong SH, Bellosi A, et al. Nuocytes represent a new innate effector leukocyte that mediates type-2 immunity. *Nature* 2010;**464**:1367–70.
36. Saenz SA, Siracusa MC, Perrigoue JG, et al. IL25 elicits a multipotent progenitor cell population that promotes T(H)2 cytokine responses. *Nature* 2010;**464**:1362–6.
37. Brabin BJ, Kelly Y. Prevalence of childhood asthma in the tropics. *Ann Trop Paediat* 1998;**18**(Suppl.):S33–9.

38. Caraballo L, Cadavid A, Mendoza J. Prevalence of asthma in a tropical city of Colombia. *Ann Allergy* 1992;**68**:525–9.
39. Dennis R, Caraballo L, Garcia E, et al. Asthma and other allergic conditions in Colombia: a study in 6 cities. *Ann Allergy Asthma Immunol* 2004;**93**:568–74.
40. Wordemann M, Polman K, Diaz RJ, et al. The challenge of diagnosing atopic diseases: outcomes in Cuban children depend on definition and methodology. *Allergy* 2006;**61**: 1125–31.
41. Sharma SK, Banga A. Prevalence and risk factors for wheezing in children from rural areas of north India. *Allergy Asthma Proc* 2007;**28**:647–53.
42. Dennis R, Caraballo L, Garcia E, et al. Prevalence of asthma and other allergic conditions in Colombia 2009–2010: a cross-sectional study. *BMC Pulm Med* 2012;**12**:17.
43. Chew FT, Lim SH, Goh DY, Lee BW. Sensitization to local dust-mite fauna in Singapore. *Allergy* 1999;**54**:1150–9.
44. Caraballo L, Puerta L, Fernandez-Caldas E, Lockey RF, Martinez B. Sensitization to mite allergens and acute asthma in a tropical environment. *J Investig Allergol Clin Immunol* 1998;**8**:281–4.
45. Acevedo N, Sanchez J, Zakzuk J, et al. Particular characteristics of allergic symptoms in tropical environments: follow up to 24 months in the FRAAT birth cohort study. *BMC Pulm Med* 2012;**12**:13.
46. Nicolaou N, Siddique N, Custovic A. Allergic disease in urban and rural populations: increasing prevalence with increasing urbanization. *Allergy* 2005;**60**:1357–60.
47. Cooper PJ. Interactions between helminth parasites and allergy. *Curr Opin Allergy Clin Immunol* 2009;**9**:29–37.
48. Caraballo L, Acevedo N. Allergy in the tropics: the impact of cross-reactivity between mites and ascaris. *Front Biosci (Elite Ed)* 2011;**3**:51–64.
49. Acevedo N, Caraballo L. IgE cross-reactivity between *Ascaris lumbricoides* and mite allergens: possible influences on allergic sensitization and asthma. *Parasite Immunol* 2011;**33**:309–21.
50. Daschner A, Alonso-Gomez A, Cabanas R, Suarez-de-Parga JM, Lopez-Serrano MC. Gastroallergic anisakiasis: borderline between food allergy and parasitic disease—clinical and allergologic evaluation of 20 patients with confirmed acute parasitism by Anisakis simplex. *J Allergy Clin Immunol* 2000;**105**:176–81.
51. Nieuwenhuizen N, Lopata AL, Jeebhay MF, Herbert DR, Robins TG, Brombacher F. Exposure to the fish parasite *Anisakis* causes allergic airway hyperreactivity and dermatitis. *J Allergy Clin Immunol* 2006;**117**:1098–105.
52. Pump KK. Echinococcosis (hydatid disease): a review and report of a case of secondary echinococcosis. *Can Med Assoc J* 1963;**89**:73–8.
53. Vuitton DA. Echinococcosis and allergy. *Clin Rev Allergy Immunol* 2004;**26**:93–104.
54. Qualizza R, Megali R, Incorvaia C. Toxocariasis resulting in seeming allergy. *Iran J Allergy Asthma Immunol* 2009;**8**:161–4.
55. Neva FA, Gam AA, Maxwell C, Pelletier LL. Skin test antigens for immediate hypersensitivity prepared from infective larvae of *Strongyloides stercoralis*. *Am J Trop Med Hyg* 2001;**65**:567–72.
56. Leighton PM, MacSween HM. *Strongyloides stercoralis*. The cause of an urticarial-like eruption of 65 years' duration. *Arch Intern Med* 1990;**150**:1747–8.
57. Buijs J, Borsboom G, Renting M, et al. Relationship between allergic manifestations and *Toxocara* seropositivity: a cross-sectional study among elementary school children. *Eur Respir J* 1997;**10**:1467–75.
58. Diemert DJ, Pinto AG, Freire J, et al. Generalized urticaria induced by the Na-ASP-2 hookworm vaccine: implications for the development of vaccines against helminths. *J Allergy Clin Immunol* 2012;**130**:169–76. e6.

59. Spillmann RK. Pulmonary ascariasis in tropical communities. *Am J Trop Med Hyg* 1975;**24**:791–800.
60. Joubert JR, de Klerk HC, Malan C. *Ascaris lumbricoides* and allergic asthma: a new perspective. *S Afr Med J* 1979;**56**:599–602.
61. Joubert JR, van Schalkwyk DJ, Turner KJ. *Ascaris lumbricoides* and the human immunogenic response: enhanced IgE-mediated reactivity to common inhaled allergens. *S Afr Med J* 1980;**57**:409–12.
62. Patterson R, Harris KE, Pruzansky JJ. Induction of IgE-mediated cutaneous, cellular, and airway reactivity in rhesus monkeys by *Ascaris suum* infection. *J Lab Clin Med* 1983;**101**:864–72.
63. Patterson R, Harris KE. IgE-mediated rhesus monkey asthma: natural history and individual animal variation. *Int Arch Allergy Immunol* 1992;**97**:154–9.
64. Tsuji M, Hayashi T, Yamamoto S, Sakata Y, Toshida T. IgE-type antibodies to *Ascaris* antigens in man. *Int Arch Allergy Appl Immunol* 1977;**55**:78–81.
65. Cooper PJ, Chico ME, Rodrigues LC, et al. Reduced risk of atopy among school-age children infected with geohelminth parasites in a rural area of the tropics. *J Allergy Clin Immunol* 2003;**111**:995–1000.
66. Leonardi-Bee J, Pritchard D, Britton J. Asthma and current intestinal parasite infection: systematic review and meta-analysis. *Am J Respir Crit Care Med* 2006;**174**: 514–23.
67. Palmer LJ, Celedon JC, Weiss ST, Wang B, Fang Z, Xu X. *Ascaris lumbricoides* infection is associated with increased risk of childhood asthma and atopy in rural China. *Am J Respir Crit Care Med* 2002;**165**:1489–93.
68. Takeuchi H, Zaman K, Takahashi J, et al. High titre of anti-*Ascaris* immunoglobulin E associated with bronchial asthma symptoms in 5-year-old rural Bangladeshi children. *Clin Exp Allergy* 2008;**38**:276–82.
69. Hunninghake GM, Soto-Quiros ME, Avila L, et al. Sensitization to *Ascaris lumbricoides* and severity of childhood asthma in Costa Rica. *J Allergy Clin Immunol* 2007;**119**: 654–61.
70. Hagel I, Cabrera M, Hurtado MA, et al. Infection by *Ascaris lumbricoides* and bronchial hyper reactivity: an outstanding association in Venezuelan school children from endemic areas. *Acta Trop* 2007;**103**:231–41.
71. Obihara CC, Beyers N, Gie RP, et al. Respiratory atopic disease, *Ascaris*-immunoglobulin E and tuberculin testing in urban South African children. *Clin Exp Allergy* 2006;**36**:640–8.
72. Dold S, Heinrich J, Wichmann HE, Wjst M. *Ascaris*-specific IgE and allergic sensitization in a cohort of school children in the former East Germany. *J Allergy Clin Immunol* 1998;**102**:414–20.
73. Alcantara-Neves NM, Badaro SJ, dos Santos MC, Pontes-de-Carvalho L, Barreto ML. The presence of serum anti-*Ascaris lumbricoides* IgE antibodies and of Trichuris trichiura infection are risk factors for wheezing and/or atopy in preschool-aged Brazilian children. *Respir Res* 2010;**11**:114.
74. Pinelli E, Willers SM, Hoek D, et al. Prevalence of antibodies against *Ascaris suum* and its association with allergic manifestations in 4-year-old children in The Netherlands: the PIAMA birth cohort study. *Eur J Clin Microbiol Infect Dis: official publication of the European Society of Clinical Microbiology* 2009;**28**:1327–34.
75. Scrivener S, Yemaneberhan H, Zebenigus M, et al. Independent effects of intestinal parasite infection and domestic allergen exposure on risk of wheeze in Ethiopia: a nested case–control study. *Lancet* 2001;**358**:1493–9.
76. Flohr C, Tuyen LN, Lewis S, et al. Poor sanitation and helminth infection protect against skin sensitization in Vietnamese children: a cross-sectional study. *J Allergy Clin Immunol* 2006;**118**:1305–11.

77. Schafer T, Meyer T, Ring J, Wichmann HE, Heinrich J. Worm infestation and the negative association with eczema (atopic/nonatopic) and allergic sensitization. *Allergy* 2005;**60**:1014–20.
78. Selassie FG, Stevens RH, Cullinan P, et al. Total and specific IgE (house dust mite and intestinal helminths) in asthmatics and controls from Gondar, Ethiopia. *Clin Exp Allergy* 2000;**30**:356–8.
79. van Riet E, Hartgers FC, Yazdanbakhsh M. Chronic helminth infections induce immunomodulation: consequences and mechanisms. *Immunobiology* 2007;**212**:475–90.
80. Erb KJ. Can helminths or helminth-derived products be used in humans to prevent or treat allergic diseases? *Trends Immunol* 2009;**30**:75–82.
81. Maizels RM, Pearce EJ, Artis D, Yazdanbakhsh M, Wynn TA. Regulation of pathogenesis and immunity in helminth infections. *J Exp Med* 2009;**206**:2059–66.
82. Lynch NR, Lopez RI, Di Prisco-Fuenmayor MC, et al. Allergic reactivity and socio-economic level in a tropical environment. *Clin Allergy* 1987;**17**:199–207.
83. Fincham JE, Markus MB, van der Merwe L, Adams VJ, van Stuijvenberg ME, Dhansay MA. *Ascaris*, co-infection and allergy: the importance of analysis based on immunological variables rather than egg excretion. *Trans R Soc Trop Med Hyg* 2007;**101**: 680–2.
84. Acevedo N, Sanchez J, Erler A, et al. IgE cross-reactivity between *Ascaris* and domestic mite allergens: the role of tropomyosin and the nematode polyprotein ABA-1. *Allergy* 2009;**64**:1635–43.
85. Koskinen JP, Laatikainen T, von Hertzen L, Vartiainen E, Haahtela T. IgE response to *Ascaris lumbricoides* in Russian children indicates IgE responses to common environmental allergens. *Allergy* 2011;**66**:1122–3.
86. Calvert J, Burney P. *Ascaris*, atopy, and exercise-induced bronchoconstriction in rural and urban South African children. *J Allergy Clin Immunol* 2010;**125**:100–5. e1–5.
87. Levin M, Muloiwa R, Le Souef P, Motala C. *Ascaris* sensitization is associated with aeroallergen sensitization and airway hyperresponsiveness but not allergic disease in urban Africa. *J Allergy Clin Immunol* 2012;**130**:265–7.
88. Carvalho EM, Bastos LS, Araujo MI. Worms and allergy. *Parasite Immunol* 2006;**28**: 525–34.
89. Johansson SG, Mellbin T, Vahlquist B. Immunoglobulin levels in Ethiopian preschool children with special reference to high concentrations of immunoglobulin E (IgND). *Lancet* 1968;**1**:1118–21.
90. Lee TD, Xie CY. IgE regulation by nematodes: the body fluid of *Ascaris* contains a B-cell mitogen. *J Allergy Clin Immunol* 1995;**95**:1246–54.
91. Dold C, Cassidy JP, Stafford P, Behnke JM, Holland CV. Genetic influence on the kinetics and associated pathology of the early stage (intestinal-hepatic) migration of Ascaris suum in mice. *Parasitology* 2010;**137**:173–85.
92. Enobe CS, Araujo CA, Perini A, Martins MA, Macedo MS, Macedo-Soares MF. Early stages of *Ascaris suum* induce airway inflammation and hyperreactivity in a mouse model. *Parasite Immunol* 2006;**28**:453–61.
93. Phills JA, Harrold AJ, Whiteman GV, Perelmutter L. Pulmonary infiltrates, asthma and eosinophilia due to *Ascaris suum* infestation in man. *N Engl J Med* 1972;**286**: 965–70.
94. Cooper PJ. Can intestinal helminth infections (geohelminths) affect the development and expression of asthma and allergic disease? *Clin Exp Immunol* 2002;**128**:398–404.
95. Cooper PJ, Chico ME, Sandoval C, et al. Human infection with *Ascaris lumbricoides* is associated with a polarized cytokine response. *J Infect Dis* 2000;**182**:1207–13.
96. Stromberg BE. Potentiation of the reaginic (IgE) antibody response to ovalbumin in the guinea pig with a soluble metabolic product from *Ascaris suum*. *J Immunol* 1980;**125**: 833–6.

97. Marretta J, Casey FB. Effect of *Ascaris suum* and other adjuvants on the potentiation of the IgE response in guinea-pigs. *Immunology* 1979;**37**:609–13.
98. Lee TD, McGibbon A. Potentiation of IgE responses to third-party antigens mediated by *Ascaris suum* soluble products. *Int Arch Allergy Immunol* 1993;**102**:185–90.
99. Holland MJ, Harcus YM, Riches PL, Maizels RM. Proteins secreted by the parasitic nematode *Nippostrongylus brasiliensis* act as adjuvants for Th2 responses. *Eur J Immunol* 2000;**30**:1977–87.
100. Paterson JC, Garside P, Kennedy MW, Lawrence CE. Modulation of a heterologous immune response by the products of *Ascaris suum*. *Infect Immun* 2002;**70**:6058–67.
101. Cooper PJ, Barreto ML, Rodrigues LC. Human allergy and geohelminth infections: a review of the literature and a proposed conceptual model to guide the investigation of possible causal associations. *Br Med Bull* 2006;**79–80**:203–18.
102. Araujo CA, Perini A, Martins MA, Macedo MS, Macedo-Soares MF. PAS-1, a protein from *Ascaris suum*, modulates allergic inflammation via IL-10 and IFN-gamma, but not IL-12. *Cytokine* 2008;**44**:335–41.
103. Tsuji N, Miyoshi T, Islam MK, et al. Recombinant *Ascaris* 16-kilodalton protein-induced protection against Ascaris suum larval migration after intranasal vaccination in pigs. *J Infect Dis* 2004;**190**:1812–20.
104. Tsuji N, Suzuki K, Kasuga-Aoki H, et al. Intranasal immunization with recombinant *Ascaris suum* 14-kilodalton antigen coupled with cholera toxin B subunit induces protective immunity to A. suum infection in mice. *Infect Immun* 2001;**69**:7285–92.
105. Tsuji N, Kasuga-Aoki H, Isobe T, Arakawa T, Matsumoto Y. Cloning and characterisation of a highly immunoreactive 37 kDa antigen with multi-immunoglobulin domains from the swine roundworm *Ascaris suum*. *Int J Parasitol* 2002;**32**:1739–46.
106. Islam MK, Miyoshi T, Yokomizo Y, Tsuji N. Molecular cloning and partial characterization of a nematode-specific 24 kDa protein from *Ascaris suum*. *Parasitology* 2005;**130**: 131–9.
107. van Riet E, Wuhrer M, Wahyuni S, et al. Antibody responses to *Ascaris*-derived proteins and glycolipids: the role of phosphorylcholine. *Parasite Immunol* 2006;**28**:363–71.
108. Perrigoue JG, Marshall FA, Artis D. On the hunt for helminths: innate immune cells in the recognition and response to helminth parasites. *Cell Microbiol* 2008;**10**:1757–64.
109. Da Silva CA, Pochard P, Lee CG, Elias JA. Chitin particles are multifaceted immune adjuvants. *Am J Respir Crit Care Med* 2010;**182**:1482–91.
110. Acevedo NEA, Briza P, Puccio F, Ferreira F, Caraballo L. Allergenicity of *Ascaris lumbricoides* tropomyosin and IgE sensitization among asthmatic patients in a tropical environment. *Int Arch of Allergy Immunol* 2011;**154**:195–206.
111. Caraballo LAN. Allergy in the tropics: the impact of cross-reactivity between mites and ascaris. *Frontiers in Bioscience* 2011;**3**:51–64.
112. Lopez N, de Barros-Mazon S, Vilela MM, Condino Neto A, Ribeiro JD. Are immunoglobulin E levels associated with early wheezing? A prospective study in Brazilian infants. *Eur Respir J* 2002;**20**:640–5.
113. Cooper PJ, Moncayo AL, Guadalupe I, et al. Repeated treatments with albendazole enhance Th2 responses to *Ascaris lumbricoides*, but not to aeroallergens, in children from rural communities in the tropics. *J Infect Dis* 2008;**198**:1237–42.
114. Endara P, Vaca M, Chico ME, et al. Long-term periodic anthelmintic treatments are associated with increased allergen skin reactivity. *Clin Exp Allergy* 2010;**40**:1669–77.
115. Watanabe K, Mwinzi PN, Black CL, et al. T regulatory cell levels decrease in people infected with *Schistosoma mansoni* on effective treatment. *Am J Trop Med Hyg* 2007;**77**: 676–82.
116. Mutapi F, Hagan P, Woolhouse ME, Mduluza T, Ndhlovu PD. Chemotherapy-induced, age-related changes in antischistosome antibody responses. *Parasite Immunol* 2003;**25**: 87–97.

117. Mutapi F, Ndhlovu PD, Hagan P, Woolhouse ME. Changes in specific anti-egg antibody levels following treatment with praziquantel for *Schistosoma haematobium* infection in children. *Parasite Immunol* 1998;**20**:595–600.
118. Maizels RM, Yazdanbakhsh M. Immune regulation by helminth parasites: cellular and molecular mechanisms. *Nat Rev Immunol* 2003;**3**:733–44.
119. Woolhouse ME, Hagan P. Seeking the ghost of worms past. *Nat Med* 1999;**5**:1225–7.
120. Oliver-Gonzalez J, Hurlbrink P, Conde E, Kagan IG. Serologic activity of antigen isolated from the body fluid of *Ascaris suum*. *J Immunol* 1969;**103**:15–9.
121. Pritchard DI, Quinnell RJ, McKean PG, et al. Antigenic cross-reactivity between *Necator americanus* and *Ascaris lumbricoides* in a community in Papua New Guinea infected predominantly with hookworm. *Trans R Soc Trop Med Hyg* 1991;**85**:511–4.
122. McWilliam AS, Stewart GA, Turner KJ. An immunochemical investigation of the allergens from *Ascaris suum* perienteric fluid. Cross-reactivity, molecular weight distribution and correlation with phosphorylcholine-containing components. *Int Arch Allergy Appl Immunol* 1987;**82**:125–32.
123. Lozano MJ, Martin HL, Diaz SV, Manas AI, Valero LA, Campos BM. Cross-reactivity between antigens of *Anisakis simplex* s.l. and other ascarid nematodes. *Parasite* 2004;**11**: 219–23.
124. Bhattacharyya T, Santra A, Majumder DN, Chatterjee BP. Possible approach for serodiagnosis of ascariasis by evaluation of immunoglobulin G4 response using *Ascaris lumbricoides* somatic antigen. *J Clin Microbiol* 2001;**39**:2991–4.
125. Jenkins RE, Taylor MJ, Gilvary NJ, Bianco AE. Tropomyosin implicated in host protective responses to microfilariae in onchocerciasis. *Proc Natl Acad Sci USA* 1998;**95**: 7550–5.
126. Burrows B, Martinez FD, Halonen M, Barbee RA, Cline MG. Association of asthma with serum IgE levels and skin-test reactivity to allergens. *N Engl J Med* 1989;**320**: 271–7.
127. Sunyer J, Anto JM, Castellsague J, Soriano JB, Roca J. Total serum IgE is associated with asthma independently of specific IgE levels. The Spanish Group of the European Study of Asthma. *Eur Respir J* 1996;**9**:1880–4.
128. Cooper PJ, Alexander N, Moncayo AL, et al. Environmental determinants of total IgE among school children living in the rural tropics: importance of geohelminth infections and effect of anthelmintic treatment. *BMC Immunol* 2008;**9**:33.
129. Klink M, Cline MG, Halonen M, Burrows B. Problems in defining normal limits for serum IgE. *J Allergy Clin Immunol* 1990;**85**:440–4.
130. Kerkhof M, Dubois AE, Postma DS, Schouten JP, de Monchy JG. Role and interpretation of total serum IgE measurements in the diagnosis of allergic airway disease in adults. *Allergy* 2003;**58**:905–11.
131. Arkestal K, Sibanda E, Thors C, et al. Impaired allergy diagnostics among parasite-infected patients caused by IgE antibodies to the carbohydrate epitope galactose-alpha 1,3-galactose. *J Allergy Clin Immunol* 2011;**127**:1024–8.
132. Commins SP, Kelly LA, Ronmark E, et al. Galactose-alpha-1,3-galactose-specific IgE is associated with anaphylaxis but not asthma. *Am J Respir Crit Care Med* 2012;**185**: 723–30.
133. Melendez AJ, Harnett MM, Pushparaj PN, et al. Inhibition of Fc epsilon RI-mediated mast cell responses by ES-62, a product of parasitic filarial nematodes. *Nat Med* 2007;**13**:1375–81.
134. Larson D, Cooper PJ, Hubner MP, et al. Helminth infection is associated with decreased basophil responsiveness in human beings. *J Allergy Clin Immunol* 2012;**130**: 270–2.
135. Larson D, Hubner MP, Torrero MN, et al. Chronic helminth infection reduces basophil responsiveness in an IL-10-dependent manner. *J Immunol* 2012;**188**:4188–99.

136. Lynch NR, Palenque M, Hagel I, DiPrisco MC. Clinical improvement of asthma after anthelminthic treatment in a tropical situation. *Am J Respir Crit Care Med* 1997;**156**:50–4.
137. Berry MA, Hargadon B, Shelley M, et al. Evidence of a role of tumor necrosis factor alpha in refractory asthma. *N Engl J Med* 2006;**354**:697–708.
138. Lajoie S, Lewkowich IP, Suzuki Y, et al. Complement-mediated regulation of the IL-17A axis is a central genetic determinant of the severity of experimental allergic asthma. *Nat Immunol* 2010;**11**:928–35.
139. Corren J, Lemanske RF, Hanania NA, et al. Lebrikizumab treatment in adults with asthma. *N Engl J Med* 2011;**365**:1088–98.
140. Artis D, Humphreys NE, Bancroft AJ, Rothwell NJ, Potten CS, Grencis RK. Tumor necrosis factor alpha is a critical component of interleukin 13-mediated protective T helper cell type 2 responses during helminth infection. *J Exp Med* 1999;**190**:953–62.
141. Chen F, Liu Z, Wu W, et al. An essential role for TH2-type responses in limiting acute tissue damage during experimental helminth infection. *Nat Med* 2012;**18**:260–6.
142. Strachan DP. Hay fever, hygiene, and household size. *BMJ* 1989;**299**:1259–60.
143. Flohr C, Nagel G, Weinmayr G, et al. Tuberculosis, bacillus Calmette–Guerin vaccination, and allergic disease: findings from the International Study of Asthma and Allergies in Childhood Phase Two. *Pediatr Allergy Immunol* 2012;**23**:324–31.
144. Holt PG. Parasites, atopy, and the hygiene hypothesis: resolution of a paradox? *Lancet* 2000;**356**:1699–701.
145. Chatkin MN, Menezes AM, Victora CG, Barros FC. High prevalence of asthma in preschool children in Southern Brazil: a population-based study. *Pediatr Pulmonol* 2003;**35**:296–301.
146. Pitrez PM, Stein RT. Asthma in Latin America: the dawn of a new epidemic. *Curr Opin Allergy Clin Immunol* 2008;**8**:378–83.
147. Pearce N, Ait-Khaled N, Beasley R, et al. Worldwide trends in the prevalence of asthma symptoms: phase III of the International Study of Asthma and Allergies in Childhood (ISAAC). *Thorax* 2007;**62**:758–66.
148. Sole D, Melo KC, Camelo-Nunes IC, et al. Changes in the prevalence of asthma and allergic diseases among Brazilian schoolchildren (13–14 years old): comparison between ISAAC Phases One and Three. *J Trop Pediatr* 2007;**53**:13–21.
149. Kuschnir FC, Alves da Cunha AJ. Environmental and socio-demographic factors associated to asthma in adolescents in Rio de Janeiro, Brazil. *Pediatr Allergy Immunol* 2007;**18**:142–8.
150. Garcia E, Aristizabal G, Vasquez C, Rodriguez-Martinez CE, Sarmiento OL, Satizabal CL. Prevalence of and factors associated with current asthma symptoms in school children aged 6–7 and 13–14 yr old in Bogota, Colombia. *Pediatr Allergy Immunol* 2008;**19**:307–14.
151. Van Dellen RG, Thompson Jr JH. Absence of intestinal parasites in asthma. *N Engl J Med* 1971;**285**:146–8.
152. Gerrard JW, Geddes CA, Reggin PL, Gerrard CD, Horne S. Serum IgE levels in white and metis communities in Saskatchewan. *Ann Allergy* 1976;**37**:91–100.
153. Yazdanbakhsh M, Kremsner PG, van Ree R. Allergy, parasites, and the hygiene hypothesis. *Science* 2002;**296**:490–4.
154. Yazdanbakhsh M, Matricardi PM. Parasites and the hygiene hypothesis: regulating the immune system? *Clin Rev Allergy Immunol* 2004;**26**:15–24.
155. Platts-Mills TA, Cooper PJ. Differences in asthma between rural and urban communities in South Africa and other developing countries. *J Allergy Clin Immunol* 2010;**125**: 106–7.
156. Rook GA. 99th Dahlem conference on infection, inflammation and chronic inflammatory disorders: darwinian medicine and the "hygiene" or "old friends" hypothesis. *Clin Exp Immunol* 2010;**160**:70–9.

157. Sironi M, Clerici M. The hygiene hypothesis: an evolutionary perspective. *Microbes Infect* 2010;**12**:421–7.
158. Jackson JA, Friberg IM, Little S, Bradley JE. Review series on helminths, immune modulation and the hygiene hypothesis: immunity against helminths and immunological phenomena in modern human populations: coevolutionary legacies? *Immunology* 2009;**126**:18–27.
159. Fumagalli M, Pozzoli U, Cagliani R, et al. Parasites represent a major selective force for interleukin genes and shape the genetic predisposition to autoimmune conditions. *J Exp Med* 2009;**206**:1395–408.
160. Fumagalli M, Pozzoli U, Cagliani R, et al. The landscape of human genes involved in the immune response to parasitic worms. *BMC Evol Biol* 2010;**10**:264.
161. Barreiro LB, Quintana-Murci L. From evolutionary genetics to human immunology: how selection shapes host defence genes. *Nat Rev Genet* 2010;**11**:17–30.
162. Lynch NR, Hagel IA, Palenque ME, et al. Relationship between helminthic infection and IgE response in atopic and nonatopic children in a tropical environment. *J Allergy Clin Immunol* 1998;**101**:217–21.
163. Cooper PJ, Chico ME, Sandoval C, Nutman TB. Atopic phenotype is an important determinant of immunoglobulin E-mediated inflammation and expression of T helper cell type 2 cytokines to ascaris antigens in children exposed to ascariasis. *J Infect Dis* 2004;**190**:1338–46.
164. Cooper PJ. Intestinal worms and human allergy. *Parasite Immunol* 2004;**26**:455–67.
165. Le Souef PN, Goldblatt J, Lynch NR. Evolutionary adaptation of inflammatory immune responses in human beings. *Lancet* 2000;**356**:242–4.
166. Hopkin J. Immune and genetic aspects of asthma, allergy and parasitic worm infections: evolutionary links. *Parasite Immunol* 2009;**31**:267–73.
167. May RM, Anderson RM. Epidemiology and genetics in the coevolution of parasites and hosts. *Proc R Soc Lond Series B, containing papers of a biological character* 1983;**219**: 281–313.
168. Dunne DW, Cooke A. A worm's eye view of the immune system: consequences for evolution of human autoimmune disease. *Nat Rev Immunol* 2005;**5**:420–6.
169. Peisong G, Yamasaki A, Mao XQ, et al. An asthma-associated genetic variant of STAT6 predicts low burden of *Ascaris* worm infestation. *Genes Immun* 2004;**5**:58–62.
170. Moller M, Gravenor MB, Roberts SE, Sun D, Gao P, Hopkin JM. Genetic haplotypes of Th-2 immune signalling link allergy to enhanced protection to parasitic worms. *Hum Mol Genet* 2007;**16**:1828–36.
171. Acevedo N, Mercado D, Vergara C, et al. Association between total immunoglobulin E and antibody responses to naturally acquired *Ascaris lumbricoides* infection and polymorphisms of immune system-related LIG4, TNFSF13B and IRS2 genes. *Clin Exp Immunol* 2009;**157**:282–90.
172. Williams-Blangero S, VandeBerg JL, Subedi J, et al. Genes on chromosomes 1 and 13 have significant effects on *Ascaris* infection. *Proc Natl Acad Sci USA* 2002;**99**:5533–8.
173. Williams-Blangero S, Vandeberg JL, Subedi J, Jha B, Correa-Oliveira R, Blangero J. Localization of multiple quantitative trait loci influencing susceptibility to infection with *Ascaris lumbricoides*. *J Infect Dis* 2008;**197**:66–71.
174. Feary JR, Venn AJ, Mortimer K, et al. Experimental hookworm infection: a randomized placebo-controlled trial in asthma. *Clin Exp Allergy* 2010;**40**:299–306.
175. Bager P, Kapel C, Roepstorff A, et al. Symptoms after ingestion of pig whipworm *Trichuris suis* eggs in a randomized placebo-controlled double-blind clinical trial. *PLoS One* 2011;**6**:e22346.
176. Oshiro TM, Enobe CS, Araujo CA, Macedo MS, Macedo-Soares MF. PAS-1, a protein affinity purified from *Ascaris suum* worms, maintains the ability to modulate the immune response to a bystander antigen. *Immunol Cell Biol* 2006;**84**:138–44.

177. Nutman TB, Hussain R, Ottesen EA. IgE production *in vitro* by peripheral blood mononuclear cells of patients with parasitic helminth infections. *Clin Exp Immunol* 1984;**58**:174–82.
178. Kojima S, Yokogawa M, Tada T. Raised levels of serum IgE in human helminthiases. *Am J Trop Med Hyg* 1972;**21**:913–8.
179. Turner KJ, Feddema L, Quinn EH. Non-specific potentiation of IgE by parasitic infections in man. *Int Arch Allergy Appl Immunol* 1979;**58**:232–6.
180. Turner KJ, Baldo BA, Anderson HR. Asthma in the highlands of New Guinea. Total IgE levels and incidence of IgE antibodies to house dust mite and *Ascaris lumbricoides*. *Int Arch Allergy Appl Immunol* 1975;**48**:784–99.
181. Lynch NR, Lopez R, Isturiz G, Tenias-Salazar E. Allergic reactivity and helminthic infection in Amerindians of the Amazon Basin. *Int Arch Allergy Appl Immunol* 1983;**72**: 369–72.
182. Gelpi AP, Mustafa A. Seasonal pneumonitis with eosinophilia. A study of larval ascariasis in Saudi Arabs. *Am J Trop Med Hyg* 1967;**16**:646–57.
183. Loffler W. Transient lung infiltrations with blood eosinophilia. *Int Arch Allergy Appl Immunol* 1956;**8**:54–9.

CHAPTER

3

Ascaris – Antigens, Allergens, Immunogenetics, Protein Structures

Malcolm W. Kennedy
University of Glasgow, Scotland, UK

OUTLINE

Ascaris: The Neglected Parasite
http://dx.doi.org/10.1016/B978-0-12-396978-1.00003-3

Copyright © 2013 Elsevier Inc. All rights reserved.

INTRODUCTION

Despite the global health impact posed by *Ascaris lumbricoides*, there have been relatively few laboratories that have been devoted to understanding the biochemical, physiological, and immunological interactions between the parasite and its host. The fact that there exists a closely related species, *Ascaris suum*, in a similarly-sized host species, the pig, has stimulated research on understanding the human infection, and has also led to the advent of materials and reagents, and the development of ideas that may be relevant to human populations exposed to both species. Genomics, transcriptomics, and proteomics are set to enhance the power of investigations into *Ascaris* and ascariasis enormously, and will hopefully extend the discoveries and associated questions that have arisen already. This chapter is primarily intended to summarize findings on the proteins that are produced by *Ascaris* in its interaction with its hosts, but also to explore and extend ideas that may have relevance to this and other nematodiases.

STAGE-SPECIFIC SURFACE AND SECRETED ANTIGENS

The antigens of *Ascaris* and all parasitic nematodes could be crudely placed into four categories – those presented on the surface of the parasites, those that are secreted (excretory/secretory; ES), those released during molting, and those usually confined to the interior of the parasites and only released upon death of a worm. For species with tissue-parasitic larval stages, such as *A. suum* and *A. lumbricoides*, all four classes are potentially important to protective immune responses, evasion thereof, and immunopathology.

The secreted proteins of the infective and later tissue-migratory larvae are different from those of other genera of nematodes,[1,2] although they do cross-react immunologically with the antigens of other ascaridids such as *Toxocara canis* and *Anisakis simplex*.[3,4] The original work, some of which used radio-labeled materials, showed that ES comprises a set of secreted (glyco)proteins that change as the larvae migrate through the tissues to arrive at the lungs.[1,5] Our unpublished work shows that the most abundant secreted proteins are glycosylated, that some occur as disulfide-bonded heterodimers (Figure 3.1), and that larvae in *in vitro* culture can be made to synthesize and secrete the same set of antigens as observed by radio-iodination of secreted materials (F. Qureshi and M.W. Kennedy, unpublished). The surface antigens of *Ascaris* larvae have also been found to change during their migration and development from infective larvae (formerly designated second stage larvae, L2, now recognized to be third

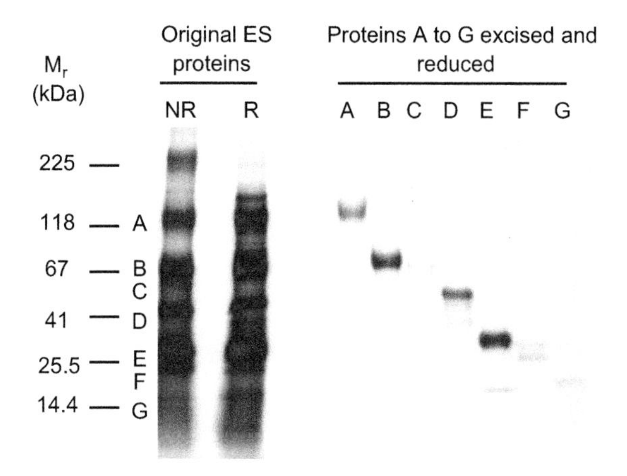

FIGURE 3.1 **The secreted antigens of the lung-stage (L3/4) larvae of *Ascaris suum* and their subunits.** Excretory-secretory material was collected from culture fluid, radio-iodinated, and then separated by protein gel electrophoresis under reducing (R) conditions (in which a reducing agent is added in order to break intermolecular disulfide bonds) or non-reducing conditions (NR) and visualized by autoradiography. This shows the complexity of the materials released by the larvae into lung tissue. Many of these proteins were found to bind to lectins, so are glycosylated (not shown). Selected proteins (labeled A to G) were removed from a previously-run preparative gel in which separation was carried out without reduction, and then run on this gel under reducing conditions. This would show whether any comprised disulfide-linked dimers or heterodimers, which would reveal themselves by a reduction from their original molecular mass or the appearance of more than one band. Most of the proteins exist as non-disulfide linked monomers, but some show that they may exist as more complex entities. The size of the proteins is indicated as relative migration (M_r) expressed in kilodaltons (kDa).

stage, L3[6]) to lung stage larvae (usually designated as L3/4, meaning a mixture of L3 and L4 larvae and those in transition).[3,6,7] Interestingly, a comparison between the ES antigens released by the tissue-invasive larvae of *A. suum* and *A. lumbricoides* revealed that the two species could be distinguished biochemically.[5] Thankfully, however, DNA-based methods have since been established that are highly discriminatory and easy to use[8–10] (see also Chapter 10). It remains too early to say whether the differences between the ES of the two species of *Ascaris* relate in any way to their host preference, although the possibility is clear (see[11,12] for ideas on host specificity and proteinase inhibitors).

It is now clear that both surface and secreted antigens of *Ascaris* larvae change during their migration. Why this should be is not known, other than the obvious guess that the parasite is passing through different tissues, thereby facing different challenges and physiological environments that require the synthesis and release of different materials during their migration. This might, for example, apply to the proteinases released by tissue-invasive *Ascaris* larvae, which can be inactivated by antibody.[13] The other idea that is often raised is that the larvae are keeping ahead of the immune response by ceasing to expose antigens to which the immune response has already begun to respond at an earlier stage of infection. But this would require no overlap in antigenicity of the different developmental stages, which may be unlikely. Also, if the "keeping ahead" hypothesis were valid, and immune responses to early larval stages were effective, how then could *Ascaris* infections be so long-lived, and involve continuous recruitment for decades[14]? Some ideas on this are presented later in this chapter.

ALLERGENS

One reason that *Ascaris* has consistently attracted attention is the allergic-like reactions that it causes. These appear to be of two broad types. The first is severe and rapid-onset reactions (allergic rhinitis, tears, etc.) that are occasionally reported by laboratory workers exposed to the adult worms or extracts thereof. These cases are apocryphal and the nature of the inciting agent is unknown. Given that a highly volatile material appears to be involved, there is no particular reason to believe that it could be due to a protein allergen, which would not be expected to diffuse through air so readily unless volatilized or droplet-dispersed. Also, there is no evidence yet that this reaction is strictly immunological. The other type of allergy-related reaction is those associated with infections with *Ascaris*, the classic example being Loeffler's syndrome.[15] This is considered to be a pulmonary type I immune hypersensitivity reaction that involves pulmonary eosinophilia and probably also IgE antibody-mediated events.[16] Loeffler's syndrome has long been associated with ascariasis, but has not been consistently researched, despite the morbidity known in arid areas with periodic, high-level seasonal transmission resulting in large numbers of larvae passing through the lungs.[17,18] The syndrome has possibly become such a routine feature of ascariasis in human populations that it is seldom reported, but, when asked, physicians working in endemic areas say it is common and harmful (L. Savioli, personal communication). Whether mortality associated with Loeffler's syndrome is significant is simply not known, though an indication of the potential is indicted by "Mass treatment for ascariasis of aboriginal children in Queensland, Australia, has reduced the mortality (mostly attributed to 'pneumonia') from 29–54% of the total deaths to 8% in the last few years" (quoted in[19]). *Ascaris*-associated pulmonary hypersensitivity reactions, and liver white spot, are also known in animals such as cattle and sheep that are exposed to *A. suum* eggs.[20] So, *Ascaris* has long been infamous for the pulmonary immune hypersensitivities it can cause, yet people can present with chronic, high intensity infections with the adult parasites yet without exhibiting disabling allergic manifestations in their lungs or intestines.[18] It has been speculated that allergic manifestations of hypersensitivity reactions, such as Loeffler's syndrome, are the consequence of seasonal infections, in which hypersensitivity responses build up during periods when transmission is not occurring, then are made manifest when transmission resumes.[17,18] The assumption then is that under conditions of continuous, low-level, trickle infections, humans do not become hypersensitized and then respond adversely to migrating larvae.

On considering *Ascaris* allergens, the one protein allergen that comes to mind is ABA-1. This is one, if not the, major component of the pseudocoelomic fluid of adult worms, and is consequently the major protein in allergen preparations (which tend to be crude extracts from adult worms) used to test for allergic hypersensitivity to *Ascaris*.[21–24] ABA-1 is also present in the larvae of *Ascaris*, although we could not find evidence that the protein is synthesized either by tissue-migrating larvae during *in vitro* culture, possibly because such culture conditions do not sufficiently mimic conditions encountered by the larvae as they migrate (F. Qureshi and M.W. Kennedy, unpublished), although cDNA encoding ABA-1 was isolated from larvae.[25]

While there are other proteins produced by *Ascaris* that are targets of IgE responses in infections (and therefore, by definition, allergens),[26,27] ABA-1 will be a recurrent theme throughout the rest of this chapter because of the detail now available on it from immune responses and immunogenetics, to its molecular structure.

CORRELATION BETWEEN IGE ANTIBODY TO ABA-1 AND ACQUIRED IMMUNITY

One of the principal reasons for characterizing the antigens of parasitic nematodes was the hope that naturally-acquired immunity could be found to associate with reactivities to single or a small set of antigen types in order to understand protective immunity and to identify protective antigens for inclusion in vaccines. These ventures began in the 1980s, but have met with limited if any success. One of the few examples in which a correlation between immune reactivity (IgE antibody specifically) in humans and naturally-acquired immunity was observed with ABA-1.[28,29] Even here, however, an effect was only observed with any confidence when the study population was subselected for only those individuals that reacted to the protein, excluding those that did not even if they reacted to other antigens of the parasite (see below for immunogenetic explanations for non-reactivity).[28] When infected people in Nigeria or

Cameroon were divided into those who persistently presented with no or few worms, and those persistently presenting with high-level infections, and screened for antibody responses to a crude mixture of antigens of *Ascaris*, no good correlations could be found for either IgG or IgE antibody responses with either phenotype.[28,29] But when only those that responded to ABA-1 were taken, then a relationship did emerge between higher IgE antibody responses to the protein and relative resistance to the infection (Figure 3.2). This remains only an association, though there are reports that immune responses to ABA-1-like proteins in other species of nematode parasites may be protective.[30] A now widely-accepted interpretation of this would be that T helper type 2 (Th2)-mediated inflammatory responses (e.g. IgE, eosinophil, and/or mast cell-mediated) are operative in ascariasis.[29,31,32] The Nigerian study also found that elevated levels of serum proteins associated with inflammatory responses correlated with natural resistance.[28] It could therefore be that the protective mechanism involves direct attack by IgE and eosinophils, and/or local allergic

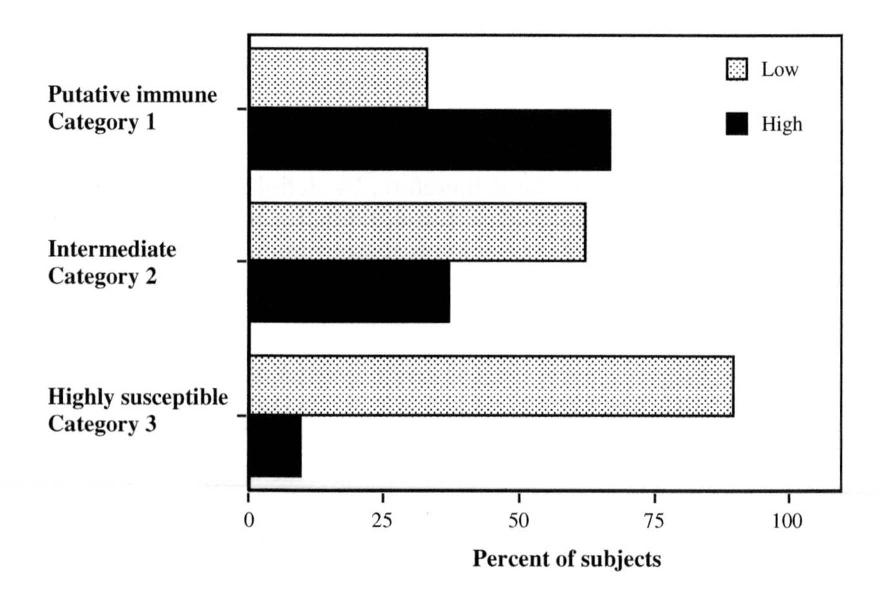

FIGURE 3.2 **Association between relative resistance to *Ascaris lumbricoides* infection and IgE antibody against ABA-1.** Infected patients (ages 5–15 years) in Nigeria were categorized as positive for IgG antibody to ABA-1 by radioimmunoassay and then subdivided into those responding with high, intermediate or low levels of IgE antibody to the recombinant form of the protein produced by *A. lumbricoides*. The proportion of individuals with the higher levels of IgE antibody to ABA-1 within the three infection classes showed pronounced differences (66.7% in category 1, 37.5% in category 2, and only 10% in category 3), with more individuals in the putatively immune group having higher levels. For full details including statistical analysis see.[28] A similar effect was observed in a study carried out in Cameroon.[29] *Figure taken from[28] with permission.*

responses to materials released by the parasite, resulting in a non-specific inflammatory response that traps and kills the larvae. In which organ and tissue this might occur in humans is not known, but experiments in mice indicate that the site of resistance to migrating larvae is probably the liver,[33] as it may also be in pigs,[34] although other sites, such as the lungs, where hypersensitivity reactions are possibly most deleterious, must remain under consideration (see also Chapter 5).

IMMUNOGENETICS AND UNPREDICTABLE IMMUNE REPERTOIRES

A puzzling early observation in people infected with nematodes (e.g. *Brugia malayi* and *Ascaris lumbricoides* itself) was that, when screened against complex antigen preparations such as worm homogenates or excretory/secretory (ES) materials, there was enormous diversity in protein antigens to which individuals produced antibody (Figure 3.3).[22,26,28,35,36] This was particularly noticeable in ascariasis, in which people known to be chronically infected, even when related and living in close proximity, produced diverse antibody repertoires.[35] Thus, most, if not all, individuals only respond to a subset of the potential

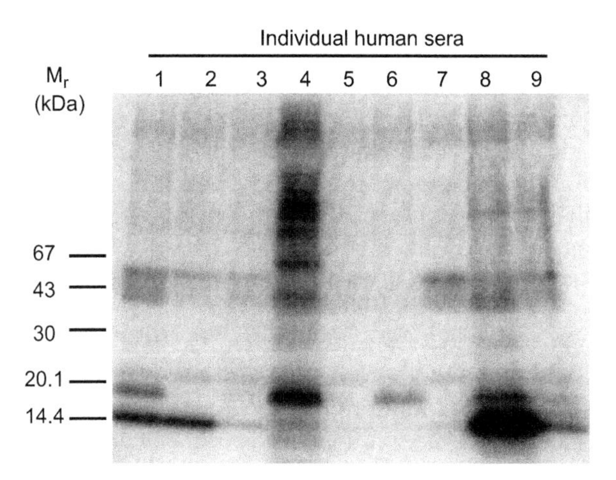

FIGURE 3.3 Diversity of immune responses to *Ascaris* antigens by infected humans. Serum from people known to have long-standing infections with *Ascaris lumbricoides* were reacted with radio-labeled antigens of the parasite and then the proteins to which they produced antibody separated by gel electrophoresis in order to reveal which antigens of the parasite different people respond to immunologically. This typifies the dramatic diversity in antibody repertoire between people, even when living in the same environment or even related. How this arises is not precisely known, but it is most likely to be under the control of the major histocompatibilty complex (MHC) (see Figure 3.4).

antigens of the parasite, and this subset varies from person to person. Moreover, this applied to both IgG and IgE antibody responses.[26,28,37] Surprisingly, this was observed in (outbred) populations of humans in which the load of foreign parasite biomass probably amounted to hundreds of grams, and also under conditions of continuous reinfection. This reduction/selectivity in the potential immune repertoire was modeled in inbred strains of mice and rats, in which the antibody repertoires were found to be under genetic control by the class II region of the major histocompatibility complex (MHC).[23,38–40] This phenomenon was soon demonstrated to be generally applicable to nematode infections.[41–43] The effect was particularly clear in experiments using radiolabeled antigens, which presumably remain in their natural conformation with only minor modification from the radiolabeling procedure (either directly by radioiodination, or biosynthetically with ^{35}S-methionine), whereas ELISA and other solid phase-based systems may deal with unfolded or denatured protein antigens in which internal epitopes are exposed.

The MHC restriction of antibody repertoires is usually due to limitations in the fitting of peptides resulting from processing of foreign proteins into the apical groove of MHC molecules presented to T cell receptors. As said above, the surprise was that restricted antibody repertoires are manifest even in outbred populations of humans, most of whom will be heterozygous across the MHC. So, if one parental MHC allele encodes an MHC protein that did not present a given antigen peptide, then the other allele might, such that the overall effect would be an almost complete (or at least expanded) repertoire of antigen recognition in chronically infected people. But, when inbred, homozygous, parental strains of mice were compared for antibody repertoires with their MHC heterozygous F_1 crosses, the anticipated simple additive repertoire was not observed (Figure 3.4).[23] This effect could be explained by little-considered immunological phenomena such as cross-tolerance with proteins encoded by background genes (discussed in[23]), but the cause of the effect as observed in *Ascaris* infections remains unknown. What it does tell us, though, is that antibody repertoires in MHC heterozygotes are not reliably predictable from the repertoires of homozygotes.[23,42] Moreover, even within inbred strains of mouse or rat exposed to the same number and scale of infections, the repertoires can vary slightly between individual animals, possibly due to the stochastic nature of the immune system.[42,43] It therefore becomes less surprising that the repertoire of an individual human is not fully predictable from that of their parents, thereby contributing to the unexpected diversity of repertoires in humans infected with pathogens such as nematodes.

Given the attention being paid here to the ABA-1 of *Ascaris* spp., it is worth mentioning that MHC restriction of the antibody response to this

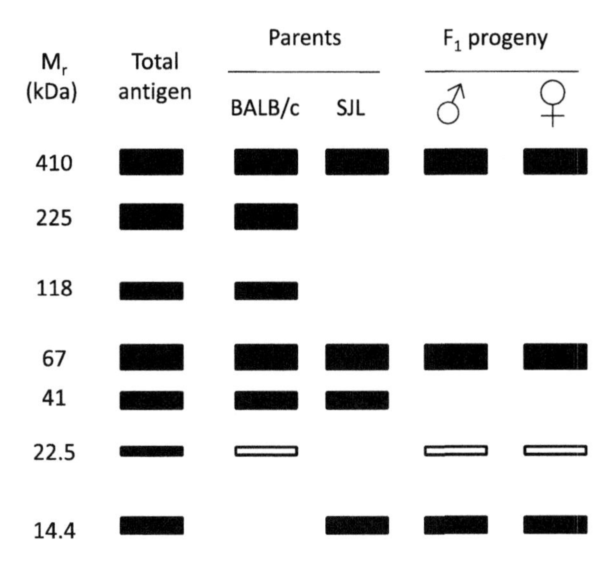

FIGURE 3.4 **Genetic control of the immune repertoire in *Ascaris* infection and its unpredictability.** Inbred mice of two strains, BALB/c and SJL, were infected with *Ascaris suum*, and their serum collected. The antibody in the serum was reacted with radio-labeled materials secreted by the lung-stage (L3/4) larvae of the parasite, and then analyzed by protein gel electrophoresis to establish the antigens of the parasite to which the different strains responded. This diagrammatic representation of the result shows that the two strains respond to different sets of protein antigens, despite being infected with the same species of parasite. The density of tone and width of the bands represent strength of recognition of each component. When the strains of mice are crossed, and their progeny infected and serum-sampled, the antigen recognition repertoire of their progeny was not predictable from the responses of their parents. The 14.4 kDa protein is ABA-1. See[23] for further details and analysis of this phenomenon. See[38] for demonstration that the repertoire is under the control of the class II region of the major histocompatibility complex (MHC).

protein is particularly stark, in that only one MHC haplotype in mice and rats (H-2^s and RT1^u, respectively) has so far been found to respond in infection,[23,24,39,43,44] and responsiveness in humans appears to be similarly infrequent (Figure 3.3).[22,28]

Interest in the immune repertoire to infection is not merely academic, but can be of direct importance for the effectiveness of immunization. For instance, it is probable that immunization against nematodes of human importance will comprise single or a limited cocktail of recombinant proteins. If people in the immunized population respond heterogeneously, then it could be that a significant proportion of vaccinees will fail to respond sufficiently to be immune. This is recognized as a significant problem for immunizations that comprise only a single type of protein (such as for hepatitis, see references in[45]). What happens in cases where the same or different pairs of MHC alleles come together in different individuals is also important and intriguing. This is emphasized by an unpublished study in which a population typed for MHC as part of a large study on genetic resistance to malaria in the Gambia found that heterozygotes at MHC loci were more likely to respond to ABA-1 than expected for their genotypes (A.V. Hill and M.W. Kennedy, unpublished). Thus, two non-responder MHC allele sets (haplotypes) could come together in an individual that then exhibits a responder phenotype. The finding with humans that MHC heterozygotes may be more likely to respond to a given type of protein antigen than expected should be explored for pure curiosity if nothing else.

Whether the genetic control of the immune repertoire, so clearly demonstrated with *Ascaris*, is of relevance to the development of natural resistance to the infection remains far from clear. Other chapters in this book relate to other correlates with immunity by following the genetics of humans that are relatively susceptible or resistant to natural infection with the parasite (see Chapter 12). For instance, whole genome screens have revealed genetic loci that are associated with natural resistance or susceptibility, but those genes are not classical MHC loci, although some may be immune-related.[46–50] But, as argued above, MHC-controlled immune repertoires are very likely to be important in immunizations using one or a small number of recombinant protein types, such that certain individuals may not respond at all to a particular antigen protein.

POLYMORPHISMS IN *ASCARIS* ANTIGENS

It is not yet known whether *Ascaris* worms are polymorphic in the antigens they produce. If they were, then this would have important implications for the acquisition and persistence of immunity in humans and even for the segregation of *A. suum* and *A. lumbricoides* into their separate host species. Detecting differences in antigen expression in ES materials of individual larvae would be difficult, though new high resolution techniques will soon be able to analyze the genome, transcriptome, and proteome of individual larvae. For the moment, it remains worthwhile to consider evidence gained some time ago, using quantitative immunofluorescence, that *A. lumbricoides* larvae are diverse in their surface-exposed antigens.[51] This work used serum antibody from infected people and quantitative immunofluorescence (which has much more discriminating power for differences in fluorescence emission than the human eye) on live larvae, and found that there were differences in the level of binding to the surfaces of larvae of *A. lumbricoides* raised from eggs collected from adult worms recovered from the same community in Nigeria (Figure 3.5). Thus, while antibody from one person would bind to larvae from the same pool with a normal distribution (donor 18), antibody from another person would have quite a different distribution (donor 20), and, significantly, in some cases their serum antibody would bind to some larvae strongly and to others weakly or not at all (donor 13). These experiments required careful controls because the pool of larvae used would have been undergoing development, but the results nevertheless indicated a high degree of diversity in surface antigens.

These observations clearly have important implications for the immunobiology of ascariasis, but have not yet been pursued. One way to look more closely at antigenic diversity of *Ascaris* larvae would be to use monoclonal antibodies. Being monospecific, monoclonal antibodies should discriminate between larvae that do and do not express a particular antigenic determinant (epitope) more effectively than can polyspecific serum antibody from infected people. Just such an approach has already revealed that larvae of *Trichostrongylus colubriformis* also exhibit antigenic diversity.[52]

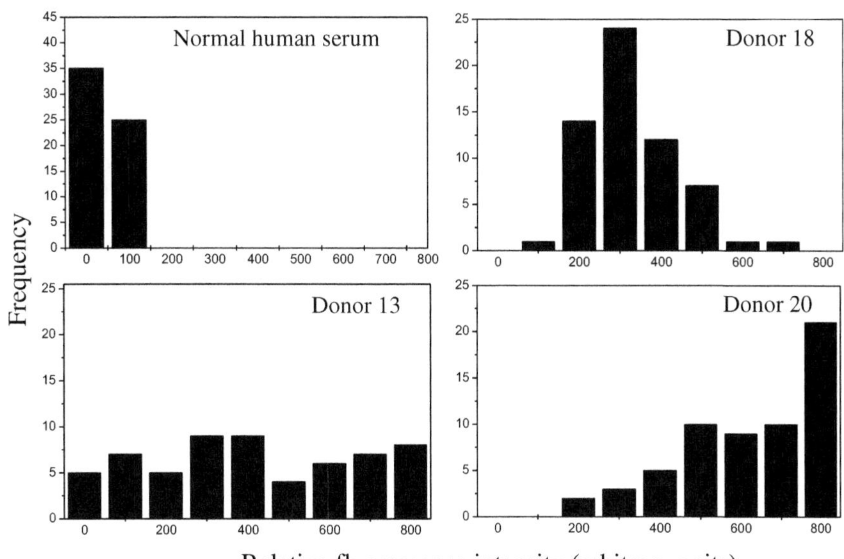

FIGURE 3.5 **Diversity of the antigens exposed on the surface of *Ascaris* larvae.** Infective larvae hatched from the same pool of eggs were incubated with serum from each of three people infected with *A. lumbricoides* and the degree of antibody binding to individual larvae measured using a microscope-based quantitative immunofluorescence method. This showed that some people produce antibody that bind similarly to the larvae (donor 18), some produce high levels of antibody to the larval surface (donor 20), but antibody from others (donor 13) bind to some larvae intensely, some intermediately, but to others not at all. The inference from this is that the surface antigens of the larvae are diverse. It should be noted that the larvae used are cultured *in vitro* and undergoing development, so careful controls had to be carried out. See.[51]

If such polymorphism in *Ascaris* were found, how could it be investigated at the genetic level? The sequencing of the *A. suum* genome[53] (see Chapter 11) provides a base from which to search for polymorphisms in the genomes of that species but also of *A. lumbricoides.* This can be done in a number of ways using genetic material from samples from an outbred population of the parasite. But a more directed approach might be to focus on genes involved in surface and secreted materials using gene transcription surveys that focus on mRNAs encoding polypeptides that exhibit secretory signal peptides, although this might not include surface antigens or carbohydrate epitopes. Genetic polymorphisms may appear as differences in the sequences of proteins that would be under immune selection (and the frequency of synonymous and non-synonymous mutations), but also in the activity and specificity of enzymes involved in the glycosylation of proteins and glycolipids, as is the case for the evolution of human ABO blood groups.[54] Apposite to this are the elegant studies showing that surface antigens of *C. elegans* are diverse and under the control of identifiable loci which, importantly, may encode glycosylating enzymes.[55–57] Intriguingly, it was also shown that surface antigens expressed by *C. elegans* can change under different culture conditions.[58,59] This is a fascinating observation, and does mean that parasitic species can also alter their antigens in response to the immune responses they encounter in order to evade attack. The classic experiments carried out in the 1970s on *Nippostrongylus brasiliensis* illustrated that the secreted materials of a parasitic nematode can alter according to whether or not a large infection is initiated at one time, or instead accumulates as a trickle infection, or whether an immune response has been mounted.[60–62] The possibility that parasitic nematodes may alter their surface antigens and secretions in response to immunological and other environmental cues is clearly one demanding detailed investigation. Those original experiments with *N. brasilieinsis* concerned secreted enzymes, but very little is known of the enzymes released by tissue-invading *Ascaris* larvae, yet they could also be both diverse and their release dependent on both innate and acquired immune responses by the host. The proteinases released by infective and lung-stage larvae, and to which neutralizing antibodies are produced, may be another good place to start.[13]

When one considers that the unequaled degree of polymorphism of immunity-related genes in vertebrates is considered to be an evolutionary response to a diverse range of pathogens that are themselves evolving to circumvent immunity (discussed in ref.[63]) then there seems to be no reason why a sexually reproducing parasite such as *Ascaris* will not itself have evolved polymorphisms in order that its population will succeed in polymorphic hosts. This would seem to be a simple prediction from the Red Queen hypothesis.[64,65] An intriguing extra possibility would be that metapopulations of parasitic nematodes such as *Ascaris* may evolve a set of polymorphisms in important antigens and immune evasion genes that suit success in a given metapopulation of humans. Also, with time, a host population may become reactive to the range of antigens expressed by its parasite population, and those parasites will then be replaced by immigrant worms bearing different antigens or immune evasion factors.

At risk of overextending the possible existence and importance of genetic polymorphisms in parasites, the next section discusses and challenges some ideas about concomitant immunity in helminth infections.

POLYMORPHISMS IN PARASITE ANTIGENS AND CONCOMITANT IMMUNITY

One of the puzzles in helminth infections is that they are so persistent and without strong evidence of protective immunity, while in the face of continual recruitment by new parasites (the average lifetime of an *Ascaris* adult worm is estimated to be about 1 year, yet infections can persist for decades).[14,37] This could be because of immunomodulation and immunosuppression by the parasites themselves, but another aspect could be, as introduced above, polymorphisms in the parasite population in their antigens and/or immunomodulatory factors. While diversity within populations of parasitic nematodes has been noticed before,[66] it has not been established yet to what degree antigenic polymorphism contributes to the slow or zero acquisition of immunity in people infected with nematodes. Another problem is that, with parasites such as *Ascaris* that contaminate the environment so effectively with their transmission stages (eggs), how is it that hosts do not more frequently become lethally overpopulated with worms?

One concept that arose to explain this latter problem in the context of schistosome infections is concomitant immunity.[67] By this hypothesis, adult parasites that are established somehow immunize against incoming larvae so as to avoid overcrowding and death of the host. Mathematical consideration of concomitant immunity falls in favor of this being a successful reproductive strategy for the parasite population.[68] But does not concomitant immunity essentially infer altruism between parasites? That is, for the concomitant immunity system to work, adult worms would be eliciting immune responses that would be lethal to their own offspring but to the benefit of unrelated adults that are already present. Also, a larva invading a host is that parasite's one and only chance to reproduce, so selection would favor the evolution of larvae that evade the system. Another problem is that concomitant immunity would limit the colonization of a host and thereby reduce the outbreeding potential of a sexually reproducing parasite. Why should we expect parasites to cooperate suicidally when selection would favor those that compete successfully with conspecifics – crudely called "cheats?" In other words, there should be a strong selective advantage for parents of infective larvae to produce offspring that would defeat concomitant immunity and get through to breed, even if that risks overcrowding the host. In the original theoretical approaches to concomitant immunity, polymorphism in the antigens or counter-immunity activities of invading larva had not been considered.[68] Thus, it would be selectively advantageous to parents to produce offspring that are different to others, resulting in the evolution of polymorphisms in antigens. That would have the added effect of increasing the genetic and antigenic diversity of those parasites that make it through to adulthood because new incoming larvae that are antigenically similar to their predecessors will be selected against, thereby favoring larvae that are different. Most nematode and schistosome parasites reproduce sexually, so there is a mechanism for the evolution and maintenance of high levels of polymorphisms within their populations, just as is so evident in immunity genes in their hosts. See[69,70] for discussion of similar points for schistosomes based on infections with parasites of differing known genotypes.

INNATE IMMUNITY AND COMPLEMENT

Protective immune responses against helminths such as *Ascaris* are frequently attributed to Th2-dependent responses, IgE in particular. But an observation made during a study that came to this conclusion also found that inflammatory markers correlated with the development of natural immunity in humans.[28] One such inflammatory-associated response is complement, which is rarely investigated in nematode infections (similarly with clotting). It has been established for a long time that nematode surfaces can activate complement non-specifically through the alternative pathway, and that their resistance to complement attack could lie partly in their rapid shedding of surface materials.[71–74] It has also been found that some nematode surfaces bind complement control proteins that can disable activation of the full complement pathway and consequent cellular inflammatory responses.[71,75] The complement system in humans is highly polymorphic, and it could be that the apparent differences in susceptibility in humans could be related in part to inheritance of differing complement alleles and their interactions.

Activation of innate responses also calls into mind pathogen-associated patterns (PAMPs) that activate toll-like receptors (TLRs). The latter are known to be activated by chemical signatures that are not present in mammals (e.g. flagellin, LPS, mannan, bacterial lipoproteins) or genetic material being present in an inappropriate compartment of cells (such as viral nucleic acid in endosomes).[76] With nematodes, there are several candidates for chemical PAMPs that are also not found in mammals, such as cuticulin proteins, ascarosides, and a plethora of protein structures (such as the NPAs) that are unknown beyond nematodes, and there is already some evidence of unusual TLR stimulation activity by certain specialized nematode molecules.[77]

THE STRUCTURE AND FUNCTION OF THE MAJOR ALLERGEN OF *ASCARIS* – ABA-1

ABA-1 was first identified as a small (~14.4 kDa; 14,400 molecular mass) protein that is particularly abundant in the pseudocoelomic fluid of adult *Ascaris*, and also appears in culture fluid of the infective and lung stage larvae.[1,24,78] It is among the antigens of the parasite that are under strong MHC-mediated genetic control of the immune repertoire in rodents (see above). As also detailed above, humans infected with the parasite are heterogeneous in their immune responses to ABA-1, such that some people respond strongly, and some not at all, despite attested infection and strong immune responses to other components of the parasite (Figure 3.3). Interest in the protein grew with the demonstration that it is an allergen in the context of infection (i.e. that it is the target of IgE antibody),[23,26,28,29] so could be relevant to allergic manifestations such as Loeffler's syndrome and IgE-based immunity. Direct amino acid sequencing of ABA-1 purified from the parasite showed that it was unrelated to any type of protein then known.[24] When cDNA encoding ABA-1 was isolated[25] it was found that the encoded protein was enriched in charged amino acids (see Figure 2 of,[79]) which has been proposed as a characteristic of allergens in general.[80,81] Whether or not ABA-1 is intrinsically allergenic, or is only so in the context of infection, remains to be determined, although there is evidence that its allergenicity is conditional on the means by which it is presented to the immune system – animals immunized with the protein in Freund's adjuvant develop strong IgG antibodies to it, but no IgE antibody.[44]

Around the time that amino acid sequence information for ABA-1 began to emerge, other laboratories working on a range of nematode parasites were focusing on an unusual class of proteins produced as large, repetitive polyprotein precursors that are post-translationally cleaved into multiple copies of ABA-1-like molecules.[82–86] Moreover, like ABA-1, these tended to be targets for IgE antibody responses.[85] ABA-1 was also found to be synthesized as a long, precursor polypeptide comprising head-to-tail, tandemly repeated regions that are post-translationally cleaved at regularly-spaced proteinase cleavage sites into about ten small proteins of approximately 14.4 kDa.[25] Some of these repeated units have identical amino acid sequences, some slightly different, and at least one exists that is only about 49% identical (Figure 3.6).[87,88]

ABA-1 is the founding member of the nematode polyprotein allergens (NPAs), and its formal systematic name is As-NPA-1 for the *A. suum* form, and Al-NPA-1 for that of *A. lumbricoides*.[82] The individual repeated units are designated by a terminal letter such as As-NPA-1A. NPAs have been found widely in many parasitic and free-living species of nematode within the Secernentean nematodes at least (which includes most of the major nematode parasites of humans and plants except for groups such as *Trichinella*, *Trichuris*, and *Capillaria*). Nematodes vary greatly in the quantities of NPAs they produce, ascaridids, especially *Ascaris*, being particularly notable in the amounts they make. In both *Ascaris*, *C. elegans*,

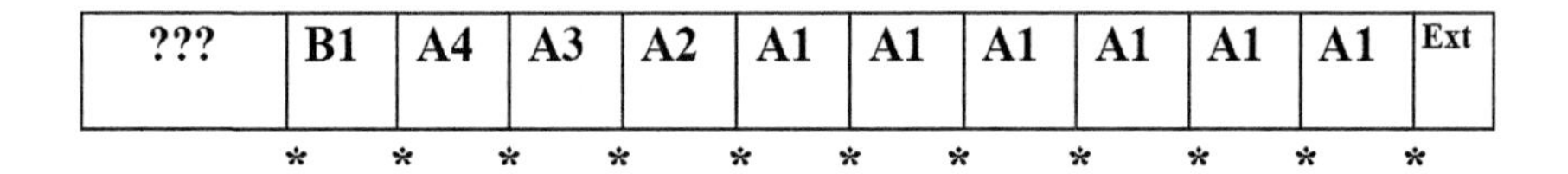


```
ABA-1A1        HHFTLESSLDTHLKWLSQEQKDELLKMKKDGKAKKELEAKILHYYDELEG
ABA-1B1        ---TMEHYLKTYLSWLTEEQKEKLKEMKEAGKTKAEIQHEVMHYYDQLHG
                  *:*  *.*:*.**::***::* :**: **:* *:: :::****:*.*

ABA-1A1        DAKKEATEHLKGGCREILKHVVGEEKAAELKNLKDSGASKEELKAKVEEA
ABA-1B1        EEKQQATEKLKVGCKMLLKGIIGEEKVVELRNMKEAGADIQELQQKVEKM
               : *::***:** **: :** ::****..**:*:*::**. :**: ***:

ABA-1A1        LHAVTDEEKKQYIADFGPACKKIYGVHTSIYGVHTSRRRR
ABA-1B1        LSEVTDEKQKEKVHEYGPACKKIFGATTLQHHRRRR
               *  ****::*: : ::*******:*. *
```

FIGURE 3.6 **Organization of the precursor polyprotein of the ABA-1 allergen of *Ascaris lumbricoides* and *Ascaris suum* and amino acid sequence heterogeneity between units.** Upper panel: The polyprotein comprises a tandemly repeated array, each unit of which encodes an ABA-1-like protein of about 14.4 kDa in size. The polyprotein is cleaved post-translationally at regularly-spaced proteinase cleavage sites (indicated by *) into multiple copies of ABA-1 proteins. At the carboxy terminal of the pre-processed polyprotein there is a short extension that is removed (Ext). The complete sequence of the polyprotein remains to be determined, and is still incomplete at the N-terminal end (as indicated by ???), but is likely to comprise a short secretory signal peptide, as found in other species.[83] Those units that have amino acid sequences that are identical or closely similar are designated units A1 to 4, and one designated B1 that is only about 49% identical to the A-types. Lower panel: Alignment of an A1- and a B1-type unit. Identical amino acids are indicated by an asterisk (*) and those that are similar in biophysical properties are indicated by a colon (:). The single tryptophan (W; single letter amino acid code), and the two cysteines (C) that are cross-linked internally in the protein, are absolutely conserved in NPAs from all species of nematode for which the sequences are known, and are double underlined in the sequences. The terminal runs of four arginines (RRRR) are where the precursor protein is cut and the arginines then removed. Regardless of the substantial differences in amino acid sequences, the A1 and B1 types have very similar lipid binding properties (see Figure 3.7 and[87]).

and *Globodera pallida*, the NPA-encoding gene is transcribed in the cells of the intestine, and in all life-cycle stages.[82,89,90] The biochemical activity of the NPAs remained a mystery until it was found by accident that ABA-1 and the other NPAs bind fatty acid and retinol (vitamin A).[83,91,92] They may therefore have a function similar to that of vertebrate serum albumin in the bulk transport of sparingly soluble lipids within the worms. Lipid binding by ABA-1 was discovered and analyzed using fluorescence-based techniques, which not only allowed the range of lipids it can bind to be investigated very quickly, but also showed that the lipid binding site in the molecule is unusually apolar (Figure 3.7).[91]

So, there arose many reasons to understand the structure of ABA-1 and its ilk, namely that (as a tandemly repetitive polyprotein) NPAs are a type

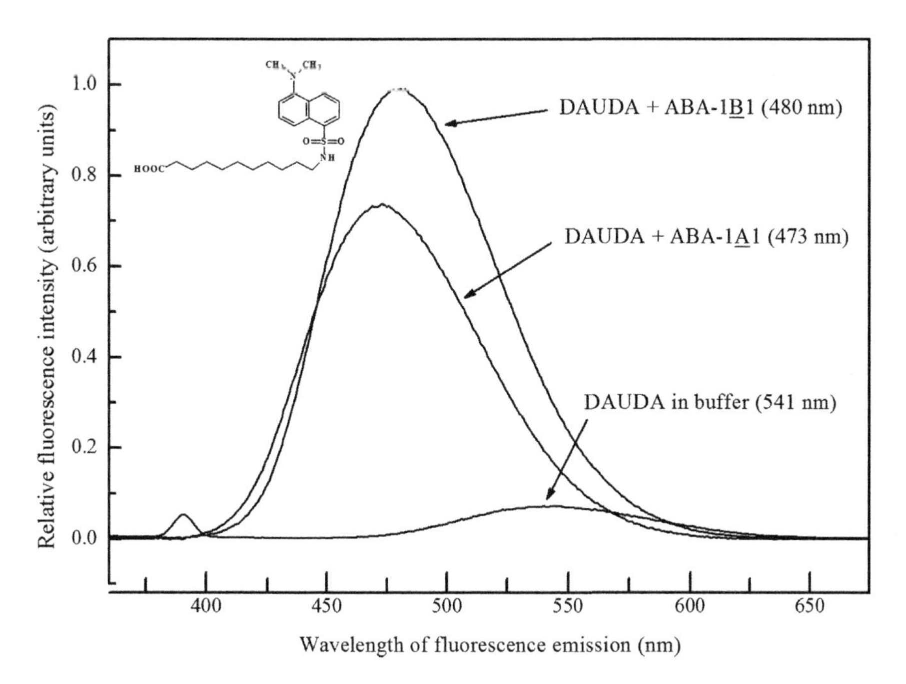

FIGURE 3.7 **Lipid binding by ABA-1.** Lipid binding activity was examined using a fatty acid attached to a dansyl fluorescent group that is environment sensitive (dansyl undecanoic acid, DAUDA, see inset). DAUDA has a low fluorescence emission in water, but its emission increases dramatically and shifts to a shorter wavelength (blue shift) when it enters a protein binding site. The degree of this shift is taken to indicate the degree of apolarity of its new environment, and the blue shift with ABA-1 is unusually large. The experiment shown compared a recombinant A-type repeat and a recombinant B-type repeat, the characteristics of which were similar. It is not shown here, but when a natural ligand, for example oleic acid, is added to the mixture of ABA-1 and DAUDA, it displaces DAUDA from the binding site back into water and its fluorescence decreases. This allows the rapid screening of natural lipids and other compounds for binding to these proteins.

of protein that is very rare in nature; NPAs are the only known lipid binding proteins to be produced as polyproteins; ABA-1 and other NPAs are frequently found to be important antigens and allergens in nematode infections and have even been described as being immunodominant (i.e. antibody responses to them are often stronger than against other antigens, and antibody screening of cDNA expression libraries from nematode parasites frequently led to the isolation of NPAs); NPAs have biophysical characteristics unlike any other lipid binding protein of their size; and, if ABA-1 proved to have unusual structural features, then it and other NPAs might comprise new drug targets. A final reason is, of course, the simple curiosity of a nematode biologist.

Solving the structure of ABA-1 required several tricks. First, the protein as purified from the parasite comes in several isoforms (here, proteins of similar structures and functions but with some of the tandem repeats having different amino acid sequences and co-purify). This heterogeneity was probably the reason for the failure of crystallization trials. A similar problem arose when a recombinant form was produced in bacteria, representing a single repeat (ABA-1A), but this too did not crystallize properly. In this case the failure may have been due to the fact that the protein molecules, although identical in amino acid sequence, were produced in bacteria, and consequently were heterogeneously loaded with different lipid ligands, which could result in slight differences in overall structure between molecules of the protein. ABA-1A's structure was finally solved by protein nuclear magnetic resonance (NMR), using stable isotope-labeled protein that had been stripped of bacterial cell-derived ligands and then reloaded with a single species of lipid (oleic acid). This revealed ABA-1A to be a helix-rich protein with a type of overall tertiary structure not known before (Figure 3.8).[93] It has two (lipid) ligand binding sites, and is completely different in structure from other fatty acid binding proteins of its size. Simple "homology" modeling using the amino acid sequences of NPAs from other nematodes provides acceptable theoretical structures (M.W. Kennedy, unpublished), so ABA-1A's structure could be valuable in understanding and elucidating the structures of other NPAs. The ABA-1A structure also confirmed an idea that arose from when the sequence of a single unit of ABA-1 was first elucidated – that each unit of an NPA is itself derived from an ancient duplication event.[25,93] Comparison of the fold of the N- and C-terminal halves of the protein showed that they were indeed similar enough to be the result of the suspected duplication (Figure 3.8).[93] If so, then the NPAs may have evolved from very small (~8 kDa) lipid binding proteins, which is about the size of the small helix-rich lipid binding proteins of plants.

Other than probably being a major lipid transporter between the tissues of nematodes, knowledge of the structure has not yet contributed

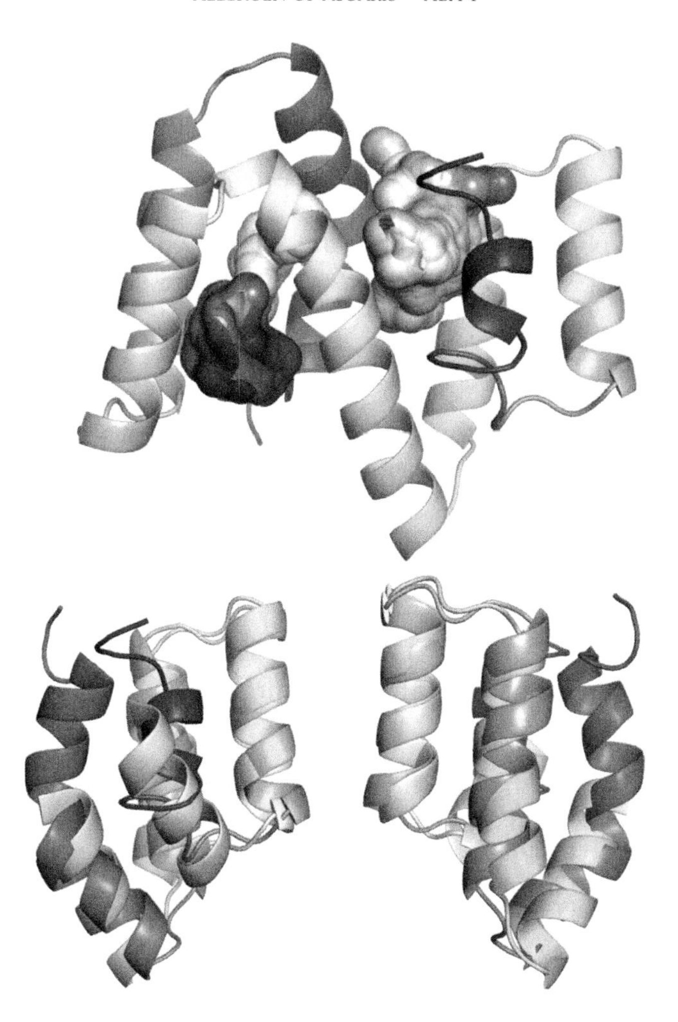

FIGURE 3.8 **The molecular structure of ABA-1. A ribbon representation of ABA-1A as solved by protein nuclear magnetic resonance.** The protein is helix-rich, and has two binding cavities, the internal surfaces of which are predominantly apolar as befits a hydrophobic binding pocket. The surfaces of the cavities are shown colored according to electrostatic potential (blue, positive, graded through white, apolar, to red, negative). The protein structure comprises two domains joined by the long central helix, and the structures of these two halves of the molecule are virtually superimposable, supporting the idea that each unit of these polyproteins is itself derived from an ancient duplication event. The lower two panels show a superimposition of the two halves of the protein from different viewpoints, illustrating the similarities in the structures of the two domains. *For a full color version of this figure go to www.gla.ac.uk/nematodes*

to a fuller understanding of what NPAs do. The question is most pertinent in parasitic nematodes if NPAs are released by the parasites into the tissues they occupy (although this is not thought to be the case for *Globodera pallida*,[94] in which case another group of unusual lipid binding proteins released by nematodes, the FAR proteins, may be important).[95,96] One possibility that is often raised, albeit so far untested, is that NPAs could be involved in interacting with host cells by delivering pharmacologically active lipids, or sequestering those produced by the host (such as retinol, platelet activating factor, eicosanoids such as arachidonic acid and leukotrienes, all of which they appear to bind[88]) in order to modulate local and systemic immune and inflammatory responses. These ideas are difficult to test, and remain to be addressed, as with other lipid binding proteins released by nematode parasites, but knowledge of NPA structures should facilitate the design of mutant proteins that do not bind selected lipids.[97]

AN UNUSUAL LIPID BINDING PROTEIN IN THE PERIVITELLINE FLUID OF *ASCARIS* EGGS

Most of this chapter has dealt with proteins of direct relevance to immune responses to *Ascaris*, and the structure of its major allergen, but structures of other interesting proteins are now emerging. These include an unusual protein found in the perivitelline fluid of the eggs of *Ascaris* (the fluid surrounding the developing larva),[89] which may be involved in maintenance of the lipid layer of the shell (e.g. by removing and replacing oxidized lipids), scavenging lipids in the fluid itself, or in maintaining the lipid balance within the whole egg/larva system. This is As-p18, which also has homologues in the periviteline fluid of other species of nematode, such as the agent of lymphatic filariasis *Brugia malayi*.[98] When originally sequenced, it was clear that As-p18 is a member of a family of small lipid binding proteins that is ubiquitous in the animal phyla, but also that it has unusual features that had not been found before. This family is the cytosolic lipid binding proteins such as the nine fatty acid binding proteins (FABPs), and the cellular retinol and retinoic acid binding proteins (CRBPs, CRABPs) of humans.[99,100] These are thought to have many functions inside cells, including central involvement in transport of lipids into and out of cells, maintaining the balance of lipid stores, and transporting small lipids involved in cell–cell signaling in regulating cellular activation and differentiation. Such proteins are needed because the lipids involved are relatively insoluble, potentially damaging to cellular membranes, and in need of protection from chemical damage, the latter applying to signaling lipids in particular. The structures of these proteins from mammals are well known, and all of them are so similar as to be virtually superimposable, despite diversity in their amino acid sequences.[99,100] Nematodes also possess cytosol-confined FABPs that appear to be similar to those found elsewhere in the animal phyla, but the unusual subfamily typified by As-p18 appears to be confined to nematodes (hence the name nemFABPs).[90] Significantly, only in nematodes do their genes encode a secretory leader peptide and the proteins are secreted from the synthesizing cells.[89,90,98] And only in nematodes is the mature polypeptide (after removal of the secretory signal peptide) longer than in other FABPs, which was speculated to mean that the mature proteins have extra loop regions unique to nemFABPs.[89] If their structures do indeed have the anticipated extra features, then they may have biological functions not seen in other phyla.

The structure of As-p18 was solved using both NMR and x-ray crystallography.[101,102] As with ABA-1, the protein was produced in bacteria and needed to be stripped of bacterial-derived lipids, then loaded uniformly with a fatty acid (oleic) so as to produce a homogeneously conformed preparation of protein. This revealed, as anticipated, that As-p18 folds as a β-barrel surrounding a central cavity (Figure 3.9), the walls of which are lined predominantly with apolar amino acid side chains. This is consistent with a hydrophobic lipid binding pocket into which a small lipid could partition. There are also some charged side chains projecting into the cavity that could, as with other FABPs, be involved in tethering a lipid's charged head group, although the orientation of lipid within As-p18 remains to be determined. As with other FABPs, the opening of the cavity is capped by two short antiparallel sections of α-helix. So, As-p18 does indeed conform to the general structure of cytosolic lipid binding proteins of the FABP/CRBP/CRABP superfamily. The atypical loop regions that had been predicted from earlier modeling of As-p18 are present, but not where predicted.[89] As-p18 is instead found to have an extended loop emerging from the opposite end of the molecule from the presumptive portal region for entry and exit of ligands beneath the pair of helices. Such a loop has not been found in any other member of this protein superfamily, structures for which are now known from a wide range of animal groups including vertebrates, insects, arachnids, and cestodes. A short loop adjacent to the portal region is also longer than in other FABPs. Thus, As-p18, and presumably other nemFABPs, does indeed exhibit features that are so far confined to nematodes.

Another interesting feature of As-p18 that may provide a clue to its function is the existence of a cluster of apolar amino acid side chains projecting into solvent from the short helix adjacent to the portal (Figure 3.9). A cluster similar to this is also seen in mammalian FABPs that have been found to interact and exchange lipids with artificial membranes by a direct contact, collisional mechanisms,[103] by which a lipid ligand can transfer between a protein and a membrane without entry into the

FIGURE 3.9 Structure of the As-p18 protein of the perivitelline fluid of *Ascaris* eggs. The crystal structure of As-p18, which is essentially identical to the NMR structure of the protein, showing the extra loop region (lower circle) at the opposite end of the molecule from the presumed portal of entry for lipid ligands.[108] The position of the presumed portal of entry of ligands into the binding cavity underneath the two short helices is indicated by the arrow. The slightly enlarged loop adjacent to the portal is indicated by the upper circle. The unusually exposed apolar side chains projecting into solvent near the portal region are shown in magenta and with the side chains added (tryptophan, methionine, isoleucine, valine, and a leucine); clusters of this kind are typical of those cytosolic fatty acid binding proteins that interact collisionally with membranes in the process of exchanging lipid ligands.[99,100,103] *Structural coordinates made available by Dr Mads Gabrielsen. For full color version of this figure go to www.gla.ac.uk/nematodes*

surrounding solvent water. The only FABP that does not perform this way (liver FABP) instead has a cluster of charged amino acids in this relative position.[103] This could mean that As-p18 also interacts with membranes by direct contact in order to collect or deliver lipids, although it remains possible that such a "sticky finger" could also be involved in its interaction with other proteins.

nemFABPs probably have functions in nematodes beyond merely being exported from the cell to transport lipids. It is clearly tempting to suppose that the existence of the loop at the opposite side of the molecule from the portal pertains to a previously unknown function of nemFABPs, and that As-p18 and other nemFABPs are present in perivitelline fluid of the egg (though seemingly also present elsewhere[90]) to perform an important role in nematode reproduction. That may be of more than general interest given that it is the reproductive stages of many nematode infections that cause the major part of associated pathologies (as in onchocerciasis). It would be interesting to know whether inactivation or deletion of nemFABP genes produce abnormal or reproduction-disabled phenotypes, *C. elegans* being the obvious system within which to test the idea.

CONCLUDING REMARKS

This chapter concentrated on proteins produced by both species of *Ascaris*, and how immune responses to them, and direct structural analysis of the proteins themselves, have illuminated principles that apply both to ascariasis and other nematodiases. The concentration on proteins is partly because they are the easiest products to investigate in relation to biochemistry, immune responses, and immunogenetics. But other types of molecule may be important in immunological phenomena, such as the modulation of responses by T cells, B cells, and other immune-associated cells that have been found with crude preparations from the worms, the active principles of which remain to be identified.[104–108] Compounds such as glycolipids and other entities could trigger specific toll-like receptors and other innate responses because their chemistries are of types that do not occur in mammals, so could therefore act as generic signals. Equally, it is clear from examples, such as ABA-1 and other NPAs, that nematodes produce proteins of types unique to them, the structures of which could also act as generic signals of nematode infection.[107] Given that NPAs are binding proteins for a range of hydrophobic ligands, some of which may be unique to nematodes, then they may yet provide important clues. It is conceivable that NPAs are not allergenic by dint of their amino acid sequences or overall tertiary structure, but only in association with the lipid ligands that accompany them. For the moment, we have good information on the ligand binding propensities of NPAs arising from simple binding assays using lipids that are of known relevance to mammalian physiology and immune signaling, but not what compounds the parasites themselves actually use them to transport.

Pertinent to this is that the two ligand binding sites of ABA-1 are different and have characteristics that may mean that they carry different classes of compound with potentially different biological activities[93]. *Ascaris* may yet have much to teach us about how nematodes modulate the immune system of their hosts. Why, for instance, do *Ascaris* larvae undergo a prolonged, hazardous, and seemingly unnecessary tissue migration from the intestine back to the intestine? Is it merely because their ancestors had intermediate hosts and their lineage cannot escape from having a compulsory migratory stage? Or is it in order to modify the immune response in their favor for when they settle down to reproduce, or to prepare the way for their successor larvae?

Acknowledgments

Most if not all of the research reported here was supported by the Wellcome Trust, UK.

References

1. Kennedy MW, Qureshi F. Stage-specific secreted antigens of the parasitic larval stages of the nematode *Ascaris*. *Immunology* 1986;**58**(3):515–22.
2. Urban Jr JF, Romanowski RD. *Ascaris suum*: protective immunity in pigs immunized with products from eggs and larvae. *Exp Parasitol* 1985;**60**(2):245–54.
3. Kennedy MW, et al. Antigenic relationships between the surface-exposed, secreted and somatic materials of the nematode parasites *Ascaris lumbricoides, Ascaris suum,* and *Toxocara canis*. *Clin Exp Immunol* 1989;**75**(3):493–500.
4. Kennedy MW, et al. The secreted and somatic antigens of the 3rd stage larva of *Anisakis simplex,* and antigenic relationship with *Ascaris suum, Ascaris lumbricoides,* and *Toxocara canis*. *Mol Biochem Parasitol* 1988;**31**(1):35–46.
5. Kennedy MW, et al. Homology and heterology between the secreted antigens of the parasitic larval stages of *Ascaris lumbricoides* and *Ascaris suum*. *Exp Immunol* 1987; **67**(1):20–30.
6. Fagerholm HP, et al. Differentiation of cuticular structures during the growth of the third-stage larva of *Ascaris suum* (Nematoda, Ascaridoidea) after emerging from the egg. *J Parasitol* 2000;**86**(3):421–7.
7. Geenen PL, et al. The morphogenesis of *Ascaris suum* to the infective third-stage larvae within the egg. *J Parasitol* 1999;**85**(4):616–22.
8. Anderson TJC, Jaenike J. Host specificity, evolutionary relationships and macrogeographic differentiation among *Ascaris* populations from humans and pigs. *Parasitology* 1997;**115**:325–42.
9. Criscione CD, et al. Microsatellite markers for the human nematode parasite *Ascaris lumbricoides:* development and assessment of utility. *J Parasitol* 2007;**93**(3):704–8.
10. Peng WD, et al. Genetic variation in sympatric *Ascaris* populations from humans and pigs in China. *Parasitology* 1998;**117**:355–61.
11. Hawley JH, Peanasky RJ. *Ascaris suum* – are trypsin-inhibitors involved in species specificity of ascarid nematodes. *Exp Parasitol* 1992;**75**(1):112–8.
12. Ng KKS, et al. Structural basis for the inhibition of porcine pepsin by *Ascaris* pepsin inhibitor-3. *Nat Struct Biol* 2000;**7**(8):653–7.

13. Knox DP, Kennedy MW. Proteinases released by the parasitic larval stages of *Ascaris suum*, and their inhibition by antibody. *Mol Biochem Parasitol* 1988;**28**(3):207–16.
14. Croll NA, et al. The population biology and control of *Ascaris lumbricoides* in a rural-community in Iran. *Trans R Soc Trop Med Hyg* 1982;**76**(2):187–97.
15. Loeffler W. Transient lung infiltrations with blood eosinophilia. *Int Arch Allergy Appl Immunol* 1956;**8**(1–2):54–9.
16. Ogilvie BM, De Savigny D. Immune response to nematodes. In: Cohen S, Warren KS, editors. *Immunology of Parasitic Infections*. 2nd ed. Oxford: Blackwell Scientific Publications; 1982. p. 715–57.
17. Gelpi AP, Mustafa A. Seasonal pneumonitis with eosinophilia – a study of larval ascariasis in Saudi Arabs. *Am J Trop Med Hyg* 1967;**16**(5):646–57.
18. Spillmann RK. Pulmonary ascariasis in tropical communities. *Am J Trop Med Hyg* 1975; **24**(5):791–800.
19. Muller R. *Worms and Diseases: A Manual of Medical Helminthology*. Heinemann; 1975. 1–161.
20. Urquhart GM, et al. *Veterinary Parasitology. Longman Scientific & Technical*. New York: Churchill Livingstone; 1987. vii + 286pp.
21. McGibbon AM, et al. Identification of the major *Ascaris* allergen and its purification to homogeneity by high-performance liquid-chromatography. *Mol Biochem Parasitol* 1990; **39**(2):163–72.
22. Kennedy MW, et al. The specificity of the antibody-response to internal antigens of *Ascaris* – heterogeneity in infected humans, and MHC (H-2) control of the repertoire in mice. *Clin Exp Immunol* 1990;**80**(2):219–24.
23. Tomlinson LA, et al. MHC Restriction of the antibody repertoire to secretory antigens, and a major allergen, of the nematode parasite *Ascaris*. *J Immunol* 1989;**143**(7):2349–56.
24. Christie JF, et al. N-terminal amino-acid-sequence identity between a major allergen of *Ascaris lumbricoides* and *Ascaris suum*, and MHC-restricted IgE responses to it. *Immunology* 1990;**69**(4):596–602.
25. Spence HJ, et al. A CDNA-encoding repeating units of the ABA-1 allergen of Ascaris. *Mol Biochem Parasitol* 1993;**57**(2):339–44.
26. Fraser EM, Christie JF, Kennedy MW. Heterogeneity amongst infected children in IgE antibody repertoire to the antigens of the parasitic nematode *Ascaris*. *Int Arch Allergy Appl Immunol* 1993;**100**(3):283–6.
27. Acevedo N, et al. Allergenicity of *Ascaris lumbricoides* tropomyosin and IgE sensitization among asthmatic patients in a tropical environment. *Int Arch Allergy Appl Immunol* 2011;**154**(3):195–206.
28. McSharry C, et al. Natural immunity to *Ascaris lumbricoides* associated with immunoglobulin E antibody to ABA-1 allergen and inflammation indicators in children. *Infect Immun* 1999;**67**(2):484–9.
29. Turner JD, et al. Allergen-specific IgE and IgG4 are markers of resistance and susceptibility in a human intestinal nematode infection. *Microbes Infect* 2005; **7**(7–8):990–6.
30. Vercauteren I, et al. Vaccination with an *Ostertagia ostertagi* polyprotein allergen protects calves against homologous challenge infection. *Infect Immun* 2004; **72**(5):2995–3001.
31. Cooper PJ, et al. Human infection with *Ascaris lumbricoides* is associated with a polarized cytokine response. *J Infect Dis* 2000;**182**(4):1207–13.
32. Dold C, Holland CV. Investigating the underlying mechanism of resistance to *Ascaris* infection. *Microbes Infect* 2011;**13**(7):624–31.
33. Dold C, et al. Genetic influence on the kinetics and associated pathology of the early stage (intestinal-hepatic) migration of *Ascaris suum* in mice. *Parasitology* 2010; **137**(1):173–85.

34. Perez J, et al. Immunohistochemical characterization of hepatic lesions associated with migrating larvae of *Ascaris suum* in pigs. *J Comp Pathol* 2001;**124**(2–3):200–6.
35. Haswell-Elkins MR, et al. The antibody recognition profiles of humans naturally Infected with *Ascaris lumbricoides. Parasite Immunol* 1989;**11**(6):615–27.
36. Kwanlim GE, Forsyth KP, Maizels RM. Filarial-specific IgG4 response correlates with active *Wuchereria bancrofti* infection. *J Immunol* 1990;**145**(12):4298–305.
37. Elkins DBM, Haswellelkins M, Anderson RM. The epidemiology and control of intestinal helminths in the Pulicat Lake region of southern India .1. Study design and pretreatment and posttreatment observations on *Ascaris lumbricoides* infection. *Trans R Soc Trop Med Hyg* 1986;**80**(5):774–92.
38. Kennedy MW, Fraser EM, Christie JF. MHC class-II (I-A) region control of the IgE antibody repertoire to the ABA-1 allergen of the nematode *Ascaris. Immunology* 1991; **72**(4):577–9.
39. Kennedy MW, et al. Genetic (major histocompatibility complex) control of the antibody repertoire to the secreted antigens of *Ascaris. Parasite Immunol* 1987;**9**(2):269–73.
40. Else KJ, et al. MHC-restricted antibody responses to *Trichuris muris* excretory secretory (E/S) antigen. *Parasite Immunol* 1990;**12**(5):509–27.
41. Kwanlim GE, Maizels RM. MHC and non-MHC-restricted recognition of filarial surface-antigens in mice transplanted with adult *Brugia malayi* parasites. *J Immunol* 1990; **145**(6):1912–20.
42. Kennedy MW, et al. MHC (RT1) restriction of the antibody repertoire to infection with the nematode *Nippostrongylus brasiliensis* in the rat. *Immunology* 1990;**71**(3):317–22.
43. Kennedy MW, et al. H-2 (I-A) control of the antibody repertoire to secreted antigens of *Trichinella spiralis* in infection and its relevance to resistance and susceptibility. *Immunology* 1991;**73**(1):36–43.
44. Christie JF, Fraser EM, Kennedy MW. Comparison between the MHC-restricted antibody repertoire to *Ascaris* antigens in adjuvant-assisted immunization or infection. *Parasite Immunol* 1992;**14**(1):59–73.
45. Wang Y, et al. Association of polymorphisms of cytokine and TLR-2 genes with long-term immunity to hepatitis B in children vaccinated early in life. *Vaccine* 2012; **30**(39):5708–13.
46. Acevedo N, et al. Association between total immunoglobulin E and antibody responses to naturally acquired *Ascaris lumbricoides* infection and polymorphisms of immune system-related LIG4, TNFSF13B and IRS2 genes. *Clin Exp Immunol* 2009; **157**(2):282–90.
47. Holland CV, et al. A possible genetic-factor influencing protection from infection with *Ascaris lumbricoides* in Nigerian children. *J Parasitol* 1992;**78**(5):915–6.
48. Peisong G, et al. An asthma-associated genetic variant of STAT6 predicts low burden of *Ascaris* worm infestation. *Genes Immun* 2004;**5**(1):58–62.
49. Skallerup P, et al. Detection of a quantitative trait locus associated with resistance to *Ascaris suum* infection in pigs. *J Parasitol* 2012;**42**(4):383–91.
50. Williams-Blangero S, et al. Localization of multiple quantitative trait loci influencing susceptibility to infection with *Ascaris lumbricoides. J Infect Dis* 2008;**197**(1):66–71.
51. Fraser EM, Kennedy MW. Heterogeneity in the expression of surface-exposed epitopes among larvae of *Ascaris lumbricoides. Parasite Immunol* 1991;**13**(2):219–25.
52. Maass DR, et al. Intraspecific epitopic variation in a carbohydrate antigen exposed on the surface of *Trichostrongylus colubriformis* infective L3 larvae. *PloS Pathog* 2009;**5**(9).
53. Jex AR, et al. *Ascaris suum* draft genome. *Nature* 2011;**479**(7374):529–33.
54. Ségurel L, et al. The ABO blood group is a trans-species polymorphism in primates. *Proc Natl Acad Sci* 2012;**109**(45):18493–8.

55. Politz SM, et al. Genes that can be mutated to unmask hidden antigenic determinants in the cuticle of the nematode *Caenorhabditis elegans*. *Proc Natl Acad Sci USA* 1990; **87**(8):2901–5.
56. Hemmer RM, et al. Altered expression of an L1-specific, O-linked cuticle surface glycoprotein in mutants of the nematode *Caenorhabditis elegans*. *J Cell Biol* 1991; **115**(5):1237–47.
57. Politz SM, Chin KJ, Herman DL. Genetic-analysis of adult-specific surface antigenic differences between varieties of the nematode *Caenorhabditis elegans*. *Genetics* 1987; **117**(3):467–76.
58. Grenache DG, et al. Environmental induction and genetic control of surface antigen switching in the nematode *Caenorhabditis elegans*. *Proc Natl Acad Sci USA* 1996; **93**(22):12388–93.
59. Olsen DP, et al. Chemosensory control of surface antigen switching in the nematode *Caenorhabditis elegans*. *Genes Brain Behav* 2007;**6**(3):240–52.
60. Ogilvie BM, Edwards AJ. Changes in the isoenzymes of acetyl cholinesterase in adult *Nippostrongylus brasiliensis* affected by immunity. *J Parasitol* 1970;**56**(4 Sec II Pt 1):253.
61. Edwards AJ, Burt JS, Ogilvie BM. Effect of immunity upon some enzymes of parasitic nematode, *Nippostrongylus brasiliensis*. *Parasitology* 1971;**62**(Apr):339–47.
62. Jones VE, Ogilvie BM. Protective immunity to *Nippostrongylus brasiliensis* in rat. 3. Modulation of worm acetylcholinesterase by antibodies. *Immunology* 1972; **22**(1):119–29.
63. Beltran S, et al. Vertebrate host protective immunity drives genetic diversity and antigenic polymorphism in *Schistosoma mansoni*. *J Evol Biol* 2011;**24**(3):554–72.
64. Van Valen L. A new evolutionary law. *Evol Theor* 1973;**1**(1):1–30.
65. Van Valen L. Molecular evolution as predicted by natural selection. *J Mol Evol* 1974; **3**(2):89–101.
66. Maizels RM, Kurniawan-Atmadja A. Variation and polymorphism in helminth parasites. *Parasitology* 2002;**125**:S25–37.
67. Smithers SR, Terry RJ. The immunology of schistosomiasis. *Adv Parasitol* 1969;**7**:41–93.
68. Brown SP, Grenfell BT. An unlikely partnership: parasites, concomitant immunity and host defence. *P Roy Soc B-Biol Sci* 2001;**268**(1485):2543–9.
69. Beltran S, Boissier J. Schistosome monogamy: who, how, and why? *Trends Parasitol* 2008;**24**(9):386–91.
70. Beltran S, Cezilly F, Boissier J. Genetic dissimilarity between mates, but not male heterozygosity, influences divorce in schistosomes. *PloS One* 2008;**3**:10.
71. Kennedy MW, Kuo YM. The surfaces of the parasitic nematodes *Trichinella spiralis* and *Toxocara canis* differ in the binding of post-C3 components of human-complement by the alternative pathway. *Parasite Immunol* 1988;**10**(4):459–63.
72. Mackenzie CD, et al. Activation of complement, the induction of antibodies to the surface of nematodes and the effect of these factors and cells on worm survival in vitro. *Eur J Immunol* 1980;**10**(8):594–601.
73. Maizels RM, Desavigny D, Ogilvie BM. Characterization of surface and excretory-secretory antigens of *Toxocara canis* infective larvae. *Parasite Immunol* 1984;**6**(1):23–37.
74. Smith HV, et al. The effect of temperature and antimetabolites on antibody-binding to the outer surface of 2nd stage *Toxocara canis* larvae. *Mol Biochem Parasitol* 1981; **4**(3–4):183–93.
75. Haapasalo K, Meri T, Jokiranta TS. *Loa loa* microfilariae evade complement attack in vivo by acquiring regulatory proteins from host plasma. *Infect Immun* 2009; **77**(9):3886–93.
76. Netea MG, Wijmenga C, O'Neill LAJ. Genetic variation in Toll-like receptors and disease susceptibility. *Nat Immunol* 2012;**13**(6):535–42.

77. Goodridge HS, et al. Immunomodulation via novel use of TLR4 by the filarial nematode phosphorylcholine-containing secreted product, ES-62. *J Immunol* 2005; **174**(1):284–93.
78. McGibbon AM, et al. Identification of the major *Ascaris* allergen and its purification to homogeneity by high-performance liquid chromatography. *Mol Biochem Parasitol* 1990; **39**(2):163–72.
79. Kennedy MW. The polyprotein allergens of nematodes (NPAs) – structure at last, but still mysterious. *Exp Parasitol* 2011;**129**(2):81–4.
80. Razzera G, et al. Mapping the interactions between a major pollen allergen and human IgE antibodies. *Structure* 2010;**18**(8):1011–21.
81. Thomas WR, Hales BJ, Smith WA. Structural biology of allergens. *Curr Allergy Asthma Rep* 2005;**5**(5):388–93.
82. Kennedy MW. The nematode polyprotein allergens/antigens. *Parasitol Today* 2000; **16**(9):373–80.
83. Britton C, et al. Extensive diversity in repeat unit sequences of the cDNA-encoding the polyprotein antigen allergen from the bovine lungworm *Dictyocaulus viviparus*. *Mol Biochem Parasitol* 1995;**72**(1–2):77–88.
84. Culpepper J, et al. Molecular characterization of a *Dirofilaria immitis* cDNA-encoding a highly immunoreactive antigen. *Mol Biochem Parasitol* 1992;**54**(1):51–62.
85. Paxton WA, et al. Primary structure of and immunoglobulin-E response to the repeat subunit of GP15/400 from human lymphatic filarial parasites. *Infect Immun* 1993; **61**(7):2827–33.
86. Poole CB, et al. Cloning of a cuticular antigen that contains multiple tandem repeats from the filarial parasite *Dirofilaria immitis*. *Proc Natl Acad Sci USA* 1992; **89**(13):5986–90.
87. Moore J, et al. Sequence-divergent units of the ABA-1 polyprotein array of the nematode *Ascaris suum* have similar fatty-acid- and retinol-binding properties but different binding-site environments. *Biochem J* 1999;**340**:337–43.
88. Xia Y, et al. The ABA-1 allergen of *Ascaris lumbricoides:* sequence polymorphism, stage and tissue-specific expression, lipid binding function, and protein biophysical properties. *Parasitology* 2000;**120**:211–24.
89. Mei BS, et al. Secretion of a novel, developmentally regulated fatty acid-binding protein into the perivitelline fluid of the parasitic nematode, *Ascaris suum*. *J Biol Chem* 1997;**272**(15):9933–41.
90. Plenefisch J, et al. Secretion of a novel class of iFABPs in nematodes: coordinate use of the *Ascaris/Caenorhabditis* model systems. *Mol Biochem Parasitol* 2000;**105**(2):223–36.
91. Kennedy MW, et al. The ABA-1 allergen of the parasitic nematode *Ascaris suum* – fatty-acid and retinoid-binding function and structural characterization. *Biochemistry* 1995;**34**(20):6700–10.
92. Kennedy MW, et al. The GP15/400 polyprotein antigen of *Brugia malayi* binds fatty-acids and retinoids. *Mol Biochem Parasitol* 1995;**71**(1):41–50.
93. Meenan NAG, et al. Solution structure of a repeated unit of the ABA-1 nematode polyprotein allergen of *Ascaris* reveals a novel fold and two discrete lipid-binding sites. *PloS Neglect Trop D* 2011;**5**(4).
94. Jones JT, et al. Identification and functional characterization of effectors in expressed sequence tags from various life cycle stages of the potato cyst nematode *Globodera pallida*. *Mol Plant Pathol* 2009;**10**(6):815–28.
95. Garofalo A, et al. The FAR protein family of the nematode *Caenorhabditis elegans* – differential lipid binding properties, structural characteristics, and developmental regulation. *J Biol Chem* 2003;**278**(10):8065–74.

96. Kennedy MW, et al. The Ov20 protein of the parasitic nematode *Onchocerca volvulus* – a structurally novel class of small helix-rich retinol-binding proteins. *J Biol Chem* 1997; **272**(47):29442–8.
97. McDermott L, et al. Mutagenic and chemical modification of the ABA-1 allergen of the nematode *Ascaris:* consequences for structure and lipid binding properties. *Biochemistry* 2001;**40**(33):9918–26.
98. Michalski ML, et al. An embryo-associated fatty acid-binding protein in the filarial nematode *Brugia malayi. Mol Biochem Parasitol* 2002;**124**(1–2):1–10.
99. Storch J, McDermott L. Structural and functional analysis of fatty acid-binding proteins. *J Lipid Res* 2009;**50**:S126–31.
100. Storch J, Thumser AE. Tissue-specific functions in the fatty acid-binding protein family. *J Biol Chem* 2010;**285**(43):32679–83.
101. Gabrielsen M, et al. Usable diffraction data from a multiple microdomain-containing crystal of *Ascaris suum* As-p18 fatty-acid-binding protein using a microfocus beamline. *Acta Crystallogr Sect F Struct Biol Cryst Commun* 2012;**68**:939–41.
102. Ibáñez-Shimabukuro M, et al. Resonance assignment of As-p18, a fatty acid binding protein secreted by developing larvae of the parasitic nematode *Ascaris suum. J Biomol NMR*; 2012. In press.
103. Kennedy MW, Beauchamp J. Sticky-finger interaction sites on cytosolic lipid-binding proteins? *Cell Mol Life Sci* 2000;**57**(10):1379–87.
104. Falcone FH, et al. *Ascaris suum*-derived products induce human neutrophil activation via a G protein-coupled receptor that interacts with the interleukin-8 receptor pathway. *Infect Immun* 2001;**69**(6):4007–18.
105. Lee TDG. Xie CY IgE regulation by nematodes – the body-fluid of *Ascaris* contains a B-cell mitogen. *J Allergy Clin Immunol* 1995;**95**(6):1246–54.
106. Paterson JCM, et al. Modulation of a heterologous immune response by the products of *Ascaris suum. Infect Immun* 2002;**70**(11):6058–67.
107. Fitzsimmons CM, Dunne DW. Survival of the fittest: allergology or parasitology? *Trends Parasitol* 2009;**25**(10):447–51.
108. Gabrielsen M, et al. Usable diffraction data from a multiple micro-domain containing crystal of *Ascaris suum* As-p18 fatty acid binding protein, using a micro-focus beam--line. *Acta Crystallographica Section F*; 2012. In press.

CHAPTER

4

Implications of *Ascaris* Co-infection

Francisca Abanyie, Tracey J. Lamb

Emory University School of Medicine, Emory Children's Center, Atlanta, GA, USA

OUTLINE

Ascaris: The Neglected Parasite
http://dx.doi.org/10.1016/B978-0-12-396978-1.00004-5

Copyright © 2013 Elsevier Inc. All rights reserved.

INTRODUCTION

The geographic distribution of gastrointestinal (GI) geohelminths is widespread, with various geographical areas of public health concern including Africa, Southeast Asia, and Central and South America. It is estimated that over one billion people are infected with *Ascaris lumbricoides* worldwide[1,2] with a notable prevalence in tropical areas where lack of safe drinking water and poor sanitation practices lead to increased exposure to infective eggs. *Ascaris* is found largely in developing countries and the epidemiological spread overlaps with that of several important diseases of the tropics including human immunodeficiency virus (HIV) and malaria (Figure 4.1). In areas where *Ascaris* is prevalent "co-infection is the rule rather than the exception."[3]

Co-infection studies have steadily become more common although funding has tended to be limited to the examination of *Ascaris* co-infection on pathogens causing high levels of mortality. In this chapter, we outline some of the general considerations in co-infection biology. We also review available data looking at interactions and effects of *Ascaris* infection on systemic co-infections with malaria or HIV. Possible effects of co-infecting pathogens on *Ascaris* biology or pathology are seldom reported or discussed. In the third part of the chapter, we consider some of the hypothetical interactions between *Ascaris* and co-infections that are more localized to specific organs of the body, specifically the liver and lungs (effects on *Ascaris* larval migration) and the gut (effects on egg hatching and adult *Ascaris* survival and fecundity).

GENERAL CONSIDERATIONS FOR CO-INFECTION INTERACTIONS IN *ASCARIS LUMBRICOIDES*-INFECTED INDIVIDUALS

Disease severity in any infection is determined by a complex relationship between several different host and pathogen traits (Table 4.1). Since pathogens have evolved to establish infection in hosts that are often co-infected with another pathogen, the study of co-infection can shed light on the biology and immunology of single infections that, in many respects, are not the norm. This point is particularly relevant when considering co-infections with *Ascaris* because of the high prevalence of this nematode infection in areas where many other diseases thrive.

It is not known whether *Ascaris* infection increases the likelihood of co-infection in a host. One study has reported that individuals who harbor higher (>5000 egg/g feces) intensities of *A. lumbricoides* infections

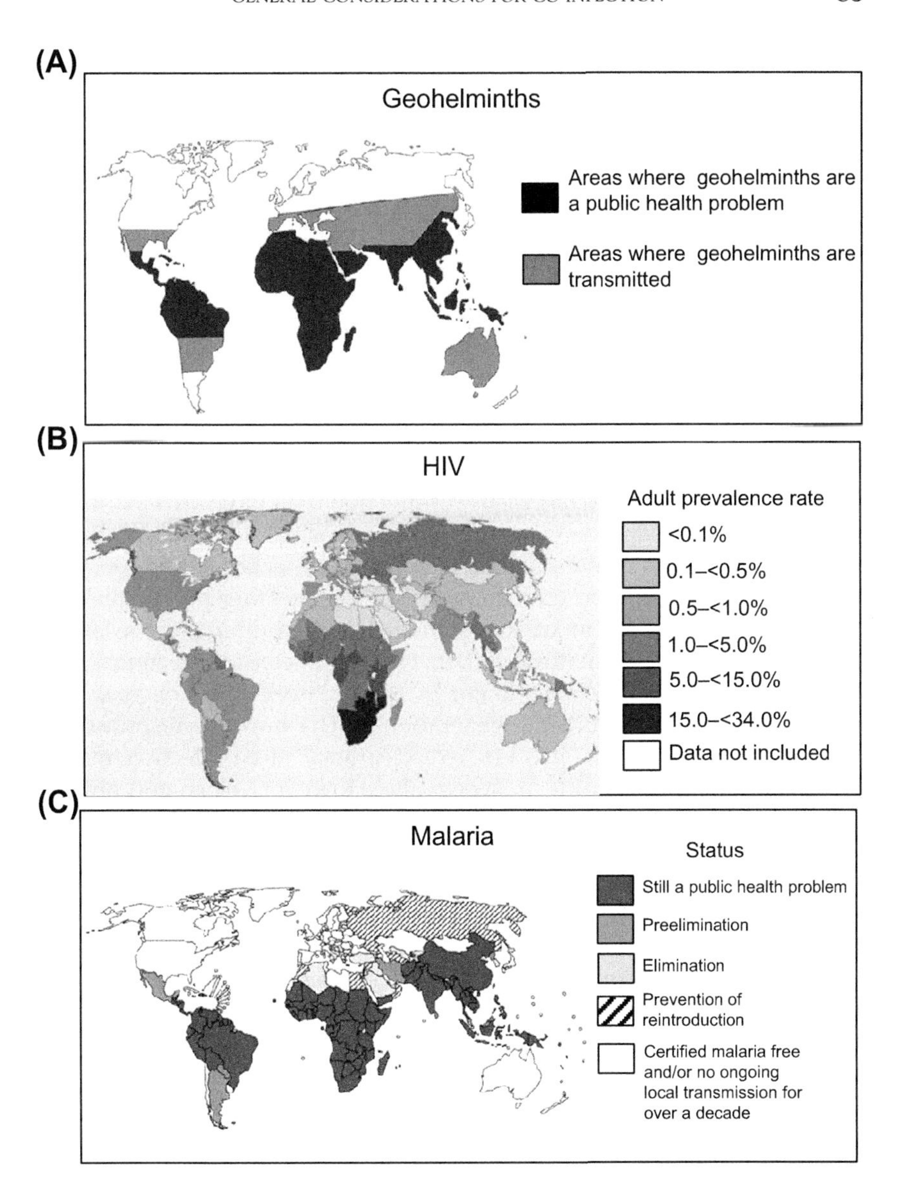

FIGURE 4.1 **Overlapping distributions of soil-transmitted helminths, HIV and malaria.** All maps are modified from the WHO in the following publications: (A) http://www.who.int/wormcontrol/documents/fact_sheets/soil_transmitted_helminths/en/index.html; (B) 2008 Report on the Global HIV epidemic; and (C) World Malaria Report 2009.

TABLE 4.1 Host and pathogen traits that may influence disease severity and outcome

Host traits	Pathogen traits
Host polymorphisms (e.g. cytokine promoter polymorphisms shaping the immune response)	Parasite polymorphisms (e.g. virulence genes)
Nutritional status	Infection frequency (e.g. biting rates of infected mosquitos)
Age and sex	Infection intensity (e.g. intense acute infection or low grade chronic infection) (see Chapter 13)
Presence of co-infection generating pre-existing host immune responses	Presence of co-infection utilizing host resources

are more likely to have *G. duodenalis* co-infection than individuals infected with *Ascaris* at a lower (<5000 egg/g feces) intensity[4] suggesting that this could be the case in some situations. *Ascaris* co-infection could alter the establishment and/or the transmission of a co-infecting pathogen. This could occur if disruption of the epithelial barrier lining the GI tract facilitates the establishment of certain viral or bacterial infections. It is also possible that the establishment of a regulatory immune response by *Ascaris* infection inhibits the generation of Th1 immune responses or Th2 responses leading to less efficient clearance of co-infecting micro-pathogens or GI-helminths, respectively. However, it is not always the case that *Ascaris*-generated immune responses are dominant in a co-infection setting[5] and the level of immune interaction is likely to differ depending on the particular co-infecting pathogen.

From the perspective of the co-infected host, it has been reported that disease severity can be altered in *Ascaris*-co-infected individuals. There are situations where this could be beneficial (for example, protection against cerebral malaria[6,7]) or detrimental (for example, lowering of CD4+ T cell numbers in HIV infection[8] which may contribute to the progression of AIDS). Both of these examples are discussed in more detail later in this chapter. Current hypotheses on the mechanisms behind such observations center on interactions mediated by the immune system. *Ascaris* can induce an immune response that can act systemically and potentially interact with the generation and efficacy of immune responses against most other pathogens in a co-infection setting. However, localized immune interactions could also occur in the liver or lungs where immune responses are generated by migrating *Ascaris* larvae in the context of pre-existing responses already active against co-infecting pathogens residing in these organs. In the gut, long-lived adult *Ascaris* nematodes persist by maintaining an anti-inflammatory environment that may interact with immune responses generated by common infections of the GI tract, particularly diarrheal-causing pathogens or co-infecting GI helminths such as hookworm or *Trichuris*.

The interaction of co-infecting pathogens within the body can be driven by resource alterations as well as immune-mediated mechanisms.[9] For example, in the gut, *Ascaris* adult nematodes feed on intestinal contents and co-infection with diarrhea-causing pathogens may change the composition or limit the availability of nutrients for *Ascaris* adult nematodes. This could lead to stunted nematode growth or fecundity.[10] Immune-mediated mechanisms and alteration in the availability of resources are not necessarily mutually exclusive considerations in co-infection biology. When considering *Ascaris*/HIV co-infection, HIV multiplies in CD4+ T cells and alterations in the capability of CD4+ T cells to become activated or proliferate during *Ascaris* co-infection may limit the main resource for HIV viral replication.

In general, magnitude of the interaction between *Ascaris* co-infection and an incoming pathogen is likely to be determined by a number of factors which include the type of infecting pathogen, the organs of the body infected, the frequency and intensity of infection by *Ascaris* and the co-infecting pathogen, the timing and order of the co-infection relative to *Ascaris* infection, and the immune responses already present from the existing *Ascaris* infection.

IMMUNE RESPONSES TO *ASCARIS LUMBRICOIDES*

To discuss possible immune-mediated interactions between *Ascaris* and co-infecting pathogens it is first necessary to review what is known of the immune responses generated by *Ascaris* infection (see also Chapter 1). The generation and maintenance of immune responses during *A. lumbricoides* infection has not yet been elucidated in fine detail. However, it is known that *Ascaris* is associated with a generic Th2 response typified by the secretion of IL-4[11–13] and circulation of IgE and IgG antibody isotypes to *Ascaris* antigens.[14–16] The ability of the body to mount Th2 responses is positively correlated with resistance to this nematode infection.[12,13] *Ascaris* infection can also induce a pro-inflammatory immune response. The inflammatory immune response may benefit one or more stages of the *Ascaris* life-cycle because the ability of monocytes to produce the acute phase cytokines tumor necrosis factor (TNF) and interleukin (IL)-1β in response to stimulation with Toll-like receptor (TLR) ligands has been positively correlated with levels of fecal *Ascaris*-egg counts.[17] A pro-inflammatory immune response could contribute to the maintenance of *Ascaris* infection by inhibiting the magnitude of the Th2 immune response and associated effector mechanisms.

Ascaris infection is thought to circumvent damaging Th2 effector mechanisms by the induction of strong immunoregulatory responses in the GI tract. Immunoregulatory cytokines IL-10 and transforming growth factor (TGF)-β can be measured systemically in *Ascaris*-infected individuals[18,19] although the particular cellular source(s) of these cytokines has not been fully characterized. *Ascaris* body extracts contain several potent immunosuppressive antigens, some of which are decorated with phosphorylcholine, a well-characterized inducer of immunosuppressive immune responses.[20] It has been demonstrated by Silva et al.[21] that the function of antigen presenting cells (such as dendritic cells) exposed to *Ascaris* antigens is suppressed in an IL-10-dependent manner. The protein from *Ascaris suum* (PAS-1) has been shown to have the ability to suppress the immune response to an unrelated protein in mice, a phenomenon that was associated with increases in IL-10 production[22] demonstrating the potential strength and systemic nature of this response. However, IL-10 production may not be strictly necessary to mediate this effect. Mucosal inflammatory immune responses stemming from induced allergy in mice was suppressed by administration of *Ascaris suum* body extracts in an IL-10 independent manner suggesting that other factors such as TGF-β can mediate this effect.[23] This immunoregulatory immune response may help to facilitate the longevity of *Ascaris* adult nematodes in the intestine.

After hatching, *Ascaris* larvae undergo migratory phases from the GI tract through the liver and lungs before being re-swallowed and establishing infection in the GI tract as adult nematodes. Not much is known regarding the immune responses generated by migrating *Ascaris* larvae in humans, but porcine models of ascariasis using *A. suum* infection suggest that migration is accompanied by an inflammatory immune response in the liver and lungs (see also Chapters 1 and 14). In the porcine liver this response manifests itself as "white spots" of cellular infiltrates, similar to granulomatous lesions, which are initially composed of eosinophils and neutrophils with the subsequent infiltration of macrophages, T and B cells, and fibroblasts.[24,25] Larval debris can also be found in mature white spots and white spot formation is associated with resistance to the establishment of *A. suum* infection in pigs.[26]

White spot formation has also been observed in the lungs, and the most prominent syndrome caused by larvae migrating through the lungs is a respiratory syndrome known as Loeffler's syndrome. This syndrome results in eosinophilic pneumonia whereby the eosinophil granule release alters the normal air spaces leading to breathing difficulties.[27] Eosinophil responses are hallmarks of Th2 responses, in particular IL-5 production,[28] suggesting that larval migration can be a strong inducer of Th2 responses and white spots are the result of type 2 inflammatory immune responses. Th2 responses may be induced by innate lymphoid cells, a recently described immune cell type that is thought to maintain epithelial barriers, such as those found in the lungs and intestine,[29] particularly if larval migration leads to physical damage of the epithelium of the lung tissue.

EFFECTS OF *ASCARIS* ON SYSTEMIC CO-INFECTIONS

The most likely explanation for interactions with *Ascaris* and co-infecting systemic pathogens is via immune modulation. Here we focus on the two most common systemic infections likely to be found in *Ascaris*-infected individuals, HIV and malaria.

HIV Infection

Although there is a growing body of reports confirming the presence of individuals co-infected with HIV and *Ascaris*, there has been relatively little work done into the effects of co-infection on HIV and *Ascaris* infection. Over 30 million people are infected with HIV globally, and 22 million people are estimated to be co-infected with helminth infections and HIV.[30,31] HIV is a retrovirus that causes acquired immunodeficiency syndrome (AIDS). There are two related retroviruses, HIV-1 and HIV-2; HIV-1 is the most predominant strain and HIV-2 is more localized to West Africa. Infection with HIV occurs via transmission in bodily fluids, in particular semen, vaginal secretions, and blood. HIV causes inflammation of the GI tract leading to destruction of enterocytes and a breakdown in the integrity of the epithelial surface.[32]

The main site of HIV replication in the body is CD4+ T cells expressing CCR5, although other cell types including monocytes/macrophages and dendritic cells can be infected and contribute to reservoirs of latent virus in the body.[33] HIV is an enveloped retrovirus that contains two copies of the viral genome and transcription factors required for reverse transcription of RNA into cDNA, a necessary prerequisite for integration into the host DNA. The amount of detectable viral load in the plasma is correlated with the proliferation of the virus in CD4+ T cells. HIV is integrated into the genome of infected cells as a provirus and the replication is controlled by a number of transcription factors which bind to the long terminal repeat (LTR) region of the virus upregulating or downregulating viral gene expression and replication.[34] HIV is capable of replicating in proliferating CD4+ T cells, including undifferentiated Th0 cells and Th2 cells,[35] utilizing a plethora of transcription factors involved in transcribing genes required for T cell proliferation to replicate.

In general a Th1 response is necessary to provide help for full activation of viral-specific CD8+ cytotoxic T cells to induce apoptosis in HIV-infected CD4+ T cells, controlling viral loads and infection (for review see

Hersperger et al.[36]). During acute HIV infection, most CD4+ T cells are depleted from the GI tract, a phenomenon that persists throughout chronic infection. In the periphery CD4+ T cell numbers initially recover but during the chronic phase of HIV infection CD4+ T cells are gradually depleted with continued viral replication over a number of years.

The impact of *Ascaris* infection in HIV-infected individuals is currently unclear. Immune activation is essential for CD8+ T cell-dependent viral control but it is also associated with viral replication and progression to AIDS. Given this dichotomy, possible effects of *Ascaris* co-infection on HIV infection are numerous, and likely to depend on the order and timing of co-infection.

The immune response generated by *Ascaris* infection biases CD4+ T cells towards a Th2/immunoregulatory (Treg) phenotype. Both Th2 cells (a source of IL-4) and T reg cells (a source of IL-10 and TGF-β) antagonize the development of Th1 cells.[37] It could be argued that established Th2 and immunoregulatory immune responses generated by an *Ascaris* infection[11] may be associated with HIV progression in *Ascaris*/HIV-co-infected individuals by antagonizing the development of Th1 cells that can provide help for CD8+ T cell-mediated control of viral loads (Figure 4.2). In support of this theory anthelmintic treatment of individuals with intestinal helminth/HIV co-infection has been associated with a decreased viral load in one study[38] and in another study a trend towards decreased viral load was associated with a statistically significant increase in CD4+ T cell counts in albendazole-treated individuals compared with untreated individuals.[8] Although no obvious changes in serum cytokine responses were apparent before and after treatment in the latter study[39] there was a marginal, but statistically significant, reduction in IL-10 production after treatment for intestinal helminth infection. This suggests that IL-10 generated by an *Ascaris* infection could be detrimental to HIV control.

However, it is possible that immunosuppressive immune responses generated in *Ascaris* infection may have a profound effect on viral loads and HIV progression without effects of immune control mechanisms. Th1 cells have been shown to be less permissible to viral replication than Th0 or Th2 cells.[35,40] It is therefore feasible to speculate that increased viral loads described in the above study[8] may have arisen because activation and expansion of Th2 cells by *Ascaris* co-infection in the gut-associated lymphoid tissue increased resources for viral replication.

Not all studies agree that helminth infection exacerbates viral loads in HIV co-infected individuals. Modjarrad et al.[41] did not find any effect of intestinal helminth treatment on viral loads in an HIV-infected population co-infected with *Ascaris* and hookworm. Helminth infection may even, in some cases, protect against the effects of HIV infection. In one study Schistosome/HIV co-infected individuals had higher CD4+ T cell counts compared with HIV singly-infected individuals, a finding that was

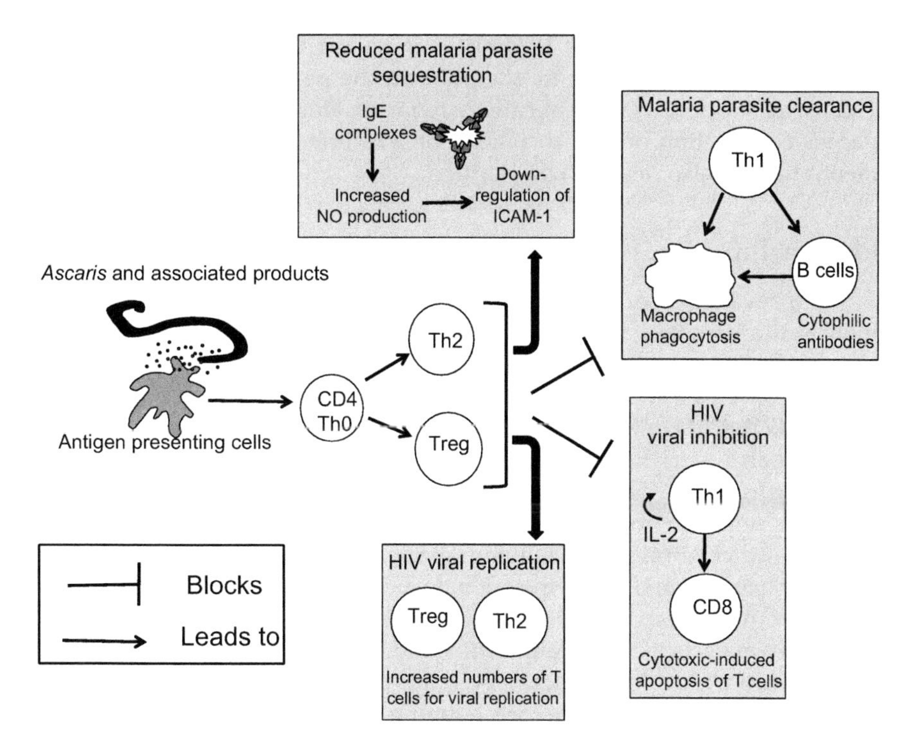

FIGURE 4.2 Possible effects of *Ascaris*-induced immune responses on HIV and malaria infection.

accompanied by a non-significant trend towards decreased viral loads.[42] The extent to which a similar effect may be observed in some populations of *Ascaris*/HIV-co-infected individuals remains to be seen. Activated CD4+ T cells are essential for HIV replication. Decreased viral loads in *Ascaris*/HIV-co-infected individuals could occur if generalized immunosuppression during *Ascaris* infection limits CD4+ T cell activation and proliferation and limits the main resource for HIV replication. IL-2 is a key cytokine required for T cell proliferation during T cell expansion in response to stimulation through the T cell receptor/CD3 complex.[43] It has been reported that *Ascaris* infection can suppress vaccine-induced immune response to oral cholera vaccine, and that this effect is associated with a deficiency in IL-2 production in response to the vaccine.[44] Since IL-2 is a key cytokine that facilitates CD4+ T cell activation it is possible that, with respect to IL-2 production and IL-2 receptor responsiveness, an *Ascaris*-induced suppression of CD4+ T cells may lead to lower viral loads in HIV-co-infected individuals.

It is currently difficult to generalize about the effects of *Ascaris* co-infection in HIV-infected individuals from current data. The relative importance of interactions between immune responses generated by

Ascaris and HIV infections leading to impaired CD8+ T cell-mediated control of viral loads, and an alteration in the permissiveness of CD4+ T cells to support viral replication, requires elucidation. The effect of *Ascaris* co-infection on the chronicity of HIV infection and the development of AIDS also needs to be clarified.

Malaria Infection

The prevalence of malaria has declined over the last decade due to increased awareness and efforts to control and ultimately eradicate this devastating disease. Despite these efforts, in 2010 there were an estimated 216 million cases and 655,000 deaths as a result of malaria infection.[45] Eighty-six percent of these deaths were in children less than 5 years of age, who are also at risk for infection and the resultant adverse effects of *Ascaris* infection.[45] The pathogenesis of malaria involves a number of different sequelae including anaemia, metabolic acidosis, respiratory distress and cerebral malaria, a neuropathy that is often fatal.[46]

Malaria infection occurs when female mosquitos deposit the sporozoite stages of malaria parasite in the dermis of the skin upon feeding. Although malaria is a systemic infection with malaria parasites replicating within red blood cells (RBC), the parasites undergo a developmental stage in the liver before entering the erythrocytes, replicating asexually every 48–72 hours depending on malaria parasite species. Malaria modifies the membrane of parasitized red blood cells (pRBCs),[47] a process that is thought to enhance the adhesive properties of the pRBCs for endothelial cells and prevent removal of pRBCs from the blood circulation by the spleen.[48] This process, known as sequestration, can occur in different organs of the body including the liver, lungs, and the brain resulting in localized inflammatory immune responses that are thought to contribute to the pathogenic symptoms of malaria.

Malaria parasites induce pro-inflammatory immune responses typified by interferon-γ (IFN-γ) and driven by CD4+ Th1 cells.[49,50] This pro-inflammatory immune response increases the phagocytic activity of macrophages and drives the production of malaria-specific cytophilic antibody isotypes, processes that enhance phagocytosis of pRBCs. When prolonged and/or overexuberant, this inflammatory immune response can contribute to the pathogenesis of malaria. However, asymptomatic malaria infections are common in endemic areas and are thought to occur because pro-inflammatory immune responses are balanced by the induction of the immunoregulatory cytokines (IL-10 and TGF-β).[49]

In Asia, *Ascaris* co-infection has been reported to have protective effects against cerebral malaria[6,7] as well as jaundice and renal failure in malaria-infected individuals.[51] Animal models have shown that immunoregulatory cytokines are essential for protecting against the pathogenesis of malaria infection, particularly when produced by CD4+ T cells.[52,53] Thus in an *Ascaris*-infected individual, protective effects may be observed if the *Ascaris* infection pre-empts or supplements the production of immunoregulatory cytokines generated by the malaria infection. However, the role of *Ascaris* co-infection in modulating malarial anaemia is less clear (Abanyie et al., unpublished observations).[54–56] In situations where the immunoregulatory immune responses generated by malaria infection are sufficiently robust, it may be that additional *Ascaris*-mediated immune regulation has no effect on modulating the pathogenesis of malaria infection.

An alternative mechanism has been proposed to explain the protective effects of *Ascaris* co-infection on cerebral malaria and involves the effects of IgE complexes induced by *Ascaris* infection. IgE complexes may stimulate the production of parasiticidal nitric oxide from endothelial cells by ligating the IgE receptor FcεRII reducing localized pRBC sequestration by downregulating the expression of intercellular adhesion molecule-1 (ICAM-1), an adhesin for RBCs infected with *Plasmodium falcaiprum*.[57] To support this hypothesis it has been observed that *Ascaris*/malaria-co-infected individuals have higher levels of reactive nitrogen intermediates (the end products of nitric oxide) and a lower number of circulating schizonts when compared with those infected with malaria alone.[7] In addition during malaria infection one study showed a positive correlation between circulating IgE levels and reactive nitrogen intermediates.[58] Although some evidence from experiments carried out on lung endothelial cells *in vitro* lends weight to the validity of this hypothesis,[59] no correlation between total IgE levels and protection from cerebral malaria has been found[60]and it has not yet been formally shown that IgE complexes induced in *Ascaris* infection lessen the sequestration of pRBCs in malaria infection. However, with the development of luciferase-expressing parasite lines in rodent models of malaria[61] it will now be possible to assess the effect of IgE complexes on the sequestration of pRBCs using whole body scanning of malaria-infected mice co-infected with model GI helminth infections or artificially injected with IgE complexes.

While preliminary evidence suggests that de-worming programs may in some situations adversely affect the pathogenesis of malaria infections as observed in the late 1970s by Murray et al.[62,63] reduced transmission of malaria within a malaria-endemic area may be a benefit of de-worming *Ascaris*-infected individuals. Co-infection has been associated with increased carriage of gametocytes, the transmissible form of malaria in the blood stream.[64] Mixed infections with *P. falciparum* and *P. vivax* were also elevated in co-infected individuals in this same study area compared to individuals with just malaria infection.[65] It is possible that *Ascaris*-generated immunoregulatory responses may impact on the development of effective parasite clearance mechanisms (Figure 4.2). In particular, the Th2 environment generated in *Ascaris* infection may favor the generation of non-cytophilic antibody isotypes from B cells responding to a malaria challenge (Figure 4.2), possibly reducing the clearance capacity of antibodies in the circulation and predisposing the host to malaria symptoms in some situations.[66]

Much remains to be learned about the interactions of malaria and *Ascaris* during co-infection. Defining the conditions under which *Ascaris* protects against malaria pathogenesis or enhance transmissibility of malaria infection is of great importance in the context of mass de-worming programs.

HYPOTHETICAL EFFECTS OF LOCALIZED CO-INFECTIONS ON *ASCARIS* CO-INFECTION

The majority of studies examining co-infection biology in patients with *Ascaris* infection and other pathogens almost exclusively report results with respect to the "other pathogen." Although almost nothing is known about the effects of co-infection on *Ascaris* biology and pathogenesis, it is possible to theorize about possible effects and associated mechanisms (Figure 4.3). For the final part of this book chapter we consider how co-infecting pathogens localized to specific organs of the body may have an impact on *Ascaris* infection. Specifically we will discuss possible effects

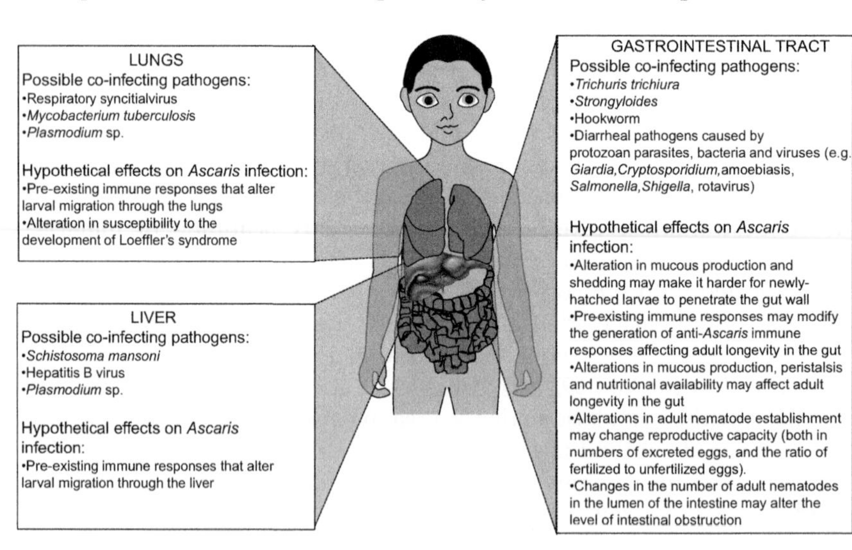

LUNGS
Possible co-infecting pathogens:
•Respiratory syncitialvirus
•Mycobacterium tuberculosis
•Plasmodium sp.
Hypothetical effects on Ascaris infection:
•Pre-existing immune responses that alter larval migration through the lungs
•Alteration in susceptibility to the development of Loeffler's syndrome
LIVER
Possible co-infecting pathogens:
•Schistosoma mansoni
•Hepatitis B virus
•Plasmodium sp.
Hypothetical effects on Ascaris infection:
•Pre-existing immune responses that alter larval migration through the liver
GASTROINTESTINAL TRACT
Possible co-infecting pathogens:
•Trichuris trichiura
•Strongyloides
•Hookworm
•Diarrheal pathogens caused by protozoan parasites, bacteria and viruses (e.g. Giardia, Cryptosporidium, amoebiasis, Salmonella, Shigella, rotavirus)
Hypothetical effects on Ascaris infection:
•Alteration in mucous production and shedding may make it harder for newly-hatched larvae to penetrate the gut wall
•Pre-existing immune responses may modify the generation of anti-Ascaris immune responses affecting adult longevity in the gut
•Alterations in mucous production, peristalsis and nutritional availability may affect adult longevity in the gut
•Alterations in adult nematode establishment may change reproductive capacity (both in numbers of excreted eggs, and the ratio of fertilized to unfertilized eggs).
•Changes in the number of adult nematodes in the lumen of the intestine may alter the level of intestinal obstruction

FIGURE 4.3 Hypothetical effects of co-infecting pathogens on *Ascaris* migration, establishment, fecundity, and pathogenesis.

of pre-existing immune response to localized co-infections in the liver (for example, *Schistosoma* eggs or hepatitis B virus (HBV)) or lungs (for example, respiratory syncitial virus (RSV) or *Mycobacterium tuberculosis*) on white spot formation in response to migrating larvae. Second, we will discuss possible effects of diarrheal diseases and co-infecting GI helminth infections on adult *Ascaris* nematodes in the gut.

Larval Migration

After migrating through the lungs, *Ascaris* larve migrate up the trachea and are re-swallowed allowing re-entry into the GI tract. Migration of *Ascaris* larvae may be a beneficial life history trait offering a window without the harsh environment of the GI tract to maximize growth leading to adult nematodes with larger body sizes and greater fecundity than non-migrating counterparts.[67] However, *Ascaris* larval migration is not necessarily benign with respect to host health. Mice infected with *Ascaris* lose body weight[68] suggesting that this phase of the life-cycle can be detrimental to the host. White spot formation in the liver of pigs experimentally infected with *A. suum* is associated with resistance to the establishment of *Ascaris* infection.[24] Prevention of *Ascaris* migration via white spot formations in turn prevents establishment of infection with adult nematodes in the gut. Since white spots are generated by an immune response triggered by larval migration, it is logical to hypothesize that pre-existing immune responses generated by co-infecting pathogens, particularly in the liver, may alter the formation of white spots, in turn impacting on the successful migration of *Ascaris* larvae and the establishment of *Ascaris* infection in a co-infected host.

White spot formation in *A. suum* infections of pigs begins with infiltration of eosinophils, as well as neutrophils,[25] suggesting that defense against migrating *Ascaris* larvae contains a strong Th2 element. Schistosomes are trematode parasites that reside in the portal vein as paired adult parasites. Some of these eggs, shed upon mating, do not successfully exit the body for excretion into the environment, but are swept by the bloodstream into the liver where they elicit a strong Th2 response[69] by virtue of the omega oils present on the egg surface.[70,71] The infiltration of immune cells to contain the egg leads to the formation of granulomas (Figure 4.4A) and heavy egg deposition is often correlated with hepatomegaly. From the perspective of migrating *Ascaris* larvae, the challenge to successfully migrate back to the GI tract could be made significantly more challenging in a host with Schistosome granulomas present in the liver. It is possible that mature white spots form more rapidly in the context of a pre-existing Th2 environment.

In the lungs, infection with RSV is common in young children causing wheezing, bronchiolitis, hypoxia, and airway dysfunction in a subset of

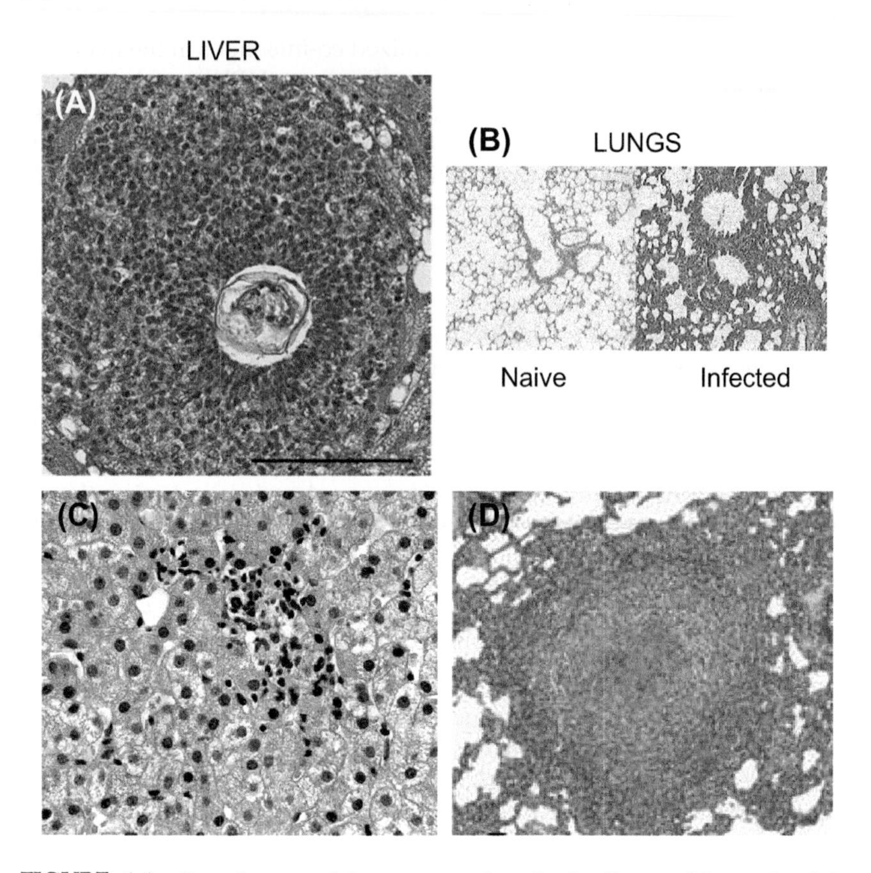

FIGURE 4.4 **Granulomas and immune reactions in the liver and lungs faced by migrating larvae in a co-infected host.** (A) shows a liver section from a C57BL/6 mouse infected with *Schistosoma japonicum* containing a liver granuloma. The section was stained with Leder stain and the scale bar represents 100 μm (adapted from Burke et al. (2010) PLOS neglected diseases e598). (B) shows lung sections from naïve (left panel) and day 15 RSV infected (right panel) mice and stained with Periodic Acid Schiff stain to highlight mucous production (adapted from[74]). (C) shows a section from a liver biopsy stained with hematoxylin-eosin (40× objective) from a patient with chronic hepatitis B with immune cell infiltration (adapted from Tokin et al. (2011) Quantitative morphometric analysis of liver biopsy: problems and perspectives, Liver Biopsy, Hirokazu Takahashi (Ed.). (D) shows a lung granuloma from a minipig 20 weeks post-infection with *Mycobacterium tuberculosis.* The lung section was stained with hemotoxylin and eosin and image taken with 40× objective (Gil et al. (2012) PLOS One 5(4):e10030).

infected individuals. In asymptomatic RSV infection a Th1 response prevails[72,73] with vigorous anti-viral CD8+ T cell responses that inhibit viral replication. Disease is associated with the development of Th2 responses, infiltration of eosinophils into the lungs and mucous production (Figure 4.4B; for review see Graham et al.[74]). The impact of RSV infection on the migration of *Ascaris* larvae is likely to depend on whether the co-infected host is resistant or susceptible to the pathogenesis of RSV infection. A diseased individual may prevent larval migration through the lung because prevailing anti-viral Th2 responses may potentiate quicker formation of white spots and overproduction of mucous (Figure 4.4B) may trap migrating larvae. On the other hand, the lungs of an individual who has been challenged with RSV infection but mounts a strong Th1 response to the virus may facilitate passage of migrating *Ascaris* larvae through the lungs by antagonizing the Th2 responses triggered by the migrating larvae, and dampening the infiltration of eosinophils and neutrophils leading to white spot formation.

Pre-existing Th1 responses affecting *Ascaris* larval migration in a similar way may also be encountered in individuals infected with hepatitis B virus (HBV), a common infection in African populations. This virus can be passed from mother to child and chronic infection can lead to cirrhosis of the liver with infiltration of immune cells (Figure 4.4C) in response to Th1 and Th17 immune responses.[75] In the lungs, *Mycobacterium tuberculosis* infection leads to the formation of granulomas (Figure 4.4D) that contain, but do not completely eliminate, the *M. tuberculosis* bacilli.[76] Lung granulomas that form in response to *M. tuberculosis* do so in the context of ongoing Th1 responses maintained by chronic low-level antigen stimulation and the cellular composition is quite different to that of granulomas surrounding Schistosome eggs.[77] Pre-existing Th1 responses present in the lungs of a person with tuberculosis may reduce Th2-mediated white spot formation in response to migrating *Ascaris* larvae potentiating *Ascaris* infection.

Gastrointestinal Tract

Co-infection with Other Geohelminths

The GI tract is likely to be inhabited by nematode infections in addition to *Ascaris* and these include hookworm (*Necator americanus* or *Ancylostoma duodenale*), *Trichuris trichiura* or *Strongyloides stercoralis*. Several studies have shown that co-infection with two or more geohelminth infections is a common finding in populations living in endemic areas[78] demonstrating that GI nematode infections such as *Ascaris* do not require exclusive use of the GI tract to survive. The immune responses generated by other GI-dwelling nematodes are of a similar skew as that generally described for *Ascaris* infections in that Th2 and immunoregulatory immune responses are dominant.[79] The extent to which antigen-specific immune responses generated by one GI nematode infection are able to influence responses to a second GI nematode infection is currently unknown. However, it can be hypothesized that in multi-geohelminth co-infections mutually beneficial immune responses are generated via the bystander effects of immunoregulatory cytokines, particularly to cross-reactive antigens.

Co-infection with Diarrhea-causing Pathogens

Diarrheal diseases are one of the most common categories of life-threatening diseases for children in developing countries. In some settings the attack rate can be as high as 12 episodes per year[80] and diarrhea can be caused by several different pathogenic organisms (Figure 4.3). Despite the high prevalence of *Ascaris* infection in areas where diarrhea is common, there are surprisingly few reports on the effects of co-infection with *Ascaris* on diarrheal diseases. Diarrheal pathogens could potentially alter the impact of *Ascaris* infection through modifications of the luminal environment of the GI tract, and through the generation of strong pro-inflammatory immune responses.

In the GI tract, there are a number of factors that could influence *Ascaris* infection in a host co-infected with a diarrheal-causing pathogen. In general mucous production is increased in diarrheal disease[81] and this may hinder the penetration of the gut wall by newly-hatched L3 larvae. *Ascaris* adult nematodes are large organisms requiring significant levels of nutrients to maintain their body size. Enteric pathogens causing diarrhea may reduce the availability of nutritional sources such as microflora for consumption by adult nematodes, a consequence of mechanisms activated to flush out the invading diarrheal pathogen.

Ascaris nematodes do not attach to the gut, and instead swim against the peristaltic flow to remain in position. Anything that disrupts smooth muscle contractability controlling gut motility (peristalsis) may affect the ability of *Ascaris* nematodes to maintain their position in the gut. T cells exert some control over peristalsis in *Trichinella spiralis*-infected mice and smooth muscle contractability is positively correlated with *Trichinella* expulsion.[82] If a similar mechanism applies to *Ascaris* infection, any immunosuppressive immune response that reduces T cell responses in *Ascaris* (for example, in HIV infection) may help *Ascaris* remain in the GI tract.

Immune Responses and Nematode Development

As described for migrating larvae, co-infection with Th1-inducing diarrheal pathogens may benefit an established *Ascaris* co-infection by helping to reduce detrimental Th2-associated effector mechanisms. However, it is possible that an established immune response generated by a co-infecting pathogen may benefit *Ascaris* beyond dampening the immune clearance mechanisms. Scattered evidence from mouse models of other medically important helminth infections suggests that immune responsiveness may provide cues for helminth development.[83–86]

In Schistosome infections, a lack of immune responsiveness, particularly from monocytes/macrophages and CD4+ T cells,[83,84] was associated with poor development of adult Schistosome parasites. There is also evidence in mouse experiments with the filarial nematode *Brugia malayi* that T cells[86] and NK cells[85] facilitate larval development. Although there is currently no evidence that the cues governing the development of nematodes such as *Ascaris* behave in a similar fashion, it is possible that pre-existing immune responses in the GI tract may enhance establishment of an incoming *Ascaris* via the provision of important growth signals for newly hatched larvae or larvae undergoing a final molt to the adult form upon re-entry into the GI tract.

CONCLUDING REMARKS

In 2001, the World Health Assembly resolved to decrease the rates of morbidity and mortality associated with geohelminths worldwide by recommending that member states endemic for these infections institute regular administration of chemotherapeutics to school-aged children. The goal of this resolution was to achieve consistent treatment of at least 75% of all school children at risk by 2010[87] (see Chapter 15). However, the impact of this strategy on the pathogenesis of co-infecting pathogens was not a main consideration.

Over time there have been many articles published discussing the possible implications of eradication of geohelminths such as *Ascaris* on the pathogenesis of co-infecting organisms,[88–90] but a comprehensive

BOX 4.1

SOME AREAS REQUIRING ELUCIDATION

Infection intensity: Are heavy *Ascaris* burdens associated with the presence of co-infecting organisms?

HIV co-infection: Does *Ascaris* co-infection alter viral loads and/or the progression of AIDS?

Malaria co-infection: Are protective effects of *Ascaris* on the development of cerebral malaria observed in geographical locations outside Asia?

Tissue migration: Does tissue containing granulomas from co-infection with other pathogens (e.g. *Schistosoma* or tuberculosis) impair or enhance *Ascaris* larval migration?

Adult longevity: Do diarrheal-causing pathogens adversely influence the longevity of adult *Ascaris* nematodes?

Egg output: Does gastrointestinal immune suppression by HIV enhance survival or fecundity of *Ascaris* adult nematodes?

body of primary data defining the interactions that occur in endemic populations is currently lacking. There are very few good quality intervention studies rather than correlational studies to help establish causality and many questions remain to be answered about the implications of *Ascaris* co-infection on life threatening diseases, a selection of which are outlined in Box 4.1. More funding is required to tackle some of these issues and shed light on the "rules" of co-infection with *Ascaris* rather than studying just the "exceptions" of singly-infected hosts.

Acknowledgments

We thank Dr. Jason Hammonds for helpful discussion, Dr. Patrice Mimche for discussion and critical reading of the manuscript, and Children's Healthcare of Atlanta for funding.

References

1. World Health Organization. *Soil-transmitted helminthiases: estimates of the number of children needing preventative chemotherapy and number treated, 2009*. Geneva: World Health Organization; 2011.
2. Cappello M, Hotez PJ. Intestinal nematodes: *Ascaris lumbricoides*. In: Long SS, Pickering LK, Prober CG, editors. *Principles and Practice of Infectious Diseases*. 3rd ed. Elsevier; 2008. p. 1296–8.
3. Keusch GT, Migasena P. Biological implications of polyparasitism. *Rev Infect Dis* 1982; **4**(4):880–2.
4. Hagel I, Cabrera M, Puccio F, Santaella C, Buvat E, Infante B, et al. Co-infection with Ascaris lumbricoides modulates protective immune responses against *Giardia duodenalis* in school Venezuelan rural children. *Acta Tropica* 2011;**117**(3):189–95.
5. Bahia-Oliveira LM, Silva JA, Peixoto-Rangel AL, Boechat MS, Oliveira AM, Massara CL, et al. Host immune response to *Toxoplasma gondii* and *Ascaris lumbricoides* in a highly endemic area: evidence of parasite co-immunomodulation properties influencing the outcome of both infections. *Mem Inst Oswaldo Cruz* 2009;**104**(2):273–80.
6. Nacher M, Gay F, Singhasivanon P, Krudsood S, Treeprasertsuk S, Mazier D, et al. *Ascaris lumbricoides* infection is associated with protection from cerebral malaria. *Parasite Immunol* 2000;**22**(3):107–13.
7. Nacher M, Singhasivanon P, Traore B, Vannaphan S, Gay F, Chindanond D, et al. Helminth infections are associated with protection from cerebral malaria and increased nitrogen derivatives concentrations in Thailand. *Am J Trop Med Hyg* 2002; **66**(3):304–9.
8. Walson JL, Otieno PA, Mbuchi M, Richardson BA, Lohman-Payne B, Macharia SW, et al. Albendazole treatment of HIV-1 and helminth co-infection: a randomized, double-blind, placebo-controlled trial. *Aids* 2008;**22**(13):1601–9.
9. Graham AL. Ecological rules governing helminth-microparasite coinfection. *Proc Nat Acad Sci USA* 2008;**105**(2):566–70.
10. Bundy DA, Golden MH. The impact of host nutrition on gastrointestinal helminth populations. *Parasitology* 1987;**95**(Pt 3):623–35.
11. Cooper PJ, Chico ME, Sandoval C, Espinel I, Guevara A, Kennedy MW, et al. Human infection with *Ascaris lumbricoides* is associated with a polarized cytokine response. *J Infect Dis* 2000;**182**(4):1207–13.

12. Jackson JA, Turner JD, Rentoul L, Faulkner H, Behnke JM, Hoyle M, et al. T helper cell type 2 responsiveness predicts future susceptibility to gastrointestinal nematodes in humans. *J Infect Dis* 2004;**190**(10):1804–11.
13. Turner JD, Faulkner H, Kamgno J, Cormont F, Van Snick J, Else KJ, et al. Th2 cytokines are associated with reduced worm burdens in a human intestinal helminth infection. *J Infect Dis* 2003;**188**(11):1768–75.
14. King EM, Kim HT, Dang NT, Michael E, Drake L, Needham C, et al. Immuno-epidemiology of *Ascaris lumbricoides* infection in a high transmission community: antibody responses and their impact on current and future infection intensity. *Parasite Immunol* 2005;**27**(3):89–96.
15. McSharry C, Xia Y, Holland CV, Kennedy MW. Natural immunity to *Ascaris lumbricoides* associated with immunoglobulin E antibody to ABA-1 allergen and inflammation indicators in children. *Infect Immun* 1999;**67**(2):484–9.
16. Haswell-Elkins MR, Kennedy MW, Maizels RM, Elkins DB, Anderson RM. The antibody recognition profiles of humans naturally infected with *Ascaris lumbricoides*. *Parasite Immunol* 1989;**11**(6):615–27.
17. Jackson JA, Turner JD, Kamal M, Wright V, Bickle Q, Else KJ, et al. Gastrointestinal nematode infection is associated with variation in innate immune responsiveness. *Microbes Infect* 2006;**8**(2):487–92.
18. Turner JD, Jackson JA, Faulkner H, Behnke J, Else KJ, Kamgno J, et al. Intensity of intestinal infection with multiple worm species is related to regulatory cytokine output and immune hyporesponsiveness. *J Infect Dis* 2008;**197**(8):1204–12.
19. Figueiredo CA, Barreto ML, Rodrigues LC, Cooper PJ, Silva NB, Amorim LD, et al. Chronic intestinal helminth infections are associated with immune hyporesponsiveness and induction of a regulatory network. *Infect Immun* 2010;**78**(7):3160–7.
20. Goodridge HS, McGuiness S, Houston KM, Egan CA, Al-Riyami L, Alcocer MJ, et al. Phosphorylcholine mimics the effects of ES-62 on macrophages and dendritic cells. *Parasite Immunol* 2007;**29**(3):127–37.
21. Silva SR, Jacysyn JF, Macedo MS, Faquim-Mauro EL. Immunosuppressive components of *Ascaris suum* down-regulate expression of costimulatory molecules and function of antigen-presenting cells via an IL-10-mediated mechanism. *Eur J Immunol* 2006; **36**(12):3227–37.
22. Oshiro TM, Enobe CS, Araujo CA, Macedo MS, Macedo-Soares MF. PAS-1, a protein affinity purified from *Ascaris suum* worms, maintains the ability to modulate the immune response to a bystander antigen. *Immunol Cell Biol* 2006; **84**(2):138–44.
23. McConchie BW, Norris HH, Bundoc VG, Trivedi S, Boesen A, Urban Jr JF, et al. Ascaris suum-derived products suppress mucosal allergic inflammation in an interleukin-10-independent manner via interference with dendritic cell function. *Infect Immun* 2006; **74**(12):6632–41.
24. Frontera E, Roepstorff A, Gazquez A, Reina D, Serrano FJ, Navarrete I. Immunohistochemical distribution of antigens in liver of infected and immunized pigs with *Ascaris suum*. *Vet Parasitol* 2003;**111**(1):9–18.
25. Nakagawa M, Yoshihara S, Suda H, Ikeda K. Pathological studies on white spots of the liver in fattening pigs. *Natl Inst Anim Health Q (Tokyo)* 1983;**23**(4):138–49.
26. Eriksen L, Lind P, Nansen P, Roepstorff A, Urban J. Resistance to *Ascaris suum* in parasite naive and naturally exposed growers, finishers and sows. *Vet Parasitol* 1992; **41**(1–2):137–49.
27. Loffler W. Transient lung infiltrations with blood eosinophilia. *Int Arch Allergy Appl Immunol* 1956;**8**(1–2):54–9.
28. Kouro T, Takatsu K. IL-5- and eosinophil-mediated inflammation: from discovery to therapy. *Int Immunol* 2009;**21**(12):1303–9.

29. Monticelli LA, Sonnenberg GF, Artis D. Innate lymphoid cells: critical regulators of allergic inflammation and tissue repair in the lung. *Curr Opin Immunol* 2012;**24**(3):284–9.
30. Fincham JE, Markus MB, Adams VJ. Could control of soil-transmitted helminthic infection influence the HIV/AIDS pandemic. *Acta Tropica* 2003;**86**(2–3):315–33.
31. UNAIDS. AIDS *epidemic update*. Geneva: World Health Organization; 2007.
32. Moir S, Chun TW, Fauci AS. Pathogenic mechanisms of HIV disease. *Annu Rev Pathol* 2011;**6**:223–48.
33. Pierson T, McArthur J, Siliciano RF. Reservoirs for HIV-1: mechanisms for viral persistence in the presence of antiviral immune responses and antiretroviral therapy. *Annu Rev Immunol* 2000;**18**:665–708.
34. Victoriano AF, Okamoto T. Transcriptional control of HIV replication by multiple modulators and their implication for a novel antiviral therapy. *AIDS Res Hum Retroviruses* 2012;**28**(2):125–38.
35. Maggi E, Mazzetti M, Ravina A, Annunziato F, de Carli M, Piccinni MP, et al. Ability of HIV to promote a TH1 to TH0 shift and to replicate preferentially in TH2 and TH0 cells. *Science* 1994;**265**(5169):244–8.
36. Hersperger AR, Migueles SA, Betts MR, Connors M. Qualitative features of the HIV-specific CD8+ T-cell response associated with immunologic control. *Curr Opin HIV AIDS* 2011;**6**(3):169–73.
37. Zhu J, Paul WE. Peripheral CD4+ T-cell differentiation regulated by networks of cytokines and transcription factors. *Immunol Rev* 2010;**238**(1):247–62.
38. Wolday D, Mayaan S, Mariam ZG, Berhe N, Seboxa T, Britton S, et al. Treatment of intestinal worms is associated with decreased HIV plasma viral load. *J Acquir Immune Defic Syndr* 2002;**31**(1):56–62.
39. Blish CA, Sangare L, Herrin BR, Richardson BA, John-Stewart G, Walson JL. Changes in plasma cytokines after treatment of *Ascaris lumbricoides* infection in individuals with HIV-1 infection. *J Infect Dis* 2010;**201**(12):1816–21.
40. Romagnani S, Maggi E, Del Prete G. HIV can induce a TH1 to TH0 shift, and preferentially replicates in CD4+ T-cell clones producing TH2-type cytokines. *Res Immunol* 1994;**145**(8–9):611–7. discussion 7–8.
41. Modjarrad K, Zulu I, Redden DT, Njobvu L, Lane HC, Bentwich Z, et al. Treatment of intestinal helminths does not reduce plasma concentrations of HIV-1 RNA in coinfected Zambian adults. *J Infect Dis* 2005;**192**(7):1277–83.
42. Elliott AM, Mawa PA, Joseph S, Namujju PB, Kizza M, Nakiyingi JS, et al. Associations between helminth infection and CD4+ T cell count, viral load and cytokine responses in HIV-1-infected Ugandan adults. *Trans R Soc Trop Med Hyg* 2003;**97**(1):103–8.
43. Malek TR. The biology of interleukin-2. *Annu Rev Immunol* 2008;**26**:453–79.
44. Cooper PJ, Chico M, Sandoval C, Espinel I, Guevara A, Levine MM, et al. Human infection with *Ascaris lumbricoides* is associated with suppression of the interleukin-2 response to recombinant cholera toxin B subunit following vaccination with the live oral cholera vaccine CVD 103-HgR. *Infect Immun* 2001;**69**(3):1574–80.
45. World Health Organization. *World Malaria Report, 2011*. Geneva: World Health Organization; 2011.
46. Mackintosh CL, Beeson JG, Marsh K. Clinical features and pathogenesis of severe malaria. *Trends Parasitol* 2004;**20**(12):597–603.
47. Cooke BM, Mohandas N, Coppel RL. The malaria-infected red blood cell: structural and functional changes. *Adv Parasitol* 2001;**50**:1–86.
48. Fonager J, Pasini EM, Braks JA, Klop O, Ramesar J, Remarque EJ, et al. Reduced CD36-dependent tissue sequestration of *Plasmodium*-infected erythrocytes is detrimental to malaria parasite growth in vivo. *J Exp Med* 2012;**209**(1):93–107.
49. Walther M, Jeffries D, Finney OC, Njie M, Ebonyi A, Deininger S, et al. Distinct roles for FOXP3 and FOXP3 CD4 T cells in regulating cellular immunity to

uncomplicated and severe *Plasmodium falciparum* malaria. *PLoS Pathog* 2009;**5**(4): e1000364.

50. Winkler S, Willheim M, Baier K, Schmid D, Aichelburg A, Graninger W, et al. Reciprocal regulation of Th1- and Th2-cytokine-producing T cells during clearance of parasitemia in *Plasmodium falciparum* malaria. *Infect Immun* 1998;**66**(12):6040–4.
51. Nacher M, Singhasivanon P, Silachamroon U, Treeprasertsuk S, Vannaphan S, Traore B, et al. Helminth infections are associated with protection from malaria-related acute renal failure and jaundice in Thailand. *Am J Trop Med Hyg* 2001;**65**(6): 834–6.
52. Freitas do Rosario AP, Lamb T, Spence P, Stephens R, Lang A, Roers A, et al. IL-27 promotes IL-10 production by effector Th1 CD4+ T cells: a critical mechanism for protection from severe immunopathology during malaria infection. *J Immunol* 2012; **188**(3):1178–90.
53. Couper KN, Blount DG, Wilson MS, Hafalla JC, Belkaid Y, Kamanaka M, et al. IL-10 from CD4CD25Foxp3CD127 adaptive regulatory T cells modulates parasite clearance and pathology during malaria infection. *PLoS Pathog* 2008;**4**(2):e1000004.
54. Green HK, Sousa-Figueiredo JC, Basanez MG, Betson M, Kabatereine NB, Fenwick A, et al. Anaemia in Ugandan preschool-aged children: the relative contribution of intestinal parasites and malaria. *Parasitology* 2011;**138**(12):1534–45.
55. Nacher M, Singhasivanon P, Gay F, Phumratanaprapin W, Silachamroon U, Looareesuwan S. Association of helminth infection with decreased reticulocyte counts and hemoglobin concentration in Thai falciparum malaria. *Am J Trop Med Hyg* 2001; **65**(4):335–7.
56. Abanyie FA, McCracken C, Kirwan P, Molloy SF, Asaolu SO, Holland CV, et al. Ascaris co-infection does not alter malaria-induced anaemia in a cohort of Nigerian preschool children. *Malar J* 2013;**12**:1.
57. Nacher M. Interactions between worm infections and malaria. *Clin Rev Allergy Immunol* 2004;**26**(2):85–92.
58. Nacher M, Singhasivanon P, Kaewkungwal J, Silachamroon U, Treeprasertsuk S, Tosukhowong T, et al. Relationship between reactive nitrogen intermediates and total immunoglobulin E, soluble CD21 and soluble CD23: comparison between cerebral malaria and nonsevere malaria. *Parasite Immunol* 2002;**24**(8):395–9.
59. Pino P, Vouldoukis I, Dugas N, Conti M, Nitcheu J, Traore B, et al. Induction of the CD23/nitric oxide pathway in endothelial cells downregulates ICAM-1 expression and decreases cytoadherence of *Plasmodium falciparum*-infected erythrocytes. *Cell Microbiol* 2004;**6**(9):839–48.
60. Duarte J, Deshpande P, Guiyedi V, Mecheri S, Fesel C, Cazenave PA, et al. Total and functional parasite specific IgE responses in *Plasmodium falciparum*-infected patients exhibiting different clinical status. *Malar J* 2007;**6**:1.
61. Franke-Fayard B, Janse CJ, Cunha-Rodrigues M, Ramesar J, Buscher P, Que I, et al. Murine malaria parasite sequestration: CD36 is the major receptor, but cerebral pathology is unlinked to sequestration. *Proc Natl Acad Sci USA* 2005;**102**(32): 11468–73.
62. Murray J, Murray A, Murray M, Murray C. The biological suppression of malaria: an ecological and nutritional interrelationship of a host and two parasites. *Am J Clin Nutr* 1978;**31**(8):1363–6.
63. Murray MJ, Murray AB, Murray MB, Murray CJ. Parotid enlargement, forehead edema, and suppression of malaria as nutritional consequences of ascariasis. *Am J Clin Nutr* 1977;**30**(12):2117–21.
64. Nacher M, Singhasivanon P, Silachamroon U, Treeprasertsu S, Krudsood S, Gay F, et al. Association of helminth infections with increased gametocyte carriage during mild falciparum malaria in Thailand. *Am J Trop Med Hyg* 2001;**65**(5):644–7.

65. Nacher M, Singhasivanon P, Gay F, Silachomroon U, Phumratanaprapin W, Looareesuwan S. Contemporaneous and successive mixed *Plasmodium falciparum* and *Plasmodium vivax* infections are associated with *Ascaris lumbricoides*: an immunomodulating effect? *J Parasitol* 2001;**87**(4):912–5.
66. Ndungu FM, Bull PC, Ross A, Lowe BS, Kabiru E, Marsh K. Naturally acquired immunoglobulin (Ig)G subclass antibodies to crude asexual *Plasmodium falciparum* lysates: evidence for association with protection for IgG1 and disease for IgG2. *Parasite Immunol* 2002;**24**(2):77–82.
67. Read AF, Skorping A. The evolution of tissue migration by parasitic nematode larvae. *Parasitology* 1995;**111**(Pt 3):359–71.
68. Lewis R, Behnke JM, Stafford P, Holland CV. Dose-dependent impact of larval *Ascaris suum* on host body weight in the mouse model. *J Helminthol* 2009;**83**(1):1–5.
69. Cheever AW, Jankovic D, Yap GS, Kullberg MC, Sher A, Wynn TA. Role of cytokines in the formation and downregulation of hepatic circumoval granulomas and hepatic fibrosis in *Schistosoma mansoni*-infected mice. *Mem Inst Oswaldo Cruz* 1998; **93**(Suppl. 1):25–32.
70. Everts B, Perona-Wright G, Smits HH, Hokke CH, van der Ham AJ, Fitzsimmons CM, et al. Omega-1, a glycoprotein secreted by *Schistosoma mansoni* eggs, drives Th2 responses. *J Experimental Med* 2009;**206**(8):1673–80.
71. Steinfelder S, Andersen JF, Cannons JL, Feng CG, Joshi M, Dwyer D, et al. The major component in schistosome eggs responsible for conditioning dendritic cells for Th2 polarization is a T2 ribonuclease (omega-1). *J Experimental Med* 2009;**206**(8): 1681–90.
72. Tang YW, Graham BS. T cell source of type 1 cytokines determines illness patterns in respiratory syncytial virus-infected mice. *J Clin Invest* 1997;**99**(9):2183–91.
73. Spender LC, Hussell T, Openshaw PJ. Abundant IFN-gamma production by local T cells in respiratory syncytial virus-induced eosinophilic lung disease. *J Gen Virol* 1998;**79** (Pt 7):1751–8.
74. Graham BS, Johnson TR, Peebles RS. Immune-mediated disease pathogenesis in respiratory syncytial virus infection. *Immunopharmacology* 2000;**48**(3):237–47.
75. Ye Y, Xie X, Yu J, Zhou L, Xie H, Jiang G, et al. Involvement of Th17 and Th1 effector responses in patients with Hepatitis B. *J Clin Immunol* 2010;**30**(4):546–55.
76. Saunders BM, Britton WJ. Life and death in the granuloma: immunopathology of tuberculosis. *Immunol Cell Biol* 2007;**85**(2):103–11.
77. Sandor M, Weinstock JV, Wynn TA. Granulomas in schistosome and mycobacterial infections: a model of local immune responses. *Trends Immunol* 2003;**24**(1):44–52.
78. Raso G, Luginbuhl A, Adjoua CA, Tian-Bi NT, Silue KD, Matthys B, et al. Multiple parasite infections and their relationship to self-reported morbidity in a community of rural Cote d'Ivoire. *Int J Epidemiol* 2004;**33**(5):1092–102.
79. Allen JE, Maizels RM. Diversity and dialogue in immunity to helminths. *Nat Rev Immunol* 2011;**11**(6):375–88.
80. Kosek M, Bern C, Guerrant RL. The global burden of diarrhoeal disease, as estimated from studies published between 1992 and 2000. *Bulletin of the World Health Organization* 2003;**81**(3):197–204.
81. Wood JD. Effects of bacteria on the enteric nervous system: implications for the irritable bowel syndrome. *J Clin Gastroenterol* 2007;**41**(Suppl. 1):S7–19.
82. Vallance BA, Croitoru K, Collins SM. T lymphocyte-dependent and -independent intestinal smooth muscle dysfunction in the T. spiralis-infected mouse. *Am J Physiol* 1998;**275**(5 Pt 1):G1157–65.
83. Davies SJ, Grogan JL, Blank RB, Lim KC, Locksley RM, McKerrow JH. Modulation of blood fluke development in the liver by hepatic CD4+ lymphocytes. *Science* 2001; **294**(5545):1358–61.

84. Lamb EW, Walls CD, Pesce JT, Riner DK, Maynard SK, Crow ET, et al. Blood fluke exploitation of non-cognate CD4+ T cell help to facilitate parasite development. *PLoS Pathog* 2010;**6**(4):e1000892.
85. Babu S, Porte P, Klei TR, Shultz LD, Rajan TV. Host NK cells are required for the growth of the human filarial parasite *Brugia malayi* in mice. *J Immunol* 1998;**161**(3):1428–32.
86. Babu S, Shultz LD, Rajan TV. T cells facilitate *Brugia malayi* development in TCRalpha(null) mice. *Experimental Parasitol* 1999;**93**(1):55–7.
87. WHO. *Schistosomiasis and soil-transmitted helminth infections.*. Geneva: World Health Organization; 2001.
88. Eziefula AC, Brown M. Intestinal nematodes: disease burden, deworming and the potential importance of co-infection. *Curr Opin Infect Dis* 2008;**21**(5):516–22.
89. Borkow G, Bentwich Z. HIV and helminth co-infection: is deworming necessary? *Parasite Immunol* 2006;**28**(11):605–12.
90. Nacher M. Interactions between worms and malaria: good worms or bad worms? *Malaria J* 2011;**10**:259.

PART II

MODEL SYSTEMS

CHAPTER

5

Larval Ascariasis: Impact, Significance, and Model Organisms

Celia V. Holland, Jerzy M. Behnke*†*, Christina Dold*‡

*Trinity College, Dublin, Ireland
†University of Nottingham, Nottingham, UK
‡University of Oxford, Oxford, UK

OUTLINE

Ascaris: The Neglected Parasite
http://dx.doi.org/10.1016/B978-0-12-396978-1.00005-7

Copyright © 2013 Elsevier Inc. All rights reserved.

INTRODUCTION

The life-cycles of *Ascaris lumbricoides*, the human roundworm and *Ascaris suum*, its counterpart in the pig, consist of three very important elements. The adult male, female and immature worms that live in the small intestine, the resistant egg stages that are produced by adult worms and pass out into the environment via the host feces and the larval parasites that emerge from the hatched eggs into the host intestine and undergo a migratory pathway within the tissues of the human or porcine host. Each of these phases of infection contributes to the unique biology of the parasite, to its public health impact and to a myriad of questions that arise in the mind of the researcher interested in *Ascaris*.

However, it is clear that our understanding of the properties of the adult worms and the resistant free-living egg stages of the parasite far outweigh that of larval infection. The reasons for this are mainly due to the fact that adult worms can be recovered from humans and pigs after the administration of anthelmintic drugs that result in the expulsion of worms from the intestine and eggs can be dissected from the uteri of adult worms, detected in the feces using diagnostic procedures and recovered from soil and pasture in the environment. Tissue-resident and migrating larvae, however, undergo a more cryptic manifestation within the parenteral tissues of humans and pigs and can only be recovered at postmortem. This has resulted in a paucity of information on the public health significance of larval ascariasis in humans due to ethical and logistic constraints. Data on larval ascariasis have mainly been accrued from animal model studies including the natural host of *A. suum*, the pig, and a range of abnormal hosts including cattle, rabbits, and small rodents such as guinea-pigs and mice (Table 5.1). This chapter will focus upon our existing knowledge of larval ascariasis derived from model organisms and the potential advantages of the use of a mouse model of resistance and susceptibility to ascariasis.

THE HEPATO-TRACHEAL MIGRATION IN ASCARIASIS

Ascaris is a parasite that exists not only as an adult worm in the intestine but also has a migratory pathway, known as a hepato-trachael migration,[1] that progresses from an infective egg (containing an L3 larva covered by an L2 cuticle)[2,3] hatching in the intestine, to a larva migrating via the portal blood vessels to the liver. After migration within the liver, and some growth, the larvae advance to the lungs, penetrate the alveolar spaces and move to the pharynx where they are coughed up and

TABLE 5.1 Parameter(s) investigated in experimentally infected model organisms excluding mice (see Table 5.2) and pigs (see Chapter 14).

Model organism	Parameter(s) investigated	Reference number
Guinea-pig	Larval migration Immunological response	52,75,79,80,82–83,91–95
Rabbit	Immunological response Larval migration Hepatic pathology Pulmonary pathology	52,75,81,96–98
Gerbil	Larval migration	99
Rat	Larval migration	93,100
Cow	Larval migration Immunological response Pathology	101–104
Lamb	Larval migration	92
Goat	Larval migration	92

swallowed, thereby returning to their former location within the small intestine where they mature into adults.[4]

The explanation for the perpetuation of this perilous journey is far from understood. Smyth[5] speculated that such an extra-intestinal migration might represent "evolutionary baggage" left over from a previous life-cycle, perhaps originally based on skin penetration. However, other authors have argued that migration may confer significant fitness benefits given the associated risks of immune-mediated damage and mortality within the tissues[6,7] and therefore by adopting this complex route, rather than just developing in the lumen of the small intestine where eggs originally hatch, the parasite gains a significant advantage. There is some evidence to suggest that migration confers enhanced parasitic growth[6] and Jungersen et al.[8] demonstrated that *A. suum* larvae infected intravenously in pigs, do not undergo larval migration, and develop more slowly.

THE IMPACT OF LARVAL MIGRATION IN NATURAL HOSTS

The public health significance of larval migration by *Ascaris* in humans remains poorly understood. The difficulty in designing appropriate and ethical studies of larval migration in humans has hindered research into this phase of the life-cycle.[9,10] Considerable evidence from both human and porcine studies points to the presence of adult worms in the intestine contributing to nutritional impairment, including growth retardation.[11] In contrast, there is a paucity of data on the impact of larval ascariasis on host fitness, including nutritional status and body weight. Hale et al.[12] concluded that the effect of migrating larvae on pigs was less pronounced than that of adult worms and Stephenson et al.[13] infected pigs with fourth-stage larvae and demonstrated that adult worms depressed the growth rate in the absence of larval migration. Nevertheless, fenbendazole treatment during the liver migration phase in experimentally infected pigs improved feed conversion rates by 22%, and treatment of pigs during the pulmonary migration phase improved these rates by 8% in comparison to controls.[14]

The initial stages of the migratory pathway involve newly hatched L3 larvae penetrating the walls of the large intestine and migrating via the bloodstream to the hepatic circulation. Despite the interest in protective intestinal immunity and a pre-hepatic barrier originally proposed by Urban et al.,[15] there is a lack of published work on the local response to this short-lived stage of the early migratory pathway. Trickle infections of *A. suum* in pigs revealed a reduction in liver white spots and larvae getting through to the lungs over time, suggesting a slow build-up of the efficiency of the pre-hepatic barrier in blocking access of larvae to the liver.[16]

The liver phase in humans remains particularly subterranean for obvious reasons. In an extensive prospective study from India, Javid and colleagues[17] described 510 patients over a 10-year period that were admitted to hospital with liver abscess. Of these, 74 (14.5%) had biliary *Ascaris* as the cause and 11 patients had intact *Ascaris* within the liver abscess. Both Sakakibara et al.[18] and Kakihara et al.[19] also reported liver abscesses in patients infected with *A. suum*. Increased levels of hepatic-originating acute phase reactants, such as C-reactive protein, ferritin, and eosinophilic cationic protein, found in putatively immune Nigerian children, indicate that an ongoing inflammatory response may represent an anti-parasite effector mechanism targeting migrating and developing liver-stage and possibly lung-stage larvae.[20]

In pigs, the consequences of *Ascaris* infection in the liver are well documented as white spots (WS), pathological lesions, composed of leucocyte infiltrations, formed as a consequence of the damage inflicted by migratory larvae and the associated inflammatory response.[21–24] There are three types of WS observed in *A. suum*-infected pigs (see Figure 14.1, Chapter 14). The granulation type WS (GT-WS) have been suggested to form along larval migration routes (small GT-WS) or encapsulate trapped larvae (large GT-WS)[23,25,26] and have been proposed to play a role in immunity to *A. suum* in pigs.[26,27] GT-WS have been proposed as precursors of the lymphonodular type WS (LN-WS);[25,26] as the appearance of the latter on 10 days post-infection coincides with the healing of GT-WS.[28] Coupled with the characteristic WS, necrotic and hemorrhagic migratory trajectories are observed in porcine livers. Certainly the potential for liver damage in human subjects is likely to be underestimated.[29]

Larval migration through the host lung tissues is known to produce a range of symptoms including asthma, dyspnoea, cough, and substernal pain that contribute to respiratory distress in both humans[30–33] and pigs.[34] While studying tuberculosis, Loeffler[35] described a transient or seasonal syndrome of pulmonary infiltrates, mild to marked respiratory symptoms, and peripheral eosinophilia that he subsequently attributed to larval *Ascaris* in the lungs, later termed "Loeffler's syndrome."[36] Gelpi and Mustafa[37] described a temporally constrained pneumonitis and eosinophilia that co-occurs with the seasonally-transmitted pattern of ascariasis in Saudi Arabia.

In humans, dyspnoea and bronchospasm may be severe[38] and have also been documented in pigs.[22,39,40] Moreover, bronchovascular damage caused by the parasite may result in secondary infections by opportunistic bacteria that may proliferate in the inflamed tissue and invade the vasculature, adding further complications associated with the migratory stages of this parasite.[41–43]

ANIMAL MODELS OF LARVAL ASCARIASIS

Modeling the *Ascaris* life-cycle and its epidemiology remains challenging and plays a part in its status as a neglected disease. As defined by Boes and Helwigh,[44] an appropriate animal model of helminth infection should mimic the human host, the parasite, and the human host–parasite system or the way in which the host and parasite interact. *A. suum* infection in the pig represents the naturally occurring host–parasite system whereby the parasite completes its entire life-cycle. Furthermore, *A. suum* is closely related to the human parasite *A. lumbricoides* (see Chapter 10) and therefore, not surprisingly, *A. suum* infections in pigs have played a major role as a model for both human and porcine *Ascaris* infection and ascariasis (see Chapters 14 and 16).

However, despite the undoubted advantages of this natural host–parasite system, the pig model has some disadvantages including its large size, cost, husbandry challenges, a lack of inbred strains, and an associated broad range of immunological reagents that could be used to dissect the immune response.

Trichuris trichiura, another important human geohelminth infection has the advantage of a rodent counterpart, *Trichuris muris*, the closely related

TABLE 5.2 Selection of studies in which the mouse was used as a model for early *Ascaris* migration and the host immunological response to *Ascaris* infection.

Study type	Reference number
Intestinal-hepatic migratory pattern	48–51,68,92,105
Hepatic-pulmonary migratory pattern	52–53,68,78,106–109
Hepatic pathology	22,68,110–112
Pulmonary pathology	22,60,74,111
Immunological response	22,59,66,76,110, 113–119
Host and parasite genetics	59,73,120

murine species in which the parasite completes its life-cycle in a manner similar to that in humans. This has enabled considerable dissection of the immune response to *T. muris* in both inbred and gene knockout strains of mice.[45] *Heligmosomoides bakeri* has proved to be an excellent model organism for nematodes of importance in livestock.[46] Tanguay and Scott[47] explored the generative mechanisms of aggregation and predisposition to an adult worm using both inbred and outbred mice infected with *H. bakeri*.

In the case of *Ascaris* infection, a wide range of model organisms have been experimentally infected, including mice, guinea-pigs, rabbits, gerbils, rats, cows, lambs, goats, and pigs (see Table 5.1 and Chapters 14 and 16). In all model organisms except pigs, the life-cycle is incomplete and models only the early stages of infection (i.e. migration from the small intestine to the liver and then from the liver to the lungs). Larval stages of *Ascaris* infection do not return successfully to the small intestine to mature and complete their life-cycle in these hosts and they are therefore referred to as abnormal hosts. It should be noted that in the vast majority of these models, susceptibility and resistance to *Ascaris* infection in either the liver or the lungs has not been clearly established. The particular focus of this chapter is upon the investigation of ascariasis in mice (see Table 5.2), and, more specifically, the development of a murine model for resistance and susceptibility to early infection. The use of mice to investigate vaccine candidates for ascariasis is described in Chapter 16.

MICE AS MODEL ORGANISMS FOR EARLY *ASCARIS* INFECTION

The first published investigation of *Ascaris* infection in mice was by Ransom and Foster[48] who described the intestinal–hepatic migratory pattern. Early studies[49,50] documented the preferential penetration of the cecal and colon wall during early migration. This migratory stage was recognized in mice prior to its discovery in pigs, highlighting the important role abnormal hosts can play in exploring fundamental aspects of the *Ascaris* life-cycle and basic parasite biology. As early as 1920, Ransom and Foster[48] detected *Ascaris* larvae in the portal vein on day 1 post-infection, the first researchers to document the role of a blood vessel in larval migration.

The timing of *Ascaris* larval migration is similar in both mice and pigs and extends over an approximately 10- to 14-day period. When infective embryonated eggs are administered to a mouse by gastric intubation (see Figure 5.1), larvae hatch in the small intestine, migrate to the large intestine and within 2 to 3 hours post-infection can be recovered from the cecal and colonic tissues.[49–51] From there, larvae migrate to the liver as early as 4 hours post-infection.[51] A few larvae can be detected in the lungs very quickly but the vast majority does not appear until days 4 to 5 post-infection.[51–53] Larval numbers then peak in this organ between

FIGURE 5.1 In experiments involving the mouse as a model organism, embryonated ova (1) are typically administered by means of gastric intubation into the stomach (2). Larvae hatch in the small intestine (3) and migrate to the large intestine (4), where they penetrate the cecal and colonic wall (5). From there, larvae migrate to the liver (6) and then to the lungs (7).

day 5 and day 7[50–53]. Slotved and colleagues[51] concluded in their important comparative work between pigs and mice that the migratory pattern of *A. suum* is similar in murine and porcine hosts and on these grounds established that mice are a suitable model organism for the study of early *Ascaris* infection.

Working with mice has a number of practical advantages including ease of manipulation and the availability of a wide range of mouse strains and immunological reagents. Behnke et al.[46] highlighted how the use of mouse model systems can enable the dissection of the mechanism of parasite resistance at a variety of sophisticated levels. Furthermore, the recovery of larvae has been shown to be higher in mice when compared to rabbits, guinea-pigs, rats,[52] and even pigs.[28] Undoubtedly, the relative host size and parasite size contribute to this higher recovery.[53] Lewis et al.[53] demonstrated increased efficiency of recovery of *Ascaris* larvae, expressed as a percentage of inoculi in mice, utilizing a modified version of the Baermann method.

THE MOUSE AS A MODEL OF *ASCARIS* AGGREGATION

One of the defining parameters that contribute to our understanding of *Ascaris* epidemiology is the intensity of infection or the worm burden of an individual host[54] (see also Chapter 7). This variable influences parasite transmission, population dynamics, and the degree of individual and community morbidity. It also determines the most effective way in which a helminth parasite can be controlled. The terms aggregation and predisposition are used to describe the fact that not all human hosts are infected equally, that some hosts carry considerably heavier burdens than most, and that there is some degree of constancy in reinfection patterns after chemotherapeutic treatment.[55–57] Nevertheless, our understanding of the mechanism behind the observed patterns of *Ascaris* infection remains incomplete. Whereas some progress has been made in assessing the role of susceptibility, both immunological and genetic, and quantitative trait locus (QTL) studies have even identified genetic loci associated with susceptibility and resistance to infection in humans[58] (see also Chapter 12), exposure to ascariasis has been difficult to measure and when it has been attempted, it has been quantified only indirectly.[57]

A major incentive to exploit mice as models is the availability of a wide range of strains with well-known genotypes. For obvious reasons, early studies did not utilize inbred strains of mice (see Table 5.2) and it was only in the 1960s and 1970s that mouse strains began to be utilized in parasitology. A seminal paper by Mitchell and colleagues[59] demonstrated variability in susceptibility to *Ascaris* in the lungs of inbred strains of mice but variations in infectivity of batches of eggs and doses resulted in lack of comparability between groups. Despite this, it was also established that resistance in resistant strains of mice did not depend on T cells, since nude athymic strains on resistant backgrounds resisted infection just as well as the wild-type parental strains and was not MHC haplotype dependent.[59] Moreover, much later Lewis et al.[60] found that even treatment with the highly anti-inflammatory and immunodepressive steroid cortisone made no impact on the capacity of resistant strains to impair larval migration to the lungs.

Prior to this, Kennedy and Qureshi[61] and Tomlinson et al.[62] had concluded that *Ascaris* in the mouse was a useful model for understanding genetic control of the immune repertoire to defined antigens of the parasite in relation to selective immune responses to antigens of *Ascaris* in infected humans (see Chapter 3).[63–65] These studies showed that there was sequential recognition of antigens with time after infection,[62,66] but concentrated on antibody produced once the immune response was mature, and hence days after the worm burdens would have declined to zero. Hence, they did not focus on the critical time when larvae were still present in the host and still migrating, and when effective resistance might be expected to operate. It is unlikely therefore that the antibody responses they described played any significant role in host-protective immunity to the larval migratory stage of *A. suum* infection in their genetically high responder mouse strains.

In an *Ascaris* mouse model, comprising a susceptible and resistant strain, a significant reduction in body weight was observed in mice that received higher doses of *Ascaris* ova indicating that larval migration and accumulation of larvae in the lungs has a significant impact upon host body condition even in abnormal hosts that sustain only the migratory and tissue-resident phase of infection.[67]

After optimization of the methodologies for infection and recovery, and having compared a panel of nine inbred strains of mice with respect to larval recovery in the lungs on day 7 post-infection, we identified a mouse model for susceptibility and resistance to early infection.[53] C57BL/6j mice are consistently highly susceptible to *Ascaris* larvae in the lungs, considerably more so than any of the other strains tested, with worm burdens peaking at day 7 post-infection, in contrast, for example, to CBA/Ca that remain consistently relatively resistant to lung infection (see Figure 5.2). Therefore, these two mouse strains with highly consistent and diverging larval burdens in their lungs represent the extremes of the host phenotype displayed in the aggregated distribution and provide an opportunity to explore the genetic and mechanistic basis of this difference in response phenotypes.[68] Early events in infection are likely to play a key role in the determination of an adult worm infection. More specifically, it has been suggested that pulmonary migration by *Ascaris* larvae in humans may create a highly polarized Th2 immune

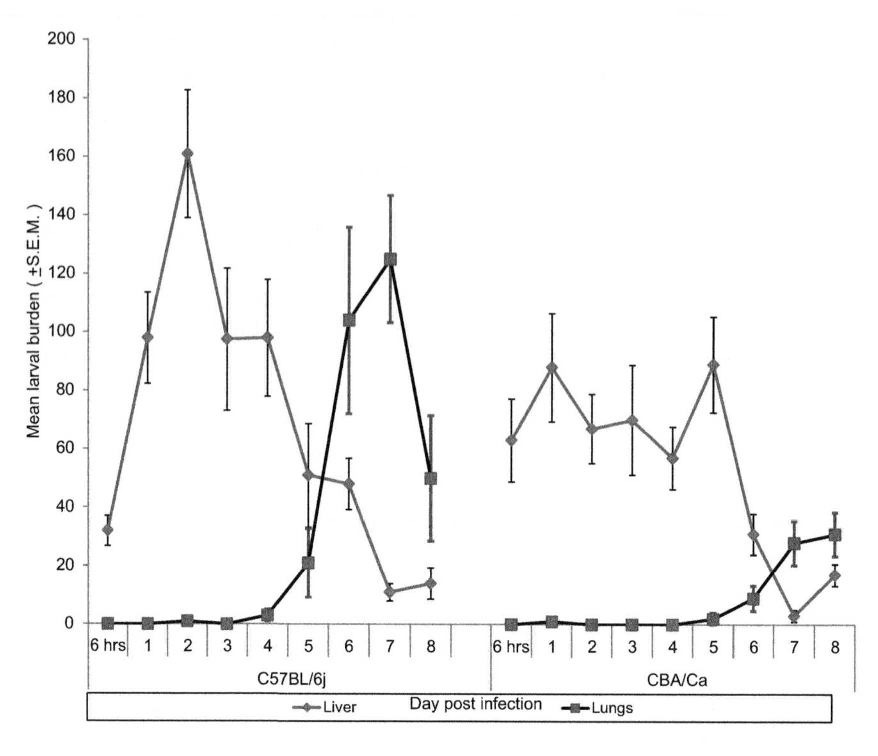

FIGURE 5.2 Changes in the mean larval burden (±S.E.M.) in the liver and lungs in C57BL/6j and CBA/Ca mice, following inoculation with 1000 *Ascaris suum* larvae.[68]

environment in the lung mucosa, and genetic variation in host response at this level may influence the subsequent course of adult worm infection.[69]

As stated above, C57BL mice appear to be considerably more susceptible to *Ascaris* larval infection than any of the other strains tested and in contrast to all the other mouse strains tested are known also to lack the *Itln*-2 gene which has been linked with expulsion of intestinal stages of *Trichinella spiralis* and *Trichuris muris*.[70–72] This suggested a simple genetic explanation for their greater susceptibility to migrating larvae and inability to control infection at the liver stage, but breeding experiments based on F_1 crosses between C57BL/6j and CBA/Ca mice, and F_2 backcrosses to both parental strains found no evidence that susceptibility segregated with absence of this gene and resistance increased with the number of copies of the *Itln*-2 gene.[73] The unusually high susceptibility of C57BL mice therefore still warrants explanation.

As Mitchell et al. found,[59] mice show acquired resistance to infection and greater numbers of lesions have been observed with more larvae surrounded by inflammatory cells in experimentally-infected mice and guinea-pigs considered resistant due to previous exposure to *A. suum*.[74,75] Therefore, since WS are associated with trapped larvae and are more abundant in resistant animals, the liver has been consistently identified as a key site in the immobilization of migrating larvae in mice,[39,59,76–78] guinea-pigs,[79,80] and rabbits.[81]

Subsequently, we used the previously described *Ascaris*[53] mouse model system to explore the significance of inflammatory processes within the murine lung (utilizing the leucocyte population in bronchalveoloar lavage (BAL) fluid and lung histopathology.[60] We concluded that the pulmonary responses reflected larval intensity, intense responses being associated with high worm burdens rather than with low worm burdens (which might have provided support for their involvement in host-protective immunity) and hence we concluded that the inflammatory response was not prominently involved in primary protection of mice to *Ascaris* infection in the lungs. However, as suggested by Cooper and colleagues,[69] the intensity of the response to this stage may have consequences for subsequent adult worm infections in hosts that enable the worms to complete their maturation (i.e. humans and pigs for *A. lumbricoides* and *A. suum*, respectively).

The lack of support for a pulmonary mechanism of resistance led us to revisit the idea that resistance operates at the level of the liver. We focused on the possibility that a hepatic or post-hepatic factor, which varies between the susceptible and resistant mouse strain, comes into play rapidly within 4 days of infection, and is non-T cell or MHC haplotype dependent, may play a critical role in successful/unsuccessful tissue migration. A study of the comparative histopathological hepatic inflammatory response in C57BL/6j and CBA/Ca mice revealed an important difference between the two strains.[68] In resistant CBA mice, the most pronounced response occurred on day 4 post-infection, this coinciding with the migration of larvae from the liver to the lungs. In contrast, the most severe inflammatory response that we detected in susceptible C57BL/6j mice occurred on day 6 post-infection when the majority of the larvae are known to have already successfully migrated to the lungs. We concluded that CBA mice handle the parasitic insult more effectively and economically in terms of an earlier response that restricts the degree of larval invasion at the liver stage, and effecting a more rapid tissue recovery without triggering additional inflammation.[68] Significantly, earlier work on guinea-pigs had identified day 4 as an important time-point with respect to encapsulation by inflammatory cells in immunized animals with an associated reduction in larval numbers noted in the lungs.[82,83] Perhaps the most intriguing aspect of all these experiments, and as yet still not explained, is that the rapid inflammatory response in the livers of resistant animals must be innately driven and is non-T cell or MHC haplotype dependent.

FUTURE DIRECTIONS

Evidence has now accumulated to confirm that selective phases of infection and their impact, specifically the role of the liver migratory phase, is tractable within a murine model of ascariasis,[29,57] and the effective hepatic response observed in the resistant CBA/Ca mice can be further dissected.[84] The liver is an important regulator of the systemic innate immune response. Hepatocytes are the primary synthesizers of all acute phase proteins (APPs), numerous soluble pathogen recognition receptors (PRRs), and 80–90% of complement components.[85] While hepatocytes play a key role in the biosynthesis of mediators of the innate immune system, principal cells of innate immunity are also resident in the liver[86] and the liver has been described as "an organ with predominant innate immunity."[86] The liver's resident monocytes are macrophage-like cells, known as Kupffer cells, accounting for 80–90% of the total population of fixed tissue macrophages in the body. Kupffer cells are localized in the liver sinusoids and therefore they are the first hepatic cells to be exposed to the antigens carried to the liver from the GI tract. It will be interesting to determine clearly which of these signals for the initiation of the innate response leading to encapsulation is absent in susceptible strains and which in particular drive the more intense host-protective response of resistant animals.

Jungersen et al.[87] recognized the importance of assessing the immunological response to *A. suum* infection at the site at which immunity is manifested. The lymphocyte population which recognized antigens in his model system was dependent upon the draining area of the lymph nodes that were assessed, each of which may be exposed to different stage-specific antigens[61] associated with the presence of larvae in the surrounding organs or may be dependent upon or constrained by the effector mechanisms of the immune response that operate in that particular organ. In our system, conventional T helper cells are unlikely to be driving the primary response that limits infection in the livers in resistant strains of mice, but clearly an innate response is involved. Therefore, we suggest that an investigation of the local production of selected hepatic cytokines as well as expansion of particular cell populations at certain key days post-infection in the susceptible and resistant strain of mice would be a useful line of enquiry, given the differential hepatic inflammatory response observed between strains. Crucial to the understanding of primary resistance, and its failure in susceptible strains of mice, is the identification of the cells that are primarily responsible for actually driving this innate inflammatory response, and of the spectrum of cytokines that they must secrete to marshal host resistance to liver stage larvae.

The mouse model of larval ascariasis could also be used to explore potential differences between *A. lumbricoides* and *A. suum* with respect to larval migration and the accompanying host response, and this could be extended to differing parasite genotypes within a species. Nesjum et al.[88] infected pigs with *A. suum* worms with unique mtDNA haplotypes and revealed an uneven distribution of the different genotypes within the host intestine. The mouse model could also be a useful and cost-effective model system to explore the host and parasite transcriptome. As described by Jex et al. [89] (see Chapter 11), the extensive and complex migration undertaken by *Ascaris* requires tightly regulated transcriptional changes in the parasite. Transcription profiles of infective L3 larvae from eggs and from the liver and lungs revealed various secreted peptidases linked to tissue migration and degradation during feeding and/or migration. Proteinases secreted by *Ascaris* larvae are known to be inhibitable by antibody,[90] and it would be interesting to discover whether proteinase inhibitors released by the liver during inflammatory and acute phase responses also inhibit them, and whether variation in the intensity and quality of secreted host proteinase inhibitors in turn is linked to the differential success of migrating larvae in resistant and susceptible strains of mice.

To conclude, the use of an animal model that provides two contrasting response phenotypes as reflected in two inbred mouse strains is an asset in teasing apart the mechanisms that confer resistance and predisposition to light and heavy *Ascaris* infection, and there is still much yet to be done to fully understand the nature of this genetic resistance. Focus upon the innate cellular immune response to migrating larvae within the liver would seem to be a particularly fruitful line of enquiry. Once detailed knowledge of the cellular hepatic response at the molecular level is available, this may enhance our ability to develop immunomodulatory therapies to elicit resistance to infection in both people and pigs. More generally, increasing our knowledge of larval ascariasis will contribute to the development of intervention strategies for early stages of the life-cycle, aimed at preventing parasite establishment and thus chronic disease.

Acknowledgments

We thank Kieran Dold for creating Figure 5.1. Celia Holland acknowledges the financial support of the Irish Research Council and Science Foundation Ireland.

References

1. Sprent JF. The life cycles of nematodes in the family Ascarididae Blanchard 1896. *J Parasitol* 1954;**40**:608–17.
2. Maung M. The occurrence of the second moult of *Ascaris lumbricoides* and *Ascaris suum*. *In J Parasitol* 1978;**8**:371–8.

3. Geenen PL, Bresciani J, Boes J, Pedersen A, Eriksen L, Fagerholm H-P, et al. The morphogenesis of *Ascaris suum* to the infective third-stage larvae within the egg. *J Parasitol* 1999;**85**:616–22.
4. Dold C, Holland CV. Ascaris and ascariasis. *Microbes Infect* 2011a;**13**(7):632–7.
5. Smyth JD. *Introduction to Animal Parasitology.* 3rd ed. Cambridge University Press; 1994. pp. 549.
6. Read AF, Skorping A. The evolution of tissue migration by parasitic nematode larvae. *Parasitology* 1995;**111**:359–71.
7. Mulcahy G, O'Neill S, Fanning J, McCarthy E, Sekiya M. Tissue migration by parasitic helminthes – an immunoevasive strategy? *Trends Parasitol* 2005;**21**:273–7.
8. Jungersen G, Fagerholm HP, Nansen P, Eriksen L. Development of patent *Ascaris suum* infections in pigs following intravenous administration of larvae hatched *in vitro*. *Parasitology* 1999;**119**:503–8.
9. Cooper ES, Whyte-Alleng CAM, Finzi-Smith JS, MacDonald TT. Intestinal nematode infections in children: the pathophysiological price paid. *Parasitology* 1992;**104**: 91–103.
10. Stephenson LS. Ascariasis. In: Stephenson LS, editor. *Impact of Helminth Infections on Human Nutrition*. London: Taylor & Francis; 1987. p. 89–118.
11. Hall A, Hewitt G, Tuffrey V, deSilva N. A review and meta-analysis of the impact of intestinal worms on child growth and nutrition. *Mat Child Nutr* 2008;**4**:118–236.
12. Hale OM, Stewart TB, Marti OG. Influence of experimental infection of *Ascaris suum* on performance in pigs. *J Anim Sci* 1985;**60**:220–5.
13. Stephenson LS, Pond WG, Nesheim MC, Krook LP, Crompton DWT. *Ascaris suum*: nutrient absorption, growth and intestinal pathology in young pigs experimentally infected with 15-day old larvae. *Exp Parasitol* 1980;**49**:15–25.
14. Stewart TB, Bidner TD, Southern LL, Simmons LA. Efficacy of fenbendazole against migrating *Ascaris suum* larvae in pigs. *Am J Vet Res* 1984;**45**:984–6.
15. Urban JF, Alizadeh H, Romanowski RD. *Ascaris suum*: development of intestinal immunity to infective second-stage larvae in swine. *Exp Parasitol* 1988;**66**:66–77.
16. Nejsum P, Thamsborg SM, Petersen HH, Kringel H, Fredholm M, Roepstorff A. Population dynamics of *Ascaris suum* in trickle-infected pigs. *J Parasitol* 2009;**95**(5):1048–53.
17. Javid G, Wani NA, Gulzar GM, Khan BA, Shah AH, Shah OJ, et al. *Ascaris*-induced liver abscess. *World J Surg* 1999;**23**:1191–4.
18. Sakakibara A, Baba K, Niwa S, Yagi T, Wakayama H, Yoshida K, et al. Visceral larva migrans due to *Ascaris suum* which presented with eosinophilic pneumonia and multiple intra-hepatic lesions with severe eosinophil infiltration. *Int Med* 2002;**41**: 574–9.
19. Kakihara D, Yoshimitsu K, Ishigami K, et al. Liver legions of visceral larva migrans due to *Ascaris suum* infection. *Abdom Imag* 2004;**29**:598–602.
20. McSharry C, Xia Y, Holland CV, Kennedy MW. Natural immunity to *Ascaris lumbricoides* associated with immunoglobulin E antibody to ABA-1 allergen and inflammation indicators in children. *Infect Immun* 1999;**67**:484–9.
21. Schwartz B, Alicata JE. *Ascaris* larvae as a cause of liver and lung lesions in swine. *J Parasitol* 1932;**19**:17–24.
22. Eriksen L. *Host parasite relations in* Ascaris suum *infection in pigs and mice*. PhD thesis. Copenhagen, Denmark: Royal Veterinary and Agricultural University; 1981.
23. Perez J, Garcia PM, Mozos E, Bautista MJ, Carrasco L. Immunohistochemical characterization of hepatic lesions associated with migrating larvae of *A. suum* in pigs. *J Comp Path* 2001;**124**:200–6.
24. Frontera E, Roepstorff A, Gazquez A, et al. Immunohistochemical distribution of antigens in liver of infected and immunized pigs with *Ascaris suum*. *Vet Parasitol* 2003;**111**:9–18.

25. Roneus O. Studies on the aetiology and pathogenesis of white spots in the liver of pigs. *Acta Vet Scand* 1966;**7**:1–112.
26. Copeman DB, Gaafar SM. Sequential development of hepatic lesions of ascaridosis in colostrum-derived pigs. *Aust Vet J* 1972;**48**:263–8.
27. Eriksen L, Anderson S, Nielsen K, Pedersen A, et al. Experimental *Ascaris suum* infection in pigs. Serological response, eosinophilia in peripheral blood, occurrence of white spots in the liver and worm recovery from the intestine. *Nor Vet Med* 1980;**32**: 233–42.
28. Roepstorff A, Eriksen L, Slotved HC, Nansen P. Experimental *Ascaris suum* infection in the pig: worm population kinetics following single inoculations with three doses of infective eggs. *Parasitology* 1997;**115**:443–52.
29. Dold C. Nematoda: *Ascaris lumbricoides*. In: Lamb TJ, editor. *Immunity to Parasitic Infection*. Wiley-Blackwell; 2012. p. 231–45.
30. Beaver PC, Danaraj TJ. Pulmonary ascariasis resembling eosinophilic lung: autopsy report with description of larvae in the bronchioles. *Am J Trop Med Hyg* 1958;**7**:100.
31. Spillmann RK. Pulmonary ascariasis in tropical communities. *Am J Trop Med Hyg* 1975;**24**:791.
32. Pawlowski ZS, Arfaa R. Ascariasis. In: Warren KS, Mahmoud AAF, editors. *Tropical and Geographic Medicine*. New York: McGraw Hill; 1984. p. 347–58.
33. Coles GC. Allergy and immunopathology of ascariasis. In: Crompton DWT, Nesheim MC, Pawlowsi ZS, editors. *Ascariasis and its Public Health Significance*. London: Taylor and Francis; 1985. p. 167–84.
34. Matsuyama W, Mizoquchi A, Iwami F, Kawababta M, Osame M. A case of pulmonary infiltration with eosinophilia caused by *Ascaris suum*. *Hihon Kokyuki Gakkai Zasshi* 1998;**36**:208–12.
35. Loeffler W. Zur Differentialdiagnose der Lungen-infiltrierungen. II Ueber Fluchtige Succedanininfiltrate (mit Eosinophilie). *Beitrage zur Klinik der Tuberkulose* 1932;**79**: 368–82.
36. Loeffler W. Transient lung infiltrations with blood eosinophilia. *Int Arch Allerg App Immunol* 1956;**8**:54–9.
37. Gelpi AP, Mustafa A. Seasonal pneumonitis with eosinophilia: a study of larval ascariasis in Saudi Arabia. *Am J Trop Med Hyg* 1967;**16**:646–57.
38. Ribeiro JD, Fischer GB. Eosinophilic lung diseases. *Paed Respir Rev* 2002;**3**:278–84.
39. Taffs LF. Immunological studies on experimental infection of pigs with *Ascaris suum* Goeze, 1782. IV Histopathology of the liver and lung. *J Helm* 1968;**42**:157–72.
40. Yoshihara S, Nakagawa M, Suda H, Ikeda K, Hanashiro K. White spots of the liver in pigs experimentally infected with *Ascaris suum*. *Nat Inst Animal Hlth Quart* 1983;**23**: 127–37.
41. Keller AE, Hillstrom HT, Gass RS. The lungs of children with *Ascaris*. A roentgenologic study. *J Am Med Abroad* 1932;**99**:1249–51.
42. Liljegren CH, Aalbaek B, Nielsen OL, Jensen HE. Some aspects of the pathology, pathogenesis, and aetiology of disseminated lung lesions in slaughter pigs. *Acta Parasitol Microbiol Scand* 2003;**111**:531–8.
43. Tjornehoj K, Eriksen L, Aalbaek B, Nansen P. Interaction between *Ascaris suum* and *Pastueralla multocida* in the lungs of mice. *Parasitol Res* 1992;**78**:525–8.
44. Boes J, Helwigh AB. Animal models of intestinal nematode infections of humans. *Parasitology* 2000;**121**:S97–111.
45. Cliffe LJ, Grencis RK. The *Trichuris muris* system: a paradigm of resistance and susceptibility to intestinal nematode infection. *Adv Parasitol* 2004;**57**:255–307.
46. Behnke JM, Menge DM, Noyes H. *Heligimosomoides bakeri*: a model for exploring the biology and genetics of resistance to chronic gastrointestinal nematode infections. *Parasitology* 2009;**136**:1565–80.

47. Tanguay GV, Scott ME. Factors generating aggregation of *Heligimosomoides polygyrus* (Nematoda) in laboratory mice. *Parasitology* 1992;**104**:519–29.
48. Ransom BH, Foster WD. Observations on the life history of *Ascaris lumbricoides. US Dept Agric Bull* 1920;**817**.
49. Jenkins DC. Observations on the early migration of the larvae of *Ascaris suum* Goetze, 1782 in white mice. *Parasitology* 1968;**58**:431–40.
50. Keittivuti B. *Sites and penetration of* Ascaris suum *larvae in experimentally infected mice and swine.* Lafayette, Indiana, USA: Purdue University; 1974.
51. Slotved HC, Eriksen E, Murrell K, Nansen P. Early *Ascaris suum* migration in mice as a model for pigs. *J Parasitol* 1998;**84**:16–8.
52. Douvres FW, Tromba FG. Comparative development of *Ascaris suum* in rabbits, guinea pigs, mice and swine in 11 days. *Proc Helm Soc Wash* 1971;**38**:246–52.
53. Lewis R, Behnke JM, Stafford P, Holland CV. The development of a mouse model to explore resistance and susceptibility to early *Ascaris suum* infection. *Parasitology* 2006;**132**:289–300.
54. Walker M, Hall A, Basanez M- G. Individual predisposition, household clustering and risk factors for human infection with *Ascaris lumbricoides*: new epidemiological insights. *PLOS Neg Trop Dis* 2011;**5**(4):e1047.
55. Holland CV, Asaolu SO, Crompton DWT, Stoddart R, MacDonald R, Torimiro SEA. The epidemiology of *Ascaris lumbricoides* and other soil-transmitted helminths in primary school children from Ile-Ife,. *Nigeria. Parasitology* 1989;**99**: 275–85.
56. Holland CV, Boes J. Distributions and predisposition: people and pigs. In: Holland CV, Kennedy MW, editors. *The Geohelminths: Ascaris, Trichuris and Hookworm.* Kluwer Academic Publishers; 2002. p. 1–24.
57. Holland CV. Predisposition to ascariasis: patterns, mechanisms and implications. *Parasitology* 2009;**136**:1537–47.
58. Williams-Blangero S, VandeBerg JL, Subedi J, Jha B, Correa-Oliveira R, Blangero J. Localization of multiple quantitative trait loci influencing susceptibility to infection with *Ascaris lumbricoides. JID* 2008;**197**:66–71.
59. Mitchell GF, Hogarth-Scott RS, Edwards RD, Lewers HM, Cousins G, Moore T. Studies on immune response to parasite antigens in mice. I. *Ascaris suum* larvae numbers and anti phosphophorylcholine responses in infected mice of various strains and in hypothymic *nu/nu* mice. *Int Arch Allerg App Immunol* 1976;**52**:64–78.
60. Lewis R, Behnke JM, Cassidy JP, Stafford P, Murray N, Holland C. The migration of *Ascaris suum* larvae, and the associated pulmonary inflammatory response in susceptible C57BL/6j and resistant CBA/Ca mice. *Parasitology* 2007;**134**: 1301–14.
61. Kennedy MW, Qureshi F. Stage-specific secreted antigens of the parasite larval stages of the nematode *Ascaris. Parasite Immunol* 1986;**58**(3):515–22.
62. Tomlinson LA, Christie JF, Fraser EM, McLaughlin D, McIntosh AE, Kennedy MW. MHC restriction of the antibody repertoire to secretory antigens, and a major allergen, of the nematode parasite Ascaris. *J Immunol* 1989;**143**(7):2349–56.
63. Fraser EM, Christie JF, Kennedy MW. Heterogeneity amongst infected children in IgE antibody repertoire to the antigens of the parasitic nematode *Ascaris. Int Arch Allerg Immunol* 1993;**100**(3):283–6.
64. Haswell-Elkins M, Kennedy MW, Maizels RM, Elkins DB, Anderson RM. The antibody recognition profiles of humans naturally infected with *Ascaris lumbricoides. Parasite Immunol* 1989;**11**(6):615–27.
65. Kennedy MW, Tomlinson LA, Fraser EM, Christie JF. The specificity of the antibody-response to internal antigens of *Ascaris* – heterogeneity in infected humans, and MHC (H-2) control of the repertoire in mice. *Clin Expt Immunol* 1990;**80**(2):219–24.

66. Kennedy MW, Gordon AM, Tomlinson LA, Qureshi F. Genetic (Major Histocompatibility Complex) control of the antibody repertoire to the secreted antigens of *Ascaris*. *Parasite Immunol* 1987;**9**(2):269–73.
67. Lewis R, Behnke JM, Stafford P, Holland CV. Dose-dependent impact of larval *Ascaris suum* on host body weight in the mouse model. *J Helminthol* 2009;**83**:1–5.
68. Dold C, Cassidy J, Stafford P, Behnke JM, Holland CV. Genetic influence on the kinetics and associated pathology of the early stage (intestinal-hepatic) migration of *Ascaris suum* in mice. *Parasitology* 2010;**137**(1):173–85.
69. Cooper PJ, Chico ME, Sandoval C, et al. Human infection with *Ascaris lumbricoides* is associated with a polarized cytokine response. *JID* 2000;**182**:1207–13.
70. Artis D. New weapons in the war on worms: identification of putative mechanisms of immune-mediated expulsion of gastrointestinal nematodes. *Int J Parasitol* 2006;**36**: 723–33.
71. Datta R, Deschoolmeester ML, Hedeler C, Paton NW, Brass AM, Else KJ. Identification of novel genes in intestinal tissue that are regulated after infection with an intestinal nematode parasite. *Infect Immun* 2005;**73**:4025–33.
72. Pemberton AD, Knight PA, Gamble J, Colledge WH, Lee J, Pierce M, et al. Innate BALB/c enteric epithelial responses to *Trichinella spiralis*: inducible expression of a novel lectin, intelectin-2 and its natural deletion in C57BL/10 mice. *J Immunol* 2004;**173**:1894–901.
73. Dold C, Pemberton A, Stafford P, Holland CV, Behnke JMB. The role of Intelectin-2 in resistance to *Ascaris suum* lung larval burdens in susceptible and resistant mouse strains. *Parasitology* 2011;**138**:660–9.
74. Sprent JF. On the toxic and allergic manifestations of produced by the tissues and fluids of *Ascaris*, effect of different tissues. *JID* 1949;**84**:221.
75. Taffs LF. Immunological studies on experimental infection of guinea pigs and rabbits with *Ascaris suum* Goeze, 1782. IV Histopathology of the liver and lung. *J Helm* 1965;**39**: 297–302.
76. Sprent JF, Chen HH. Immunological studies in mice infected with the larvae of *Ascaris lumbricoides*, criteria of immunity and immunizing effect of isolated worm tissues. *JID* 1949;**84**:111–24.
77. Johnstone C, Levental R, Soulsby EJL. The spin method for recovering tissue larvae and its use in evaluating C57BL/6 mice as a model for the study of resistance to infection with *Ascaris suum*. *J Parasitol* 1978;**64**:1015–20.
78. Song JS, Kim JJ, Min DY, Lee KT. Studies on the comparative migration patterns of *Ascaris suum* larvae between primary and re-infected mice. *Korean J Parasitol* 1985;**23**: 247–52.
79. Kerr KB. The cellular response in acquired resistance in guinea pigs to an infection with pig *Ascaris*. *Am J Hyg* 1938;**27**:28–51.
80. Fallis AM. *Ascaris lumbricoides* infection in guinea pigs with special reference to eosinophilia and resistance. *Can J Res Ser D* 1948;**26**:307–27.
81. Arean VM, Crandall CA. The effect of immunization on the fate of injected second stage *Ascaris lumbricoides* larvae in the rabbit. *Am J Trop Med Hyg* 1962;**11**:369–79.
82. Soulsby EJL. Immunization against *Ascaris lumbricoides* in the guinea pig. *Nature* 1957;**179**:783–4.
83. Khoury PB, Stromberg BE, Soulsby EJL. Immune mechanisms to *Ascaris suum* in inbred guinea pigs. I. Passive transfer of immunity by cells or serum. *Immunology* 1977;**32**: 405–11.
84. Dold C, Holland CV. Investigating the underlying mechanism of resistance to *Ascaris* infection. *Microbes Infect* 2011b;**13**(7):624–31.
85. Gao B, Jeong WI, Tian Z. Liver: an organ with predominant innate immunity. *Hepatology* 2008;**47**:729–36.

86. Doherty DG, O'Farrelly C. Innate and adaptive lymphoid cells in the human liver. *Immun Rev* 2000;**174**:5.
87. Jungersen G, Eriksen L, Nansen P, et al. Regional immune responses with stage-specific antigen recognition profiles develop in lymph nodes of pigs following *Ascaris suum* larval migration. *Parasite Immunol* 2001;**23**:185–94.
88. Nejsum P, Roepstorff A, Anderson TJC, Jorgensen C, Fredholm M, Thamsborg SM. The dynamics of genetically marked *Ascaris suum* infections in pigs. *Parasitology* 2008;**136**: 193–201.
89. Jex AR, et al. *Ascaris suum* draft genome. *Nature* November 2011;**479**(7374):529–33. 24.
90. Knox DP, Kennedy MW. Proteinases released by the parasitic larval stages of *Ascaris suum*, and their inhibition by antibody. *Mol Biochem Parasitol* 1988;**28**(3):207–16.
91. Stewart FH. Further experiments on *Ascaris* infection. *BMJ* 1916a;**2**:486.
92. Ransom BH, Foster WD. Recent discoveries concerning the life history of *Ascaris lumbricoides*. *J Parasitol* 1919;**5**:93–9.
93. Yoshida S. On the development of *Ascaris lumbricoides* L. *J Parasitol* 1919;**5**(3):105–15.
94. Yoshida S. On the migrating course of *Ascaris* larvae in the body of the host. *J Parasitol* 1920;**6**:19–27.
95. Beraldo WT. Dias Da Silva W, Pudles J. Antigenic properties of purified fractions from *Ascaris lumbricoides* var. *suum* on naturally sensitized guinea-pig. *BJ Pharmacol* 2012;**17**(2):236–44.
96. Galvin TJ. Development of human and pig *Ascaris* in the pig and the rabbit. *J Parasitol* 1968;**54**:1085–91.
97. Berger H. Experimentally induced patent infections of *Ascaris suum* in rabbits. *J Parasitol* 1971;**57**:344–7.
98. Yoshida A, Nagayasu E, Horii Y, Maruyama H. A novel C-type lectin identified by EST analysis in tissue migratory larvae of. *Ascaris suum*. *Parasitol Res* 2012;**110**:1583–6.
99. Cho S, Egami M, Ohnuki H, Saito Y, Chinone S, Shichinohe K, et al. Migration behaviour and pathogenesis of five ascarid nematode species in the Mongolian gerbil *Meriones unguiculatus*. *J Helminthol* 2007;**81**(1):43–7.
100. Stewart FH. On the life-history of *Ascaris lumbricoides*. *BMJ* 1916b;**2**:5.
101. Greenway JA, McCraw BM. *Ascaris suum* in calves I. Clinical signs. *Can J Com Med* 1970;**34**:227–37.
102. Greenway JA, McCraw BM. *Ascaris suum* in calves II. Circulating and marrow eosinophil responses. *Can J Com Med* 1970;**34**:238–44.
103. McCraw BM, Greenway JA. *Ascaris suum* infection in calves III. Pathology. *Can J Comp Med* 1970;**34**:247–55.
104. McCraw BM. The development of *Ascaris suum* in calves. *Can J Comp Med* 1975;**39**: 354–7.
105. Ransom BH, Cram EB. The course of migration of *Ascaris* larvae. *Am J Trop Med* 1921;**1**: 129–59.
106. Stewart FH. Note on *Ascaris* infection in man, the pig, rat and mouse. *Ind Med Gaz* 1917a;**52**:272–3.
107. Stewart FH. On the development of *Ascaris lumbricoides* Lin. and *Ascaris suilla* Duj. in the rat and mouse. *Parasitology* 1917b;**9**:1.
108. Sprent JF. On the migratory behavior of the larvae of various *Ascaris* species in white mice. I. Distribution of larvae in tissues. *JID* 1952;**90**:165.
109. Sinha BN. The migratory behaviour of the larvae of *Ascaris lumbricoides* (Linnaeus, 1758) in white mice. *Ind Vet J* 1967;**44**:292–7.
110. Bindseil E. Immunity to *Ascaris suum*. 2. Investigations of the fate of larvae in immune and non-immune mice. *Acta Path Microbiol Scand* 1969;**77**:223–34.
111. Crandall CA, Crandall RB. *Ascaris suum*: immunological responses in mice. *Exp Parasitol* 1971;**30**:426–37.

112. Bindseil E. Pathogenesis of liver lesions in mice following a primary infection with *Ascaris suum*. *Vet Pathol* 1981;**18**:804–12.
113. Bindseil E. Immunity to *Ascaris suum*. 1. Immunity induced in mice by means of material from adult worms. *Acta Path Microbiol Scand* 1969;**77**:218–22.
114. Guerrero J, Silverman PH. *Ascaris suum*: immune reactions in mice. I. Larval metabolic and somatic antiges. *Exp Parasitol* 1969;**26**:272–81.
115. Bindseil E. Immunity to *Ascaris suum*. 3. The importance of the gut for immunity in mice. *Acta Path Microbiol Scand* 1970;**78**:183–9.
116. Bindseil E. Immunity to *Ascaris suum*. 4. The effect of different stimulations upon challenge within mice. *Acta Path Microbiol Scand* 1970;**78**:191–5.
117. Brown AR, Crandall CA, Crandall RB. The immune response and acquired resistance to *Ascaris suum* infection in mice with an X-linked B lymphocyte defect. *J Parasitol* 1977;**63**:950–2.
118. Jeska EL, Stankiewicz M. Responses of NFR/N inbred mice to very low-dose infections with *Ascaris suum*. *Int J Parasitol* 1989;**19**:85–9.
119. Pineda MRP, Ramos JDA. *Ascaris suum* infective eggs upregulate IL-4, 5 and 10 in BALB/c mice. *Philipp Sci Lett* 2012;**5**(2):139–49.
120. Peng W, Yuan K, Peng G, Qiu L, Dai Z, Yuan F, et al. *Ascaris*: development of selected genotypes in mice. *Exp Parasitol* 2012;**131**:69–74.

15. [illegible] in mice following a primary infection with [illegible]
[illegible] immunity induced in mice by means [illegible]
16. [illegible] resistance to mice [illegible]
and vaccinia virus [illegible]
17. [illegible] The importance of the gap [illegible]
[illegible]
18. [illegible] effect of different [illegible]
[illegible]
19. [illegible] The [illegible] response and acquired [illegible]
[illegible]
20. [illegible]
[illegible]
21. [illegible]
[illegible]

CHAPTER

6

The Neurobiology of *Ascaris* and Other Parasitic Nematodes

*Antony O.W. Stretton**, *Aaron G. Maule*[†]

[*]University of Wisconsin-Madison, Madison, WI, USA

[†]Queen's University Belfast, Belfast, UK

OUTLINE

http://dx.doi.org/10.1016/B978-0-12-396978-1.00006-9

Copyright © 2013 Elsevier Inc. All rights reserved.

INTRODUCTION

Why would anyone interested in the nervous system of nematodes work on *Ascaris*? There are a number of disadvantages that could undermine its selection as a research model: (1) due to the success of anthelmintic treatment of pigs, *Ascaris suum*, the most common species in domestic animals, is increasingly hard to find in host small intestines at slaughterhouses; (2) it is difficult to maintain for more than a few days in good physiological condition in the laboratory; (3) it does not provide a tractable genetic system; (4) it smells bad; and (5) it exudes a powerful allergen that can lead to debilitating asthma. The property that trumps all of these disadvantages is size. Adult female *A. suum* can be up to 35 cm in length. Correspondingly, the neurons are also large, with cell bodies ranging from 5 to 60 μm in diameter, features that have made possible the electrophysiological study of neurons and their synapses by using sharp microelectrodes. In addition, large neuronal size means that the dissection of single identified neurons is relatively easy, such that the biochemistry of single neurons is possible.

At the turn of the 19th/20th centuries, Richard Hesse[1] and Richard Goldschmidt[2] used light microscopy to investigate and to generate anatomical reconstructions of the neuronal tissue of the parasitic nematode *Ascaris* (as was standard practice until the mid-1970s, these worms were designated *Ascaris lumbricoides* even though they were recovered from host pigs – they should now be referred to as *Ascaris suum*) (see Chapter 10) providing the earliest data on its organization in nematodes. Even then, the relative simplicity and consistent arrangement of the nematode nervous system between individual specimens of *Ascaris* was noted. Around 100 years later White et al. published their transmission electron microscopy-based reconstructions of the ventral nerve cord[3] and entire nervous system[4] of *Caenorhabditis elegans*, providing the first opportunities for meaningful comparisons between distinct nematode species and elevating the appeal of this free-living nematode as a model organism for research.

The technical advantages offered by the size of *Ascaris* meant that much of the early electrophysiological and pharmacological understanding of nematode nerve-muscle function was derived from research on this genus. Classical work on the electrophysiology of *Ascaris* muscle was pioneered by del Castillo and colleagues.[5,6] Importantly, Chitwood and Chitwood[7] noted the structural conservation in the nervous systems of many nematodes, including *Ascaris*. White et al.[3] noted the close similarities in the structure of the muscles, and of the nerves that innervate them, between both *C. elegans* and *Ascaris*. Further work on the nervous system of *A. suum* provided more detailed data on its structure and organization and facilitated more meaningful comparisons.[8,9]

These observations on the consistency in nematode nervous system structure between species hold firm to this day and continue to underscore the value of the baseline physiology and pharmacology work on *Ascaris* to the broader understanding of nematode neurobiology and behavior. Despite having diverged in the order of 500 million years ago,[10] the nervous systems of *Ascaris* and *C. elegans* are recognized as virtual scale models of each other.[11] Considering the evolutionary distance between them, this structural conservation is remarkable.

One quirk of nematode parasite control is the predominance of chemotherapeutics which act to disrupt normal neuromuscular function. In spite of the fact that all front-line anthelmintics were discovered empirically, the vast majority have mechanisms of action that center on the dysregulation of normal parasite behavior through the impairment of normal coordinated neuromuscular function.[12] This is most apparent in nematodes whereby drugs such as derquantel, emodepside, levamisole, the macrocyclic lactones, monepantel, piperazine, and pyrantel cause paralysis, reducing parasite ability to survive within the host.[13–19] The fact that many of these agents have broad-spectrum activities across diverse nematode species illustrates the widespread conservation of neuromuscular signaling components, suggesting that functional similarities echo the structural similarities of nematode nervous systems noted by the pioneers of nematode neurobiology. *A. suum* is now well established as a model species for the pharmacological/physiological interrogation of anthelmintic mechanisms of action.[13,20–24]

Since neuromuscular function emerged independently as a useful target resource from the empirical screening programs that identified these drugs in the first place, it holds candidature as the *prima facie* resource for targets for chemotherapeutic interventions in helminth parasites.[12,25] The portfolio of anthelmintic targets sourced from nematode neuromuscular signaling systems is a component of a relatively small number of classical signaling pathways (e.g. acetylcholine, glutamate, γ-aminobutyric acid). This encourages the view that other exploitable parasite control targets involved in distinct signaling pathways associated with neuromuscular function await discovery and provide impetus for research on the neurobiology of these organisms.[26,27] Already established as a key model species for the interrogation of nematode parasite physiology/pharmacology, the publication of the *A. suum* genome sequence has added to its appeal as a model system for nematode research (see Chapter 11).[28] Indeed, these new data elevate *Ascaris* from a surrogate model for physiology/pharmacology studies in nematodes to a tool for discovery biology and candidate target validation.

While there are many aspects of neural function yet to be understood in *Ascaris*, some of the outstanding issues will be emphasized here.

NEURONAL STRUCTURE AND ORGANIZATION

Ascaris and Other Parasites

The *Ascaris* nervous system promised to be ideal for testing the hypothesis that a comprehensive functional description of the "wiring diagram" (or the "connectome"), including the anatomical connectivity and physiological properties of neurons and synapses, would be sufficient to predict behavior. In addition to being large enough for electrophysiological and biochemical experiments, *Ascaris* neurons are few in number (298 in adult females) and exhibit an unusually simple branching pattern. However, after assembling an extensive description of the structure and function of the *A. suum* motor nervous system, and creating a model that predicted the mechanism for generating locomotory behavior, further experiments showed that the hypothesis was wrong; the structural features, together with detailed measurements of the physiological properties of neurons and their synapses, did not correctly predict behavior – something crucial was missing. At least part of what was missing was the influence of neuromodulators that alter the properties of the neurons and/or synapses thereby regulating the functional connectivity of neurons.[29–31] Indeed, *A. suum* possesses a large and complex array of neuromodulators; in addition to small molecular weight transmitters like dopamine and serotonin, there is in excess of 250 neuropeptides. One of these *Ascaris* neuropeptides (designated AF1) was the first to be tested in this way in nematodes; it exerted its paralyzing action by disrupting dorsoventral signaling in the inhibitory motor neurons. Although many other nematode neuromodulators have been studied physiologically, still the majority remain to be explored.

Neuroanatomy

A. suum has a long and distinguished history in neuroanatomy. Hesse[1] and Goldschmidt[2] made detailed studies of the *A. suum* nervous system at the turn of the last century. Hesse described general features of the system, including the head ganglia, nerve ring, and nerves. Goldschmidt studied the neurons in the head, and found that the number of neurons in different individuals was identical (162 in the head ganglia, including the retrovesicular ganglion at the anterior end of the ventral nerve cord). Furthermore, their relative position and morphology (cell body size, shape, and cytoplasmic inclusions) enabled him to recognize each of the neurons in different individuals. He concluded that their nervous systems, at least at this level of analysis, were identical, and he was the first to elaborate the concept of the "Identified Neuron," which has been so important in cellular neurobiology, especially in invertebrates. Of course, Goldschmidt used light microscopy, and illustrated his findings with drawings. The descriptions of the neuronal cell bodies are remarkably accurate and still valid. He also attempted to trace neuronal processes within the nerve ring and nerve cords. In most cases he did not connect identified nerve processes with identified cell bodies, and these diagrams are not informative. In addition, he saw areas where the processes of different cells apparently "melted" together and he interpreted these as sites of cytoplasmic continuity between the neurons, in support of the reticular theory of the nervous system (along with Golgi, in joint opposition to Ramon y Cajal and his followers who believed in the Neuron Doctrine). Ultrastructural datasets generated using electron microscopy are not consistent with this explanation.[9] Areas of apparent "melting" are sites of synaptic contact, where the membranes of neighboring neuronal processes have complex interdigitations so that the membranes appear indistinct as they are observed in the light microscope at different focal planes within a section. When these same sections are re-sectioned for electron microscopy, they have the structure of typical chemical synapses.

Other investigators reported on other aspects of the nervous system. Voltzenlogel described the 30 neurons of the tail ganglia.[32] Hesse[1] described the pattern of dorsoventral commissures in the rest of the body; Otto[33] recognized a chain of neuronal cell bodies, presumed to be motor neurons, in the ventral nerve cord. To describe the morphology of the motor neurons in the ventral and dorsal nerve cords required the analysis by light microscopy of serial sections of complete worms.[8] Even for relatively short individual worms, it still required the analysis of 10,000 serial 10 μm sections per worm. Reconstructions revealed a repeating pattern of motor neurons, comprising a basic set of five repeats of 11 motor neurons. In addition, the anterior and posterior ends of the ventral cord depart subtly from the basic repeat pattern, and "extra" neurons are present, so that the ventral cord includes a total of 72 motor neurons in females. Within the basic 11 neuron repeat, there are seven types of motor neuron, and light microscopy and subsequent electron microscopy showed that four types innervate dorsal muscle and three types innervate ventral muscle. The neuronal commissures, which connect the ventral cord and dorsal cord, have been crucial experimentally, as described below.

Finally, there are 14 neurons in the two lateral lines, and 20 neurons in the pharynx, giving a grand total of neurons in female *A. suum* of 298 (Figure 6.1), close to the total of 302 found in the *C. elegans* hermaphrodite. The four additional neurons in *C. elegans* are sensory neurons in the head.

The neurons give rise to neurites that project into the nerve ring and into several longitudinal nerve cords. The major nerve cords are the ventral and dorsal cords (VC and DC) which contain the axons that make neuromuscular synapses onto the ventral or dorsal musculature,

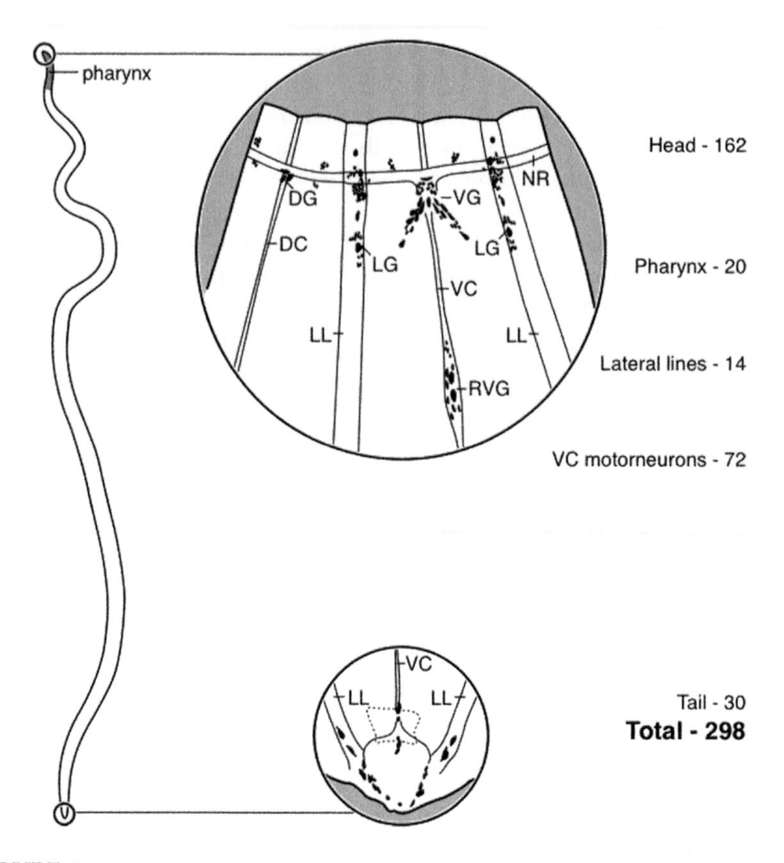

FIGURE 6.1 Female *Ascaris suum* in locomotory posture. Position of head ganglia and tail ganglia are circled. The extent of the pharynx is indicated by gray shading. Panels to right show neuronal cell bodies (filled profiles; their relative size is to scale) in anterior ganglia and in tail ganglia. Dotted area in the tail indicates the rectum. Numbers of neurons in head ganglia, pharynx, lateral lines, ventral nerve chord, and tail are indicated, for a grand total of 298 neurons. The total for *Caenorhabditis elegans* is 302[4]. NR, nerve ring; VG, ventral ganglion; LG, lateral ganglion; DG, dorsal ganglion; LL, lateral line; VC, ventral nerve cord; DC, dorsal nerve cord: RVG, retrovesicular ganglion.

respectively, and produce the mostly dorsoventral muscle contractions of the propagating body waveform during locomotion. The major nerve cords also include interneurons, just two in the DC, and about 30 in the ventral cord (depending on the position along the animal). There are also four sublateral nerve cords (two subdorsal and two subventral) that are close to the lateral lines. They give rise to dual innervation of anterior muscles,[34] and are believed to generate three-dimensional movement of the anterior part of the worm. The sublateral cords gradually taper in diameter and eventually terminate near the vulva; the density of neuromuscular connections similarly declines, corresponding with the observed gradual reduction in the amplitude of left or right movements from the head posteriorly. There are also lateral nerves that extend from head to tail along each lateral line.

The Importance of Motor Neuron Commissures

The dorsoventral commissures in the body of *A. suum* had been noticed by Hesse,[1] but it was only after reconstruction from serial sections that the individual types of neurons were recognized, and it was shown that the dorsoventral commissures were all branches of motorneurons.[8] There are seven types that occur in five repeating subunits, named "segments," along the length of the worm. Anatomical and electrophysiological experiments showed that there were both excitatory and inhibitory motor neurons, and that there were separate motor neurons controlling the dorsal and ventral muscles. The dorsal motor neurons were named DE1, DE2, DE3, and DI, and the ventral motor neurons VE1, VE2, and VI. Each segment contains 11 motor neurons, with four of the neuronal types (DE1, VE1, VE2, and VI) present twice in each segment, and DE2, DE3, and DI only once. The cell bodies of all these motor neurons are in the ventral nerve cord; five types of neurons (DE1, DE2, DE3, DI, and VI) have dorsoventral commissures, so there are seven commissures per segment. Extensive anatomical studies at both light and electron microscope levels showed that all the motor neurons had output regions where they synapsed onto muscle and other neurons, and input regions (dendrites) where they received synapses from other neurons. For neurons with commissures, these two regions were in different nerve cords. Another interesting finding was that the sole input to inhibitory neurons was from the opposite excitors (DE1, DE2, and DE3 to VI, and VE1 and VE2 to DI). These synapses mediate reciprocal inhibitory activity between dorsal and ventral musculature.

Physiology of Neurons

The early physiological characterization of the motor neurons made use of the anatomical reproducibility of the position of the commissures, so that dissected preparations could be made in which the dorsal and ventral halves of the animal were linked only by a single commissure (validated by serial section reconstruction). This allowed single neurons to be stimulated and their effects on muscle to be recorded, leading to the initial characterization of dorsal/ventral and excitatory/inhibitory motor neurons.[8] In 1989, Ralph Davis made a crucial contribution;[35] in a slit-open piece of the worm he carefully dissected away the muscle cells that overlie a commissure to allow microelectrode penetration for recording and/or stimulation. He did control experiments to show that the integrity of the neurons was not compromised by the dissection, which was important since one of his key discoveries was that the motor neurons did not

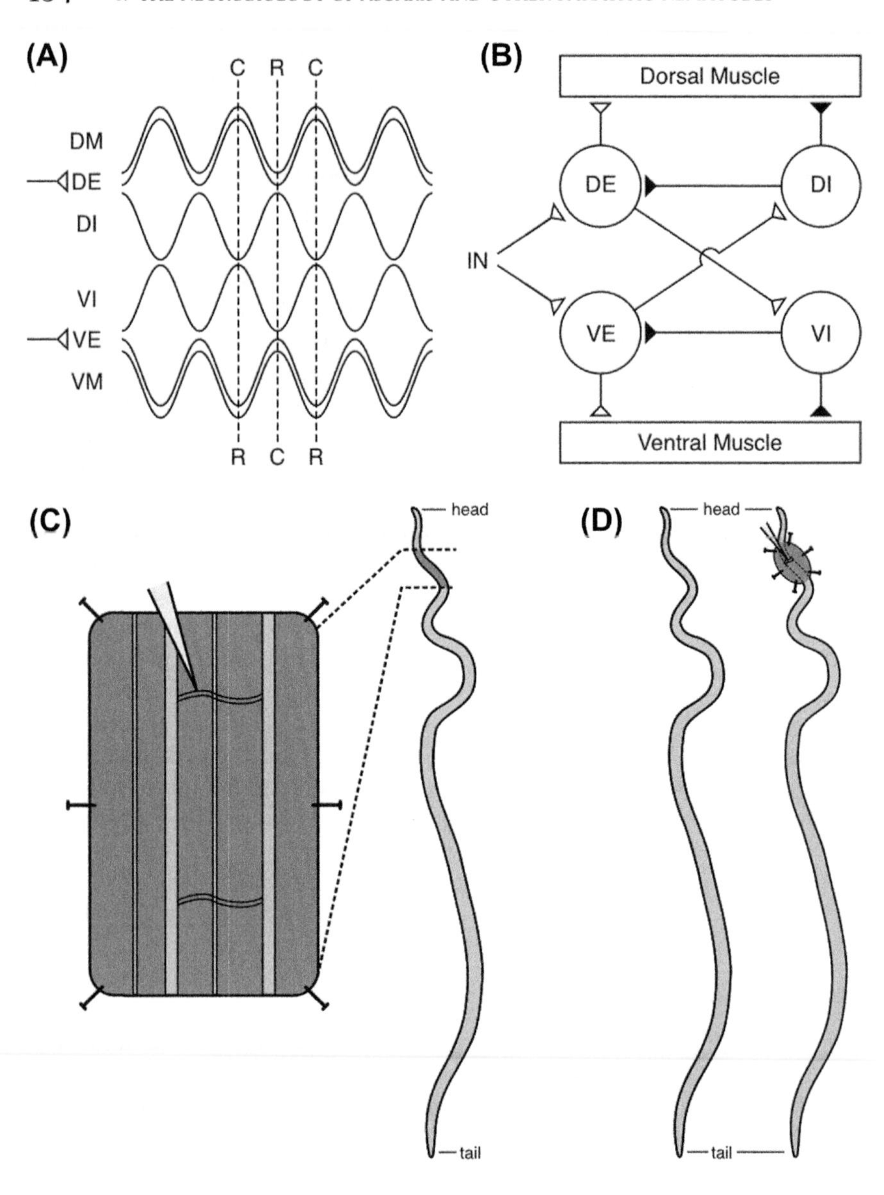

conduct action potentials, despite their controlling muscle cells several centimeters distant. Long distance signaling in these neurons is passive, and depends on a very high specific membrane resistance (between 70 and 250 $\Omega.cm^2$), so that the space constants are of the order of 1 cm (graded electrical signals decay exponentially, and the space constant determines the extent). Signaling is analog rather than digital (in most other nervous systems, information is coded digitally by the frequency of action potentials). Analog signaling is also seen in synaptic transmission, which is graded, and there is tonic release of transmitter.[36]

Recordings using patch electrodes (electrodes that form seals on the surface of the membrane rather than being inserted through it, as with sharp electrodes) in *C. elegans* show that the membrane resistance is also high; this means that in neurons as small as those in *C. elegans*, the cells are essentially iso-potential, and action potentials may not be needed.[37] Analog signaling and synaptic transmission fit well with such properties. It is perhaps more surprising that large nematodes function well without having developed action potentials for long distance signaling. This may be a factor that contributes to the upper limit in the size of nematodes.

Pairwise recordings from motor neurons showed the neuron–neuron synapses. Besides the input to inhibitory motor neurons from their opposite excitors (DE to VI, and VE to DI) mediating reciprocal inhibition as mentioned above, there are synapses from inhibitors to the same side excitors (VI to VE, and DI to DE; Figure 6.2B). Taken together with another property of VI and DI motor neurons, namely that they generate slow oscillations of the membrane potential when they are depolarized,[36,38] this led to a hypothesis of how the alternating dorsal and ventral contractions that characterize *A. suum* (and indeed nematode) locomotory behavior are generated (Figure 6.2A,B). Once a dorsal excitor is depolarized (by input from interneurons) it will depolarize VI and cause it to oscillate. The inhibitory synapse from VI to VE will make VE oscillate out of phase with VI, and the excitatory synapse from VE to DI will make DI respond with oscillations in antiphase to VI; similarly the inhibitory synapse from DI to DE will make DI and DE oscillate in antiphase. Peaks of DE oscillations should produce dorsal contractions and ventral relaxations, and peaks in VE oscillations should produce ventral contractions and dorsal relaxations, and the oscillations will produce alternating dorsal and ventral contractions, which interchange with the characteristic frequency of the

FIGURE 6.2 **Model for generation of locomotory waves, and dissections used for electrophysiological testing of model. A**. Predictions of alternation of contractions and relaxations in dorsal muscle (DM) and ventral muscle (VM), and activity in dorsal excitatory (DE) motor neurons, dorsal inhibitory motor neurons (DI), ventral excitatory motor neurons (VE), and ventral inhibitory motor neurons (VI). **B**. Synaptic connectivity between motor neurons and neuromuscular synapses. Open triangles, excitatory synapses; filled triangles, inhibitory synapses. The sole input to inhibitory motor neurons is from the opposite excitors. Excitors receive excitatory input from interneurons (IN). **C**. Fully dissected preparation for recording from ventrodorsal commissures. A short length of worm is removed (dotted lines) and slit open longitudinally and pinned flat. Intracellular microelectrode recordings are made from commissures. **D**. Semi-intact preparation for correlating electrical activity in commissures of motor neurons with body waveform propagation. A small region of the body wall is slit open and pinned flat to allow recording from commissures. The rest of the body is free to generate locomotory movements.

oscillations (Figure 6.2A). Supporting this hypothesis, pairwise recordings from DI and VI, and of DI and DE neurons, showed that they oscillated in antiphase, as predicted by the hypothesis. These recordings were made in pieces of worm 3–5 cm long, continuously perfused with *Ascaris* artificial physiological saline solution at 37°C (Figure 6.2C).

A more rigorous test was aimed at correlating the oscillatory potentials with the propagating behavioral waveform. A semi-intact preparation of a whole worm was made.[30] A small slit in the body wall allowed microelectrode penetration of chosen motor neurons, while the rest of the worm was free to move within a narrow trough (Figure 6.2D). The results were compelling in that the individual oscillations in the motor neurons do not correspond with separate body bends – the oscillations occur at about 10 times the frequency of the propagating waveform, and it is bouts of oscillations rather than individual oscillations that correlate with behavior.[30] The conclusion is that the description of the properties of the neurons and their synapses that were incorporated into the hypothesis is incomplete. The same neurons in the fully dissected and perfused preparation, and in the semi-intact preparation, have different activities. One obvious source of these differences is the presence and activity of neuromodulators, which would most likely be washed away in the fully dissected and perfused preparation. Many neuromodulators have potent activity on *Ascaris* motorneurons,[39] so it is clear that their effects must be incorporated into the next hypothetical model to be tested. Developing models that accurately encompass the diverse effects of a changeable mix of neuromodulators remains a major challenge. For example, it is already known that single neurons can contain multiple peptides or combinations of peptides with other modulators, and these are likely to be co-released to act on distinct receptors on the target cell(s). Little is known about the physiological consequences of such a mix of neuromodulators acting on individual nerve cells. Furthermore, the release of modulators into the circulating pseudocoelomic fluid (PCF), where they would act as neurohormones, is strongly supported by the morphology of certain neurons. Clearly, understanding which neuromodulators and/or neurohormones are actually released onto the target neurons and establishing the impact of these physiological interactions needs to be investigated in the future.

NEUROMODULATORS AND SIGNALING

Classical Transmitters and their Receptors

While much data on classical neurotransmitter occurrence in nematode parasites are available, it has been reviewed previously and will not be repeated here.[40–42] The largest and most diverse group of classical neurotransmitter receptors in nematodes is the cysteine-loop ligand-gated ion channels (cys-loop LGICs) which comprise both anion and cation conducting receptors responsible for fast synaptic transmission and formed by the convergence of five homologous membrane protein subunits to encase the channel pore.[43] Most knowledge on these ion channels has been garnered from work on *C. elegans* which possesses in excess of 100 LGIC subunit protein encoding genes. This remarkable subunit complexity facilitates the generation of a diverse array of receptors that can be gated by a variety of different classical transmitters. Further complexity derives from the fact that cys-loop LGICs that conduct cations can be selectively gated by either acetylcholine (ACh) or γ-aminobutyric acid (GABA) and those conducting anions can be selectively gated by ACh, GABA, glutamate or serotonin, providing for a bewildering array of LGIC receptor subtypes. It is remarkable that *C. elegans* boasts the greatest diversity of ionotropic ACh receptor subunits known in any animal. A relatively constricted, yet diverse, set of 44 putative cys-loop LGICs were reported from the filarial nematode *Brugia malayi*[44] with only 19 identified in *Trichinella spiralis*.[45]

The primary excitatory neurotransmitter in nematodes is ACh which plays a central role in nematode locomotion; the predominant receptors are muscle-based cys-loop LGICs that resemble classical ionotropic nicotinic receptors of vertebrates. While we know little of the ionotropic ACh receptor complements in parasitic nematodes, some differences in subunit expression have been reported.[46] Pharmacological studies on anthelmintic action and resistance have generated much of the impetus for research on these receptors in parasites, with the levamisole- and pyrantel-sensitive receptors attracting the most intense scrutiny.[47,48] Truncated subunits were identified in three trichostrongylid nematodes (*Haemonchus contortus*, *Teladorsagia circumcincta*, and *Trichostrongylus colubriformis*) and correlated with levamisole resistance.[47] The newest anthelmintics, the amino-acetonitrile derivative monepantel and derquantel, the semi-synthetic derivative of paraherquamide, also act at ionotropic ACh receptors and further stimulate interest in parasite ion channel biology.

Following the pioneering work of del Castillo and colleagues,[5] a primary inhibitory role in nematode locomotion has been attributed to GABA through its action on ligand-gated chloride channels (LGCCs). As with the ionotropic ACh receptors, features including distinct pharmacology and the actions of the anthelmintic piperazine stimulated interest in these inhibitory receptors which are now known to display the subunit diversity that typifies LGICs in nematodes.[43,49]

Distinct LGCCs are now believed to be gated by ACh, GABA, and glutamate as well as various biogenic amines including dopamine, octopamine, serotonin, and tyramine. A dopamine-sensitive LGCC, designated HcGGR3, was characterized from *H. contortus* and was found to be expressed in mechanosensory cells associated with the cervical papillae.[50] Curiously, quantitative polymerase chain reaction methods revealed that the transcript for this channel was downregulated in two ivermectin-selected laboratory strains of *H. contortus* compared to a genetically-related unselected strain. Although macrocyclic lactone treatment seems to apply some form of selective pressure on this channel, no direct link with anthelmintic resistance has been established. All *H. contortus* life stages were also found to express a homologue of the *C. elegans* tyramine-gated ionotropic receptor which, following functional expression, was shown to be sensitive to dopamine, octopamine, and tyramine.[51] While subunits of many different ionotropic receptors have been identified in parasites,[49] we are a long way from understanding the roles they play in parasite biology.

Additional layers of complexity in classical transmitter signaling include their selective activation of the metabotropic seven transmembrane G-protein coupled receptors (GPCRs) that act via guanine nucleotide-binding proteins (G-proteins), so-called molecular switches that facilitate the translation of extracellular binding of ligands to intracellular second messenger signaling cascades. G-protein coupled ACh receptors in nematodes are designated GARs and three subtypes are known in *C. elegans*. Data on GARs in parasitic worms are limited to reports of atypical pharmacology and the characterization of GAR-1 from *A. suum*.[52] A number of serotonin GPCRs has been characterized in *C. elegans*, but there is little information on homologous receptors in parasites. Seven biogenic amine receptors were reported from the genome of *B. malayi* including three serotonin, two dopamine, and two tyramine/octopamine receptors.[44,53] The functional expression of a tyramine GPCR from *B. malayi* (Bm4) revealed it to be pharmacologically distinct from its closest *C. elegans* homologue (TYRA-2).[54] Although in comparison to *C. elegans*, a greatly restricted complement of GPCRs were reported from the root knot nematode *Meloidogyne incognita* genome,[55] much of this constriction appears to be due to the absence of olfactory receptors which form a greatly expanded cohort in *C. elegans*. In the related species *Meloidogyne hapla*, putative receptors for ACh, dopamine, glutamate, octopamine, serotonin, and tyramine, have been identified.[56]

Ascaris

Another important feature of the motor neuron commissures is that they are branches of identified neurons, and single commissures can be dissected for biochemical analysis. This dissection was developed by Carl Johnson,[57] who also devised an assay for choline acetyltransferase (ChAT, the biosynthetic enzyme for acetylcholine [ACh]) that was sensitive enough to measure ChAT in single commissures. Dissected commissures included a small area of surrounding hypodermis, so pieces of hypodermis of equal area were assayed as controls, and low levels of ChAT were found in hypodermis. The presence of an excitatory commissure increased the ChAT level dramatically, whereas there was no increase contributed by an inhibitory commissure. An interesting finding that came out of these experiments was that the concentration of ChAT in the hypodermis progressively increased the further anteriorly it was sampled. Analysis of serial frozen sections showed that the peak activity was at the tip of the head, anterior to the nerve ring. Tentatively this was ascribed to the arcade cells of the head. The role of ChAT in the head has not been further explored, although it may be responsible for release of ACh into the PCF, and contribute, along with ACh tonically released from excitatory motorneurons, to basic muscle tone, possibly by interacting with non-synaptic ACh receptors on muscle.[58]

The issue of whether there are chemical components of the PCF that affect the properties of neurons and/or muscle cells also arises when considering the role of neuromodulators in the control of behavior, and needs to be more fully addressed.

Since del Castillo and colleagues showed that muscle was hyperpolarized by exogenous GABA,[5] this was believed to be the transmitter in inhibitory motor neurons of *A. suum*. A GABA-specific antibody showed that GABA is concentrated in the commissures, cell bodies, and nerve cord processes of DI and VI motor neurons, in strong support of del Castillo's hypothesis.[59] There are other neurons that are GABA-immunoreactive in *A. suum*,[60] and comparison with *C. elegans* GABAergic neurons revealed a robust difference in cellular expression in the two species. In both species the dorsal and ventral inhibitory motor neurons (DI and VI in *A. suum*, and DD and VD in *C. elegans*) are GABA-immunoreactive, as are the four RME neurons of the nerve ring and the DVB neuron in the tail. In each species there are two GABA-immunoreactive neurons in the ventral ganglion. However, in *C. elegans* these neurons are the unpaired RIS and AVL neurons,[61] whereas in *A. suum* the GABA-immunoreactive neurons are the paired AIY or AIM neurons (these neuronal types have not been distinguished in *A. suum*).

There is a further feature of the GABAergic motorneurons that is not yet understood functionally, namely the presence of branches from the DI and VI commissures that spread along the sublateral cords as they cross them.[59] These branches receive synapses from excitatory axons in the sublateral cords. Remarkably, there are different connections with DI and VI motor neurons, such that VI neurons receive synapses from dorsal sublateral axons, and DI neurons from ventral sublaterals. It is not at all clear how important these connections are in affecting the control of movement by the DI and VI neurons.

Other Transmitters in Ascaris

In female *A. suum*, serotonin (5-HT) is present in one pair of neurons (NSM neurons) in the pharynx.[62,63] These neurons are very atypical of nematode neurons in that they are highly branched. They extend multiple varicose endings to the outside surface of the pharynx; by electron microscopy these varicosities are packed with dense-core vesicles, and have no postsynaptic partners. Morphologically they seem to be neurosecretory neurons, releasing serotonin into the PCF. They have an especially high density of endings immediately under the nerve ring.[63] When 5-HT is injected into intact worms, it produces a rapid paralysis. An unresolved question is whether there is tonic release of 5-HT, and therefore whether in the circulating PCF there is a basal level of 5-HT that contributes to muscle tone, or other aspects of behavior.

Glutamate is another small molecular mass transmitter that is important in *A. suum*;[64,65] DE2 motor neurons receive glutamatergic excitatory input from small interneurons that conduct calcium-mediated action potentials in the VC.[66] The pharmacology of this system explains the potency of kainic acid, domoic acid, and quisqualate as anthelmintics that have long been used in Asia.[67] These compounds are also well-known tools in the study of glutamatergic mechanisms in vertebrates.

Neuropeptides and their Receptors

Nematode neuropeptides have been divided into three groupings, two of which form distinct peptide families (the FMRFamide-like peptides or FLPs and the insulin-like peptides or INSs) and one which includes at least 11 distinct families (the neuropeptide-like proteins or NLPs). The theme of structural diversity that pervades the components of nematode classical neurotransmitter signaling is mirrored within nematode neuropeptide signaling systems with each family encompassing large numbers of distinct peptides (~70 FLPs, ~40 INSs, and ~70 NLPs). Assuming that all of these ligands are involved in intercellular communication from neurons, there is much scope for a complex signaling milieu. In contrast to the ligand diversity inherent to neuropeptide signaling in nematodes, the numbers of known neuropeptide receptors are more limited with initial trawls recovering 54 candidates from the *C. elegans* genome,[68] although the impact of splice variation (whereby a single gene encodes multiple proteins through alternative splicing) could provide for additional complexity.[69]

Ascaris Neuropeptides – Discovery and Cellular Localization

The variety of neuropeptides in *A. suum* is impressive. The first nematode peptides to be characterized were the AF peptides (*Ascaris* FMRFamide-like peptides or FLPs), isolated chemically from crude peptide extracts, and sequenced by Edman degradation.[30,70–72] Although effective, this approach required large amounts of material (e.g. from 60,000 worms) and extensive and time-consuming purification procedures. Since then, expressed sequence tag (EST) and genomic sequences have led to the prediction of precursor proteins that include many more putative peptides, including more FLPs,[73] and the NLPs which may or may not have C-terminal amidation.[74] Since the precursor proteins are cleaved by proteases in the secretory pathway, and different cells express different proteases, it is important also to detect the peptide product itself, since predictions from the common cleavage sites (typically single or pairs of basic amino acids) may be erroneous. Alternative splicing is also an issue that affects the interpretation of peptide-encoding genes found in genomic DNA.

Peptide discovery has recently been vastly aided by the use of mass spectrometry (MS). Initially, matrix assisted laser desorption ionization-time of flight (MALDI-TOF) MS was used on dissected ganglia to detect peptides from single ganglia.[75] In this technique, there is the simultaneous vaporization and ionization of the molecules in the sample; the released ions are then accelerated by a voltage gradient and the time taken to drift down a flight tube to a target that detects each ion is proportional to the momentum of the ion, which is in turn dependent on the mass to charge ratio. Tandem MS, in which selected ions are fragmented, also allows *de novo* sequencing, which can also be carried out on the peptides from single ganglia. In other systems, it was already possible to carry out MS on single neurons,[76,77] but in *A. suum*, the isolation of single neurons was difficult because of their tight association with surrounding hypodermal tissue, which acts as the glia (the non-neuronal tissue that surrounds neurons in most higher organisms) in nematode nervous systems. Finally, the dissection technique for isolating single neurons was perfected, and subsequent MS and MS/MS together facilitated the discovery of many novel peptides.[78] About 25% of the different types of neurons in *A. suum* have been examined so far, and all contain neuropeptides; when the transcripts encoding these peptides are examined, often it is clear that the peptides in a single neuron are the products of multiple genes. Many of the peptides are novel, and in many cases the previously unknown peptides outnumber the known peptides. One of the additional advantages of single cell MS is that besides the chemical identification of the processed peptides in a neuron, the cellular localization is simultaneously described. This is usually extended and validated by other techniques such as immunocytochemistry (if specific antibodies can be generated) and/or *in situ* hybridization, so that the full expression pattern of peptides and their encoding transcripts can be determined. With very rare exceptions, the three techniques give identical results. Although none of these three techniques is quantitative, there are consistent differences in the observed results, so that in a particular neuron, some peptides and transcripts are strongly and consistently expressed, and others are weakly but consistently expressed. In other cases a peptide or transcript may be very weakly expressed to the extent that there are some samples of that neuron that do not reveal any detectable gene product. The significance of this variability is unclear. It is possible that the differences generate differences in behavior, but it is also possible that the neural circuitry is robust against such variations – comparable variability, yet stability in overall function, has been reported in the stomatogastric ganglion of crustacea.[79]

Bioactivity of Ascaris Neuropeptides

Several different assays have been used to assess the physiological effects of peptides in *A. suum*: (1) injection of peptide into the intact worm, and observation of effects on locomotory behavior and body length, either in unconstrained worms,[39] or in worms in a glass tube (18 mm internal diameter) after the worms were ligatured approximately 5 mm posterior to the pharynx – this procedure induced regular anterior propagating waveforms (the basis for this response is not understood; similar effects are induced by decapitation, so perhaps removal of the head ganglia and/or the pharynx and constriction of the "neck" have mechanisms in common);[80] (2) muscle strip assays where the effects of peptides on muscle contraction are either due to the peptide alone or on the amplitude of ACh-induced contraction; (3) assays of peptide-induced changes in cyclic AMP and GMP;[80,81] (4) electrophysiological changes in motorneurons;[39] (5) contraction of the *A. suum* ovipositor/ovijector;[82,83] and (6) effects on spike activity and contraction in the *A. suum* pharynx. The currently available results of these assays have been very recently reviewed and will not be further summarized here.[84]

An additional imaginative assay was developed by Martin, who noticed blebs (irregular bulges of the plasma membrane) that appeared after collagenase treatment of *A. suum*, and showed that they were large vesicles of muscle cell membrane that were suitable for patch recording, and characterization of single channel properties.[85] These preparations, or preparations in which the membranes of muscle cell bags were cleaned by enzyme digestion with collagenase, have been used to show the effects of AF2 (KHEYLRFamide) on voltage-activated channels in *A. suum* muscle.[86]

The majority of the peptides tested for their effects on behavior lead to paralysis, or drastic reduction in the amplitude and frequency of locomotory waveforms. One set of peptides has very different effects – the six peptides AF3, AF4, AF10, AF13, AF14, and AF20, all products of the *afp-1* (*As-flp-1*) transcript, and sharing a PGVLRFamide C-terminus, produce exaggerated high amplitude waveforms, and the waveforms extend further posteriorly than in a control worm.[80] When extended to the effects of peptides on the electrophysiology of motor neurons, the number of categories of different effects is raised to about seven.[39] However, taken together, the total number of categories of different biological effects observed is considerably less than the number of peptides, suggesting either that there may be functional redundancy, or that the number of biological tests is still too small. There are many potential targets that have not yet been tested electrophysiologically, and to test all neurons with all known peptides is scarcely feasible. It is also certain that the concentration of the ligand and the affinity of receptor(s) are important functionally.

Among the many aspects of neuropeptide function that are currently unknown are the time course of the release of peptides from the cells that synthesize them. In some cases, e.g. the effects of AF1 (KNEFIRFamide) on inhibitory motor neurons, there is close contact (synapse-like) between the interneurons in the DC and VC that contain AF1 and the inhibitory motor neurons that are short-circuited by the peptide.[87] In other cases there is apparently secretion into the PCF, as mentioned above for the serotonin-containing neurons of the pharynx. It would be interesting to assay the PCF for its levels of neuromodulators, including peptides and biogenic amines, and to determine whether they change with behavioral state, or developmental state.

The locomotory behavior of the L3 larva and the adult are very different, with the L3 larva behaving like *C. elegans*: posteriorly propagating waves drive the animal forwards, and anteriorly propagating waves drive it backwards. This is the opposite in adult *A. suum*; in a tube of dimensions like that of a pig small intestine, it is anteriorly propagating waves that lead to anterior locomotion. In addition, the locomotory waveform of adult *A. suum* contains three waves, whereas that of *C. elegans* has a fraction more than a single wave. The *A. suum* L3 has only a single wave, although the number of neurons is similar to that of the adult. One possibility is that there are changes in modulator levels that alter the physiological properties of parts of the motor nervous system or of muscle. Another enigma is that the anteriorly propagating waves in adult *A. suum* are initiated near the vulva; the mechanism is unclear.

Neuropeptides in Other Parasites

The peptide extraction and biochemical characterization approach that worked so well in *A. suum* was much less evident in other parasites, with discovery confined to a total of three FLPs (KHEYLRFamide, KSAYMRFamide, and LQPNFLRFamide) that were structurally characterized from the third larval stage of *H. contortus* with KHEYLRFamide being the most abundant peptide detected.[88–90]

A second method of parasite neuropeptide discovery relied on the identification of predicted peptides from cloned and sequenced gene transcripts. Five *flp* genes have been characterized from second (infective) stage larval *Globodera pallida* (potato cyst nematode) including *Gp-flp-1*, *Gp-flp-6*, *Gp-flp-12*, *Gp-flp-14*, and *Gp-flp-18*.[91] Another four *flp* genes were characterized from the root knot nematode *Meloidogyne incognita* (*Mi-flp-1*, *Mi-flp-7*, *Mi-flp-12*, *Mi-flp-14*); *Mm-flp-12* was also identified from *Meloidogyne minor*.[92] While these efforts yielded novel information on neuropeptide genes in parasites, they did not allow meaningful comparisons between the neuropeptide complements of parasites and *C. elegans*. Broader guidance on the diversity of *flp* genes in nematode parasites was provided from a bioinformatics screen for *flp* transcripts from expressed sequence tag (EST) data for 32 nematode parasites. These efforts identified 31 *flp* genes in parasites, 29 of which had orthologs in *C. elegans*.[73] While the *flp* gene complement recorded for any one parasite was markedly smaller than *C. elegans*, the vast majority of *C. elegans flp* genes had orthologs that occurred in some nematode parasite species.

More recently, genome-wide neuropeptide gene surveys have been completed for *M. incognita* and the migratory endoparasite *Bursaphelenchus xylophilus* which causes pine wilt disease.[55,93] The former reported the *Mi-flp* and *Mi-ins* gene complements whereas the latter reported the *Bx-flp*, *Bx-ins*, and *Bx-nlp* gene complements. Both of these plant parasitic nematodes display constricted *flp* gene diversity with 19 (*M. incognita*) and 21 (*B. xylophilus*) *flp* genes compared to the 29 *flp* genes in *C. elegans*.

While no NLPs have been structurally characterized from parasites other than *Ascaris*, a screen of parasite transcriptomic data uncovered orthologs of 26 of the 46 *C. elegans nlp* genes.[73] The *Tc-nlp-12* transcript has been characterized from the sheep parasite *Trichostrongylus colubriformis*, with expression identified in the tail and appearing to match that for the *C. elegans* ortholog.[94] Genomic screens of *M. incognita* and *B. xylophilus* recovered sequence homologs for 21 and 17 of the *C. elegans nlp* genes, respectively. While these datasets reveal complexity in the parasitic nematode *flp* and *nlp* gene complements, they are similarly reduced in comparison to *C. elegans*. Data on parasitic nematode *ins* genes is restricted to the genomic survey of *B. xylophilus* which identified seven *ins* gene orthologs, dramatically fewer than the 40 genes (39 *ins* genes and *daf-28*) identified in *C. elegans* and suggesting even more profound constriction of *ins* family genes or structural divergence that prevented discovery using *C. elegans*-derived search strings.

The Role of Neuropeptides in Other Parasites

As with *Ascaris*, most data on neuropeptide function in other parasites relate to FLPs. Two *flp-18* peptides (AF3, AVPGVLRFamide, and AF4, GDVPGVLRFamide) increased spontaneous phasic activity, induced body wall muscle contractions and decreased cyclic AMP levels in *A. galli*, effects that were independent of cholinergic signaling.[95] Spontaneous and

ACh-induced contractions of *H. contortus* were found to be enhanced by injecting KSAYMRFamide (AF8) whereas KHEYLRFamide (AF2) injections inhibited ACh-induced contraction. An additional observation which involved testing AF8 on *H. contortus* isolates that were susceptible or resistant to the cholinergic agonist levamisole revealed differential sensitivity to the peptide, suggesting that its actions involve cholinergic signaling.[90] FLRFamide was reported to increase the body movements of infective stage (J2) *M. incognita*.[96]

The application of reverse genetic approaches to parasitic nematodes through the application of RNA interference (RNAi) has provided platforms to interrogate the consequences of silencing the expression of neuronal targets such as neuropeptides. Using RNAi, a small number of *flp* genes have been silenced in infective stage *G. pallida* J2s (*Gp-flp-1, Gp-flp-6, Gp-flp-12, Gp-flp-14,* and *Gp-flp-18*) and *M. incognita* J2s (*Mi-flp-18*), in every case resulting in locomotory phenotypes including various forms of reduced migrational competence or paralysis;[97,98] after 24 hours of RNAi-induced *flp* gene silencing, J2 stage migration through sand columns was reduced between 71 and 100%. Recently, RNAi of the *Gp-flp-32* gene phenocopied RNAi of a putative *Gp-flp-32* receptor gene in infective stage worms, making them migrate faster and appearing to have no detrimental impact on plant infection.[99] While it seems logical to hypothesize that the silencing of a neuropeptide (or its cognate receptor) that inhibits normal movement would result in "faster" worms, the rationale for a parasite using a peptide in this way whenever there were no observed detrimental effects of being faster is less obvious. This was the first study to functionally deorphanize a parasite neuropeptide receptor using reverse genetic approaches in whole worms.

FUTURE PERSPECTIVES AND PARASITE CONTROL

The control of nematode parasites in humans and animals has relied heavily on drugs that compromise normal neuromuscular function. Most typically, these anthelmintics bind to ionotropic receptors that are gated by classical neurotransmitters such as ACh, GABA or glutamate. Overwhelming evidence demonstrates that the disruption of normal neuromuscular function in nematode parasites is devastating for these worms, compromising their viability as parasites and pathogens.

The large numbers of different nematode parasites that cause disease mean that another expectation of drugs for the treatment of nematode parasite infections is broad spectrum activity. Fortunately, many key components of nematode neuromuscular function are well conserved across diverse parasite species, providing opportunities for drugs which disrupt this facet of nematode biology to have broad spectrum activities.

Indeed, any doubt in the value of nematode neuromuscular function as a resource for anthelmintic targets is dispelled by the fact that the three most recent additions to this drug arsenal (derquantel, emodepside, and monepantel) are embedded within it.

While recent additions to the anthelmintic portfolio are extremely welcome, this cannot be allowed to encourage complacency in efforts to diversify the parasiticide arsenal. In addition to remaining gaps in available anthelmintic therapies, an inevitable consequence of the repeated use of drugs of this kind is resistance. Drug resistance in nematodes of livestock is well established and continues to undermine animal health and food security. Further, the expanding mass distribution and treatment approaches to the control of human helminthiases heighten the threat of resistance in nematode parasites of medical importance (see Chapter 15). While vaccines fail to meet nematode parasite control needs, anthelmintics will continue to shoulder the control burden and drug resistance will ensure the relentless erosion of these control options. Therefore, sustaining effective control of nematode parasites demands the continual addition of new mechanism of action anthelmintics to the pharmaceutical arsenal. The potential of LGCCs and of diverse elements of neuropeptide signaling in nematode parasites as targets for future anthelmintics is particularly compelling and has been reviewed recently.[27,49]

As the leading model for research on nematode neuromuscular physiology, *A. suum* is well placed to continue to inform research on anthelmintics. The recent completion of its genome[28] and published reports of its amenability to RNAi[100,101] provide clear opportunities to expand the utility of *Ascaris* beyond that of a model to inform basic neuroscience to a platform for the discovery and validation of new anthelmintic targets. RNAi-based reverse genetic approaches in animal parasitic nematodes have proven unreliable[102–106], undermining the exploitation of this technology to inform gene function and target validation. Several studies report successful RNAi in larval stage *A. suum* indicating that the gene silencing effectors are functional in *Ascaris*, at least in the larval stages.[100,101] Recent work has demonstrated robust RNAi of gene targets in neuronal and non-neuronal tissues of adult female *A. suum* (unpublished observations, Neil Warnock, Ciaran McCoy, Nikki Marks, and Angela Mousley, Queen's University Belfast), providing much hope that *Ascaris* can become a powerful tool for the functional genomics of animal parasitic nematodes.

It seems highly likely that the key advantages offered to neurobiology research by *Ascaris* will ensure that it remains at the forefront of neurobiology research for some considerable time. Among parasites, *Ascaris* provides a totally unique platform for the interrogation of new components of neuromuscular signaling, informing the mechanisms of action of new anthelmintics, facilitating the discovery of novel signaling molecules and enabling single neuron physiology and biochemistry. Combining the advantages of being able to examine the electrophysiology and biochemistry of single neurons with the opportunities to silence selected gene transcripts using RNAi would provide a powerful platform for parasite gene function studies at the single cell level. Exploiting these unique opportunities will ensure that *Ascaris* continues to inform basic neurobiology and parasite control.

Acknowledgments

The authors would like to thank Philippa Claude for critical reading of the chapter and Bill Feeny for help with the illustrations. AOWS acknowledges support from US Public Health Service grant RO1-AI15429, National Science Foundation grant IOS1145721, and a John Bascom Professorship from the University of Wisconsin-Madison. AGM acknowledges support from the Biotechnology and Biological Sciences Research Council, Merial, and the Department of Agriculture and Rural Development for Northern Ireland.

References

1. Hesse R. Uber das Nervensystem von *Ascaris lumbricoides* und *Ascaris megalocephala. Z Wiss Zool Abt A* 1892;**90**:73–136.
2. Goldschmidt R. Das Nervensystem von *Ascaris lumbricoides* und Megalocephala. Ein versuch, in den aufbau eines einfachen nervensystems einzudringen, Zweiter Teil. *Z Wiss Zool Abt A* 1908;**90**:73–136.
3. White JG, Southgate E, Thomson JN, et al. The structure of the ventral nerve cord of *Caenorhabditis elegans. Philos Trans R Soc Lond B Biol Sci* 1976;**275**(938):327–48.
4. White JG, Southgate E, Thomson JN, et al. The structure of the nervous system of the nematode *Caenorhabditis elegans. Philos Trans R Soc Lond B Biol Sci* 1986;**314**(1165):1–340.
5. Del Castillo J, De Mello WC, Morales T. Inhibitory action of gamma-aminobutyric acid (GABA) on *Ascaris* muscle. *Experientia* 1964;**20**(3):141–3.
6. Del Castillo J, De Mello WC, Morales T. The initiation of action potentials in the somatic musculature of *Ascaris lumbricoides. J Expl Biol* 1967;**46**(2):263–79.
7. Chitwood BG, Chitwood MB. *Introduction to Nematology.* University Park Press; 1974.
8. Stretton AO, Fishpool RM, Southgate E, et al. Structure and physiological activity of the motoneurons of the nematode *Ascaris. Proc Natl Acad Sci USA* 1978;**75**(7):3493–7.
9. Angstadt JD, Donmoyer JE, Stretton AO. Retrovesicular ganglion of the nematode *Ascaris. J Comp Neurol* 1989;**284**(3):374–88.
10. Vanfleteren JR, Van de Peer Y, Blaxter ML, et al. Molecular genealogy of some nematode taxa as based on cytochrome c and globin amino acid sequences. *Mol Phylogenet Evol* 1994;**3**(2):92–101.
11. Nanda JC, Stretton AOW. In situ hybridization of neuropeptide-encoding transcripts afp-1, afp-3, and afp-4 in neurons of the nematode *Ascaris suum. J Comp Neurol* 2010; **518**:896–910.
12. Geary TG, Klein RD, Vanover L, et al. The nervous systems of helminths as targets for drugs. *J Parasitol* 1992;**78**(2):215–30.
13. Martin RJ. Neuromuscular transmission in nematode parasites and antinematodal drug action. *Pharmacol Ther* 1993;**58**(1):13–50.
14. Martin RJ. Modes of action of anthelmintic drugs. *Vet J* 1997;**154**(1):11–34.

15. Rufener L, Mäser P, Roditi I, et al. *Haemonchus contortus* acetylcholine receptors of the DEG-3 subfamily and their role in sensitivity to monepantel. *PLoS Pathog* 2009; **5**(4):e1000380.
16. Rufener L, Keiser J, Kaminsky R, et al. Phylogenomics of ligand-gated ion channels predicts monepantel effect. *PLoS Pathog* 2010;**6**(9). e1001091.
17. Geary TG, Moreno Y. Macrocyclic lactone anthelmintics: spectrum of activity and mechanism of action. *Curr Pharm Biotechnol* 2012;**13**(6):866–72.
18. Krücken J, Harder A, Jeschke P, et al. Anthelmintic cyclooctadepsipeptides: complex in structure and mode of action. *Trends Parasitol* 2012;**28**(9):385–94.
19. Martin RJ, Buxton SK, Neveu C, et al. Emodepside and SL0-1 potassium channels: a review. *Exp Parasitol* 2012;**132**(1):40–6.
20. Kass IS, Wang CC, Walrond JP, et al. Avermectin B1a, a paralyzing anthelmintic that affects interneurons and inhibitory motoneurons in *Ascaris*. *Proc Natl Acad Sci USA* 1980;**77**(10):6211–5.
21. Kass IS, Stretton AO, Wang CC. The effects of avermectin and drugs related to acetylcholine and 4-aminobutyric acid on neurotransmission in *Ascaris suum*. *Mol Biochem Parasitol* 1984;**13**(2):213–25.
22. Martin RJ, Valkanov MA, Dale VM, et al. Electrophysiology of *Ascaris* muscle and anti nematodal drug action. *Parasitology* 1996;**113**:S137–56.
23. Zinser EW, Wolf ML, Alexander-Bowman SJ, et al. Anthelmintic paraherquamides are cholinergic antagonists in gastrointestinal nematodes and mammals. *J Vet Pharmacol Ther* 2002;**25**(4):241–50.
24. Holden-Dye L, Crisford A, Welz C, et al. Worms take to the slow lane: a perspective on the mode of action of emodepside. *Invert Neurosci* 2012;**12**(1):29–36.
25. Maule AG, Mousley A, Marks NJ, et al. Neuropeptide signaling systems – potential drug targets for parasite and pest control. *Curr Top Med Chem* 2002;**2**(7):733–58.
26. Raymond V, Sattelle DB. Novel animal-health drug targets from ligand-gated chloride channels. *Nat Rev Drug Discov* 2002;**1**(6):427–36.
27. McVeigh P, Atkinson L, Marks NJ, et al. Parasite neuropeptide biology: seeding rationale drug target selection. *Int J Parasitol Drugs Drug Resist* 2012;**2**:76–91.
28. Jex AR, Liu S, Li B, et al. *Ascaris suum* draft genome. *Nature* 2011;**479**(7374): 529–33.
29. Weimann JM, Marder E. Switching neurons are integral members of multiple oscillatory networks. *Curr Biol* 1994;**4**:896–902.
30. Davis RE, Stretton AO. The motor nervous system of *Ascaris*: electrophysiology and anatomy of the neurons and their control by neuromodulators. *Parasitology* 1996; **113**:S97–117.
31. Bargmann CI. Beyond the connectome: how neuromodulators shape neural circuits. *Bioessays* 2012;**34**(6):458–65.
32. Voltzenlogel E. Untersuchen über den anatomischen under histologischen bau des hinterendes von *Ascaris megalocephala* und *Ascaris lumbricoides*. *Zool Jahrb Abt Anat* 1902; **16**:481–510.
33. Otto A. Ueber das Nervensystem der Eingeweidewürmer. *Mag Entdeck Ges Naturk* 1816;**7**:223–33.
34. Stretton AO. Anatomy and development of the somatic musculature of the nematode *Ascaris*. *J Exp Biol* 1976;**64**(3):773–88.
35. Davis RE, Stretton AO. Passive membrane properties of motorneurons and their role in long-distance signaling in the nematode *Ascaris*. *J Neurosci* 1989;**9**(2): 403–14.
36. Davis RE, Stretton AO. Signaling properties of *Ascaris* motorneurons: graded active responses, graded synaptic transmission, and tonic transmitter release. *J Neurosci* 1989; **9**(2):415–25.

37. Goodman MB, Hall DH, Avery L, et al. Active currents regulate sensitivity and dynamic range in *C. elegans* neurons. *Neuron* 1998;**20**(4):763–72.
38. Angstadt JD, Stretton AO. Slow active potentials in ventral inhibitory motor neurons of the nematode *Ascaris*. *J Comp Physiol A* 1989;**166**(2):165–77.
39. Davis RE, Stretton AO. Structure–activity relationships of 18 endogenous neuropeptides on the motor nervous system of the nematode *Ascaris suum*. *Peptides* 2001; **22**(1):7–23.
40. Brownlee DJ, Fairweather I. Exploring the neurotransmitter labyrinth in nematodes. *Trends Neurosci* 1999;**22**(1):16–24.
41. Brownlee D, Holden-Dye L, Walker R. The range and biological activity of FMRFamide-related peptides and classical neurotransmitters in nematodes. *Adv Parasitol* 2000;**45**: 109–80.
42. Holden-Dye L, Walker RJ. Neurobiology of plant parasitic nematodes. *Invert Neurosci* 2011;**11**(1):9–19.
43. Jones AK, Sattelle DB. The cys-loop ligand-gated ion channel gene superfamily of the nematode, *Caenorhabditis elegans*. *Invert Neurosci* 2008;**8**(1):41–7.
44. Ghedin E, Wang S, Spiro D, et al. Draft genome of the filarial nematode parasite *Brugia malayi*. *Science* 2007;**317**(5845):1756–60.
45. Williamson SM, Walsh TK, Wolstenholme AJ. The cys-loop ligand-gated ion channel gene family of *Brugia malayi* and *Trichinella spiralis*: a comparison with *Caenorhabditis elegans*. *Invert Neurosci* 2007;**7**(4):219–26.
46. Bennett HM, Williamson SM, Walsh TK, et al. ACR-26: a novel nicotinic receptor subunit of parasitic nematodes. *Mol Biochem Parasitol* 2012;**183**(2):151–7.
47. Neveu C, Charvet CL, Fauvin A, et al. Genetic diversity of levamisole receptor subunits in parasitic nematode species and abbreviated transcripts associated with resistance. *Pharmacogenet Genomics* 2010;**20**(7):414–25.
48. Martin RJ, Robertson AP, Buxton SK, et al. Levamisole receptors: a second awakening. *Trends Parasitol* 2012;**28**(7):289–96.
49. Accardi MV, Beech RN, Forrester SG. Nematode cys-loop GABA receptors: biological function, pharmacology and sites of action for anthelmintics. *Invert Neurosci* 2012; **12**(1):3–12.
50. Rao VT, Siddiqui SZ, Prichard RK, et al. A dopamine-gated ion channel (HcGGR3*) from *Haemonchus contortus* is expressed in the cervical papillae and is associated with macrocyclic lactone resistance. *Mol Biochem Parasitol* 2009;**166**(1):54–61.
51. Rao VT, Accardi MV, Siddiqui SZ, et al. Characterization of a novel tyramine-gated chloride channel from *Haemonchus contortus*. *Mol Biochem Parasitol* 2010;**173**(2):64–8.
52. Kimber MJ, Sayegh L, El-Shehabi F, et al. Identification of an *Ascaris* G protein-coupled acetylcholine receptor with atypical muscarinic pharmacology. *Int J Parasitol* 2009; **39**(11):1215–22.
53. Smith KA, Komuniecki RW, Ghedin E, et al. Genes encoding putative biogenic amine receptors in the parasitic nematode *Brugia malayi*. *Invert Neurosci* 2007;**7**(4):227–44.
54. Smith KA, Rex EB, Komuniecki RW. Are *Caenorhabditis elegans* receptors useful targets for drug discovery: pharmacological comparison of tyramine receptors with high identity from *C. elegans* (TYRA-2) and *Brugia malayi* (Bm4)? *Mol Biochem Parasitol* 2007; **154**(1):52–61.
55. Abad P, Gouzy J, Aury JM, et al. Genome sequence of the metazoan plant-parasitic nematode *Meloidogyne incognita*. *Nat Biotechnol* 2008;**26**(8):909–15.
56. Opperman CH, Bird DM, Williamson VM, et al. Sequence and genetic map of *Meloidogyne hapla*: a compact nematode genome for plant parasitism. *Proc Natl Acad Sci USA* 2008;**105**(39):14802–7.
57. Johnson CD, Stretton AO. Localization of choline acetyltransferase within identified motoneurons of the nematode *Ascaris*. *J Neurosci* 1985;**5**(8):1984–92.

58. Johnson CD, Stretton AOW. Neural control of locomotion in *Ascaris*: anatomy, physiology and biochemistry. In: Zuckerman BM, editor. *Nematodes as Bological Models.* New York: Academic Press, Inc.; 1980. p. 159–95.
59. Johnson CD, Stretton AO. GABA-immunoreactivity in inhibitory motor neurons of the nematode *Ascaris. J Neurosci* 1987;**7**(1):223–35.
60. Guastella J, Johnson CD, Stretton AO. GABA-immunoreactive neurons in the nematode *Ascaris. J Comp Neurol* 1991;**307**(4):584–97.
61. Schuske K, Beg AA, Jorgensen EM. The GABA nervous system in *C. elegans. Trends Neurosci* 2004;**27**(7):407–14.
62. Brownlee DJ, Fairweather I, Johnston CF, et al. Immunocytochemical demonstration of peptidergic and serotoninergic components in the enteric nervous system of the roundworm, *Ascaris suum* (Nematoda, Ascaroidea). *Parasitology* 1994;**108**(1): 89–103.
63. Johnson CD, Reinitz CA, Sithigorngul P, et al. Neuronal localization of serotonin in the nematode *Ascaris suum. J Comp Neurol* 1996;**367**(3):352–60.
64. Davis RE. Neurophysiology of glutamatergic signaling and anthelmintic action in *Ascaris suum*: pharmacological evidence for a kainate receptor. *Parasitology* 1998;**116**: 471–86.
65. Davis RE. Action of excitatory amino acids on hypodermis and the motornervous system of *Ascaris suum*: pharmacological evidence for a glutamate transporter. *Parasitology* 1998;**116**:487–500.
66. Davis RE, Stretton AO. Extracellular recordings from the motor nervous system of the nematode, *Ascaris suum. J Comp Physiol A* 1992;**171**(1):17–28.
67. Takemoto T. Isolation and structural identification of naturally occurring excitatory amino acids. In: McGeer EG, Olney JW, McGeer P, editors. *Kainic Acid as a Tool in Neurobiology.* New York: Raven Press; 1978. p. 1–15.
68. Bargmann CI. Neurobiology of the *Caenorhabditis elegans* genome. *Science* 1998; **282**(5396):2028–33.
69. Greenwood K, Williams T, Geary T. Nematode neuropeptide receptors and their development as anthelmintic screens. *Parasitology* 2005;**131**. S169–77.
70. Cowden C, Stretton AOW, Davis RE. AF1, a sequenced bioactive neuropeptide isolated from the nematode *Ascaris suum. Neuron* 1989;**2**:1465–73.
71. Cowden C, Stretton AOW. AF2, an *Ascaris* neuropeptide: isolation, sequence, and bioactivity. *Peptides* 1993;**14**:423–30.
72. Cowden C, Stretton AOW. Eight novel FMRFamide-like neuropeptides isolated from the nematode *Ascaris suum. Peptides* 1995;**16**:491–500.
73. McVeigh P, Leech S, Mair GR, et al. Analysis of FMRFamide-like peptide (FLP) diversity in phylum Nematoda. *Int J Parasitol* 2005;**35**:1043–60.
74. McVeigh P, Alexander-Bowman S, Veal E, et al. Neuropeptide-like protein diversity in phylum Nematoda. *Int J Parasitol* 2008;**38**(13):1493–503.
75. Yew JY, Kutz KK, Dikler S, et al. Mass spectrometric map of neuropeptide expression in *Ascaris suum. J Comp Neurol* 2005;**8**:396–413.
76. Li L, Garden RW, Romanova EV, et al. In situ sequencing of peptides from biological tissues and single cells using MALDI-PSD/CID analysis. *Anal Chem* 1999; **71**:5451–8.
77. Neupert S, Predel R. Mass spectrometric analysis of single identified neurons of an insect. *Biochem Biophys Res Commun* 2005;**327**(3):640–5.
78. Jarecki JL, Andersen K, Konop CJ, et al. Mapping neuropeptide expression by spectrometry in single dissected identified neurons from the dorsal ganglion of the nematode *Ascaris suum. ACS Chem Neurosci* 2010;**1**:505–19.
79. Marder E. From biophysics to models of network function. *Ann Rev Neurosci* 1998;**21**: 25–45.

80. Reinitz CA, Herfel HG, Messinger LA, et al. Changes in locomotory behavior and cAMP produced in *Ascaris suum* by neuropeptides from *Ascaris suum* or *Caenorhabditis elegans*. *Mol Biochem Parasitol* 2000;**111**:185–97.
81. Reinitz CA, Pleva AE, Stretton AOW. Changes in cyclic nucleotides, locomotory behavior, and body length produced by novel endogenous neuropeptides in the parasitic nematode *Ascaris suum*. *Mol Biochem Parasitol* 2011;**180**:119–32.
82. Fellowes RA, Maule AG, Marks NJ, et al. Modulation of the motility of the vagina vera of *Ascaris suum* in vitro by FMRF amide-related peptides. *Parasitology* 1998; **116**(3):277–87.
83. Fellowes RA, Maule AG, Martin RJ, et al. Classical neurotransmitters in the ovijector of *Ascaris suum*: localization and modulation of muscle activity. *Parasitology* 2000; **121**(3):325–36.
84. Marks NJ, Maule AG. Parasitic nematode peptides. In: Kastin A, editor. *Handbook of Biologically Active Peptides*. 2nd ed. Elsevier Inc.; 2012.
85. Martin RJ. Gamma-aminobutyric acid- and piperazine-activated single channel currents form *Ascaris suum* body muscle. *Br J Pharmacol* 1985;**84**:445–61.
86. Verma S, Robertson AP, Martin RJ. The nematode neuropeptide AF2 (KHEYLRF-NH2) increases voltage-activated calcium currents in *Ascaris suum* muscle. *Br J Pharmacol* 2007;**151**:888–99.
87. Sithigorngul P, Jarecki JL, Stretton AOW. A specific antibody to neuropeptide AF1 (KNEFIRFamide) recognizes a small subset of neurons in *Ascaris suum*: differences from *Caenorhabditis elegans*. *J Comp Neurol* 2011;**519**:1546–61.
88. Keating CD, Holden-Dye L, Thorndyke MC, et al. The FMRFamide-like neuropeptide AF2 is present in the parasitic nematode *Haemonchus contortus*. *Parasitology* 1995; **111**(4):515–21.
89. Geary TG, Marks NJ, Maule AG, et al. Pharmacology of FMRFamide-related peptides in helminths. *Ann NY Acad Sci* 1999;**897**:212–27.
90. Marks NJ, Sangster NC, Maule AG, et al. Structural characterisation and pharmacology of KHEYLRFamide (AF2) and KSAYMRFamide (PF3/AF8) from *Haemonchus contortus*. *Mol Biochem Parasitol* 1999;**100**:185–94.
91. Kimber MJ, Fleming CC, Bjourson A, et al. FMRFamide-related peptides in potato cyst nematodes. *Mol Biochem Parasitol* 2001;**116**:199–208.
92. Johnston MJ, McVeigh P, McMaster S, et al. FMRFamide-like peptides in root knot nematodes and their potential role in nematode physiology. *J Helminthol* 2010;**21**:1–13.
93. Kikuchi T, Cotton JA, Dalzell JJ, et al. Genomic insights into the origin of parasitism in the emerging plant pathogen *Bursaphelenchus xylophilus*. *PLoS Path* 2011;**7**(9): e1002219.
94. McVeigh P, Leech S, Marks NJ, et al. Gene expression and pharmacology of nematode NLP-12 neuropeptides. *Int J Parasitol* 2006;**36**(6):633–40.
95. Trim N, Boorman JE, Holden-Dye L, et al. The role of cAMP in the actions of the peptide AF3 in the parasitic nematodes *Ascaris suum* and *Ascaridia galli*. *Mol Biochem Parasitol* 1998;**93**:263–71.
96. Masler EP. Responses of *Heterodera glycines* and *Meloidogyne incognita* to exogenously applied neuromodulators. *J Helminthol* 2007;**81**(4):421–7.
97. Kimber MJ, McKinney S, McMaster SM, et al. *flp* Gene disruption in a parasitic nematode reveals motor dysfunction and neuronal sensitivity to RNA interference. *FASEB J* 2007;**21**(4):1233–43.
98. Dalzell JJ, McMaster S, Fleming CC, et al. Short interfering RNA-mediated gene silencing in *Globodera pallida* and *Meloidogyne incognita* infective stage juvenilles. *Int J Parasitol* 2010;**40**:91–100.
99. Atkinson L, Stevenson M, McCoy C, et al. *flp-32* ligand/receptor silencing phenocopy a faster phenotype in plant pathogenic nematodes. *PLoS Path*. In press.

100. Xu MJ, Chen N, Song HQ, et al. RNAi-mediated silencing of a novel *Ascaris suum* gene expression in infective larvae. *Parasitol Res* 2010;**107**(6):1499–503.
101. Chen N, Xu MJ, Nisbet AJ, et al. *Ascaris suum*: RNAi mediated silencing of enolase gene expression in infective larvae. *Exp Parasitol* 2011;**127**(1):142–6.
102. Geldhof P, Visser A, Clark D, et al. RNA interference in parasitic helminths: current situation, potential pitfalls and future prospects. *Parasitology* 2007;**134**:609–19.
103. Knox DP, Geldhof P, Visser A, et al. RNA interference in parasitic nematodes of animals: a reality check? *Trends Parasitol* 2007;**23**:105–7.
104. Lendner M, Doligalska M, Lucius R, et al. Attempts to establish RNA interference in the parasitic nematode *Heligmosomoides polygyrus*. *Mol Biochem Parasitol* 2008; **161**:21–31.
105. Maule AG, McVeigh P, Dalzell JJ, et al. An eye on RNAi in nematode parasites. *Trends Parasitol* 2011;**27**(11):505–13.
106. Viney ME, Thompson FJ. Two hypotheses to explain why RNA interference does not work in animal parasitic nematodes. *Int J Parasitol* 2007;**38**:43–7.

PART III

EPIDEMIOLOGY OF ASCARIASIS

CHAPTER

7

Ascaris lumbricoides: New Epidemiological Insights and Mathematical Approaches

*Martin Walker**, *Andrew Hall*†, *María-Gloria Basáñez**

* Imperial College London, London, UK

† University of Westminster, London, UK

OUTLINE

Ascaris: The Neglected Parasite

http://dx.doi.org/10.1016/B978-0-12-396978-1.00007-0

Copyright © 2013 Elsevier Inc. All rights reserved.

INTRODUCTION

Ascaris lumbricoides (Nematoda: Ascaridida), commonly called the large intestinal roundworm, parasitizes the gastrointestinal tract of humans. It is often referred to as a helminth, a term which encompasses a large number of nematode and platyhelminth parasites. *A. lumbricoides* is also a soil-transmitted helminth (STH), a group of human gastrointestinal nematodes transmitted via direct contact with eggs or larvae in soil.[1] In addition to *A. lumbricoides*, the STHs of greatest public health importance are *Trichuris trichiura* (whipworm) and the hookworm species *Ancylostoma duodenale* and *Necator americanus*. Approximately 1 billion people are infected with at least one of these parasites.[1,2] The burden of *A. lumbricoides* infection alone has been estimated as to cost anywhere between 1.8 and 10.5 million disability-adjusted life years (DALYs) (see Chapter 13).[3]

A. lumbricoides, together with *T. trichiura* and hookworm, comprise three of the 13 neglected tropical diseases (NTDs) of greatest public health importance.[4] The term neglected refers to the lack of attention that has traditionally been paid to these infections by the research, medical and international funding communities alike. Reasons for this, in the case of the STHs, include that the majority of disease burden is borne by the world's most impoverished people, acute pathology is uncommon and the physical, cognitive and economic effects of chronic long-term infection are difficult to assess.[1] However, in the context of the current unprecedented momentum and commitment to NTD control,[5,6] embodied by the recent London Declaration against NTDs,[7] there are renewed calls for research and development to be embedded at the core of control and intervention strategies (see Chapter 15).[8] Research priorities include: fundamental biology and transmission dynamics,[9] diagnostics[10] and mathematical modeling.[11]

Biology and Life-Cycle

A. lumbricoides is the largest intestinal nematode to infect humans. Females typically weigh 4–7 grams and measure 20–30 cm long. Males are smaller, weighing 2–3 grams. Adult worms tend to inhabit the jejunum – the middle section of the small intestine – feeding on ingesta and swimming against the flow to maintain their position. Reproductively mature females mate with mature males enabling production of fertile eggs. Unfertilized mature females release infertile eggs which can be distinguished from their fertile counterparts under microscopic examination. Eggs are egested into the environment with the host's feces.[12]

Shaded moist soil and temperatures of 28–32°C provide optimum conditions for embryonation, under which L1 larvae will develop within the egg approximately 10 days after release from the host.[12] This process takes around 50 days at a less favorable 17°C.[13] Embryonation is followed by the first molt and formation of an L2 larva a few days later, a process that again occurs within the egg, giving rise to the infectious stage; an egg containing an L3 larva covered by an L2 cuticle.[14,15] In this state, and under favorable conditions, eggs are thought to be able to survive for up to 15 years,[12,13] although their life expectancy is stated as 28–84 days[16] reflecting high variability among survival times.

Humans ingest infectious eggs through fecal contamination of soil, foodstuffs, vegetable crops (for instance, fertilized with "night soil") and water supplies. Eggs hatch within the duodenum and larvae penetrate the intestinal mucosa, migrate towards the liver, where the L2 cuticle is shed, and from there pass to the lungs. This takes approximately 10 days during which L3 larvae molt to L4 before returning to the small intestine via the bronchi, trachea and esophagus. Here the final molt occurs and immature adults are formed.[12,17] It takes 7–12 weeks for worms to reach reproductive maturity, females maturing slightly faster than males.[18] The life expectancy of an adult worm is thought to be 1–2 years.[19,20]

There are debatably 13 or 16 known species of *Ascaris* which exclusively infect mammalian hosts (see Chapter 10).[21] *Ascaris suum* preferentially parasitizes the domestic pig but is very closely related to *A. lumbricoides*. Indeed the two possibly represent strains of the same species;[12,22] both cross-infection[22,23] and hybridization[24] have been demonstrated in sympatric populations, although the former not consistently across geographical locations.[25,26] This genetic closeness, combined with the anatomical, physiological, immunological, metabolic and nutritional similarities between humans and pigs, as well as the economic considerations of *Ascaris* infections in livestock, has prompted use of the pig as a model for *A. lumbricoides* infections in humans (see Chapters 14 and 16).[27,28]

Morbidity and Mortality due to Ascariasis

The pathology associated with *A. lumbricoides* infection is predominantly chronic, mainly related to malnutrition caused by perpetual infestation with adult worms, but severe acute effects may result from particularly heavy infections.[13,21] The chronic effects of infection are of much greater public health significance, as is the case for STHs in general[29] and indeed for many other helminth parasites of humans.[4]

Malnutrition is common to all STH infections[29] although particular species are associated with specific nutrient deficiencies. For example, hookworm is strongly associated with anemia caused by iron deficiency because hookworms suck blood from the intestinal mucosa and secrete anticoagulants to prevent clotting.[30] By contrast, *A. lumbricoides* impairs the absorption of fats, vitamin A and iodine and the digestion of lactose by reducing the appetite of infected individuals and damaging their intestinal villi.[31] The effects of STH-associated malnutrition are most pronounced in children and are broadly similar: impaired growth, physical ability, work capacity and cognition.[13,29]

Larval migration can lead to acute but short-lived pulmonary ascariasis or the more severe pneumonitis, both of which are rare, although their public health significance is not fully understood (Chapter 5).[13] Acute disease caused by adult worms is also uncommon,[32] but tends to be associated with complications arising from a large number of worms (worm burden) harbored by the host. The most common causes of such morbidity are intestinal or bile duct obstruction but many other complications have been documented.[21] Mortality caused by either larval or adult *A. lumbricoides* is exceptional[21,32] and, in general, acute pathology is of much less public health significance than chronic morbidity.[13]

Treatment

A. lumbricoides is extremely susceptible to a single dose of a wide range of anthelmintic drugs, but particularly to the bezimidazoles, which are used in mass treatment programs aimed specifically at STHs and also as part of combination therapy against lymphatic filariasis.[33] *A. lumbricoides* is also susceptible to ivermectin, the mainstay of onchocerciasis (and lymphatic filariasis in Africa) control.[34] The estimated range of efficacies of the anthelmintic drugs against *A. lumbricoides, T. trichiura* and hookworms are given in Table 7.1.

TABLE 7.1 Ranges of efficacies (%) of albendazole (ABZ), mebendazole (MBZ), pyrantel (PYR), ivermectin (IVM) and diethylcarbamazine (DEC)

Species	ABZ (≥400 mg)	MBZ (500 mg)	PYR (10 mg/kg)	IVM (150–200 μg/kg)	ABZ + IVM ABZ + DEC	DEC (150 mg) (6 mg/kg)
Ascaris lumbricoides	CR 67–100[241,242] ERR 87–100[242,243]	CR 91–100[241,242] ERR 96–99[242]	CR 79–100[241,242]	CR 50–78[241,243] ERR 94[243]	CR 78; ERR 100[243]	CR 24; ERR 34[243]
					CR 78; ERR 97[243]	CR 31; ERR 77[244]
Trichuris trichiura	CR 10–77[241,242] ERR 0–90[242,243]	CR 16–100[241,242] ERR 81–93[242]	CR 0–56[241]	CR 11–80[241,243] ERR 87[243]	CR 65; ERR 98[243]	CR 3; ERR 20[243]
					CR 19; ERR 79[243]	CR 77; ERR 86[244]
Necator americanus, Ancylostoma duodenale	CR 33–95[241,242] ERR 64–100[3]	CR 95–100[241] CR 1–27; ERR 0–98[242]	CR 19–88[241,242]	CR 0–20[241]		
					CR 90; ERR 94[244]	CR 26; ERR 36[244]

Abbreviations: CR, cure rate (proportion of treated individuals who became parasitologically negative); ERR, egg reduction rate (proportion reduction in egg count from baseline). *Adapted from Basáñez et al.*[113]

TABLE 7.2 Summary of chemo-expulsion studies that have collected adult *Ascaris lumbricoides* from humans by study starting date

Location	Study period	Participants*	Age range (years)	Drug	Study type	Reference
Iran	1972–1973	652	5–40+	PYR	CX	245
Iran	1973–1974	252	0–45+	PYR	CX	205
South Korea	1975–1978	853	NA	PYR	CX	246
South Korea	1977–1980	NA	NA	PYR	LG	247,248
Bangladesh	1982	203	0.5–15	PYR	CX	249
Burma (Myanmar)	1982	239	5–39+	LEV	LG	112,58
Philippines	1983	308	primary school	FLU	CX	250
Panama	1983–1984	203	3–5	LEV	CX	251
India	1984–1985	224	2–65+	PYR	LG	19
Philippines	1985–1986	150	0–14	PYR	CX	252
Mexico	1986	118	2–10	MBZ	LG	81
Nigeria	1987	808	5–16	LEV	LG	59
St. Lucia	NA†	113	1–30+	MBZ	LG	158
Bangladesh	1988–1989	1,765	1–98	PYR	LG	98
Madagascar	1990–1992	428	5–11	PYR	LG	253
China	1994–1995	222	0–55+	PYR	CX	254
Nepal	1998–2003	1,007	3–85	ABZ	LG	77

* *The number of participants whose worm burden was successfully assessed after anthelmintic treatment. For longitudinal studies, this is the number after the first round of treatment.*

† *Not reported in the published literature.*

Drug abbreviations: ABZ, albendazole; FLY, flubendazole; LEV, levamisole; MBZ, mebendazole; PYR, pyrantel.

Study type abbreviations: CX, cross-sectional; LG, longitudinal.

An extremely important research application of anthelmintics, particularly pyrantel pamoate, is their use as chemo-expulsive agents for collecting adult worms (Table 7.2), the gold-standard data for estimation of infection intensity. Pyrantel pamoate is particularly suitable for this purpose since it paralyzes adult worms – rather than killing and potentially damaging them – which are then expelled intact from the gut by peristalsis.

DIAGNOSIS OF INFECTION

Current infections with *A. lumbricoides* can be diagnosed in five main ways: (1) by seeing the worms in the gut either during endoscopy, in sonographic images,[35] in X-ray photographs or in tomographic images;[36] (2) by seeing worms after they are expelled, either naturally or after treatment; (3) by detecting the worms' metabolites in urine;[37] (4) by detecting the worms' DNA in feces;[38,39] and (5) by seeing the worms' eggs in feces under a microscope. Although the last method is the most simple, direct and most widely used in developing countries, methods using polymerase chain reactions followed by electrophoresis or fluorescence of the DNA[38,39] may well become more commonplace in developed countries as microscopy skills are lost, or not trusted, and as "multiplex" methods to detect several intestinal parasites at once are developed.

Two diagnostic methods are not useful: clinical signs or symptoms and antibody tests. There are no specific clinical signs or symptoms of infection, and if the worm load is light the effects may be negligible. Antibodies to parasite-specific antigens can be detected in blood[40] or saliva[41] but are not useful for diagnosis as they cannot distinguish between past and current infections. However, they may be useful to estimate the proportion of the population that is or has been infected with *A. lumbricoides*.

The metabolites of *A. lumbricoides* detected in the urine of infected people are 2-methyl-butyramide and 2-methyl-valeramide, and their concentration is proportional to the worm burden.[37] However, gas–liquid chromatography is required to detect these substances after they have been extracted from urine in a solvent, so it is not a simple or practicable method.

The recent development of methods involving amplifying worm DNA by polymerase chain reaction has included primers for *A. lumbricoides*, and several species of worms can be detected in the same assay using fluorescent probes.[39] However, it is not clear how sensitive this method is in comparison with microscopy.

Compared with these highly technical methods, the most simple, reliable and accurate diagnosis of infection is still achieved by seeing the worms' eggs under a microscope at a magnification of 100×, as they are large (55–75 μm by 35–55 μm) and are orange, yellow or brown,[42] so are easy to identify. A slide can be prepared and examined in a few minutes and the only equipment needed is a microscope. However, the eggs of *A. lumbricoides* cannot be distinguished from those of *A. suum*, a species that can infect humans in addition to its normal host, the pig. *A. lumbricoides* and A. *suum* are identical in terms of their method of transmission, infection, pathophysiology and treatment.[43]

Although the specificity of diagnosing *A. lumbricoides*/*A. suum* should usually be 100% as the eggs are pathognomonic, infections can be missed if the concentration of eggs in feces is very low. The sensitivity of diagnosis depends on three main factors, namely, the presence of mature female worms, as males do not produce eggs; the number of eggs produced by mature female worms; and the dilution of eggs in the feces. The number of eggs released by female worms into the gut depends on several factors. First, whether the worms have been fertilized by a male, although infertile "decorticated" eggs can be released in small numbers by mature, unfertilized female worms. Infertile eggs tend to be longer and thinner than normal.[42] Second, the presence of other *A. lumbricoides* in the gut, which act to inhibit the production of eggs in a process called density-dependent fecundity so that the number of worms per female decreases in proportion to the worm burden.[44,45] This is a non-linear effect as the presence of a few worms appears substantially to affect the concentration of eggs in feces. Third, there is evidence of a wide range variation in the *per capita* fecundity of worms in different parts of the world.[46] This means that the relationship between the concentration of eggs in feces and the number of worms in the gut is specific to the locality, but may also affect the diagnosis of infections. However, a single female *A. lumbricoides* can produce up to 200,000 eggs per day,[47] which is 1 egg per milligram in 200 g of stool, if thoroughly mixed.

The probability of seeing an egg in a sample of feces depends on their dilution and dispersion in the fecal mass, which is smaller in children than adults. There is evidence from examining multiple samples from the same stool that eggs may be clumped,[48] a characteristic that may result from the aggregation of worms within hosts.[49] Ideally every fecal sample should be thoroughly mixed before a subsample is taken for diagnosis, perhaps with a known volume of 0.9% saline containing a surfactant to separate and disperse the eggs. The probability of seeing an egg will depend on the amount of feces examined and on whether or not any fecal detritus is removed or clarified.

There are several alternative methods of microscopical diagnosis. The simplest is to examine a direct smear of feces (typically 5–10 mg) on a glass slide. A study in Brazil found this method to be more sensitive than the Kato-Katz method, which samples 3–6 times as much feces.[50] The Kato-Katz method applies glycerol to a standard volume of filtered feces deposited in a template on a slide; the glycerol serves to clarify the fecal detritus so that after 30 minutes the eggs of worms become easier to see.[51] The amount of filtered feces placed on the slide depends on the diameter of the hole in the template. The number of eggs on the slide are counted and then multiplied by a standard factor depending on the template used. However, in an experiment a template 6 mm in diameter and 1.0 mm thick delivered between 12 and 34 mg of feces in 95% of samples,[52] which should have led to multiplication factors of between 29 and 83× to calculate the concentration of eggs per gram (epg) of feces.

The multiplication factor for such a template should be 35, presuming that feces have a specific gravity of 1.0.

The eggs of some helminths can be separated and concentrated from feces by flotation or sedimentation, or flotation followed by sedimentation. In a simple flotation process, feces are mixed with a solution of $ZnSO_4$ with a specific gravity (SG) of 1.18[53] or sugar with an SG of 1.27[54] and then centrifuged. The surface fluid is removed with either a loop or a cover slip and then examined under a microscope.

Sedimentation methods use gravity to separate eggs from light fecal detritus either in water[54] or by adding ether or ethyl-acetate and then spinning the vigorously shaken suspension in a centrifuge, which traps detritus in the upper, immiscible ether layer. A small volume of the sediment is then examined for eggs under a microscope.[48] If the bottle of fixative is weighed before and after the feces are added, this method can give a quantitative egg count expressed in epg of feces. This provides an indicator of the worm burden, although this is specific to the site because of differences in female fecundity between sites.[46]

A comparative study in Zanzibar of 56 infants infected with *A. lumbricoides* found that a simple gravity sedimentation method using 2.0 g of feces was more sensitive respectively than the Kato-Katz method using an estimated 0.42 mg, a modified ethyl acetate sedimentation method using 1.0 g, and a dual sedimentation and flotation method using 5.0 g.[54]

TRANSMISSION IN COMMUNITIES

Quantifying Infection

Patterns of infection within communities are best studied using intensity data because infection intensity relates to an individual's contribution to transmission,[20] and serves as an indicator of probable morbidity.[55,56] Prevalence (presence or absence of infection) data are much less informative, particularly in highly endemic communities where the non-linear relationship between prevalence and intensity[57] results in the vast majority of people being infected.

Worm burdens are the "gold standard" intensity data which have shaped much of our understanding of the population biology and epidemiology of *A. lumbricoides*. Worms can be collected by chemoexpulsion, a technique that exploits the action of anthelmintic drugs which kill or paralyze worms so that they are expelled intact from the gut by peristalsis (Table 7.2). The high efficacy of the anthelmintics used for this purpose (Table 7.1) means that the number of worms counted in the stools passed by a patient for 48–72 hours post-treatment[19,58,59] provides a reliable and accurate measure of their infection intensity.

In contrast to worm counts, fecal egg counts vary considerably from day to day and among samples taken from the same stool.[47,48] Moreover, because the distribution of worms among individuals within a community is ubiquitously overdispersed relative to the Poisson or random distribution,[20] egg counts are highly variable (and zero-inflated, see "Statistical models," below), providing a very inaccurate measure of infection intensity (Figure 7.1). That said, epidemiological surveys using egg counts are fast, affordable and relatively easy to undertake, making it the most widely abundant available data for estimation of infection intensity.[60]

Heterogeneities in Infection

The distribution of worms among hosts adheres to the Pareto principle, also known as the 80/20 rule; approximately 80% of the worm population tends to be harbored by about 20% of hosts. This reflects a high degree of heterogeneity in infection rates among individuals.[20,61] Such heterogeneity may be caused by variability in exposure; innate (genetic) susceptibility; acquired immunity or by a so-called "clumped" infection process whereby multiple adult worms establish simultaneously, presumably because multiple infectious larvae are acquired per infection event.[62–64] In a highly endemic urban community in Bangladesh, correlation observed among the weights of individual adult *A. lumbricoides* within infra-populations (the population of worms within a host) has been presented as evidence for a clumped infection process.[49,65] Similar conclusions have been drawn from a study conducted in Guatemala, where worm mitochondrial DNA sequences were shown to be clustered (genetically similar) within individuals.[66]

Measuring exposure to infectious larvae directly is notoriously difficult.[67] Consequently, the estimation of exposure has been restricted to the measurement of concentrations of fecal silica as a proxy for soil contamination of food and geophagic activity.[68–70] An alternative approach has been to infer heterogeneities in exposure from statistical identification of risk factors associated with infection intensity; factors that explain some of the observed variability in intensity among individuals. Numerous studies have identified a diverse and often inconsistent range of factors associated with *A. lumbricoides* egg output,[71] although a recent meta-analysis of prevalence-based epidemiological surveys has demonstrated the protective effect of access to sanitation facilities.[72] Just four studies have used gold-standard worm count data to study risk factors,[73–76] identifying household-, agricultural-, host sex- and poverty-related factors associated with worm burdens.

The Role of Host Genetics

There is strong evidence for a genetic component to host susceptibility (see Chapter 12). Pedigree analysis of a Nepalese community has indicated that 30–50% of the variability among individuals' worm burdens is explained by host genetic differences.[77] Furthermore, at least three loci on chromosomes 8, 11 and 13 have been associated with susceptibility to infection.[78,79] The genetic component of susceptibility explains partly the observation that worm burdens of members of the same household tend to be similar.[73,80,81] However, shared household exposures are also important. Molecular analysis of worms from the same Nepalese community showed genotypic clustering of worms infecting members of the same household, suggesting that households, at least in this community, are important transmission foci (Chapter 8).[82] The importance of the household in the transmission of *A. lumbricoides* and *T. trichiura* – and to a lesser extent hookworm – was first identified in the late 1920s to early 1930s from work conducted in China,[83] Panama[84] and the southern United States of America.[85] These ideas were revisited by Williams et al. in 1974[86] and by others in the late 1980s[81] and early 1990s,[87] and expounded in 1996 by Cairncross et al.[88] who described the household (for *Ascaris* and *Trichuris*) and the public environments (for hookworm) as "fundamental arenas of disease transmission."

The Role of Host Immunity

The role of either the innate or adaptive (acquired) immune response in protecting against infection with *A. lumbricoides* remains incompletely understood (Chapter 1).[89] Infection elicits the production of cytokines, predominantly associated with a Th2-type response, and antibodies of all isotypes, particularly IgE and *Ascaris*-specific IgE.[90,91] These immunoglobulins do not seem to be protective;[92] rather, they reflect current or past infection.[93,94] However, the concomitant rise in cytokine concentrations has been associated with reduced infection intensities,[95,96] although the elicited effector mechanisms remain incompletely understood.

An alternative (or complementary) approach to investigating the operation of acquired immunity has been to use mathematical models to demonstrate that the characteristic "peaked" or "convex" age-intensity profile of *A. lumbricoides*,[19,58,97,98] in which infection peaks in school-age (5–15-year-old) children and declines in adults (Figure 7.2), can be reproduced under the assumptions of: (1) a gradual development of protective immunity from chronic exposure to worm antigens, and (2) a slow loss of immunological memory.[99–101] Such models also predict a "peak shift," whereby infection intensity (or prevalence) peaks at a higher level and at an earlier host age in communities where transmission is more intense, causing protective immunity to develop more

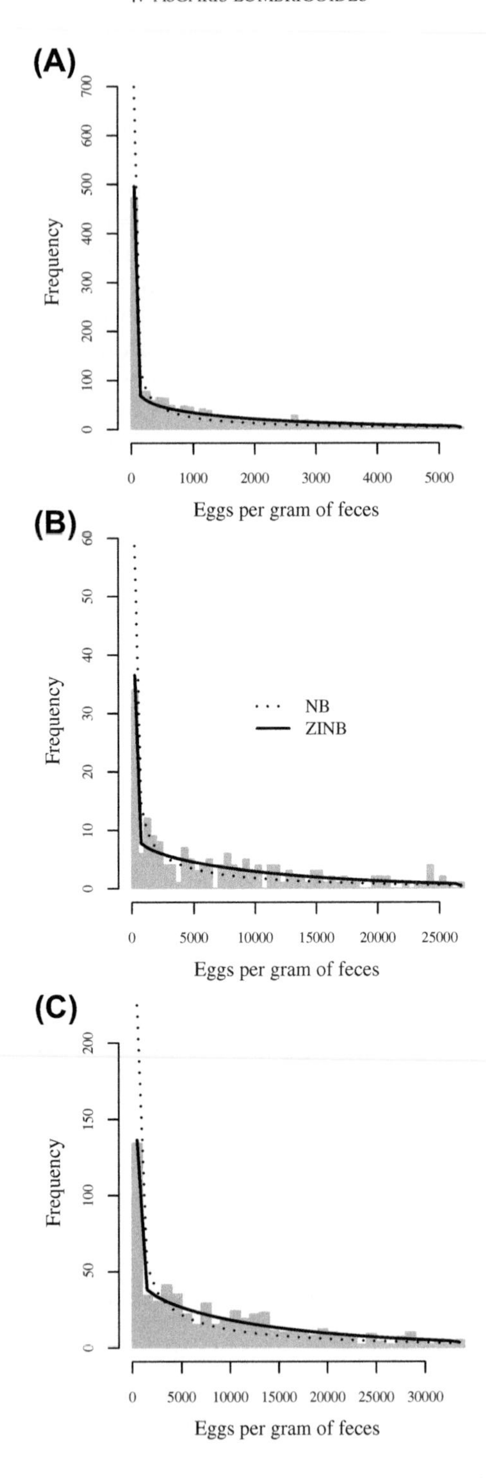
(A)
Frequency
0
100
200
300
400
500
600
700
0
1000
2000
3000
4000
5000
Eggs per gram of feces
(B)
Frequency
0
10
20
30
40
50
60
NB
ZINB
0
5000
10000
15000
20000
25000
Eggs per gram of feces
(C)
Frequency
0
50
100
150
200
0
5000
10000
15000
20000
25000
30000
Eggs per gram of feces

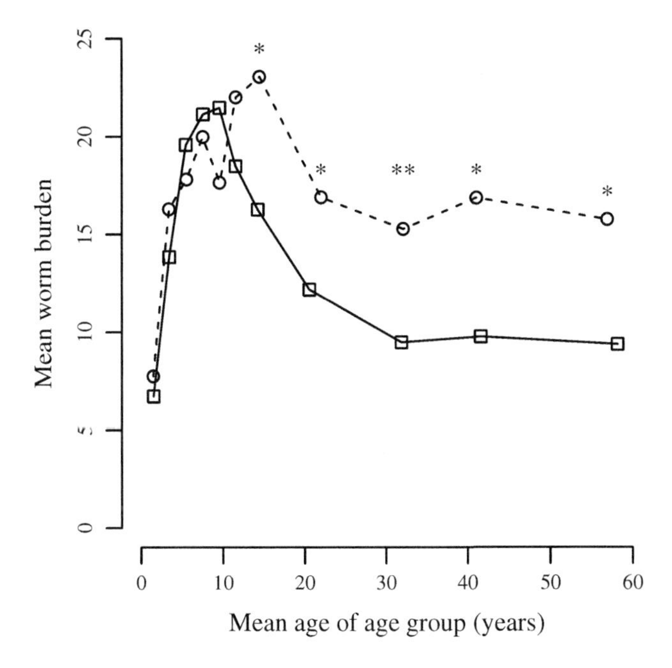

FIGURE 7.2 **Baseline age-intensity profile of *Ascaris lumbricoides* in a community in Bangladesh, stratified by host sex.**[104] Data points are arithmetic mean worm burdens from individuals in the age groups: 1–3; 4–5; 6–7; 8–9; 10–11; 12–13; 14–17; 18–27; 28–37; 38–47; 48+ years. Squares and circles represent means for males and females, respectively. Stars indicate statistically significant differences between the sexes as indicated by a two-sample *t*-test. * p-value < 0.05; ** p-value < 0.01.

rapidly.[102] There is no parasitological evidence for a peak shift in *A. lumbricoides*, although a recent study has identified a peak shift in IgE concentrations, possibly indirect evidence for protective immunity.[103] However, variability in age-intensity profiles between the sexes, particularly in adults, has been anecdotally related to different cultural and occupational practices of men and women[104] and boys and girls[75]

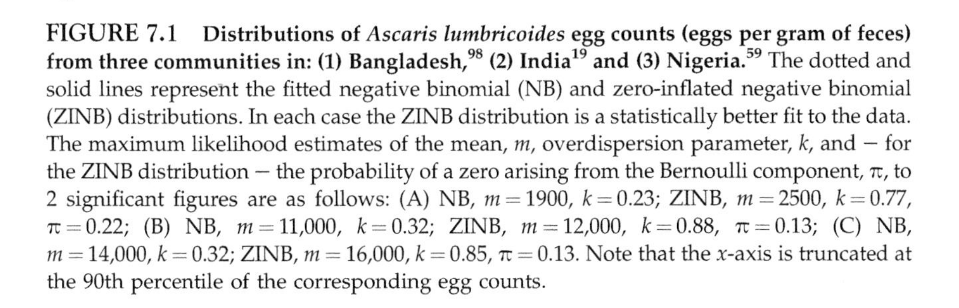

FIGURE 7.1 **Distributions of *Ascaris lumbricoides* egg counts (eggs per gram of feces) from three communities in: (1) Bangladesh,[98] (2) India[19] and (3) Nigeria.**[59] The dotted and solid lines represent the fitted negative binomial (NB) and zero-inflated negative binomial (ZINB) distributions. In each case the ZINB distribution is a statistically better fit to the data. The maximum likelihood estimates of the mean, m, overdispersion parameter, k, and – for the ZINB distribution – the probability of a zero arising from the Bernoulli component, π, to 2 significant figures are as follows: (A) NB, $m = 1900$, $k = 0.23$; ZINB, $m = 2500$, $k = 0.77$, $\pi = 0.22$; (B) NB, $m = 11{,}000$, $k = 0.32$; ZINB, $m = 12{,}000$, $k = 0.88$, $\pi = 0.13$; (C) NB, $m = 14{,}000$, $k = 0.32$; ZINB, $m = 16{,}000$, $k = 0.85$, $\pi = 0.13$. Note that the x-axis is truncated at the 90th percentile of the corresponding egg counts.

(Figure 7.2), suggesting that behaviorally mediated changes in exposure may override any putative effects of acquired immunity.

Reinfection following Curative Treatment

In endemic settings, reinfection following curative anthelmintic treatment occurs rapidly, rebounding to pre-treatment infection levels in approximately 1 year.[105] Heterogeneities in infection rates ensure that rates of reinfection also vary among individuals within a community. A striking manifestation of this heterogeneity is that individuals' worm burdens following treatment and subsequent reinfection are correlated with pre-treatment worm burden. This phenomenon, known as "predisposition" to infection, has been demonstrated in many communities, both after a single round and over multiple rounds of anthelmintic treatment.[106,107] Predisposition is also evident at the household level: worm burdens tend to be associated among members of the same household[73,80,81] and average household worm burdens tend to be similar between rounds of treatment and reinfection.[108,109]

Recent analysis of the interplay between household clustering and individual predisposition in an urban community in Bangladesh found that individual predisposition has limited epidemiological significance compared to the predisposition effect driven by the clustering of infections within households.[76] This complements the results of analyses comparing the associations between the worm burdens of parents and their (genetically related) children and between unrelated parents, from which it was concluded that any genetic basis to individual predisposition must be overwhelmed by household-related behavioral or environmental factors.[110] Furthermore, probabilistic modeling of the expected magnitude of correlation between pre- and post-treatment worm burdens has revealed that observed correlations are too small to be explained solely by long-term differences in host susceptibilities, and that short-term, transient heterogeneity, such as arising from clumped infections, is a more important driver of predisposition.[111] It thus appears that individual predisposition is generally weak,[98,107] driven largely by transient short-term heterogeneities in infection,[111] and swamped by the putative effects of the household.[76,110]

A pattern emerging from studies of communities in Bangladesh,[98] Burma (Myanmar)[112] and Nigeria[59] is that rates of reinfection post-treatment decline with (baseline) host age (Figure 7.3). This observation is based on worm counts, as opposed to egg counts, and so cannot be explained by the decreased dilution of eggs in the typically smaller mass of feces produced by children compared with adults (see "Diagnosis of infection," above). Consequently, this suggests that infection rates in

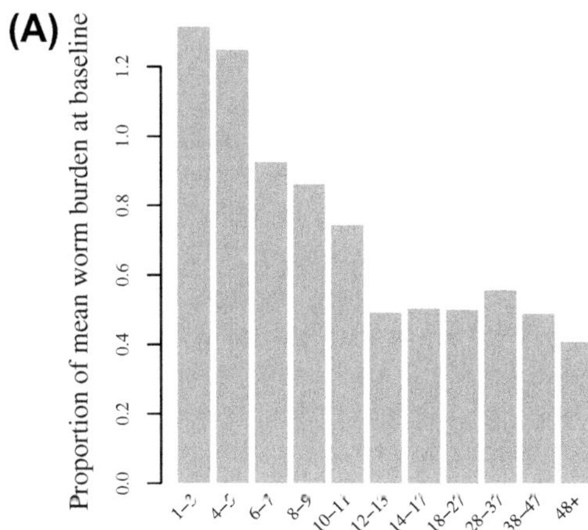

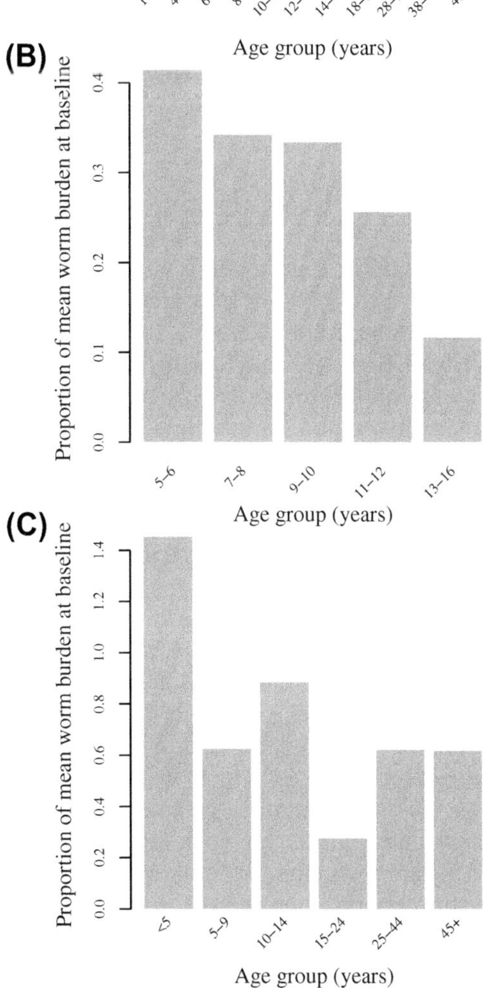

FIGURE 7.3 **Age-dependent patterns of reinfection with *Ascaris lumbricoides* following curative treatment.** Data are arithmetic mean worm burdens, by age group, after a period of reinfection expressed as a proportion of baseline worm burden. Data collected from communities in: (A) Bangladesh following 6 months' reinfection;[98] (B) Nigeria following 6 months' reinfection;[59] (C) Burma (Myanmar) following 12 months' reinfection.[112]

children change rapidly over the reinfection period; behaviorally-mediated exposure may change rapidly over the first three years of life as children learn to walk and explore their environment, which will increase their exposure to *A. lumbricoides* eggs. This changing "force of infection" (the per host rate of parasite acquisition) has important implications for control programs selecting cohorts of children to be followed up for the purpose of monitoring and evaluation.[113]

An additional observation from reinfection studies that have collected data on the size (weight/length) of individual worms[44,114] is that the size of worms is host-age dependent, probably due to crowding constraints exacerbated by the size of the gut lumen, and that after treatment newly acquired worms appear to grow rapidly, possibly due to the release of density-dependent constraints. Worm size is associated with fecundity[115] and so these observations have potentially important consequences on transmission following anthelmintic treatment, although the epidemiological significance of this is yet to be fully elucidated.[44]

MATHEMATICAL APPROACHES

The seminal work of Crofton in 1971[116] embedded the frequency distribution of parasites among hosts at the heart of quantitative parasitological research. Crofton revisited the words of Cassie,[117] who in 1963 wrote that, "The frequency distribution model may be applied at two levels, the empirical and the fundamental. Empirically it is desirable to condense the sample data, so that any given population may be described by a few parameters, which are readily comparable with the corresponding parameters of another population. [...] The fundamental model, on the other hand is based on some hypothesis of some real biological significance. [...] If it fits the data better than other possible models, it provides some justification for the hypothesis concerned."

Classical statistical inference is grounded in the "empirical," and a great deal of applied statistical analysis in the area has focused on the adequacy of distributions for describing parasitological data and on the best methods of estimating their parameters. Population dynamics models are more associated with the fundamental, being defined in terms of underlying population (biological) processes which determine the number of parasites within a host. To what extent the frequency distribution of worms among hosts is considered can vary greatly, from a simple deterministic description of changes in the (population) mean[118] – which may include "empirical" components to account for the effect of the frequency distribution on population processes (the so-called hybrid structure, see "Dynamics models of infection and transmission," below and Chapter 9)[119,120] – to fully stochastic considerations of changes in the number of parasites within individual hosts.[63]

Statistical Models

Here we highlight some statistical methods that are of broad applicability to the analysis of parasitological data in general and of *A. lumbricoides* in particular. Emphasis is placed on appropriate univariate and multivariate (regression) analysis of overdispersed count data, often with an excess of zeros; the use of hierarchical methods for clustered and longitudinal data, and the closely related and burgeoning discipline of temporal–spatial data analysis. We also discuss briefly more general methods of fitting non-linear and population dynamics models to data.

Negative Binomial and Zero-inflated Negative Binomial Distributions

In 1941, Fisher[121] successfully used the negative binomial distribution to describe the overdispersed distribution of ticks on sheep. Since this time, the distribution has become a ubiquitous and well-validated description of the distribution of parasites of humans,[16,20] wildlife[122] and of adult *A. lumbricoides*.[57] In parasitology, the negative binomial distribution is expressed in terms of its mean, m, and dispersion parameter, k, which inversely describes the degree of parasite aggregation. The Poisson distribution is obtained as $k \to \infty$ and the log series distribution arises as $k \to 0$.[123,124] The geometric distribution is a special case of the negative binomial distribution when $k = 1$.

Two methods are commonly used to estimate k. The moment estimator (so called because it is based on the first two moments, the mean and the variance s) $k = m^2/(s^2 - m)$ is derived from equating the variance of the negative binomial distribution to the *sample* variance, s^2.[125] This estimator substantially overestimates k (underestimates the degree of overdispersion) at low sample sizes.[126] A refined moment estimator, which partially corrects for this bias, is given by $k = (m^2 - s^2/n)/(s^2 - m)$.[127] The maximum likelihood estimator of k[123] is generally recommended over moment estimators[128] owing to its reduced bias and superior asymptotic efficiency.[129–131]

Since k is a critical parameter in population dynamics models of *A. lumbricoides* and of helminth parasites more generally, inaccurate or biased estimation may profoundly affect the output from such models. A straightforward method of mitigating such effects is to estimate parameters from sufficiently large sample sizes. However, this may be less straightforward when estimators are required for different population strata; obtaining sufficiently large sample sizes from different age groups typically requires increased sampling effort for older individuals, who are less frequent in the population, most obviously because of mortality. This issue has been highlighted in the context of observed age-intensity profiles for cestode (*Diphyllobothrium ditremum*) infections of Arctic char (*Salvelinus alpinus*).[128] Because the mean of overdispersed count data underestimates the true population mean to an increasing degree with declining sample size,[126] smaller observed intensities of infection in older age groups can be a statistical artifact.

The prevalence of infection for a negative binomial parasite distribution is given by $1 - P(0|m,k)$, where $P(0|m,k)$ is the probability a host has zero worms given the mean intensity of infection, m, and overdispersion parameter, k. The relationship between the prevalence, p, and mean intensity of infection, m, is non-linear, $p = 1 - (1 + m/k)^{-k}$; prevalence initially increases rapidly with increasing mean intensity, before saturating at a level determined by parameter k. Data on *A. lumbricoides* worm counts collated from a variety of communities suggest that the best fit to the prevalence–intensity relationship is obtained when the value of k is allowed to increase linearly with the mean.[57] It is noteworthy that although prevalence is, in general, a less informative index of infection, it is unbiased by sample size.[126]

The zero-inflated negative binomial distribution is a mixture of negative binomial and Bernoulli (binary) distributions (Figure 7.1). This permits zero counts to arise from either the Bernoulli component or the count component, which inflates the number of zeros relative to the negative binomial. Recent analysis of *A. lumbricoides* egg counts collected from a community in Bangladesh has demonstrated the superiority of a zero-inflated negative binomial over its non-zero-inflated counterpart in describing the data.[44] This is in accordance with another recent analysis of *A. lumbricoides, T. trichiura* and hookworm egg counts collected in northeast Ecuador as part of a study into the ancillary effects on STHs of long-term mass ivermectin treatments aimed principally at controlling onchocerciasis.[132] Zero inflation has also been shown in egg counts from parasites of livestock[133,134] and wildlife,[135] and so empirical evidence is building to suggest that this is a characteristic feature of fecal egg count data.

Excess zeros in *A. lumbricoides* egg count distributions are likely to arise from two predominant mechanisms: (1) hosts harboring only males (unfertilized females produce unfertilized and still contribute to positive counts), and (2) failure of the diagnostic method to detect parasite eggs. The first mechanism can be accounted for if per host worm counts are available and stratified by worm sex. That is, if the second mechanism is unimportant, removal of male-only infection should remove the majority of excess zeros. The data from Bangladesh indicate that this is not the case, which is unsurprising given the modest sensitivity of the diagnostic method. Furthermore, the fraction of zero counts declines with increasing

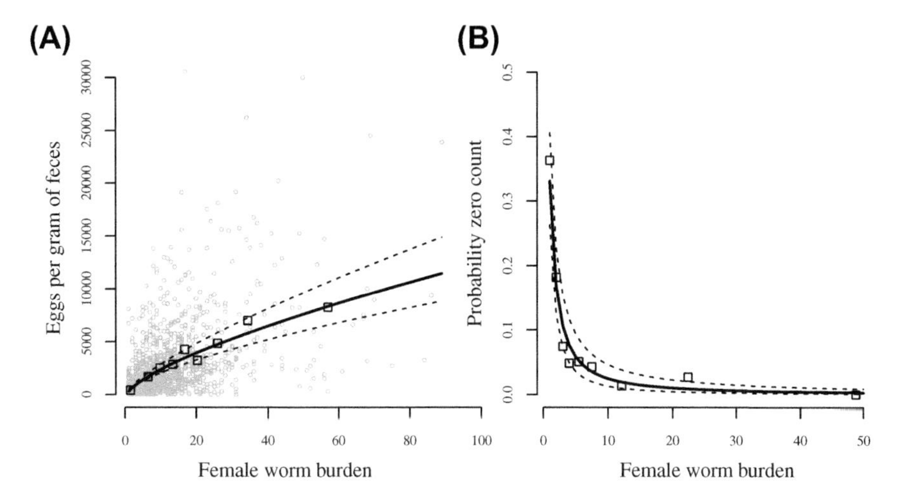

FIGURE 7.4 **The output of a regression model fitted to *Ascaris lumbricoides* egg counts (eggs per gram of feces) using a zero-inflated negative binomial (ZINB) distribution.** The mean of the negative binomial (NB) component, m, and probability of the Bernoulli component, π, were modeled as dependent on the female worm burden, x, using the following relationships: $\ln(m) = a + b \ln(x)$; $\ln[\pi / (1-\pi)] = c + d \ln(x)$. The over-dispersion parameter, k, of the NB component was assumed constant. The fitted mean, m, and probability, π, along with corresponding 95% confidence intervals (CIs) are denoted by the thick solid and thin dashed lines in (A) and (B), respectively. Maximum likelihood estimates to 2 significant figures are as follows: NB component, $a = 470$, $b = 0.71$, $k = 1.1$; Bernoulli component, $c = 0.33$, $d = -1.3$. The value of b is statistically significantly less than 1 (95% CI: 0.66–0.76) indicating density-dependent female worm fecundity. In (A) the small gray circular data points represent the individual egg count data and the large squares represent the mean egg counts in the following female worm burden groups: 1–4; 5–8; 9–11; 12–15; 16–18; 19–22; 23–30; 31–39; 40+. In (B) the large square data points represent the proportion of zero counts in the following worm burden groups: 1; 2; 3; 4; 5–6; 7–8; 9–16; 16–32; 33+.

female worm burden, pointing to a density-dependent diagnostic sensitivity (Figure 7.4).[44]

Extrapolating individual-level diagnostic sensitivity to the chance of detecting infection in a randomly sampled individual (moving from an individual to a population level) indicates that diagnostic sensitivities will vary considerably among communities with different endemicities (Figure 7.5). Furthermore, sensitivity will decline throughout the course of an effective control program, potentially leading to overoptimistic assessment of the achieved reduction in infection and premature cessation of control.

Regression Models

The preponderance of overdispersion in parasitological data means that classical regression techniques, which assume normal errors with

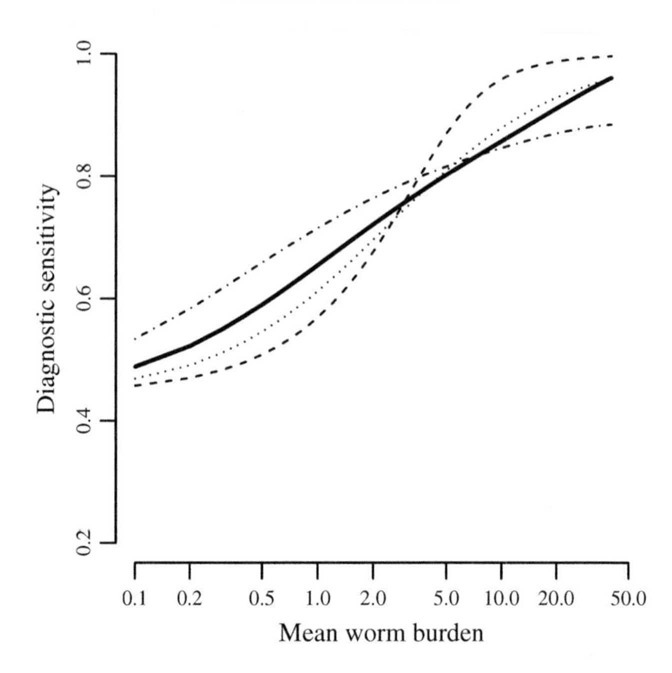

FIGURE 7.5 **The predicted relationship between the diagnostic sensitivity of an ether-sedimentation technique[48] and average community worm burden.** Predictions were derived using the fitted relationship between the probability of observing a zero egg count, π, and female worm burden, x, depicted in Figure 7.4(B) and making the assumptions that: there is an equal (1:1) sex ratio of males and females, and the distribution of males *and* females is well described by a negative binomial distribution with mean m. The over-dispersion parameter was varied: $k \rightarrow \infty$ (Poisson distribution), thin dashed line; $k = 1$, thin dotted line; $k = 0.1$, thin dot-dash line; k is a parameterized linear function of m from Guyatt et al.,[57] thick solid line.

a constant variance (homoskedasticity) are rarely appropriate. One solution is to transform the data so that they conform better to these assumptions,[136] although the often used logarithmic transformation is not recommended.[137] Rather, Anscombe[138] derived an inverse-hyperbolic sine for transforming negative binomial data (optimized transformations for the Poisson and binomial distributions are also given in Anscombe's paper).

The development of generalized linear modeling methods by Nelder and Wedderburn in 1972[139] permitted efficient regression analysis of non-normally distributed data that fell within the exponential family of distributions, a family that includes, among others, the Poisson, binomial and the negative binomial with *known* dispersion parameter k.[140,141] The generalized linear model framework has also been extended to permit simultaneous estimation of k and other regression parameters (covariate coefficients),[142] and to allow k to depend on covariates.[143] Negative binomial regression has now become a more commonly included component of statistical software packages, and a number of studies throughout the last decade have applied it to analyze egg count data on *A. lumbricoides*[144,145] and other STH infections.[146–148]

Zero-inflated Poisson[149–151] and negative binomial[152] regression techniques were developed primarily for economic applications during the mid-1980s to early 1990s.[153] More recently they have been adopted by the parasitological research community, having been applied to spatial risk models of *Schistosoma mansoni* infection[154] and *S. mansoni*–hookworm co-infections;[155] the comparison of STH infections in ivermectin-naïve and treated communities,[132] and the analysis of density- and female weight-dependent fecundity in *A. lumbricoides*.[44] Aside from providing a generally improved description of egg count data, zero-inflated models permit modeling of the count and Bernoulli components. This has been exploited to demonstrate that the preponderance of zeros in *A. lumbricoides* egg counts depends on the female worm burden, presumably arising from parasite density-dependent sensitivity of the diagnostic test.[44] Modeling zeros in this way may also be useful for random effects modeling, where diagnostic sensitivity may be heterogeneous among communities with different endemicities. A better understanding of the components of diagnostic performance of parasitological assays will become increasingly important as control programs shift their aim from morbidity control to elimination of infection.

Hierarchical, Mixed Effects Models

Hierarchical models, which are also referred to as mixed or random effects models, are used to analyze non-independent, clustered data that arise when observations are made from distinct or related units.[156,157] For example, observations made on the same individual, either at the same or at different points in time (longitudinal data) will generally be more similar than observations made from different individuals. Similarly, observations made from members of the same household, school or community will generally be more similar than observations made from different households, schools or communities. The former two examples (i.e. individuals and households) are well illustrated by predisposition and household clustering of *A. lumbricoides* infection, respectively.[106] In statistical parlance, these two phenomena simply represent the clustering or correlation typically observed among repeated observations made from on the same distinct units.

Interestingly, predisposition and household clustering have only recently been explored (by analyzing data from Bangladesh) using methods which exploit the natural three-level hierarchical structure of worm counts measured repeatedly from individuals, before and after chemo-expulsive treatment, residing in separate households.[76] Prior to this, predisposition and household clustering had been studied independently of one another. Predisposition had been demonstrated by the statistical significance of a non-parametric measure of statistical dependence between worm burdens from the same individual.[19,98,158] Household clustering had typically been demonstrated by dichotomizing individuals' worm burdens as heavy or light and by determining a statistically significant difference between the number per household observed and the number expected by chance.[108] The recent hierarchical modeling approach, however, revealed that individual predisposition was very weak and almost entirely subsumed under the clustering effect of the household.[76]

Aside from quantifying the magnitude of correlation among clustered data, hierarchical techniques permit critical adjustment for the estimated uncertainty of regression coefficients which, if ignored, can lead to erroneous statistical inference. Generally, failing to account for correlation among non-independent data leads to overoptimistic (narrow) estimates of regression coefficient standard errors (underestimating the uncertainty around such estimates). This can lead to erroneous rejection of a null hypothesis (type I error). Koukounari et al.[159] highlighted the importance of accounting for hierarchical levels of variation when analyzing the effectiveness of mass praziquantel treatments on levels of *Schistosoma mansoni* infection and morbidity in children measured pre- and post-treatment and attending different schools. Such hierarchical structures are common in data collected as part of protocols for the monitoring and evaluation of interventions.

Hierarchical generalized linear models, commonly referred to as generalized linear mixed models, may be fitted to data using an extension of the generalized linear model framework[157] which is now standard in statistical software packages. Methods for fitting hierarchical negative binomial[160] and zero-inflated models have also been developed,[161,162] although Bayesian methods which exploit the power and versatility of Markov chain Monte Carlo sampling may prove a more reliable means of fitting such models.[163,164] Bayesian methods are readily accessible through software packages such as OpenBUGS, the currently maintained and updated version of WinBUGS,[165] and JAGS.[166]

Spatial and Spatial–Temporal Models

Spatial models are hierarchical models where the clustering units are spatially structured such that the degree of correlation between pairs of observations depends on the Euclidean distance between them. Analogously, for spatial–temporal models, the correlation between observations made on the same unit but at different times may depend on their degree of temporal separation. Such correlation is termed *autocorrelation;*

the degree of correlation between a pair of observations depends on the degree of spatial or temporal separation.

Helminth spatial models typically include covariates, including remotely sensed environmental and climatic data,[167] often with the aim of enhancing the predictive accuracy of the fitted model. Bayesian Markov chain Monte Carlo methods are generally advantageous for fitting such models since they permit simultaneous estimation of the effects of covariates and spatial clustering.[168] Maximum likelihood methods are more disjointed, often involving initial fitting of a regression model followed by an assessment of spatial autocorrelation using residuals.[144,169]

Spatial models have been extensively used to construct predictive prevalence maps of STH and *Schistosoma* infections at both country[170] and regional[171,172] levels. For these infections there is an abundance of cross-sectional egg count-based community prevalence data from locations across the globe. Furthermore, these data have been collated, geo-referenced and stored in the Global Atlas of Helminth Infection database[60] which has greatly facilitated access to a rich source of data to feed geo-statistical models.[173,174] The model-predicted prevalence can be used to identify target populations for treatment[167] and, combined with demographic data, facilitate estimation of the global burden of infection and morbidity, although estimation of the latter is fraught with uncertainty.[3] The relationship between environmental covariates such as land surface temperature and precipitation and the developmental and survival rates of the free-living infective stages of the STH parasites[167] also permits an understanding of possible distributional changes in infection prevalence under a range of climate change scenarios.[175]

Spatial modeling approaches have also yielded insight into the effects that the underlying species-specific biology has on the large-scale distribution of infection. For example, Clements et al.[171] found a decreased prevalence of *A. lumbricoides* in rural compared to peri-urban or urban areas in the Great Lakes region of East Africa, in accordance with the longstanding perception that transmission of *A. lumbricoides* and *T. trichiura* is more pronounced in densely populated urban locations (in contrast to hookworm which is more associated with rural settings).[176] Brooker et al.[167] elucidated the relationship between prevalence and land surface temperature which peaks between 29 and 32°C, before declining such that in areas where the land surface temperature exceeds 36–37°C prevalence is typically <5%.[169] Saathoff et al.,[144] working in northern KwaZulu-Natal, South Africa, found that vegetation density was strongly associated with a higher prevalence of *A. lumbricoides*, results in accordance with those of a previous larger scale spatial study in Cameroon.[177] Bayesian geo-statistical models have also been developed to predict infection intensity (of *S. mansoni*) by fitting a negative binomial distribution to the data and accounting for their spatial correlation.[178]

The increase in mass drug administration (MDA, the preventive chemotherapy approach advocated by the World Health Organization)[179] poses a dilemma for geo-statistical modeling approaches as infection levels may become progressively more linked to the impact of past and ongoing local control interventions and less tied to environmental, climatic and socio-economic variables and indeed to local historical data. Consequently, it will become increasingly important to include temporal components in spatial models,[170] potentially informed by mathematical transmission models. Spatially explicit transmission models have been developed for directly-transmitted microparasitic infections, such as measles, foot-and-mouth disease and influenza,[180] and also on a community scale for schistosomiasis.[181,182]

Fitting Transmission Models to Data

It is quite common for some parameters of human transmission models (see "Dynamic models of infection and transmission," below) to be directly unobservable. For example, plausible density-dependent effects on the establishment of adult *A. lumbricoides* within the gut cannot be directly observed. In such cases, estimation relies on fitting transmission models to baseline, reinfection or longitudinal data with respect to the "free" unobservable parameters. How this is achieved, and to what degree of statistical rigor, depends on the transmission model and data in question.

The first important point to note is that fitting transmission models generally falls outside of the realms of traditional (linear) regression analysis, which rests on the assumption that parameters are multiplicative coefficients of covariates. The covariate when fitting a transmission model is typically time and/or host age, and it is common for population parameters of interest to be non-linearly related to these variables. This means that the highly efficient algorithms used for fitting regression models, such as least squares, weighted least squares or iteratively reweighted least squares, are inappropriate, and more general-purpose and computationally more intensive numerical methods, including Markov chain Monte Carlo, are required. Computational demand is also stepped up since models often require solving by numerical integration for each iterated value of the parameter(s) of interest during the fitting procedure.

Deterministic transmission models typically consider only the mean of the distribution, which is modeled with respect to time and/or age. This is sufficient to fit the model to *mean*-based data (such as the mean intensity of infection within an age category) estimated from a reasonably large sample size using a non-linear weighted least squares methodology.[183] This technique is called trajectory matching or shooting. For example, to fit a simple immigration–death transmission model (see "Dynamic models of infection and transmission," below) of *A. lumbricoides* to

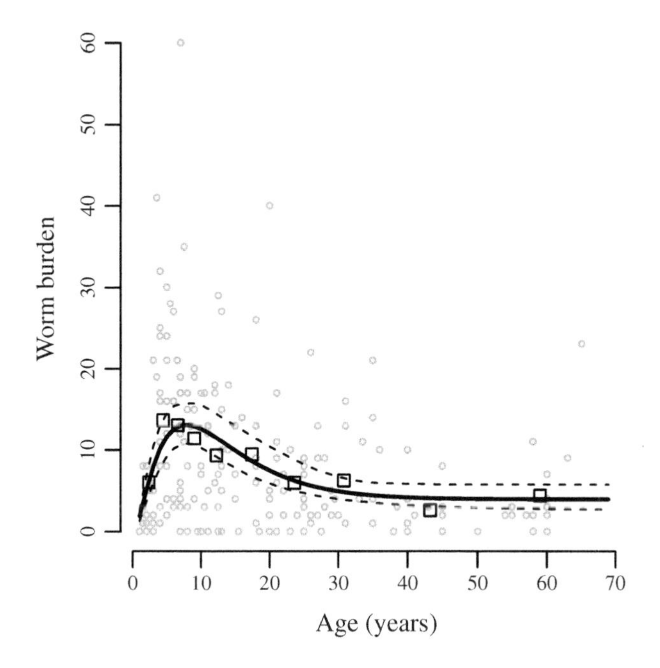

FIGURE 7.6 **An immigration–death model fitted by trajectory matching to baseline *Ascaris lumbricoides* worm burdens from a community in India.**[19] Small gray circles represent individual worm counts and the large squares represent the mean worm burden in the age categories: 1–3.5; 4–5; 6–7; 7.5–10.5; 11–13; 14–20; 21–26; 27–35; 36–50; 51+ years. The thick solid line is the fitted immigration–death model with *per capita* worm mortality rate, $\mu = 1$ year^{-1}, and age, a, -dependent force of infection (FOI), $\Lambda(a)$, given by an exponentially damped linear model[240] of the form $\Lambda(a) = (\beta_1 a - \beta_0)\exp(-\beta_2 a) + \beta_0$. The properties of $\Lambda(a)$ are such that the FOI at birth is 0 year^{-1}, $\Lambda(0) = 0$; the initial (and maximum) instantaneous rate of increase in the FOI is $\beta_0\beta_2 + \beta_1$ year^{-2}, and the long-term residual FOI as $a \to \infty$ is β_0 year^{-1}. The thin dashed lines denote 95% confidence intervals around the fitted model. The model was fitted by maximum likelihood to the individual data, assuming a negative binomial distribution with constant overdispersion parameter k. Maximum likelihood estimates to 2 significant figures are as follows: $\beta_0 = 4.0$ year^{-1}; $\beta_1 = 0.17$ year^{-2}; $\beta_2 = 5.0$ year^{-1}; $k = 0.98$.

baseline age-worm burden data one could calculate the mean worm burden in different age groups and fit the model trajectory by least squares, weighted by the sample size of each age group.

To fit a model to individual data, other distributional properties of the data must be either assumed or derived from the dynamic model. The former approach remains firmly within the domain of (traditional) empirical inference and is, in essence, analogous to trajectory matching using individual data (Figure 7.6). Revisiting the previous example, rather than grouping the worm burden data by age groups and calculating the mean, one could fit directly to the individual data assuming a negative binomial distribution with either constant k or, perhaps more realistically, k linearly dependent on the modeled mean.[57]

The latter approach of deriving distributional properties from the underlying modeled population processes is more akin to Cassie's idea of the fundamental (or mechanistic) processes that underlie observed epidemiological patterns, and requires either that the dynamic model has been formulated stochastically at an individual level or that it is possible to infer or derive distributional properties from the deterministic formulation. For example, a frequently applied interpretation of the deterministic immigration–death process is the following.[56,64,111,184,185] The population rate of parasite establishment, Λ, is considered as the average of individual rates of homogeneous or inhomogeneous Poisson processes; an individual rate, Λ_i, is given by the population (average) rate multiplied by a "susceptibility" factor, $\Lambda_i = s_i\Lambda$; susceptibility factors are assumed to be gamma distributed with mean 1 and variance $1/k$; the resulting *marginal* distribution of worms among individuals is negative binomial with mean Λ and overdispersion parameter k. Thus, the example of an immigration–death model fitted to individual *A. lumbricoides* age-worm burden data with *constant* overdispersion parameter k has plausible mechanistic (or fundamental) interpretation.

Generally, it is not possible to derive such neat analytical results from either explicitly modeled or inferred underlying stochastic processes. Indeed the marginal distribution will seldom conform to a closed-form probability distribution. This is particularly the case with longitudinal data where one has a joint distribution of repeated correlated observations made on the same individual and independent observations made from different individuals. In such circumstances fitting becomes a challenging task, reliant on computational-intensive methods. One such approach is to formulate a dynamic model in discrete time, modeling explicitly the stochastic evolution of the model using a state-space construction. Combining model predictions – conditional on current parameter values – with observed data at specific times (or host ages) permits evaluation of the likelihood. Consequently, Markov chain Monte Carlo methods can be used to evaluate the posterior distribution of the unknown parameter. For an introduction to state-space models see Chapter 11 of Bolker[186] and references therein.

Dynamic Models of Infection and Transmission

The development of mathematical, dynamic models for helminth infections was led by the pioneering contributions of Anderson and colleagues,[64,119,120,187,188] beginning in earnest in the 1970s and 1980s to understand theoretical and applied questions on helminth transmission dynamics and population biology (Chapter 9). Since this time, much progress has been made in parameterizing mathematical models using baseline (endemic equilibrium) parasitological data, and modifying basic model structures to facilitate understanding of observed epidemiological patterns which are driven by directly unobservable population processes (for a brief history of human helminth models see Basáñez et al.).[11]

Models of directly-transmitted helminths, such as the STHs, generally have an immigration–death structure at their core. The rate of immigration (establishment) of adult worms depends on the distribution of infective stages within the environment, density-dependent processes operating upon parasite establishment (including immunity), and on numerous potential host-related factors which drive heterogeneity. The rate of death (or loss) is determined by the *per capita* mortality rate of the parasite, which again may depend on density- and host-dependent factors.

Deterministic, Stochastic and Hybrid Approaches

The structure of helminth transmission models ranges from a simple deterministic description of changes in the mean number of worms per host,[118] to stochastic considerations of changes in the number of worms within individual hosts.[63] Deterministic models which ignore all stochastic elements, such as the distribution of worms among hosts, can be used to model the population mean of an underlying stochastic model under assumptions of a modestly large population size and an absence of non-linear population processes. The latter assumption is extremely restrictive given the range of non-linearities (essentially density-dependent processes) which characterize helminth population dynamics.[16,20]

Stochastic models (see, for example,[64,184,189–191]) provide a mechanistic framework with which to model the dynamic and fundamentally random processes which define transmission of infection among individuals within a community. Such models can be used to explore and predict how the dynamics of the mean and higher moments of the parasite frequency distribution are driven by underlying population processes. For example, a common application of stochastic models of *A. lumbricoides* and other STHs has been to understand how heterogeneities in host susceptibility,[61] the number of infectious eggs or larvae acquired per infection event (clumped infections) and parasite-induced host mortality can drive overdispersion[61,62,192] in the distribution of worms among hosts. Such models have also been used to predict how overdispersion will change with time (host age) under different generative mechanisms, with the aim of formulating hypotheses for testing against observed epidemiological data.[193,194]

A general conclusion arising from stochastic formulations of STH population dynamics is that a quite bewildering array of heterogeneous population processes can generate observed macro-epidemiological trends, a point noted by Cassie in 1963[117] and revisited by Crofton in 1971[116], both prior to any formal mathematical analysis. It is also apparent that without simplifying assumptions or approximations to aid analytical exploration,[62,192,195] models can become too intractable to yield useful results.

Hybrid models[196] are essentially deterministic but include, in a phenomenological manner, stochastic elements such as the distribution of parasites among hosts or the observed host demography (Chapter 9). This permits the population-level effects of non-linear density-dependent processes, and demographic variability in infection and transmission, to be accounted for. This approach has been used to great effect[16,20] and has shaped our fundamental understanding of the population dynamics of helminths in general, and of *A. lumbricoides* in particular.

Parasite Distribution and Density-dependent Processes

The size and stability of helminth populations is critically regulated by the interaction between the overdispersed distribution of parasites among hosts[119,197,198] and the predominant mechanism causing the overdispersion.[199] In general, overdispersion enhances the severity of density-dependent processes, highlighting the critical requirement for population dynamics models to account for this ubiquitous characteristic of helminth populations.

Negative density-dependent mechanisms restrict the rate of population growth at high densities and induce resilience to perturbation resulting from the relaxation of constraints when densities decrease.[16,64,200,201] The only directly observable density-dependent process in *A. lumbricoides* infections is on female worm fecundity, namely, the *per capita* egg output (eggs per gram of feces) declines with increasing worm burden (Figure 7.4), although similar but unobservable effects on mortality or establishment, possibly interacting with host immune responses, are possible.[197] Regardless of the specific density dependencies in operations, a worrying connotation for the success of *A. lumbricoides* control is that increasing degrees of overdispersion, which occur with *decreasing* average worm burdens,[57] enhance the parasite's resilience and sustainability at very low population densities.[202]

A positive, or facilitating, density-dependent effect common to all dioecious (separate sexes) obligatory parasites is the female mating probability;[203] the chance of a female worm infecting a host without a co-infecting male worm decreases with increasing population density. Facilitating density-dependent mechanisms (Allee effect[204]) act as a destabilizing force, either promoting runaway growth (though at high population densities constraining processes are likely to operate) or, at low densities, failing to facilitate enough growth leading to population collapse.[118] Just as with constraining density dependencies, positive density dependencies are enhanced by the overdispersed distribution of worms among hosts since the majority of worms reside in large infrapopulations. Consequently, the threshold density of population collapse, or breakpoint density (also known as the transmission breakpoint or unstable equilibrium), is likely to be extremely low for *A. lumbricoides*,[205] partly on account of the assumed polygamous nature of male worms (one male can mate all females within a host), but also because of the overdispersed worm distribution.

Age-structured Models

Many of the parameters included in population dynamics models are likely to vary with host age. Indeed for *A. lumbricoides*, age-related changes in exposure and (protective) acquired immunity have long been considered as the principal drivers underlying the observed convex age-infection profiles.[16,20,67] Age-related changes in contamination of the environment are equally as important, but because of a lack of independent epidemiological data, relative rates of contamination among age groups must be either assumed, or if possible, inferred indirectly from model output.[206]

In their most general form, age-structured deterministic models describe changes in the mean worm burden with respect to time *and* host age, often using partial differential equations (see Chapter 9).[16,201] Consequently, they offer greater realism in two key ways. First, they can capture adequately observed age-dependent epidemiological patterns.[16,20] Second, the interplay between age-infection profiles and the demography of the host population is modeled explicitly; the contribution to transmission of a particular age group depends on their average worm burden, the net severity of any density-dependent constraints on female worm fecundity, their contamination of the environment, and their proportional representation in the host population.[16,201]

Age-structured models may also be formulated in a less general manner, tailored to the context of their application. This can ease community-specific parameterizations, especially when limited epidemiological data are available. For example, the EpiWorm[206,207] transmission model for *A. lumbricoides*, *T. trichiura* and hookworm, which is one of three models included in the EpiDynamics software suite of age-structured helminth models (the others being EpiFil for lymphatic filariasis[208,209] and EpiSchisto for schistosomiasis[210]), is optionally divided into age groups of children and adults, or school children, other children and adults. This is because the chief aim of the model is to simulate the effect of mass treatments targeted at (school-age) children, the predominant strategy of global STH control.[179]

Modeling Treatment

The effects of anthelmintic treatment on *A. lumbricoides*, and on STHs in general, have been modeled in two ways (Chapter 9). The first assumes that treatments are administered to the population at random for a certain time period and that during this period adult worms incur an increased mortality rate which is a function of the coverage, the drug efficacy and the frequency of treatment.[188,211] This approach has yielded: (1) analytical results pertaining to a critical proportion of the population that it is necessary to treat to achieve specific reductions in mean worm burden, including reductions which reduce burdens under breakpoint densities, for different intensities of transmission (encapsulated in the basic reproduction ratio or R_0), and (2) an empirical understanding of how selectively treating specific age groups leads to collateral reductions in infection levels in untreated age groups due to the overall effect on environmental transmission.[16,20,211]

The alternative approach is to model treatment rounds explicitly by assuming the instantaneous death of a fixed[206] or random[56] fraction of worms following each treatment which is determined by the drug efficacy and the level of coverage.[64] The framework has also been elaborated to take into account the dynamic effects of treatment on the distribution of worms among hosts.[56] Together these developments permit: (1) an accurate reflection of the effects of treatment on the net severity of density-dependent processes; (2) the estimation of community morbidity using a threshold worm burden to define disease; (3) the quantification of community benefit as the area between the infection (prevalence or intensity) time curve and the equilibrium level of infection (i.e. the level of infection in the absence of intervention),[206] and (4) by linking rounds of treatments to monetary costs, cost-effectiveness analysis.[212]

Although the effect of mass treatments has largely been assessed in terms of population-level effects, such as bounce-back (reinfection) times following treatment[56] or the effectiveness of targeting specific age groups,[206] models have also been used to explore individual-based treatment effects. In particular, the so-called selective strategy, which involves selectively treating heavily infected or high risk individuals, has been shown to be extremely effective in reducing average worm burdens, so long as the degree of overdispersion is high.[201] This result is in parallel with the effectiveness of targeting "superspreaders" of directly-transmitted microparasites.[213] Further, individual-based stochastic simulations have indicated that the population-level (mean) rate of reinfection following treatment is strongly dependent on the predominant aggregation-generating mechanism; aggregation arising from differences in host susceptibilities (analogous to predisposition) causes more rapid reinfection than when aggregation is generated solely by a clumped infection process.[64]

The cost of repeated parasitological surveys means that selective treatment strategies are unlikely to be deployed on a large scale. Rather, mathematical models of school-age targeted treatments are in broad consensus with the view that this strategy is a cost-effective means of controlling morbidity.[206,212] This is in accordance with empirical evidence demonstrating that de-worming children produces improvements in growth, educational attainment[214–216] and school participation[217] (although it is noteworthy that quantifying the benefit of de-worming is notoriously difficult and is the subject of some recent controversy).[218] Moreover, the schools' infrastructure offers a convenient conduit to achieve high coverage of the most heavily infected individuals.

Model Validation

Validation of model predictions against parasitological (or other) data is imperative, particularly if output is going to be trusted and used by intervention planners and control strategists. Validation can be undertaken in a number of ways: a qualitative appraisal of model output against observed data;[206,210] the calculation of generic statistical measures of fit, such as the coefficient of determination (R^2);[219] or the calculation of (log-) likelihoods, whereby the higher the likelihood, the better the fit.[208]

Ideally, all model parameters should be estimated prior to validation using independent data. In practice, this is frequently not achieved since data used for parameter estimation are often validated against post-intervention data collected from the same individuals or population. For example, EpiWorm has been fully parameterized for *A. lumbricoides* using baseline worm burden data from Burma (Myanmar) and qualitatively validated against mean worm burdens collected over a 2-year period of 3-monthly treatments in the same community.[206,220]

For validation data to be fully independent of parameterization data is, in essence, making the somewhat optimistic assumption that the model in question is completely generalizable across heterogeneous host–parasite populations. For example, the fecundity of female *A. lumbricoides* is highly variable among geographical locations, yet the underlying processes causing this are unknown,[46] and consequently reparameterization and revalidation are necessary on a community basis. More generally, best practice should involve continual reappraisal of model output especially in the contemporary context of parasite populations being exposed to repeated and regular doses of anthelmintics.[113]

Anthelmintic Resistance

Given the continuing increase in MDA of anthelmintics,[33] population dynamics models adapted to capture the effects of evolving anthelmintic drug resistance[221] may become invaluable tools for decision support in refining treatment strategies (such as the use of combination therapies) in the face of declining single-dose drug efficacy.[11] Currently there is no evidence for anthelmintic (benzimidazole) resistant *A. lumbricoides*, although this is not the case for human hookworms,[222] and a single nucleotide polymorphism (SNP) in parasite β-tubulin – associated with widespread benzimidazole resistance in nematode species of veterinary importance[223] – has recently been shown to occur at high frequency in *T. trichiura*-infected people repeatedly treated with benzimidazoles in Panama.[224]

For transmission models to be most useful they need to be parameterized using data on the phenotypic manifestations (increased drug tolerance of worms, potential fitness costs) and genetic basis of resistance (how many loci would be involved, whether or not these are linked, whether anthelmintic resistance is conferred by recessive alleles) which should be collected as part of control program monitoring and evaluation activities.[225] It is also likely that individual-based, stochastic models will be necessary on the basis of modeling work in veterinary animals, demonstrating that the mode of infection (trickle or clumped) significantly influences the rate of spread of anthelmintic resistance.[226,227]

Elimination

In certain circumscribed foci, intervention efforts are moving from morbidity control to transmission interruption and elimination of STHs, including *A. lumbricoides*.[228] Experiences from Japan,[229] the Republic of Korea[230] and most recently from the People's Republic of China[231] suggest that such ambitious targets will only be achieved using integrated control approaches, encompassing a consolidation of MDA combined with socio-economic improvements, particularly improved access to sanitation.[72,232] Similar conclusions have been reached from mathematical models; the theoretical transmission breakpoint densities for *A. lumbricoides* and other STHs are extremely low,[16,203] meaning that mass treatments alone would need to be administered at a prohibitively high coverage and frequency.[20]

Despite this, in low endemicity settings or where combined control initiatives have resulted in prolonged heavy suppression of transmission, models may be warranted that are suitable for assessing the feasibility of elimination under specific conditions that reflect the reality of a control program. In such circumstances, individual-based stochastic modeling approaches become necessary, in essence because at low parasite densities the probability of stochastic population fade-out (extinction) becomes important, a phenomenon which is omitted from mean-based (mass action) deterministic constructs. The output from multiple runs of stochastic models can be used to construct probabilistic statements about the chance of elimination under specified control conditions.

In other helminthiases, namely filariasis and schistosomiasis, individual-based stochastic models have been developed as stand-alone computer simulation models (ONCHOSIM for onchocerciais,[233] LYMFASIM for lymphatic filariasis[234] and SCHISTOSIM[235] for schistosomiasis) and have been used for answering very applied questions on the effectiveness of helminth control strategies[236] and the feasibility of elimination.[237] Although such models offer the potential to include much more complexity and realism compared to their deterministic counterparts, adequate parameterization can be difficult unless extremely detailed, community-specific epidemiological information is available. This renders such models rather location specific and makes their applicability to the broader epidemiological context rather uncertain.

CONCLUSIONS

Ascariasis, like other helminthiases, is currently being targeted by MDA of anthelmintics on an unprecedented scale. To guide, enhance and ensure the ongoing effectiveness of MDA-based control it is imperative that the epidemiology and population biology of *A. lumbricoides* continues to be elucidated by applying and developing mathematical and statistical analytical approaches. Those applying these disciplines in research, development and data analysis should endeavor to communicate and collaborate with those involved with policy making, planning and application to maximize benefits to public health. To substantiate this ideal, we conclude by revisiting some of the parasite's epidemiology which has particular relevance to the effective implementation of MDA-based control.

Monitoring and evaluation of the effectiveness of MDA — which when undertaken is generally based on the direct smear Kato-Katz method for detecting eggs in feces — must take into account diagnostic performance under changing endemicities. In particular, because diagnostic sensitivity declines with decreasing infection intensity, egg count-based estimates of intensity and prevalence will become increasingly underestimated. Premature cessation of adequate control could result if strategic decisions on when MDA can safely be halted or reduced in intensity (treatment frequency) are based on absolute pre-defined thresholds levels of infection which do not account for variable diagnostic sensitivity. Stopping treatments too early will not only lead to recrudescence of infection but also — if alleles conferring anthelmintic resistance have been spreading through the parasite population — to widespread and catastrophic treatment failure following recommencement of control.[238]

In general, egg count-based threshold levels of infection – to define targets for control or to inform the most appropriate control strategy – should be decided locally, principally because the fecundity of female *A. lumbricoides* appears to be markedly geographically heterogeneous.[46] For example, the World Health Organization indicates that when >10% of a target population have high intensity infections with *A. lumbricoides* – defined by the threshold of 50,000 eggs per gram of feces – intensive control efforts (two or three rounds of MDA per year) should be undertaken.[239] Yet, because of geographically variable fecundity, this threshold is associated with markedly different worm burdens in different locations.

The observed variability in fecundity also has profound implications for the generality of transmission models. Female worm fecundity, in conjunction with the severity of density dependence, is critical in relating an individual's worm burden to his or her contribution to transmission. It is also a component of the basic reproductive number, R_0, and the theoretical transmission breakpoint density. Consequently, both of these quantities will be locale specific. Fecundity can only be estimated from paired egg count–worm count data, the latter obtained by chemo-expulsion techniques. Thus, for transmission models to be most useful in informing local control policy and strategy, epidemiological surveys would ideally include limited and judiciously planned chemo-expulsions.

In most settings, elimination of *A. lumbricoides* by MDA alone is unrealistic. Reasons for this include the high reproductive capacity of female worms; the overdispersed distribution of worms among hosts which enhances the parasite population's resilience to perturbation and facilitates persistence at very low densities, and the increasing degree of overdispersion that occurs with decreasing parasite density. Consequently, morbidity reduction is often a more achievable aim. It is thus rather conspicuous that the relationship between infection and disease is poorly understood. This is because the effects of chronic infection, mediated chiefly by malnutrition, are rather covert and not associated with readily measurable biomarkers; such as hemoglobin as an indicator of iron-deficiency anemia caused by hookworm infection. Attempts to quantify population levels of morbidity associated with *A. lumbricoides* have applied – without empirical justification – a threshold worm burden above which morbidity is observed.[56,57,206] More research is clearly needed in this area so that populations can be targeted for control on the basis of disease rather than infection; to facilitate the development of disease models; and to enable cost-effective analysis of morbidity control strategies.

In future, the development of spatially explicit population dynamics transmission models which could be coupled to geo-statistical models would represent an exceptional achievement of great practical relevance to the efficient implementation of helminth control programs. In theory, spatio-temporal maps could be generated and used to refine, in real-time, MDA or other intervention strategies to reflect the effects of ongoing control at a variety of spatial scales. Furthermore, in the longer term, model-based maps could be used to refine and alter strategy in response to climatic, environmental and socio-economic changes. Such aims are technically challenging and will only be achieved through committed and enduring collaborative efforts among all those involved. Yet, in these times of unprecedented international commitment and financial support, such aspirations should be embraced.

References

1. Bethony J, Brooker S, Albonico M, et al. Soil-transmitted helminth infections: ascariasis, trichuriasis, and hookworm. *Lancet* 2006;**367**:1521–32.
2. de Silva NR, Brooker S, Hotez PJ, Montresor A, Engels D, Savioli L. Soil-transmitted helminth infections: updating the global picture. *Trends Parasitol* 2003;**19**:547–51.
3. Brooker S. Estimating the global distribution and disease burden of intestinal nematode infections: adding up the numbers – a review. *Int J Parasitol* 2010;**40**:1137–44.
4. Hotez PJ, Molyneux DH, Fenwick A, et al. Control of neglected tropical diseases. *N Engl J Med* 2007;**357**:1018–27.
5. World Health Organization. *Accelerating work to overcome the global impact of neglected tropical diseases: a roadmap for implementation*. Geneva: World Health Organization; 2012.
6. World Health Organization. *Working to Overcome the Global Impact of Neglected Tropical Diseases: First WHO Report on Neglected Tropical Diseases*. Geneva: World Health Organization; 2010.
7. Department for International Development. London declaration on neglected tropical diseases, http://www.dfid.gov.uk/Documents/publications1/NTDEvent-London DeclarationonNTDs.pdf; 2012.
8. Utzinger J. A research and development agenda for the control and elimination of human helminthiases. *PLoS Negl Trop Dis* 2012;**6**:e1646.
9. Lustigman S, Geldhof P, Grant WN, Osei-Atweneboana MY, Sripa B, Basáñez M- G. A research agenda for helminth diseases of humans: basic research and enabling technologies to support control and elimination of helminthiases. *PLoS Negl Trop Dis* 2012;**6**:e1445.
10. McCarthy JS, Lustigman S, Yang G-J, et al. A research agenda for helminth diseases of humans: diagnostics for control and elimination programmes. *PLoS Negl Trop Dis* 2012;**6**:e1601.
11. Basáñez M-G, McCarthy JS, French MD, et al. A research agenda for helminth diseases of humans: modelling for control and elimination. *PLoS Negl Trop Dis* 2012;**6**:e1548.
12. Crompton DWT. Biology of *Ascaris lumbricoides*. In: Crompton DWT, Nesheim MC, Pawlowski ZS, editors. *Ascariasis and its Prevention and Control*. London: Taylor & Francis; 1989. p. 9–44.
13. O'Lorcain P, Holland CV. The public health importance of *Ascaris lumbricoides*. *Parasitology* 2000;**121**(Suppl):S51–71.
14. Geenen PL, Bresciani J, Boes J, et al. The morphogenesis of *Ascaris suum* to the infective third-stage larvae within the egg. *J Parasitol* 1999;**85**:616–22.
15. Maung M. The occurrence of the second moult of *Ascaris lumbricoides* and *Ascaris suum*. *Int J Parasitol* 1978;**8**:371–8.

16. Anderson RM, May RM. Helminth infections of humans: mathematical models, population dynamics, and control. *Adv Parasitol* 1985;**24**:1–101.
17. Dold C, Holland CV. *Ascaris* and ascariasis. *Microbes Infect* 2011;**13**:632–7.
18. Seo BS. Epidemiology and control of ascariasis in Korea. *Korean J Parasitol* 1990;**28**(Suppl):S49–61.
19. Elkins DB, Haswell-Elkins M, Anderson RM. The epidemiology and control of intestinal helminths in the Pulicat Lake region of Southern India. I. Study design and pre- and post-treatment observations on *Ascaris lumbricoides* infection. *Trans R Soc Trop Med Hyg* 1986;**80**:774–92.
20. Anderson RM, May RM. *Infectious Diseases of Humans: Dynamics and Control*. Oxford: Oxford University Press; 1992.
21. Crompton DW. *Ascaris* and ascariasis. *Adv Parasitol* 2001;**48**:285–375.
22. Nejsum Jr P, Parker ED, Frydenberg J, et al. Ascariasis is a zoonosis in Denmark. *J Clin Microbiol* 2005;**43**:1142–8.
23. Anderson TJC. *Ascaris* infections in humans from North America: molecular evidence for cross-infection. *Parasitology* 1995;**110**:215.
24. Criscione CD, Anderson JD, Sudimack D, et al. Disentangling hybridization and host colonization in parasitic roundworms of humans and pigs. *Proc Biol Sci* 2007;**274**:2669–77.
25. Peng W, Zhou X. Genetic variation in sympatric *Ascaris* populations from humans and pigs in China. *Parasitology* 1998;**117**:355–61.
26. Anderson TJC, Romero-Abal ME, Jaenike J. Genetic structure and epidemiology of *Ascaris* populations: patterns of host affiliation in Guatemala. *Parasitology* 1993;**107**:319–34.
27. Holland C, Boes J. Distributions and predisposition: people and pigs. In: Holland CV, Kennedy MW, editors. *The Geohelminths: Ascaris, Trichuris and Hookworm*. New York: Springer; 2002. p. 1–24.
28. Boes J, Medley GF, Eriksen L, Roepstorff A, Nansen P. Distribution of *Ascaris suum* in experimentally and naturally infected pigs and comparison with *Ascaris lumbricoides* infections in humans. *Parasitology* 1998;**117**:589–96.
29. Stephenson LS, Latham MC, Ottesen EA. Malnutrition and parasitic helminth infections. *Parasitology* 2000;**121**(Suppl):S23–38.
30. Hotez PJ, Cerami A. Secretion of a proteolytic anticoagulant by *Ancylostoma* hookworms. *J Exp Med* 1983;**157**:1594–603.
31. Stephenson LS. Pathophysiology of intestinal nematodes. In: Holland CV, Kennedy MW, editors. *The Geohelminths:* Ascaris, Trichuris *and Hookworm*. New York: Springer; 2002.
32. de Silva NR, Guyatt HL, Bundy DA. Morbidity and mortality due to *Ascaris*-induced intestinal obstruction. *Trans R Soc Trop Med Hyg* 1997;**91**:31–6.
33. Prichard RK, Basáñez M-G, Boatin B, et al. A research agenda for helminth diseases of humans: intervention for control and elimination. *PLoS Negl Trop Dis* 2012;**6**:e1549.
34. Basáñez M-G, Pion SDS, Churcher TS, Breitling LP, Little MP, Boussinesq M. River blindness: a success story under threat? *PLoS Med* 2006;**3**:e371.
35. Wu S. Sonographic findings of *Ascaris lumbricoides* in the gastrointestinal and biliary tracts. *Ultrasound Q* 2009;**25**:207–9.
36. Das CJ, Kumar J, Debnath J, Chaudhry A. Imaging of ascariasis. *Australas Radiol* 2007;**51**:500–6.
37. Hall A, Romanova T. *Ascaris lumbricoides*: detecting its metabolites in the urine of infected people using gas–liquid chromatography. *Exp Parasitol* 1990;**70**:35–42.
38. Basuni M, Muhi J, Othman N, et al. A pentaplex real-time polymerase chain reaction assay for detection of four species of soil-transmitted helminths. *Am J Trop Med Hyg* 2011;**84**:338–43.

39. Taniuchi M, Verweij JJ, Noor Z, et al. High throughput multiplex PCR and probe-based detection with Luminex beads for seven intestinal parasites. *Am J Trop Med Hyg* 2011;**84**:332–7.
40. Chatterjee BP, Santra A, Karmakar PR, Mazumder DN. Evaluation of IgG4 response in ascariasis by ELISA for serodiagnosis. *Trop Med Int Health* 1996;**1**:633–9.
41. Needham CS, Lillywhite JE, Beasley NMR, Didier JM, Kihamia CM, Bundy DAP. Potential detection for diagnosis in saliva of intestinal nematode infections through antibody detection in saliva. *Trans R Soc Trop Med Hyg* 1996;**90**:526–30.
42. World Health Organization. *Bench aids for the diagnosis of intestinal parasites*. Geneva: World Health Organization; 1994.
43. Leles D, Gardner SL, Reinhard K, Iñiguez A, Araujo A. Are *Ascaris lumbricoides* and *Ascaris suum* a single species? *Parasit Vectors* 2012;**5**:42.
44. Walker M, Hall A, Anderson RM, Basáñez M- G. Density-dependent effects on the weight of female *Ascaris lumbricoides* infections of humans and its impact on patterns of egg production. *Parasit Vectors* 2009;**2**:11.
45. Kotze AC, Kopp SR. The potential impact of density dependent fecundity on the use of the faecal egg count reduction test for detecting drug resistance in human hookworms. *PLoS Negl Trop Dis* 2008;**2**. e297.
46. Hall A, Holland C. Geographical variation in *Ascaris lumbricoides* fecundity and its implications for helminth control. *Parasitol Today* 2000;**16**:540–4.
47. Sinniah B. Daily egg production of *Ascaris lumbricoides*: the distribution of eggs in the faeces and the variability of egg counts. *Parasitology* 1982;**84**:167–75.
48. Hall A. Quantitative variability of nematode egg counts in faeces: a study among rural Kenyans. *Trans R Soc Trop Med Hyg* 1981;**75**:682–7.
49. Walker M, Hall A, Basáñez M- G. Trickle or clumped infection process? An analysis of aggregation in the weights of the parasitic roundworm of humans, *Ascaris lumbricoides*. *Int J Parasitol* 2010;**40**:1373–80.
50. Luciano F, Santos N, José E, Cerqueira L, Soares NM. Comparison of the thick smear and Kato-Katz techniques for diagnosis of intestinal helminth infections [Comparação das técnicas de sedimentação espontânea e Kato-Katz para diagnóstico das helmintoses intestinais]. *Rev Soc Bras Med Trop* 2005;**38**:196–8.
51. World Health Organization. *Basic laboratory methods in medical parasitology*. Geneva: World Health Organization; 1991.
52. Engels D, Nahimana S, De Vlas SJ, Gryseels B. Variation in weight of stool samples prepared by the Kato-Katz method and its implications. *Trop Med Int Health* 1997;**2**:265–71.
53. Garcia LS. *Diagnostic Medical Parasitology*. Herndon, VA: American Society for Microbiology; 2007.
54. Goodman D, Haji HJ, Bickle QD, et al. A comparison of methods for detecting the eggs of *Ascaris, Trichuris*, and hookworm in infant stool, and the epidemiology of infection in Zanzibari infants. *Am J Trop Med Hyg* 2007;**76**:725–31.
55. Guyatt HL, Bundy DA. Estimating prevalence of community morbidity due to intestinal helminths: prevalence of infection as an indicator of the prevalence of disease. *Trans R Soc Trop Med Hyg* 1991;**85**:778–82.
56. Medley GF, Guyatt HL, Bundy DA. A quantitative framework for evaluating the effect of community treatment on the morbidity due to ascariasis. *Parasitology* 1993;**106**:211–21.
57. Guyatt HL, Bundy DA, Medley GF, Grenfell BT. The relationship between the frequency distribution of *Ascaris lumbricoides* and the prevalence and intensity of infection in human communities. *Parasitology* 1990;**101**:139–43.
58. Thein-Hlaing, Than-Saw, Htay-Htay-Aye Myint-Lwin. Thein-Maung-Myint. Epidemiology and transmission dynamics of *Ascaris lumbricoides* in Okpo village, rural Burma. *Trans R Soc Trop Med Hyg* 1984;**78**:497–504.

59. Holland CV, Asaolu SO, Crompton DW, Stoddart RC, Macdonald R, Torimiro SE. The epidemiology of *Ascaris lumbricoides* and other soil-transmitted helminths in primary school children from Ile-Ife, Nigeria. *Parasitology* 1989;**99**:275–85.
60. Brooker S, Hotez PJ, Bundy DAP. The global atlas of helminth infection: mapping the way forward in neglected tropical disease control. *PLoS Negl Trop Dis* 2010;**4**:e779.
61. Anderson RM, Gordon DM. Processes influencing the distribution of parasite numbers within host populations with special emphasis on parasite-induced host mortalities. *Parasitology* 1982;**85**:373–98.
62. Isham V. Stochastic models of host-macroparasite interaction. *Ann Appl Probab* 1995;**5**:720–40.
63. Tallis GM, Leyton MK. Stochastic models of populations of helminthic parasites in the definitive host. *I. Math Biosci* 1969;**4**:39–48.
64. Anderson RM, Medley GF. Community control of helminth infections of man by mass and selective chemotherapy. *Parasitology* 1985;**90**(Pt 4):629–60.
65. Walker M, Hall A, Basáñez M- G. Trickle or clumped infection process? A stochastic model for the infection process of the parasitic roundworm of humans, *Ascaris lumbricoides. Int J Parasitol* 2010;**40**:1381–8.
66. Anderson TJC, Romero-Abal ME, Jaenike J. Mitochondrial DNA and *Ascaris* microepidemiology: the composition of parasite populations from individual hosts, families and villages. *Parasitology* 1995;**110**:221–9.
67. Bundy DA, Medley GF. Immuno-epidemiology of human geohelminthiasis: ecological and immunological determinants of worm burden. *Parasitology* 1992;**104**(Suppl): S105–19.
68. Wong MS, Bundy DA. Quantitative assessment of contamination of soil by the eggs of *Ascaris lumbricoides* and *Trichuris trichiura. Trans R Soc Trop Med Hyg* 1990;**84**:567–70.
69. Wong MS, Bundy DA, Golden MH. The rate of ingestion of *Ascaris lumbricoides* and *Trichuris trichiura* eggs in soil and its relationship to infection in two children's homes in Jamaica. *Trans R Soc Trop Med Hyg* 1991;**85**:89–91.
70. Geissler W, Mwaniki D, Thiong F, Friis H. Geophagy as a risk factor for geohelminth infections: a longitudinal study of Kenyan primary school children. *Trans R Soc Trop Med Hyg* 1998;**92**:7–11.
71. Scott ME. *Ascaris lumbricoides*: a review of its epidemiology and relationship to other infections. *Ann Nestle Eng* 2008;**66**:7–22.
72. Ziegelbauer K, Speich B, Mäusezahl D, Bos R, Keiser J, Utzinger J. Effect of sanitation on soil-transmitted helminth infection: systematic review and meta-analysis. *PLoS Med* 2012;**9**:e1001162.
73. Haswell-Elkins M, Elkins D, Anderson RM. The influence of individual, social group and household factors on the distribution of *Ascaris lumbricoides* within a community and implications for control strategies. *Parasitology* 1989;**98**:125–34.
74. Holland CV, Taren DL, Crompton DW, et al. Intestinal helminthiases in relation to the socioeconomic environment of Panamanian children. *Soc Sci Med* 1988;**26**:209–13.
75. Kightlinger LK, Seed JR, Kightlinger MB. *Ascaris lumbricoides* intensity in relation to environmental, socioeconomic, and behavioral determinants of exposure to infection in children from southeast Madagascar. *J Parasitol* 1998;**84**:480–4.
76. Walker M, Hall A, Basáñez M- G. Individual predisposition, household clustering and risk factors for human infection with *Ascaris lumbricoides*: new epidemiological insights. *PLoS Negl Trop Dis* 2011;**5**:e1047.
77. Williams-Blangero S, Subedi J, Upadhayay RP, et al. Genetic analysis of susceptibility to infection with *Ascaris lumbricoides. Am J Trop Med Hyg* 1999;**60**:921–6.
78. Williams-Blangero S, VandeBerg JL, Subedi J, et al. Genes on chromosomes 1 and 13 have significant effects on *Ascaris* infection. *Proc Natl Acad Sci USA* 2002;**99**:5533–8.

79. Williams-Blangero S, Vandeberg JL, Subedi J, Jha B, Corrêa-Oliveira R, Blangero J. Localization of multiple quantitative trait loci influencing susceptibility to infection with *Ascaris lumbricoides*. *J Infect Dis* 2008;**197**:66–71.
80. Chai JY, Seo BS, Lee SH, Cho SY. Epidemiological studies on *Ascaris lumbricoides* reinfection in rural communities in Korea II. Age-specific reinfection rates and familial aggregation of the reinfected cases. *Korean J Parasitol* 1983;**21**:142–9.
81. Forrester JE, Scott ME, Bundy DA, Golden MH. Clustering of *Ascaris lumbricoides* and *Trichuris trichiura* infections within households. *Trans R Soc Trop Med Hyg* 1988;**82**:282–8.
82. Criscione CD, Anderson JD, Sudimack D, et al. Landscape genetics reveals focal transmission of a human macroparasite. *PLoS Negl Trop Dis* 2010;**4**. e665.
83. Cort WW, Stoll NR. Studies on *Ascaris lumbricoides* and *Trichuris trichiura* in China. *Am J Epidemiol* 1931;**14**:655–89.
84. Brown H. Human *Ascaris* as a household infection. *J Parasitol* 1927;**13**:206–12.
85. Otto GF, Cort WW, Keller AE. Environmental studies of families in Tennessee infested with *Ascaris lumbricoides, Trichuris trichiura* and hookworm. *Am J Epidemiol* 1931;**14**:156–93.
86. Williams D, Burke G, Hendley JO. Ascariasis: a family disease. *J Pediatr* 1974; **84**:853–4.
87. Anderson TJ, Zizza CA, Leche GM, Scott ME, Solomons NW. The distribution of intestinal helminth infections in a rural village in Guatemala. *Mem Inst Oswaldo Cruz* 1993;**88**:53–65.
88. Cairncross S, Blumenthal U, Kolsky P, Moraes L, Tayeh A. The public and domestic domains in the transmission of disease. *Trop Med Int Health* 1996;**1**:27–34.
89. Dold C, Holland CV. Investigating the underlying mechanism of resistance to *Ascaris* infection. *Microbes Infect* 2011;**13**:624–31.
90. Cooper PJ, Chico ME, Sandoval C, et al. Human infection with *Ascaris lumbricoides* is associated with a polarized cytokine response. *J Infect Dis* 2000;**182**:1207–13.
91. Bradley JE, Jackson JA. Immunity, immunoregulation and the ecology of trichuriasis and ascariasis. *Parasite Immunol* 2005;**26**:429–41.
92. King E-M, Kim HT, Dang NT, et al. Immuno-epidemiology of *Ascaris lumbricoides* infection in a high transmission community: antibody responses and their impact on current and future infection intensity. *Parasite Immunol* 2005;**27**:89–96.
93. Palmer DR, Hall A, Haque R, Anwar KS. Antibody isotype responses to antigens of *Ascaris lumbricoides* in a case-control study of persistently heavily infected Bangladeshi children. *Parasitology* 1995;**111**:385–93.
94. Haswell-Elkins MR, Kennedy MW, Maizels RM, Elkins DB, Anderson RM. The antibody recognition profiles of humans naturally infected with *Ascaris lumbricoides*. *Parasite Immunol* 1989;**11**:615–27.
95. Turner JD, Faulkner H, Kamgno J, et al. Th2 cytokines are associated with reduced worm burdens in a human intestinal helminth infection. *J Infect Dis* 2003;**188**:1768–75.
96. Jackson JA, Turner JD, Rentoul L, et al. T helper cell type 2 responsiveness predicts future susceptibility to gastrointestinal nematodes in humans. *J Infect Dis* 2004;**190**:1804–11.
97. Bundy DA, Kan SP, Rose R. Age-related prevalence, intensity and frequency distribution of gastrointestinal helminth infection in urban slum children from Kuala Lumpur, Malaysia. *Trans R Soc Trop Med Hyg* 1988;**82**:289–94.
98. Hall A, Anwar KS, Tomkins AM. Intensity of reinfection with *Ascaris lumbricoides* and its implications for parasite control. *Lancet* 1992;**339**:1253–7.
99. Anderson RM, May RM. Herd immunity to helminth infection and implications for parasite control. *Nature* 1985;**315**:493–6.
100. Berding C, Keymer AE, Murray JD, Slater AF. The population dynamics of acquired immunity to helminth infection. *J Theor Biol* 1986;**122**:459–71.

101. Woolhouse ME. A theoretical framework for immune responses and predisposition to helminth infection. *Parasite Immunol* 1993;**15**:583–94.
102. Woolhouse ME. Patterns in parasite epidemiology: the peak shift. *Parasitol Today* 1998;**14**:428–34.
103. Blackwell AD, Gurven MD, Sugiyama LS, et al. Evidence for a peak shift in a humoral response to helminths: age profiles of IgE in the Shuar of Ecuador, the Tsimane of Bolivia, and the U.S. NHANES. *PLoS Negl Trop Dis* 2011;**5**:e1218.
104. Hall A, Anwar KS, Tomkins A. The distribution of *Ascaris lumbricoides* in human hosts: a study of 1765 people in Bangladesh. *Trans R Soc Trop Med Hyg* 1999;**93**:503–10.
105. Jia T-W, Melville S, Utzinger J, King CH, Zhou X- N. Soil-transmitted helminth reinfection after drug treatment: a systematic review and meta-analysis. *PLoS Negl Trop Dis* 2012;**6**:e1621.
106. Holland CV. Predisposition to ascariasis: patterns, mechanisms and implications. *Parasitology* 2009;**136**:1537–47.
107. Keymer A, Pagel M, Schad GA, Warren KS. Predisposition to helminth infection. In: Schad GA, Warren KS, editors. *Hookworm Disease: Current Status and New Directions*. London: Taylor & Francis; 1990. p. 177–209.
108. Forrester JE, Scott ME, Bundy DAP, Golden MHN. Predisposition of individuals and families in Mexico to heavy infection with *Ascaris lumbricoides* and *Trichuris trichiura*. *Trans R Soc Trop Med Hyg* 1990;**84**:272–6.
109. Chan L, Bundy DA, Kan SP. Aggregation and predisposition to *Ascaris lumbricoides* and *Trichuris trichiura* at the familial level. *Trans R Soc Trop Med Hyg* 1994;**88**:46–8.
110. Chan L, Bundy DA, Kan SP. Genetic relatedness as a determinant of predisposition to *Ascaris lumbricoides* and *Trichuris trichiura* infection. *Parasitology* 1994;**108**:77.
111. McCallum HI. Covariance in parasite burdens: the effect of predisposition to infection. *Parasitology* 1990;**100**:153.
112. Thein-Hlaing, Than-Saw, Htay-Htay-Aye, Myint-Lwin. Reinfection of people with *Ascaris lumbricoides* following single, 6-month and 12-month interval mass chemotherapy in Okpo village, rural Burma. *Trans R Soc Trop Med Hyg* 1987;**81**:140–6.
113. Basáñez M-G, French MD, Walker M, Churcher TS. Paradigm lost: how parasite control may alter pattern and process in human helminthiases. *Trends Parasitol* 2012;**28**:161–71.
114. Elkins DB, Haswell-Elkins M. The weight/length profiles of *Ascaris lumbricoides* within a human community before mass treatment and following reinfection. *Parasitology* 1989;**99**:293–9.
115. Sinniah B, Subramaniam K. Factors influencing the egg production of *Ascaris lumbricoides*: relationship to weight, length and diameter of worms. *J Helminthol* 1991;**65**:141–7.
116. Crofton HD. A quantitative approach to parasitism. *Parasitology* 1971;**62**:179–93.
117. Cassie RM. Frequency distribution models in the ecology of plankton and other organisms. *J Anim Ecol* 1962;**31**:65–92.
118. Macdonald G. The dynamics of helminth infections, with special reference to schistosomes. *Trans R Soc Trop Med Hyg* 1965;**59**:489–506.
119. Anderson RM, May RM. Regulation and stability of host-parasite population interactions: I. Regulatory processes. *J Anim Ecol* 1978;**47**:219–47.
120. May RM, Anderson RM. Regulation and stability of host-parasite population interactions: II. Destabilizing processes. *J Anim Ecol* 1978;**47**:249–67.
121. Fisher RA. The negative binomial distribution. *Ann Eugen* 1941;**11**:182–7.
122. Shaw DJ, Grenfell BT, Dobson AP. Patterns of macroparasite aggregation in wildlife host populations. *Parasitology* 1998;**117**:597–610.
123. Bliss C, Fisher RA. Fitting the negative binomial distribution to biological data. *Biometrics* 1953;**9**:176–200.

124. Anscombe FJ. Sampling theory of the negative binomial and logarithmic series distributions. *Biometrika* 1950;**37**:358–82.
125. Anscombe FJ. The statistical analysis of insect counts based on the negative binomial distribution. *Biometrics* 1949;**5**:165–73.
126. Gregory RD, Woolhouse ME. Quantification of parasite aggregation: a simulation study. *Acta Trop* 1993;**54**:131–9.
127. Elliott JM. *Some Methods for the Statistical Analysis of Samples of Benthic Invertebrates.* Ambleside: Freshwater Biological Association; 1977.
128. Wilson K, Bjørnstad ON, Dobson AP, et al. Heterogeneities in macroparasite infections: patterns and process. In: Hudson PJ, Rizzoli AP, Grenfell BT, Heesterbeek JAP, Dobson AP, editors. *The Ecology of Wildlife Diseases.* Oxford: Oxford University Press; 2001. p. 6–44.
129. Lloyd-Smith JO. Maximum likelihood estimation of the negative binomial dispersion parameter for highly overdispersed data, with applications to infectious diseases. *PLoS One* 2007;**2**:e180.
130. Piegorsch WW. Maximum likelihood estimation for the negative binomial dispersion parameter. *Biometrics* 1990;**46**:863–7.
131. Saha K, Paul S. Bias-corrected maximum likelihood estimator of the negative binomial dispersion parameter. *Biometrics* 2005;**61**:179–85.
132. Moncayo AL, Vaca M, Amorim L, et al. Impact of long-term treatment with ivermectin on the prevalence and intensity of soil-transmitted helminth infections. *PLoS Negl Trop Dis* 2008;**2**:e293.
133. Denwood MJ, Stear MJ, Matthews L, Reid SWJ, Toft N, Innocent GT. The distribution of the pathogenic nematode *Nematodirus battus* in lambs is zero-inflated. *Parasitology* 2008;**135**:1225–35.
134. Nødtvedt A, Dohoo I, Sanchez J, et al. The use of negative binomial modelling in a longitudinal study of gastrointestinal parasite burdens in Canadian dairy cows. *Can J Vet Res* 2002;**66**:249–57.
135. Ziadinov I, Deplazes P, Mathis A, et al. Frequency distribution of *Echinococcus multilocularis* and other helminths of foxes in Kyrgyzstan. *Vet Parasitol* 2010;**171**:286–92.
136. Box GEP, Cox DR. An analysis of transformations. *J R Stat Soc Series B Stat Methodol* 1964;**26**:211–52.
137. O'Hara RB, Kotze DJ. Do not log-transform count data. *Methods Ecol Evol* 2010;**1**:118–22.
138. Anscombe FJ. The transformation of the Poisson, binomial and negative-binomial-data. *Biometrika* 1948;**35**:246–54.
139. Nelder AJA, Wedderburn RWM. Generalized linear models. *J R Stat Soc Ser A* 1972; **135**:370–84.
140. McCullagh P, Nelder JA. *Generalized Linear Models.* London: Chapam & Hall; 1989.
141. Wilson K, Grenfell BT. Generalized linear modelling for parasitologists. *Parasitol Today* 1997;**13**:33–8.
142. Lawless JF. Negative binomial and mixed poisson regression. *Can J Stat* 1987; **15**:209–25.
143. Hilbe JM. *Negative Binomial Regression.* Cambridge: Cambridge University Press; 2007.
144. Saathoff E, Olsen A, Kvalsvig JD, Appleton CC, Sharp B, Kleinschmidt I. Ecological covariates of *Ascaris lumbricoides* infection in schoolchildren from rural KwaZulu-Natal, South Africa. *Trop Med Int Health* 2005;**10**:412–22.
145. Carneiro FF, Cifuentes E, Tellez-Rojo MM, Romieu I. The risk of *Ascaris lumbricoides* infection in children as an environmental health indicator to guide preventive activities in Caparaó and Alto Caparaó, Brazil. *Bull World Health Organ* 2002;**80**:40–6.

146. de Silva NR, Pathmeswaran A, Fernando SD, et al. Impact of mass chemotherapy for the control of filariasis on geohelminth infections in Sri Lanka. *Ann Trop Med Parasitol* 2003;**97**:421–5.
147. Brooker S, Jardim-Botelho A, Quinnell RJ, et al. Age-related changes in hookworm infection, anaemia and iron deficiency in an area of high *Necator americanus* hookworm transmission in south-eastern Brazil. *Trans R Soc Trop Med Hyg* 2007;**101**:146–54.
148. Flohr C, Tuyen LN, Lewis S, et al. Low efficacy of mebendazole against hookworm in Vietnam: two randomized controlled trials. *Am J Trop Med Hyg* 2007;**76**:732–6.
149. Lambert D. Zero-inflated Poisson regression, with an application to defects in manufacturing. *Technometrics* 1992;**34**:1–14.
150. Mullahy J. Specification and testing of some modified count data models. *J Econom* 1986;**33**:341–65.
151. Heilbron DC. Zero-altered and other regression models for count data with added zeros. *Biom J* 1994;**36**:531–47.
152. Greene W. *Accounting for excess zeros and sample selection in Poisson and negative binomial regression models.* Working Paper EC-94–10,. Department of Economics, New York University; 1994.
153. Ridout M, Demétrio CGB, Hinde J. *Models for count data with many zeros.* Cape Town: International Biometric Conference; 1998.
154. Vounatsou P, Raso G, Tanner M, N'goran EK, Utzinger J. Bayesian geostatistical modelling for mapping schistosomiasis transmission. *Parasitology* 2009;**136**:1695–705.
155. Soares Magalhães RJ, Biritwum N-K, Gyapong JO, et al. Mapping helminth co-infection and co-intensity: geostatistical prediction in Ghana. *PLoS Negl Trop Dis* 2011;**5**:e1200.
156. Paterson S, Lello J. Mixed models: getting the best use of parasitological data. *Trends Parasitol* 2003;**19**:370–5.
157. Clayton DG. Generalized linear mixed models. In: *Markov Chain Monte Carlo in Practice.* London: Chapman & Hall; 1996.
158. Bundy DAP, Cooper ES, Thompson DE, Didier JM, Simmons I. Epidemiology and population dynamics of *Ascaris lumbricoides* and *Trichuris trichiura* infection in the same community. *Trans R Soc Trop Med Hyg* 1987;**81**:987–93.
159. Koukounari A, Sacko M, Keita AD, et al. Assessment of ultrasound morbidity indicators of schistosomiasis in the context of large-scale programs illustrated with experiences from Malian children. *Am J Trop Med Hyg* 2006;**75**:1042–52.
160. Booth JG, Casella G, Friedl H, Hobert JP. Negative binomial loglinear mixed models. *Stat Modelling* 2003;**3**:179–91.
161. Hall DB. Zero-inflated Poisson and binomial regression with random effects: a case study. *Biometrics* 2000;**56**:1030–9.
162. Min Y, Agresti A. Random effect models for repeated measures of zero-inflated count data. *Stat Modelling* 2005;**5**:1–19.
163. Gelman A, Carlin JB, Stern HS, Rubin DB. *Bayesian Data Analysis.* London: Chapman & Hall; 2003.
164. Gilks WR, Richardson S, Spiegelhalter DJ. Introducing Markov chain Monte Carlo. In: Gilks WR, Richardson S, Spiegelhalter DJ, editors. *Markov Chain Monte Carlo in Practice.* London: Chapman & Hall; 1996. p. 1–20.
165. Lunn DJ, Thomas A, Best N, Spiegelhalter D. WinBUGS – a Bayesian modelling framework: concepts, structure, and extensibility. *Stat Comput* 2000;**10**:325–37.
166. Plummer M. *JAGS: a program for analysis of Bayesian graphical models using Gibbs sampling*. Vienna: Proceedings of the 3rd International Workshop on Distributed Statistical Computing; 2003.
167. Brooker S, Clements ACA, Bundy DAP. Global epidemiology, ecology and control of soil-transmitted helminth infections. *Adv Parasitol* 2006;**62**:221–61.

168. Diggle PJ, Tawn JA, Moyeed RA. Model-based geostatistics. *J R Stat Soc Ser C Appl Stat* 1998;**47**:299–350.
169. Brooker S, Kabatereine NB, Tukahebwa EM, Kazibwe F. Spatial analysis of the distribution of intestinal nematode infections in Uganda. *Epidemiol Infect* 2004; **132**:1065–71.
170. Pullan RL, Gething PW, Smith JL, et al. Spatial modelling of soil-transmitted helminth infections in Kenya: a disease control planning tool. *PLoS Negl Trop Dis* 2011;**5**:e958.
171. Clements ACA, Deville M-A, Ndayishimiye O, Brooker S, Fenwick A. Spatial co-distribution of neglected tropical diseases in the east African great lakes region: revisiting the justification for integrated control. *Trop Med Int Health* 2010; **15**:198–207.
172. Brooker S, Singhasivanon P, Waikagul J, et al. Mapping soil-transmitted helminths in Southeast Asia and implications for parasite control. *Southeast Asian J Trop Med Public Health* 2003;**34**:24–36.
173. Diggle PJ, Ribeiro PJ. *Model based Geostatistics*. New York: Springer; 2007.
174. Magalhães RJS, Clements ACA, Patil AP, Gething PW, Brooker S. The applications of model-based geostatistics in helminth epidemiology and control. *Adv Parasitol* 2011;**74**:267–96.
175. Weaver HJ, Hawdon JM, Hoberg EP. Soil-transmitted helminthiases: implications of climate change and human behavior. *Trends Parasitol* 2010;**26**:574–81.
176. Crompton DW, Savioli L. Intestinal parasitic infections and urbanization. *Bull World Health Organ* 1993;**71**:1–7.
177. Brooker S, Hay SI, Tcheum Tchuente L-A, Ratard R. Using NOAA-AVHRR data to model human helminth distributions in planning disease control in Cameroon, West Africa. *Photogramm Eng Remote Sensing* 2002;**68**:175–9.
178. Clements ACA, Moyeed R, Brooker S. Bayesian geostatistical prediction of the intensity of infection with *Schistosoma mansoni* in East Africa. *Parasitology* 2006;**133**:711–9.
179. World Health Organization. *Preventative chemotherapy in human helminthiasis. coordinated use of anthelmintic drugs in control interventions: a manual for health professionals and programme managers*. Geneva: World Health Organization; 2006.
180. Riley S. Large-scale spatial-transmission models of infectious disease. *Science* 2007;**316**:1298–301.
181. Gurarie D, Seto EYW. Connectivity sustains disease transmission in environments with low potential for endemicity: modelling schistosomiasis with hydrologic and social connectivities. *J R Soc Interface* 2009;**6**:495–508.
182. Gurarie D, King CH. Heterogeneous model of schistosomiasis transmission and long-term control: the combined influence of spatial variation and age-dependent factors on optimal allocation of drug therapy. *Parasitology* 2005;**130**:49–65.
183. Bates DM, Watts DG. *Nonlinear Regression Analysis*. New York: Wiley; 1988.
184. Hadeler KP, Dietz K. Nonlinear hyperbolic partial differential equations for the dynamics of parasite populations. *Comput Math Appl* 1982;**9**:415–30.
185. Dietz K. Overall population patterns in the transmission cycle of infectious disease agents. In: Anderson RM, May RM, editors. *Population Biology of Infectious Diseases*. New York: Springer; 1982. p. 87–102.
186. Bolker BM. *Ecological Models and Data in R*. Princeton: Princeton University Press; 2008.
187. Anderson RM. Mathematical models of host–helminth parasite interactions. In: Usher MB, Williamson MH, editors. *Ecological Stability*. London: Chapman & Hall; 1974.
188. Anderson RM. The dynamics and control of direct life-cycle helminth parasites. In: Barigozzi C, editor. *Vito Volterra Symposium on Mathematical Models in Biology*. New York: Springer; 1980. p. 278–322.

189. Kretzschmar M. Persistent solutions in a model for parasitic infections. *J Math Biol* 1989;**27**:549–73.
190. Adler FR, Kretzschmar M. Aggregation and stability in parasite–host models. *Parasitology* 1992;**104**:199–205.
191. Kretzschmar M, Adler FR. Aggregated distributions in models for patchy populations. *Theor Popul Biol* 1993;**43**:1–30.
192. Herbert J, Isham V. Stochastic host–parasite interaction models. *J Math Biol* 2000;**40**:343–71.
193. Pacala SW, Dobson AP. The relation between the number of parasites/host and host age: population dynamic causes and maximum likelihood estimation. *Parasitology* 1988;**96**:197–210.
194. Quinnell RJ, Grafen A, Woolhouse MEJ. Changes in parasite aggregation with age: a discrete infection model. *Parasitology* 1995;**111**:635–44.
195. Grenfell BT, Dietz K, Roberts MG. Modelling the immuno-epidemiology of macroparasites in naturally fluctuating host populations. In: Grenfell BT, Dobson AP, editors. *Ecology of Infectious Diseases in Natural Populations*. Cambridge: Cambridge University Press; 1995. p. 362–83.
196. Nåsell I. *Hybrid Models of Tropical Infections*. New York: Springer; 1985.
197. Keymer A. Density-dependent mechanisms in the regulation of intestinal helminth populations. *Parasitology* 1982;**84**:573–87.
198. Medley GF. Which comes first in host–parasite systems: density dependence or parasite distribution? *Parasitol Today* 1992;**8**:321–2.
199. Rosà R, Pugliese A. Aggregation, stability, and oscillations in different models for host-macroparasite interactions. *Theor Popul Biol* 2002;**61**:319–34.
200. Churcher TS, Filipe JAN, Basáñez M- G. Density dependence and the control of helminth parasites. *J Anim Ecol* 2006;**75**:1313–20.
201. Anderson RM, May RM. Population dynamics of human helminth infections: control by chemotherapy. *Nature* 1982;**297**:557–63.
202. Churcher TS, Ferguson NM, Basáñez M- G. Density dependence and overdispersion in the transmission of helminth parasites. *Parasitology* 2005;**131**:121–32.
203. May RM. Togetherness among schistosomes: its effects on the dynamics of the infection. *Math Biosci* 1977;**35**:301–43.
204. Courchamp F, Berec L, Gascoigne J. *Allee Effects in Ecology and Conservation*. Oxford: Oxford University Press; 2008.
205. Croll NA, Anderson RM, Gyorkos TW, Ghadirian E. The population biology and control of *Ascaris lumbricoides* in a rural community in Iran. *Trans R Soc Trop Med Hyg* 1982;**76**:187–97.
206. Chan MS, Guyatt HL, Bundy DA, Medley GF. The development and validation of an age-structured model for the evaluation of disease control strategies for intestinal helminths. *Parasitology* 1994;**109**:389–96.
207. Chan MS, Bradley M, Bundy DA. Transmission patterns and the epidemiology of hookworm infection. *Int J Epidemiol* 1997;**26**:1392–400.
208. Chan MS, Srividya A, Norman RA, et al. Epifil: a dynamic model of infection and disease in lymphatic filariasis. *Am J Trop Med Hyg* 1998;**59**:606–14.
209. Norman RA, Chan MS, Srividya A, et al. EPIFIL: the development of an age-structured model for describing the transmission dynamics and control of lymphatic filariasis. *Epidemiol Infect* 2000;**124**:529–41.
210. Chan MS, Guyatt HL, Bundy DA, Booth M, Fulford AJ, Medley GF. The development of an age structured model for schistosomiasis transmission dynamics and control and its validation for *Schistosoma mansoni*. *Epidemiol Infect* 1995;**115**:325.
211. Anderson RM. The population dynamics and control of hookworm and roundworm infections. In: Anderson RM, editor. *Population Dynamics of Infectious Diseases*. London: Chapman & Hall; 1982. p. 67–108.

212. Guyatt HL, Chan MS, Medley GF, Bundy DAP. Control of *Ascaris* infection by chemotherapy: which is the most cost-effective option? *Trans R Soc Trop Med Hyg* 1995;**89**:16–20.
213. Lloyd-Smith JO, Schreiber SJ, Kopp PE, Getz WM. Superspreading and the effect of individual variation on disease emergence. *Nature* 2005;**438**:355–9.
214. Stephenson LS, Latham MC, Adams EJ, Kinoti SN, Pertet A. Physical fitness, growth and appetite of Kenyan school boys with hookworm, *Trichuris trichiura* and *Ascaris lumbricoides* infections are improved four months after a single dose of albendazole. *J Nutr* 1993;**123**:1036–46.
215. Raj SM, Naing NN. Ascariasis, trichuriasis, and growth of schoolchildren in Northeastern Peninsular Malaysia. *Southeast Asian J Trop Med Public Health* 1998;**29**:729–34.
216. Hall A, Bobrow E, Brooker S, et al. Anaemia in schoolchildren in eight countries in Africa and Asia. *Public Health Nutr* 2001;**4**:749–56.
217. Miguel E, Kremer M. Worms: identifying impacts on education and health in the presence of treatment externalities. *Econometrica* 2004;**72**:159–217.
218. Bundy DAP, Kremer M, Bleakley H, Jukes MCH, Miguel E. Deworming and development: asking the right questions, asking the questions right. *PLoS Negl Trop Dis* 2009;**3**:e362.
219. Chan S, Montresor A, Savioli L, Bundy DAP. Planning chemotherapy based schistosomiasis control: validation of a mathematical model using data on *Schistosoma haematobium* from Pemba, Tanzania. *Epidemiol Infect* 1999;**123**:487–97.
220. Thein-Hlaing Than-Saw. Myat-Lay-Kyin. The impact of three-monthly age-targetted chemotherapy on *Ascaris lumbricoides* infection. *Trans R Soc Trop Med Hyg* 1991;**85**:519–22.
221. Churcher TS, Basáñez M-G. Density dependence and the spread of anthelmintic resistance. *Evolution* 2008;**62**:528–37.
222. Vercruysse J, Albonico M, Behnke JM, et al. Is anthelmintic resistance a concern for the control of human soil-transmitted helminths? *Int J Parasitol Drugs Drug Resist* 2011;**1**:14–27.
223. Wolstenholme AJ, Fairweather I, Prichard R, von Samson-Himmelstjerna G, Sangster NC. Drug resistance in veterinary helminths. *Trends Parasitol* 2004;**20**:469–76.
224. Diawara A, Drake LJ, Suswillo RR, et al. Assays to detect beta-tubulin codon 200 polymorphism in *Trichuris trichiura* and *Ascaris lumbricoides*. *PLoS Negl Trop Dis* 2009;**3**:e397.
225. Albonico M, Engels D, Savioli L. Monitoring drug efficacy and early detection of drug resistance in human soil-transmitted nematodes: a pressing public health agenda for helminth control. *Int J Parasitol* 2004;**34**:1205–10.
226. Cornell SJ, Isham VS, Grenfell BT. Drug-resistant parasites and aggregated infection – early-season dynamics. *J Math Biol* 2000;**41**:341–60.
227. Cornell SJ, Isham VS, Smith G, Grenfell BT. Spatial parasite transmission, drug resistance, and the spread of rare genes. *Proc Natl Acad Sci USA* 2003;**100**:7401–5.
228. Knopp S, Stothard JR, Rollinson D, et al. From morbidity control to transmission control: time to change tactics against helminths on Unguja Island, Zanzibar. *Acta Trop;* 2012;. http://dx.doi.org/10.1016/j.actatropica.2011.04.010. In press.
229. Kasai T, Nakatani H, Takeuchi T, Crump A. Research and control of parasitic diseases in Japan: current position and future perspectives. *Trends Parasitol* 2007;**23**:230–5.
230. Hong S, Chai J, Choi M, Huh S. A successful experience of soil-transmitted helminth control in the Republic of Korea. *Korean J Parasitol* 2006;**44**:177–85.
231. Wang L-D, Guo J-G, Wu X-H, et al. China's new strategy to block *Schistosoma japonicum* transmission: experiences and impact beyond schistosomiasis. *Trop Med Int Health* 2009;**14**:1475–83.

232. Bartram J, Cairncross S. Hygiene, sanitation, and water: forgotten foundations of health. *PLoS Med* 2010;**7**:e1000367.
233. Plaisier AP, van Oortmarssen GJ, Habbema JD, Remme J, Alley ES. ONCHOSIM: a model and computer simulation program for the transmission and control of onchocerciasis. *Comput Methods Programs Biomed* 1990;**31**:43–56.
234. Plaisier AP, Subramanian S, Das PK, et al. The LYMFASIM simulation program for modeling lymphatic filariasis and its control. *Methods Inf Med* 1998;**37**:97–108.
235. Vlas SJ, Van Oortmarssen GJ, Gryseels B, Polderman AM, Plaisier AP, Habbema JD. SCHISTOSIM: a microsimulation model for the epidemiology and control of schistosomiasis. *Am J Trop Med Hyg* 1996;**55**:170–5.
236. Stolk WA, de Vlas SJ, Borsboom GJJM, Habbema JDF. LYMFASIM, a simulation model for predicting the impact of lymphatic filariasis control: quantification for African villages. *Parasitology* 2008;**135**:1583–98.
237. Winnen M, Plaisier AP, Alley ES, et al. Can ivermectin mass treatments eliminate onchocerciasis in Africa? *Bull World Health Organ* 2002;**80**:384–91.
238. Churcher TS, Basáñez M- G. Sampling strategies to detect anthelmintic resistance: the perspective of human onchocerciasis. *Trends Parasitol* 2009;**25**:11–7.
239. Montresor A, Gyorkos TW, Crompton DWT, Bundy DAP, Savioli L. Monitoring helminth control programmes. Guidlines for monitoring the impact of control programmes aimed at reducing morbidity caused by soil-transmitted helminths and schistosomes, with particular reference to school-age children. Geneva: World Health Organization; 1999.
240. Farrington CP. Modelling forces of infection for measles, mumps and rubella. *Stat Med* 1990;**9**:953–67.
241. Stepek G, Buttle DJ, Duce IR, Behnke JM. Human gastrointestinal nematode infections: are new control methods required? *Int J Exp Pathol* 2006;**87**:325–41.
242. Keiser J, Utzinger J. Efficacy of current drugs against soil-transmitted helminth infections. *JAMA* 2008;**299**:1937–48.
243. Belizario VY, Amarillo ME, de Leon WU. de los Reyes AE, Bugayong MG, Macatangay BJC. A comparison of the efficacy of single doses of albendazole, ivermectin, and diethylcarbamazine alone or in combinations against Ascaris and Trichuris spp. *Bull World Health Organ* 2003;**81**:35–42.
244. Mani TR, Rajendran R, Munirathinam A, et al. Efficacy of co-administration of albendazole and diethylcarbamazine against geohelminthiases: a study from South India. *Trop Med Int Health* 2002;**7**:541–8.
245. Arfaa F, Ghadirian E. Epidemiology and mass-treatment of ascariasis in six rural communities in central Iran. *Am J Trop Med Hyg* 1977;**26**:866–71.
246. Seo BS, Cho SY, Chai JY. Frequency distribution of *Ascaris lumbricoides* in rural Koreans with special reference on the effect of changing endemicity. *Korean J Parasitol* 1979;**17**:105–13.
247. Seo BS, Cho SY, Chai JY, Hong ST. Comparative efficacy of interval mass treatment on *Ascaris lumbricoides* infection in Korea. *Korean J Parasitol* 1980;**18**:145–51.
248. Seo BS, Chai JY. Effect of two-month interval mass chemotherapy on the reinfection of *Ascaris lumbricoides* in Korea. *Korean J Parasitol* 1980;**18**:153–63.
249. Martin J, Keymer A, Isherwood RJ, Wainwright SM. The prevalence and intensity of *Ascaris lumbricoides* infections in Moslem children from northern Bangladesh. *Trans R Soc Trop Med Hyg* 1983;**77**:702–6.
250. Cabrera BD. Reinfection and infection rates of ascariasis in relation to seasonal variation in the Philippines. *Southeast Asian J Trop Med Public Health* 1984; **15**:394–401.
251. Holland CV, Crompton DW, Taren DL, et al. *Ascaris lumbricoides* infection in pre-school children from Chiriqui Province, Panama. *Parasitology* 1987;**95**:615–22.

252. Monzon RB, Cabrera BD, Cruz AC, Baltazar JC. The "crowding effect" phenomenon in *Ascaris lumbricoides. Southeast Asian J Trop Med Public Health* 1990;**21**:580–5.
253. Kightlinger LK, Seed JR, Kightlinger MB. The epidemiology of *Ascaris lumbricoides, Trichuris trichiura,* and hookworm in children in the Ranomafana rainforest. Madagascar. *J Parasitol* 1995;**81**:159–69.
254. Peng W, Cui X, Zhou X. Comparison of the structures of natural and re-established populations of *Ascaris* in humans in a rural community of Jiangxi, China. *Parasitology* 2002;**124**:641–7.

CHAPTER

8

Genetic Epidemiology of *Ascaris*: Cross-transmission between Humans and Pigs, Focal Transmission, and Effective Population Size

Charles D. Criscione

Texas A&M University, College Station, TX, USA

OUTLINE

INTRODUCTION

In many regards, the field of genetic epidemiology (a.k.a. molecular or evolutionary epidemiology and here defined as the use of genetic/molecular markers to infer some aspect of the parasite/pathogen's

Ascaris: The Neglected Parasite
http://dx.doi.org/10.1016/B978-0-12-396978-1.00008-2

Copyright © 2013 Elsevier Inc. All rights reserved.

population biology such as transmission, population growth, or selected traits) asks the same questions as asked in the field of conservation genetics. Is there just one species or are there cryptic evolutionary units, is the species fragmented into subpopulations, was the fragmentation the result of human perturbation, is the population declining, what facilitates connectivity/gene flow among subpopulations, what was the source of invasion (outbreak) for an exotic species (emerging pathogen), what loci are of adaptive significance? The key difference between epidemiology and conservation is the end goal. Epidemiologists try to eliminate or reduce populations of parasites/pathogens. In contrast, conservationists strive to maintain or increase population sizes and continuity of endangered species. Population genetic applications are now integral in conservation because it is well recognized that low genetic diversity, small effective population sizes, and population fragmentation (all three of which can be measured via genetic methods) can increase the chance of population extinction.[1,2] Because conservation geneticists are interested in these factors to prevent extinction, then it seems logical that epidemiologists could use similar data to help reduce or eradicate parasites/pathogens. Indeed, because of the parallel questions between the fields, much of the population genetics theory, methods, and reasoning that are used in conservation genetics could be applied to genetic epidemiology. For instance, it is recognized that low genetic diversity can reduce evolutionary potential (i.e. the ability of populations to evolve to cope with environmental change).[1,3] Chemotherapy control programs are a major environmental change for parasites. Given that drug resistance has evolved among several helminths,[4,5] it seems reasonable that reducing genetic diversity, via a reduction in effective population size (discussed below), should be an imperative epidemiological goal to help prevent drug-resistant evolution.

In this chapter, I discuss three pertinent applications of population genetics (all of which have been utilized in conservation biology) to further our understanding of *Ascaris* epidemiology in fine scale geographic studies. First, I focus on whether sympatric populations of *Ascaris* in humans and pigs constitute separate populations in order to ascertain if there is cross-transmission between human and pig hosts. Second, I discuss the use of landscape genetics to identify foci of transmission and epidemiologically relevant variables correlated to substructure of parasite populations. These first two topics correspond to a series of recently proposed hierarchical questions aimed at addressing local scale population genetics in metazoan parasites.[6] Thus, I refer readers to Gorton and colleagues[6] for a more general discussion of these topics in metazoan parasites. Also, these sections are not intended to be a comprehensive summary of the *Ascaris* population genetics literature as this was recently reviewed by Peng and Criscione.[7] The third section of this chapter proposes the novel integration of the effective population size (N_e) parameter into population monitoring and epidemiological studies of parasites. Using microsatellite data from a metapopulation of *A. lumbricoides* in Nepal, I demonstrate the utility of estimating N_e with single-sample, contemporary estimators. I also discuss assumptions and provide some guidelines for estimating N_e. My goal is to emphasize the importance of the above topics in epidemiological research, highlight the population genetic methodologies that have been used, and point to new directions that may aid the development or monitoring of *Ascaris* (and metazoan parasites in general) control programs.

A species' life history and the way samples are collected can influence interpretation of some of the genetic analyses I discuss. Thus, I first provide a brief summary of the biological characteristics of *Ascaris*. Sampling will be addressed in the context of each study that is discussed below and just note here that genotypes were always obtained from adult worms. *Ascaris* has a direct life-cycle where mature male and female adult worms reside in the lumen of the small intestine.[8] The mating system has not been extensively studied. However, recent paternity analyses indicate there is polyandry in pig *Ascaris*[9] and Hardy–Weinberg equilibrium, indicating random mating, has been observed on very local scales (i.e. within people in a single village).[10] A female can produce millions of eggs over her lifetime, which is about 1 year.[11] Eggs are released into the external environment where they can persist for 6 to 9 years.[12] Infection occurs by ingestion of eggs via fecal contaminated material. Larvae hatch in the small intestine, penetrate the intestinal wall, migrate to the lung to become fourth-stage larvae, and then migrate up the trachea back into the esophagus and ultimately the small intestine. In about 60 days from the point of infection, females will start to produce eggs.[8] Key life history aspects in terms of population genetics are that breeding worms are transiently separated into groups (i.e. hosts)[13] and that the long-lived eggs can lead to overlapping generations. As will be discussed, the latter is of significance because breeding worms that end up in the same host may be of different offspring cohorts (i.e. there is overlapping of generations).

ASCARIS CROSS-TRANSMISSION BETWEEN HUMANS AND PIGS

The subject of whether *Ascaris* in humans and pigs is one or two species (*A. lumbricoides* and *A. suum*, respectively) is still being discussed[7,14,15] and really points to an underlying question that is central for many human parasites: are there reservoir hosts (i.e. is there zoonotic transmission)? The answer to this question would clearly impact control strategies in terms of which hosts should be targeted: just humans or both humans and pigs? It is clear from mitochondrial sequence (mtDNA) data that there is strong neutral genetic differentiation between roundworms originating from sympatric host species.[16–18] These data indicate there is non-random transmission between the host species such that there is not a single source pool of infection shared by humans and pigs. However, because there were no fixed allelic genetic differences between human and pig *Ascaris* samples, these results were unable to ascertain if there were two completely independent transmission cycles (one through humans and one through pigs) or if there was limited cross-infection between the two host species. The lack of fixed sequence differences could result from incomplete lineage sorting (retention of ancestral lineages in descendent taxa) with no cross-transmission, current introgression (hybrid offspring resulting from cross-breeding between human and pig *Ascaris*), or cross-transmission, but no interbreeding (e.g. a worm is a first generation migrant from one host species to the other).[7] In areas of non-endemic human transmission (USA, Denmark, and Japan), worms obtained from humans had DNA sequences that matched those obtained from pigs.[19–21] These data clearly show cross-transmission from a pig source into humans and raise the possibility that the lack of fixed differences observed in human–pig endemic areas is also due to cross-transmission. Thus, two important questions are raised: (1) how can one detect cross-transmission in human–pig endemic sites, and (2) if there is cross-transmission, is there introgression between human and pig *Ascaris*?

Criscione and colleagues[22] addressed these questions with genetic-based assignment/model-based clustering methods.[23,24] These methods, which have a history in species management applications, use information from multilocus genotypes (commonly assuming Hardy–Weinberg equilibrium and linkage equilibrium among loci) to ascertain population membership of individuals.[25] They can also be used for identifying first generation migrants and hybrid individuals. Genetic assignment/model-based clustering methods provide several advantages for allowing one to detect hybrids. First, analyses can be conducted when no taxa-specific markers exist,[26,27] as is currently the case with *Ascaris* of humans and pigs.[7] Second, separate samples where each only contains individuals of a single parental population are not required.[26] Third, *a priori* delineation of populations is not necessary (i.e. no knowledge of underlying substructure is needed for the analyses). The latter is important as the finding of cryptic species and substructure is not uncommon among metazoan parasites.[6]

From both a village in Guatemala and a county in the Hainan Province of China, Bayesian clustering methods with genotypes of adult worms clearly delineated genetically structured parasite populations between human and pig hosts in sympatry.[22] These results were in accordance with previous mtDNA-based studies.[16–18] Moreover, the multilocus genotype data enabled the identification of hybrid worms (4% in Guatemala and 7% in China).[22] The finding of hybrids necessarily implies that there was cross-transmission between human and pig hosts because a worm of pig origin and of human origin had to meet in the same host in order to mate. This cross-transmission and interbreeding had to be recent as the methods employed can only detect hybrids going back two generations.[26] Zhou and colleagues[28] used the same methods to ascertain the frequency of cross-transmission across six provinces in China. They observed similar results and identified both first generation migrants (~7% of sampled worms) and hybrid worms (also ~7%), both of which were predominantly collected from human hosts. Notably, the authors state "The results strongly suggest pig *Ascaris* as an important source of human ascariasis in endemic area where both human and pig *Ascaris* exist. In consideration of current control measures for human ascariasis targeting only infected people, it is urgently needed to revise current control measures by adding a simultaneous treatment to infected pigs in the sympatric endemics".[28] With these new molecular tools at hand, it will be prudent to perform additional studies from sympatric populations to determine if limited cross-transmission is a global theme especially in relation to different pig-raising, cultural, or economic conditions. It will also be of interest to see if cross-transmission continues to show a largely pig to human pattern and to explore the mechanisms that generate the genetic differentiation between the host-associated populations despite the high frequencies of cross transmission.[7]

Aside from the direct inference of cross-transmission, what is the epidemiological significance of limited cross-transmission and introgression? Criscione and colleagues[22] highlight two critical aspects. First, while there is significant genetic differentiation between *Ascaris* populations in humans and pigs, the long-term ability to cross-transmit between host species remains possible. Thus, even in non-endemic sites, human infection via a pig source remains possible (as evidenced by several studies[19–21]). Also, this ability may have led to a complex evolutionary history of multiple host colonization events.[7] Second, hybridization can lead to introgression of adaptive genes[29] and hybridization itself may produce new combinations of parasite genotypes that increase parasite virulence or host range via host immune evasion.[30] Little attention has been given to these aspects of *Ascaris* epidemiology. Because parasite hybridization is of long-term epidemiological significance in terms of the evolution of novel host infectivity genes or drug-resistant genes, it will be critical to begin mapping regions of genomic introgression in relation to host species infectivity patterns in *Ascaris*.

LANDSCAPE GENETICS AS A MEANS TO INFER *ASCARIS* TRANSMISSION WITHIN A HOST POPULATION

Effective *Ascaris* control will require detailed knowledge of parasite dispersal to fully evaluate transmission patterns among individual human hosts. The extent of parasite dispersal, however, is difficult to ascertain with data based solely on infection intensities (i.e. number of worms per infected host or a surrogate such as eggs per gram of feces). This is because direct observation of parasite offspring leaving one host and subsequently infecting the same or a new host is nearly impossible.[31] Thus, while intensity data are necessary to explore factors that explain the variation in the distribution of parasites among individual hosts,[32] they do little to answer the question of where did an individual acquire their infection (i.e. are there different foci of infection in the single human population).

Identification of population subdivision via population genetics analyses of multilocus genotypic data provides a powerful means to infer macroparasite dispersal among subdivided units such as individuals or groups of hosts (e.g. households).[33–36] When using genetic data to infer transmission among individual hosts, the sampling unit should be the parasite stage that infects that host.[6] In the case of *Ascaris*, adult worms would be genotyped from human hosts. If, for example, expelled *Ascaris* eggs from humans were used, then measures of genetic differentiation could be inflated due to the possible sampling of sibling parasites. I refer readers to Steinauer and colleagues[37] for a more thorough discussion of this type of sampling. Additional insight into what controls the transmission process can be gained by using landscape genetic statistical approaches to test if epidemiological variables correlate with the observed parasite genetic structure. Landscape genetics is a multidisciplinary field that incorporates spatial statistics, landscape ecology, and population genetics to evaluate the role of landscape variables (e.g. altitude, ground cover) in shaping genetic differentiation among populations.[38] In this regard, landscape genetics has parallel goals with the field of spatial epidemiology, which examines the correlates of spatial variation in infection intensity patterns.[39] As landscape genetics is still a developing field where several methodologies are being explored, I refer readers to a special issue in *Molecular Ecology* that highlights this field in more detail.[40] Here, I demonstrate the application of landscape genetics to the epidemiology of *A. lumbricoides* from an endemic population in Jiri, Nepal.[41]

The goals of the study by Criscione and colleagues[41] were to determine if there was more than one source pool of infection (i.e. foci of infection) and, if so, to examine epidemiological variables that may correlate with these foci. If there is high mixing and dispersal of parasites across the human population, then the parasites would have a panmictic population structure. Thus, people would effectively be acquiring infections from a common parasite population (i.e. a single source pool of infection). In contrast, repeated transmission that is localized at particular foci across the human population would limit parasite mixing, leading to parasite genetic differentiation within a single human population. The finding of multiple genetic clusters of parasites, therefore, is an indication that there could be multiple infection foci (see Figure 1 in Criscione and colleagues[41]). Adult *A. lumbricoides* were collected from 320 people across 165 households that spanned an area approximately 14 km^2. In addition to spatial sampling, two temporal samples (~3 years apart, so a total of 211 household-by-year samples) were taken for some regions of the village. For logistic reasons, temporal sampling was staggered for three regions of Jiri such that one group of houses was sampled in 1998 and 2001, a second group in 1999 and 2002, and a third in 2000 and 2003. As noted below, time of collection explained less than 1% of the variance in the genetic structure of the parasite population.[41] A total of 1094 roundworms were genotyped at 23 autosomal microsatellite markers.[10] Model-based Bayesian clustering (implemented in the program STRUCTURE[23]) was used to analyze the multilocus parasite genotypes to determine if there was underlying genetic structure among the sampled worms. Importantly, no prior spatial or temporal information was included (or needed) in this analysis.

There was strong support for local-scale genetic structuring with 13 genetic clusters of parasites identified. The results of the population clustering analyses were subsequently incorporated into a non-parametric multivariate analysis of variance[42,43] to elucidate spatial, geographical, or epidemiological features associated with the partitioning of genetic variation among the sampled worms. This analysis provided a novel approach to integrating individual-based genetic assignment results with downstream statistical analyses.[41] The independent variables included a nested design (household and hosts nested within household) and eight covariates: host age, host sex, host density (number of people living in the house), elevation, geographic distance among households (latitude–longitude combined), infection intensity, parasite sex, and time of collection. When variables were analyzed independently, household explained >63% of the variance in genetic structuring whereas each covariate always accounted for <15%. When the nested design was conditioned on the eight covariates (i.e. variance due to the covariates was accounted for first), the contribution of household was still high and explained >36%. In contrast, none of the eight covariates were significant after accounting for the nested design. Interestingly, time had no impact on the underlying genetic structure even when compared pairwise between time periods for 18 households with sufficient sample sizes for testing.[41] Furthermore, a spatial autocorrelation analysis showed that parasites between households within 540 m were more genetically similar than expected by chance alone. Genetic differentiation measured as F_{CT} (hierarchical F-statistic of household to the total) was 0.023 and highly significant ($p < 0.0001$)[41].

These results[41] revealed three key insights into transmission of *A. lumbricoides* in Jiri: there were separate foci of transmission at this local scale, households and nearby houses shared genetically related parasites, and people reacquired their worms from the same source pool of infection over time. These results challenge the dogma that a single human community will correspond to a homogeneous parasite population (implicit in many classic models[44,45] of parasite transmission that measure a single basic reproduction number, R_0). In Jiri, multiple source pools of infection need to be considered when modeling parasite transmission. Thus, when using models to evaluate control strategies in Jiri, it would be more appropriate to consider incorporating parasite populations that exist in an interconnected network, i.e. metapopulation.[46]

Although I emphasized how population genetics can be used to elucidate transmission patterns, I note that I do not view landscape genetics as a panacea for epidemiological goals in general, nor do I view genetics data as a replacement for infection intensity data. Rather I see the two types of data as providing different, but complementary, information about the transmission process. For example, Walker and colleagues[32] found that in Bangladesh host age and sex explained part of the variation in worm burdens. In contrast, host age and sex were not correlated to how worm genetic variation was partitioned in Jiri, Nepal.[41] I realize that data from the two studies are not directly comparable as they were from different locations, but the point is that both parasite intensity and genetic data are needed to fully elucidate the transmission process. Thus, in this hypothetical comparison, although gender may account for differences in worm burdens within a household (females have higher intensities possibly due to peridomiciliary behaviors that increase exposure[32]), males and females are still getting their worms from the same source of infection. Lastly, it should be noted that the patterns in Jiri may not extrapolate to other locations as differences in human behavior, topography, and external environmental conditions could alter transmission patterns. For instance, a communal use of human feces for fertilizer may facilitate parasite dispersal thereby creating a single source pool of parasites. Thus, the assumption of a single infectious pool of parasites will need to be tested for each population of interest and as evidenced by the study in Jiri,[41] even on very local scales.

EFFECTIVE POPULATION SIZE: EPIDEMIOLOGICAL UTILITY AND ESTIMATION

The effective population size (N_e) is the size of an ideal population that has the same rate of genetic drift as the population under consideration. The "ideal" population follows the models of Wright[47] and Fisher[48] and, in simple terms, refers to the situation where every individual has an equal opportunity to contribute genes to the next generation.[49,50] The effects of genetic drift can be measured several ways such as by the increase in inbreeding, increase in variance in allele frequency, or loss of heterozygosity over generations. Hence, there are different definitions of N_e: inbreeding N_e, variance N_e, and eigenvalue N_e, respectively.[51] In closed populations of constant size, the different concepts have similar or identical values of N_e, but certain demographic scenarios can lead to different estimates of N_e depending on which aspect of drift is being measured.[49–52] My discussion will largely not make a distinction between the different N_e concepts; however, the estimates I provide are more closely related to inbreeding N_e. Commonly, but not always, N_e is smaller than the actual census population size (N_c) because some parents contribute many more offspring to the next generation than others.

Of what interest is parasite N_e to epidemiologists? There are both long-term (evolutionary) and short-term (ecological) utilities of N_e. Evolutionary importance stems from the fact that N_e directly determines the rate of drift where the loss of neutral genetic variation (often quantified via expected heterozygosity; H_e) each generation is expected to decline by a rate inversely dependent on N_e.[51] N_e is also needed to assess the relative importance of the three other evolutionary mechanisms (mutation, gene flow, and selection). For instance, equilibrium gene diversity in the infinite alleles model is determined by N_e and the mutation rate (u) such that

$$H_e = \frac{\theta}{\theta + 1}, \tag{8.1}$$

where $\theta = 4N_e u$.[51] Additionally, if $N_e s << 1$ (s = selection coefficient), change in allelic frequency is determined primarily by genetic drift rather than selection.[47] Given these above relationships, it is clear why N_e is an important parameter in conservation biology.[53] Indeed, conservationists are concerned about populations with small N_e because there is lower genetic variation to respond to environmental change (i.e. lower adaptive potential), the breeding of closely related individuals can reduce the fitness of an outbreeding species (inbreeding depression), and deleterious alleles can become fixed at low N_e.[1–3] All of the latter may increase the chance for population extinction.[1] Of course, the latter is the goal for epidemiologists. Consequently, from a disease management perspective, reducing parasite N_e has the long-term goal of helping to reduce parasite adaptive potential. Because drift affects loci across the genome, reducing parasite N_e may help reduce standing genetic variation at any given locus that could become of adaptive significance in the face of drastic environmental changes (e.g. application of drugs or vaccines). Moreover, the parameter N_e itself is necessary to help model the potential for drug resistance evolution in relation to the selective pressures induced by chemotherapy programs.

In ecological (epidemiological) terms, N_e is important as it is directly determined by life history variation. Demographic factors such as fluctuating population size, non-binomial variation in reproductive success and unequal sex ratios can cause N_e to deviate (likely lower) from N_c.[51] Thus, knowledge of what demographic factors impact parasite N_e might begin to help link the microevolutionary dynamics of parasites to transmission models that examine the reproductive potential and population growth of parasites.[54] Admittedly, measuring demographic variables can be difficult in parasites. Thus, I believe that more immediate applications of using N_e in epidemiological studies will stem from recent developments of single-sample, contemporary genetic estimators of N_e. In particular, the linkage disequilibrium (LD-N_e)[55,56] and sibship assignment (SA-N_e)[57] methods hold great promise to estimate N_e in parasite populations. Because these methods require only the genotyping of a sample of parasites from a single time point, they will be useful in generating estimates of N_e for the long-term applications noted above. Moreover, for short-term applications, recent simulations have shown that LD-N_e estimates from two time points can be used to detect population bottlenecks[58] or fragmentation of a population.[59] Therefore, what I envision for short-term applications is the use of N_e estimates as a genetic means to monitor parasite control programs. For instance, one can ask if a chemotherapy program not only reduces worm burdens (N_c), but also N_e. Does a control program reduce parasite dispersal across the treated population (i.e. cause population fragmentation)?

I am unaware of any study that has provided contemporary estimates of N_e in a metazoan parasite of animals much less the application of N_e estimates to monitoring a macroparasite control program. Genetic monitoring studies of parasites largely focus on levels of allelic richness (A) or H_e.[60] While it is important to report the latter two statistics, there are disadvantages to these indices of genetic diversity. First, A is subject to sample size unless rarefaction (i.e. subsampling larger samples to compare richness values among samples with different sample sizes) is used. Second, both A and H_e (or the DNA sequence data equivalent, π) are affected by mutation rate. This means these two measures provide somewhat redundant information, as A increases so does H_e (e.g. with two equally frequent alleles $H_e = 0.5$, with four $H_e = 0.75$). Being affected by mutation also means comparisons across studies that use different loci may be inhibited as different loci (e.g. SNPs vs. microsatellites) may have different mutation rates. In contrast, changes in N_e will be comparable across studies and species. Third, while A and H_e may provide an indication of immediate evolutionary potential, they have no predictive value for future levels of genetic diversity.[61] As noted above, N_e is a critical parameter in many evolutionary models including future H_e. Below I provide an example of how contemporary N_e estimates can be used to further elucidate the epidemiology and population dynamics of human parasites.

The N_e estimates in Table 8.1 were generated with genotype data from *A. lumbricoides* in Jiri (same data set as described for the landscape genetics study[41]). I note that the study by Criscione and colleagues[41] was not designed to address specific questions about N_e or the effects of chemotherapy on parasite population dynamics. Worms were originally collected to examine how human genetics may play a role in parasite infection intensities.[62–64] Thus, sampling is less than ideal for some of the questions I address below. Furthermore, I am assuredly violating certain assumptions for some of the population genetic theoretical models that I utilize below. I try to highlight where some of these assumptions may be violated. However, I encourage readers to research the references for the models as space limitations prevent an in-depth discussion of all assumptions. My main goal in going through several models is to show epidemiologically related questions one could ask with N_e and to highlight some sampling issues associated with estimating N_e. Nonetheless, despite the assumptions I make, I believe the presented data do provide a reasonable approximation for some important population dynamics of *Ascaris* in Jiri.

I am primarily interested in estimating the parasite N_e from households (i.e. subpopulations of the *Ascaris* metapopulation in Jiri). As discussed previously, there was significant parasite genetic structure across Jiri that was largely explained by households (>63%).[41] As there was focal transmission around households, this would be the scale by which one would monitor the impact of a control program on parasite population dynamics. Also, because of the genetic subdivision, the N_e of subpopulations will be of relevance in relation to adaptive potential (i.e. this is the level by which one would monitor genetic diversity or model the relative influence of genetic drift versus selection). In my data set, household N_e was estimated from a sample of adult *Ascaris* that were collected from individual people of a household after chemotherapy treatment.[41] A critical aspect to consider is what effective size is being estimated from this collected sample. This is outlined in Figure 8.1. Because *Ascaris* has long-lived egg stages in the external environment,[12] the effective number of adults breeding in year t (N_t) will have a proportion of their offspring

TABLE 8.1 Estimates of household-by-year N_e based on the linkage disequilibrium[55,56] (LD) or sibship assignment[57] (SA) methods with their lower and upper 95% confidence intervals

House ID_year[a]	Genotyped worms[b]	House intensity[c]	Allele freq. cutoff[d]	LD-N_e	LD 95% lower	LD 95% upper	SA-N_e	SA 95% lower	SA 95% upper
014_1999	11	19	0.05	18.1	12.9	27.7[f]	110	37	infinite
014_2002	13	13	0.04	21.4	15.6	31.6	52	24	447
076_1999	10	29	0.06	20.5	13.9	34.5[f]	180	57	infinite
077_2000	10	12	0.06	104	34.5	infinite	180	53	infinite
080_2000	12	18	0.05	−303.3[e]	89.3	infinite	2.15×10^9 [e]	1	infinite
092_2000	11	13	0.05	24.3	14.8	52.7	44	20	635
097_2000	22	32	0.03	42.5	34.1	55.3	116	59	444
097_2003	37	43	0.02	66.7	54.8	83.9	133	79	272
119_2002	27	28	0.02	41.3	34.3	51	117	67	321
121_1999	29	89	0.02	90.1	67.4	132.3	180	100	862
121_2002	13	15	0.04	53.3	30.2	171.8[f]	104	48	infinite
122_1999	82	173	0.02	314.8	211.5	582.6	251	183	363
122_2002	42	48	0.02	271.8	178.3	546.6	265	164	680
123_1999	33	115	0.02	59.2	48.8	74.1	92	58	170

123_2002	21	31	0.03	146.7	78.5	785.3 [f]	420	131	infinite
124_1999	11	34	0.05	447.7	43.4	infinite	2.15×10^9 [e]	1	infinite
128_1999	13	24	0.04	28	19.7	45.2 [f]	156	64	infinite
133_2002	15	20	0.04	34.7	25.5	51.9 [f]	210	83	infinite
134_1999	14	32	0.04	54.8	32.6	142.7 [f]	364	95	infinite
135_1999	65	132	0.02	240	178.1	359.2	208	147	310
135_2002	72	85	0.02	184.5	147.4	242.7	173	125	246
140_1998	23	31	0.03	40.7	31.9	54.6	67	36	195
140_2002	14	20	0.04	183.2	57.6	infinite	364	113	infinite
148_1998	13	23	0.04	−1590.5[e]	78.9	infinite	312	101	infinite
148_2002	22	24	0.03	89.1	57.2	185.3	185	89	12,788
152_1998	12	22	0.05	37.9	22.6	94.5 [f]	132	47	infinite

[a]*Household identification numbers correspond to those in Figure 2 of Criscione and colleagues.*[41]
[b]*The number of worms that were genotyped at 23 microsatellite markers per household-by-year. Raw data are from Criscione and colleagues.*[41]
[c]*The total number of worms collected per household-by-year after albendazole treatment. See Criscione and colleagues*[41] *for details of sampling.*
[d]*Alleles with frequencies below this value were omitted when estimating* N_e *with the LD method.*[56]
[e]*Negative or* 2.15×10^9 *estimates of* N_e *are regarded as infinite (see text for explanation).*[56]
[f]*The LD-*N_e *method had an upper bound for the 95% CI for the given allele frequency cutoff, but at other cutoffs, estimates typically included infinity as the upper bound. In contrast, LD-*N_e *estimates in shaded rows often provided bounded CI even at other allele frequency cutoffs.*
The jackknife method was used for the LD interval[55] and the SA interval is estimated in the program.[67] Estimates were generated with the programs LDNE[66] and COLONY,[67] respectively. The 13 shaded rows highlight where both estimators yielded N_e estimates bounded by confidence intervals.

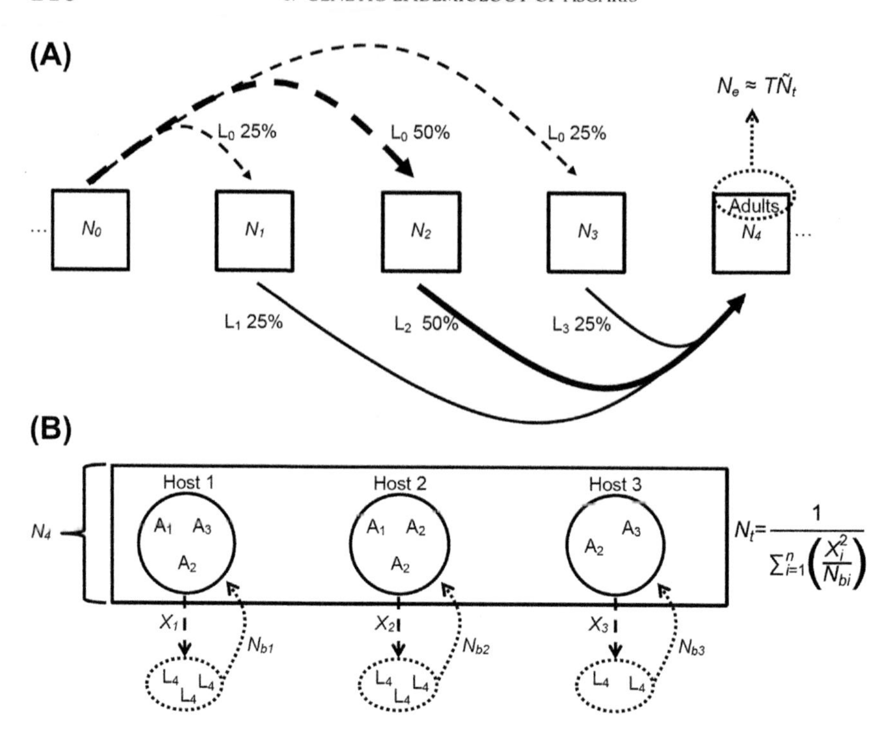

FIGURE 8.1 Diagram showing how the life history of *Ascaris* and sampling relate to the estimation of effective size parameters. (A) Illustration of how the long-lived egg stage leads to overlapping generations in a single subpopulation (e.g. a household in the current study[41]). Boxes represent the effective number of adults breeding in year t (N_t). Five breeding years are shown and an arbitrary year is chosen as year $t = 0$. Generation length (T; average age to adulthood) is not known for *Ascaris*. For demonstration, $T = 2$ is shown. As an example, 25% of the offspring from N_0 (L_0) will become adults in year 1, 50% in year 2, and 25% in year 3 (dashed curved arrows). In a given breeding year, adult worms will be of mixed ages (i.e. they originate from different temporal breeding cohorts). For instance, N_4 will be a mixture of 25% offspring from year 1 (L_1), 50% L_2, and 25% L_3 (solid curved arrows). The use of a single-sample, genetic estimator (e.g. LD-N_e or SA-N_e) on a random sample of adult worms across hosts within a subpopulation (dotted oval) provides an estimate of the generational N_e[56] (see text). Generational N_e is $\approx T\tilde{N}_t$, where $\tilde{N}_t$ is the harmonic mean of the N_t's within a generation.[65] (B) Illustration of how breeders within a given year (figure is shown for N_4) are subdivided among individual hosts. As noted above, adult worms will be of mixed ages (e.g. A_1, A_2, A_3). Eggs passed from each host are the offspring of year 4 breeders (L_4). The use of a single-sample, genetic estimator on a random sample of eggs from single host (dotted oval) such as might be obtained from a fecal sample provides an estimate of the effective number of breeders in that host (N_{bi}; dotted arrow). The equation to calculate N_t is shown on the right and is a function of the N_{bi}'s and the X_i's, where X_i is the proportional contribution of progeny from the ith host to the mixed pool that makes up the next generation of parasite breeders. Note that if the species had discrete generations, N_t is N_e. See Criscione and Blouin[13] for more thorough discussion of using a model of subdivided breeders to estimate parasite N_e.

that survive to reproduce in years $t+1$, $t+2$, and so on (Figure 8.1A). These proportions determine the average age to adulthood (i.e. generation length, T).[65] Thus, even though *Ascaris* adults may live only a year in their host,[11] generation length is likely several years longer due to the fact that eggs can persist 6–9 years in the environment.[12] Interestingly, *Ascaris* life history closely approximates that of semelparous, age-structured species such as annual plants with seed banks and Pacific salmon. A detailed theoretical treatment of estimating N_e in the latter groups of organisms is given by Waples.[65] In short, generational N_e is $\approx T\tilde{N}_{t,}$, where T is generation time in years and $\tilde{N}_t$ is the harmonic mean of the N_t's within a generation.[65] An important point to recognize is that a sample of adult worms of a given breeding year will contain individuals of mixed ages, i.e. there are overlapping generations (Figure 8.1A). With the LD-N_e and SA-N_e methods, the estimated N_e reflects that of the breeders that produced the sampled adult worms (i.e. the parents of the sampled worms) and not the sampled worms themselves (i.e. not N_t). While cautioning that testing is needed, Waples and Do[56] conjectured that a mixed-aged sample that includes a number of consecutive age classes approximately equal to generation length should produce an estimate roughly corresponding to generational N_e. Thus, throughout the chapter, I will assume that the sample of adult worms from each household provides an estimate of generational N_e of each subpopulation (Figure 8.1A). I will return to the estimation of N_t (Figure 8.1B) in my concluding remarks.

I used two single-sample, contemporary estimators, LD-N_e and SA-N_e,[55–57] as implemented in the programs LDNE[66] and COLONY v2.0.2.1[67], respectively. Both of these methods provide estimates that are related to the inbreeding N_e.[56,57] The LD-N_e method can be sensitive to rare alleles, thus I followed the recommendations of Waples and Do[56] for using alleles with frequencies above a cutoff given the sample size (see Table 8.1). The random mating system option was used. In COLONY, I selected the male and female polygamy options without inbreeding. These latter options in the two programs seem reasonable given the current state of knowledge about *Ascaris* mating systems.[9,10] Length of run and likelihood precision (full-likelihood) were set to medium in COLONY. I used the update allele frequency option and the complexity prior, which should result in a higher N_e estimate (compared to not using it) as this prior discourages complex pedigree inference.

Table 8.1 provides the estimates of N_e per household-by-year where 10 or more worms were genotyped ($n=26$). There are several important patterns and questions that emerge from these data. First, sample size matters in obtaining estimates that are not infinite or do not have an upper confidence interval of infinity. Infinity estimates (negative values in the LD-N_e method or the 2.15×10^9 values in the SA- N_e method) result when sampling error swamps the genetic signature of genetic drift in the case of LD-N_e estimates[56] or when little to no pedigree structure is found in the SA-N_e method. Of the 26 estimates, only 13 (Table 8.1, shaded rows) gave values that had bounded confidence intervals for both estimators. When looking at the other 13 estimates, it appears that several of the LD-N_e estimates had upper confidence limits when the SA-N_e method did not. However, it is important to note that these LD-N_e estimates (white rows and marked in Table 8.1) were sensitive for the allele frequency cutoff such that other cutoff values returned an infinity upper bound (data not shown). In contrast, LD-N_e estimates in the shaded rows had upper bound confidence intervals regardless of allele frequency cutoff. Thus, there was congruence between the two methods in returning estimates with uncertainty in the upper confidence limits for the same household-by-year samples. Thirteen of the 13 estimates with uncertainty in the upper confidence limits (white rows) had $n \leq 21$, whereas 11 of the 13 estimates with bounded confidence intervals had an $n > 21$ (Table 8.1). Small sample sizes will only provide bounded confidence intervals if the true N_e is small ($\leq$50), which is likely the case for houses 014_2002 and 092_2000 (Table 8.1). The reason is that the larger the true N_e and the smaller the sample size, the less likely one is to find related individuals in the sample (Table 2 in Waples and Waples[68]). Thus, if small sample sizes yield estimates with unbounded confidence limits it is difficult to ascertain whether the true N_e is large or whether it is small, but a larger sampling error is to blame. If one wants to detect populations that have a true N_e of 500–1000, sample sizes need to be around 50 with about 20 polymorphic loci.[56] It appears my current data set was able to get bounded confidence limits with $n = 22–40$ because true N_e of each subpopulation was likely much less than 500. Several studies[56–59] have used simulations to address sampling, thus I refer readers to these papers for a discussion of appropriate samples sizes and number of loci to use in relation to types of questions one might ask with N_e estimates.

Interestingly, almost all point estimates range in the mid tens to low hundreds. Even the unbounded confidence interval estimates, which still can give some indication of the lower bounds of N_e, tend to show low N_e point estimates. From here on, however, I will restrict my analyses and discussion to the 11 estimates that had $n > 21$ (Table 8.2). Even though houses 014_2002 ($n = 13$) and 092_2000 ($n = 11$) had estimates with bounded confidence intervals, I removed them from subsequent analyses to avoid bias. Bias may originate because I would be omitting the other houses with $n \leq 21$ that potentially really do have larger effective sizes, but could not get an accurate estimate due to small sample size. There was a high correlation between the point estimates of the two estimators ($r = 0.894, p = 0.0002, n = 11$; Table 8.2). These data show good congruence between the two estimators and give me high confidence I am getting

TABLE 8.2 N_e Estimates from household-by-year with $n > 21$. The "Best" estimate-N_e is the harmonic mean of the LD-N_e and SA-N_e methods

House ID_year	House intensity	LD-N_e	SA-N_e	"Best" estimate-N_e
097_2000	32	42.5	116	62.2
097_2003	43	66.7	133	88.8
119_2002	28	41.3	117	61
121_1999	89	90.1	180	120.1
122_1999	173	314.8	251	279.3
122_2002	48	271.8	265	268.3
123_1999	115	59.2	92	72
135_1999	132	240	208	222.9
135_2002	85	184.5	173	178.6
140_1998	31	40.7	67	50.6
148_2002	24	89.1	185	120.3
Harmonic mean		76.9	137.9	98.8

accurate estimates of the parental breeding population N_e that contributed to the infections in each household. This is especially true given the two methods utilize very different methods (linkage disequilibrium versus identification of pedigree structure) to estimate N_e. The harmonic means of the household N_e point estimates ($n = 11$) were 76.9 (95% CI: 55.6–116.8) and 137.9 (95% CI: 108.2–183.6) for the LD-N_e and SA-N_e, respectively (CI based on 1000 bootstraps over the point estimates of the household-by-year samples). The harmonic mean is used because the distribution of N_e can be highly skewed.[56] Waples and Do[56] also suggest that if two single-sample estimators are independent and are estimating the same parameter from a population, then a more precise or "best" estimate of N_e can be obtained by taking the harmonic mean of the two single-sample estimators. The "best" estimate-N_e's for the 11 household-by-year samples with $n > 21$ are given in Table 8.2. The harmonic mean of these "best" estimates is 98.8 (bootstrap 95% CI: 73.5–139.1).

One of the questions that can be asked with these data is whether or not drug treatment impacted N_e. Simulations have shown that LD-N_e estimates from two time points can be used to detect a population genetic bottleneck.[58] After omitting samples with small n, I only had three houses (97, 122, 135) with estimates from both time periods (people in households were treated and worms collected, then three years later this was repeated). This is a small sample size, but visual (i.e. not statistical) assessment of the values and their confidence intervals (Table 8.1) does not reveal any discernible impact of chemotherapy on the N_e of these *Ascaris* subpopulations (even if houses with small n are examined). These genetic results parallel prior epidemiological data from Jiri where after 1 year of treatment both prevalence (year 1 = 27.2%, year 2 = 24.2%) and mean number of worms expelled per individual (2.37 and 2.67) showed little change.[62]

The latter scenario also begs the question of whether N_e reflects N_c of *Ascaris* in Jiri. Intuitively, as N_c increases, so should N_e. However, I caution that the relationship between N_e and N_c under different demographic scenarios is generally not well understood and may vary considerably among species.[69] In some free-living organisms, the ratio of N_e/N_c decreases as population density increases.[69] Experimental data in flour beetles suggest this may be caused by an increase in the variation in reproductive success among individuals as N_c increases.[70] Therefore, it might be that there is an asymptotic relationship between N_e and N_c such that N_e levels off even as N_c gets larger (Figure 8.2). The latter relationship would be important in epidemiological studies because a drop in N_c may not constitute a drop in N_e until a critical N_c is reached. This would be crucial in terms of the evolution of drug resistance because a huge selection pressure via chemotherapy could be imposed on the population without a drop in N_e. Selection is more efficient with larger N_e. Thus, both worm count and genetic data are warranted in epidemiological studies if

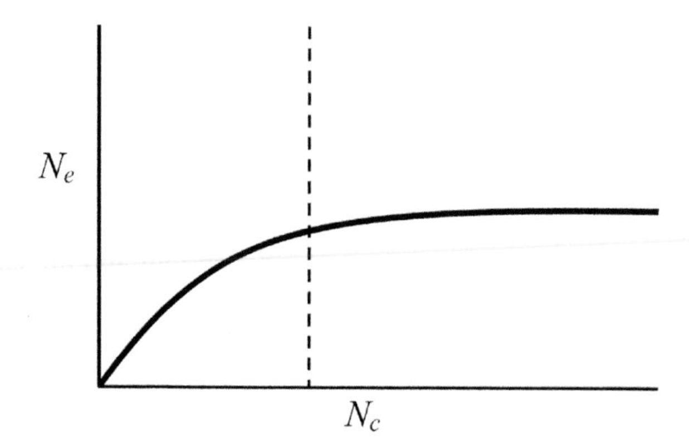

FIGURE 8.2 Hypothetical asymptotic relationship between N_e and N_c. Dotted line denotes critical N_c where N_e no longer substantially increases as N_c increases. One possible explanation for this pattern is that as population density increases, variation in reproductive success may increase considerably, thus substantially reducing N_e. An important epidemiological implication is that above this critical point, N_c could be reduced drastically without a dramatic effect on N_e. In relation to the *Ascaris* data presented, the correlation between household intensities and the single-sample N_e estimates may suggest *Ascaris* subpopulations already exist below the critical N_c value.

one of the goals is also to reduce genetic diversity/adaptive potential. I did not have a means to estimate the N_c of the parents that produced the sampled worms from each household especially since the parents are likely from different breeding years (i.e. different N_t's; Figure 8.1A). As a surrogate, I tested for a correlation between the "best" estimate-N_e's and the infection intensities recorded for each house (Table 8.2), which implicitly assumes large census populations beget large populations. There was a significant correlation between infection intensity and "best" estimate-N_e ($r = 0.61$, $p = 0.047$, $n = 11$). Thus, in these samples N_e may be a good tracker of N_c. It would be encouraging if this result holds in future studies because that means N_e estimates might be useful to monitor not only adaptive potential, but also intensity data following an *Ascaris* treatment program. More data are certainly needed, but it is interesting to speculate that these correlations suggest that the *Ascaris* subpopulations do not exist at high densities (e.g. mean intensity per person was ~2.5 in Jiri[62]) where an asymptotic relationship between N_e and N_c would be relevant (Figure 8.2). In comparison, an asymptotic relationship may be more pertinent in parasites that have high infection intensities per host (hundreds to thousands) such as several trichostrongylid nematodes of livestock. Interestingly, among nematodes, the latter group is largely where drug resistance has been reported.[4,5]

The overall metapopulation N_e (N_{eT}) is also of interest in relation to dynamics that occur among subpopulations (e.g. equal subpopulation contributions to the migrant pool versus extinction/recolonization dynamics). My goal in this section is to compare an estimate of N_{eT} using Wright's island model[71] to an estimate of N_{eT} from the single-sample estimators. I caution the combining of samples across subpopulations (and across years as in this data set) and the subsequent use of these single-sample estimators has not been quantitatively tested as a means to estimate N_{eT}. Thus, the following should be treated as a thought exercise rather than definitive conclusions. I used the entire data set of 1094 worms and obtained an LD-N_e estimate of 1062 (95% CI: 975–1161, at the 0.02 cutoff) and SA-N_e estimate of 1645 (95% CI: 1502–1789). The harmonic mean of these two estimates yields a "best" estimate of $N_{eT} = 1291$. In Wright's island model,[71] N_{eT} is a function of subpopulation N_e and genetic differentiation (F_{ST}) such that

$$N_{eT} \approx \frac{nN_e}{1 - F_{ST}}, \tag{8.2}$$

where n is the number of subpopulations and each subpopulation has the same N_e. This model assumes that subpopulations contribute equally to the migrant pool. As can be seen in Eq. (8.2), as genetic differentiation increases among subpopulations, N_{eT} can exceed the sum of the subpopulation effective sizes.[49,72] This is because while each subpopulation loses variation due to drift, each subpopulation will become fixed for different alleles. Thus, genetic variation is maintained over the entire metapopulation. However, in metapopulation models where some subpopulations have greater contributions to the migrant pool than others or where subpopulations go extinct and are recolonized via founders of another subpopulation, N_{eT} can be greatly reduced below the sum of subpopulation effective sizes.[49,61,72,73] If estimates of the three parameters in Eq. (8.2) can be obtained to estimate N_{eT}, then the island model value can be compared to the single-sample N_{eT} "best" estimate to draw on conclusions about subpopulation contributions to the migrant pool. Criscione and colleagues[41] reported that genetic differentiation among households was 0.023 (the equivalent of F_{ST}). Furthermore, using a Bayesian clustering method,[23] they identified 13 core clusters, which I will use as n. Obviously N_e was not the same across households, but for the purpose of illustration I will assume they were and use the harmonic mean (Table 8.2), 98.8 (95% CI: 73.5–139.1). Based on the latter values, the island model N_{eT} is 1314 (possible range from 979 to 1851), which is in agreement with the single-sample "best" estimate of 1291. Therefore, this comparison suggests that *Ascaris* subpopulations in Jiri reflect more of Wright island model rather than a metapopulation where subpopulations have large unequal contributions to the migrant pool or recolonization–extinction dynamics. If the latter were true, then it seems like the single-sample estimators would be producing an estimate well below that predicted from the island model. Readers are encouraged to delve into the references above[49,61,72,73] to get an understanding of all model assumptions. Here I point out two concerns in this data set. First of which is the number of subpopulations I used in Eq. (8.2). If the landscape genetics study[41] did not sample all possible subpopulations, then 13, and thus the estimate of N_{eT} from the island model, would be an underestimate. Second, I also assumed that the harmonic mean N_e of the households reflects the central tendency of the N_e of the 13 genetic clusters. This seems reasonable as households were largely composed of individuals belonging to a single cluster. However, all clusters are not represented by the houses in Table 8.2, and a few houses may represent the same cluster (i.e. there is pseudoreplication).

I did not have a means to estimate N_c for each subpopulation. However, if I assume stable human population growth and infection patterns are constant over time, I can estimate a census size for *Ascaris* across the Jiri metapopulation (N_{cT}). This enables me to get a N_{eT}/N_{cT} ratio. Using the average prevalence of 25.7% and intensity of 2.52 worms per infected host data from Williams-Blangero and colleagues,[62] and the 1991 census count of the Jiri human population of 7138, the N_{cT} of *A. lumbricoides* would be 4623. Accordingly, $N_{eT}/N_{cT} = 0.28$ when using the single-sample "best" estimator for N_{eT}. The single-sample estimators used here would reflect uneven sex ratios and variation in reproductive success of the previous breeding generation. In an extensive review by Frankham,[69] the mean N_e/N_c ratio was 0.35 (95% CI: 0.28–0.42) among species for which variation in reproductive success and uneven sex ratios were taken into account to obtain demographic estimates of N_e. Thus, the *Ascaris* value falls just on the edge for what is known from single generation N_e/N_c estimates of other species.

The following may be a bit of an extrapolation because of the restrictive assumptions of the island model,[74] but I think it is a useful exercise in what genetic data and a N_e/N_c ratio might be able to tell us. Under the assumptions of Wright's island model[47,74] genetic differentiation is a function of subpopulation N_e and migration rate (m) where

$$F_{ST} \approx \frac{1}{4N_e m + 1}. \tag{8.3}$$

As discussed above, the island model might approximate the *Ascaris* population dynamics in Jiri. Thus, it seemed reasonable to estimate the effective number of migrants per generation ($N_e m$) from Eq. (8.3). Using a F_{ST} of 0.023[41], $N_e m = 10.61$. If the N_{eT}/N_{cT} ratio of 0.28 also represents the ratio within subpopulations, then that means about 38 census worms/per generation are migrants into the foci of transmission around households. This does not mean all 38 census worms become adults or even infect a person. It would be more appropriate to say a minimum of 10 migrant census worms infect people (necessarily adult worm infections because $N_e m$ represents individuals that contribute to the gene pool), but up to 38 census worms infecting a household were acquired from another transmission focus per worm generation. A key point here is "per worm generation." *Ascaris* adult worms live about 1 year in their host.[11] Thus, one might conclude generation time is 1 year and, therefore, 10–38 migrant worms per year cause infections. However, as noted above, the long-lived egg stages of *Ascaris* will increase generation time. Thus, these 10–38 migrant worms will be spread out likely over several years.

Above I have focused on using single-sample estimators to estimate the N_e of the parents that generated the infections in the sampled households. One can also estimate long-term or coalescent N_e that reflects the historical evolutionary dynamics of a population. Such an estimate may provide a historical baseline for what the parasite's N_e was like prior to the implementation of a control program. Waples[49] provides a summary about estimating long-term N_e. Here, I illustrate estimation of long-term N_e with the Jiri *Ascaris* data while also highlighting some of the caveats discussed by Waples.[49] Long-term N_e requires an estimate of $\theta = 4N_e u$, which means an estimate of u is also needed. Importantly, an accurate estimate of N_e via an estimate of θ will be dependent on a reliable estimate of u; a 10× change in u leads a 10× change in the N_e

estimate.[49] Model-based genealogical simulations are preferable to estimate θ,[75] though these are computationally intensive. For simplicity, I estimated θ with Eq. (8.1), which has the assumption that the population under consideration is closed to immigration.[49] In comparison to samples from China and Guatemala, *Ascaris* from Jiri are highly genetically differentiated.[22] Thus, on a global scale the Jiri metapopulation of *A. lumbricoides* is likely relatively isolated. Nonetheless, sampling of locations around Jiri is needed to ascertain potential regional influences on the long-term N_e estimate provided below. To estimate the coalescent N_e of the metapopulation, I used $H_e = 0.71$, which was reported over all 1094 genotyped nematodes;[41] thus, $\theta = 2.45$. There are no estimates of u for microsatellites in *Ascaris*; therefore, I used estimates from the nematode *Caenorhabditis elegans*.[76] Repeat motif and length can affect u so I calculated the average u from the six di- and five tetra-nucleotide motif loci with lengths less than 70 repeats[76] (mean $u = 0.000542$ and 0.0000362, respectively) as this would reflect the microsatellite loci in my data set. I had 19 di- and 4-tetra microsatellites, and used a weighted average to obtain an estimate of $u = 0.000454$. Using this value of u, the coalescent $N_e = 1347$. This long-term estimate is nearly identical to the single-sample "best" estimate of N_{eT} (1291).

CONCLUDING REMARKS

Above I discussed how population genetics data can be used to identify cross-transmission and focal transmission. In addition, I introduced N_e as a means to help genetically monitor epidemiologically relevant parasites. All the methods I have used come with assumptions and require appropriate sampling. With regards to cross-transmission and focal transmission, more discussion can be found in prior studies.[7,22,41] Here, I will conclude with a discussion of using N_e estimators for parasites especially in relation to *Ascaris* biology.

Single-sample, contemporary estimators assume closed populations with discrete generations.[49] In regards to the assumption of a closed population, simulations showed that the LD-N_e estimator is little affected by migration unless $m > 0.1$, in which case an estimate from a subpopulation will approach N_{eT}.[77] The latter does not appear to be an issue in this *Ascaris* data set. Because *Ascaris* has a "seed bank" life history, it clearly does not have discrete generations. When dealing with a species with overlapping generations, generational N_e is of most significance for monitoring adaptive potential or modeling the effects of selection. How then can one estimate generational N_e for *Ascaris*? As conjectured and assumed in this chapter, the use of single-sample estimators on a sample with a mixed-age cohort (adult worms in the case of *Ascaris*) may actually provide an estimate of generational N_e (Figure 8.1A).[56] If this holds true (currently being tested by R. Waples, personal communication), one should aim for larger sample sizes than the current data set (e.g. ≥50 per subpopulation) in order to make sure that all potential cohorts making up a generation are sampled. If this does not hold true, extensive data collection will be needed to obtain an estimate of generational N_e (i.e. using the formula $N_e \approx T\tilde{N}_t$)[65] as one will need estimates of T and the N_t's. An estimate of T for *Ascaris* will likely require experiments in pigs by either monitoring infections from a cohort of eggs over years or using different aged pastures (i.e. eggs left standing 1 year, 2 years, etc.) to estimate infection efficiencies of different egg ages. For now, it must suffice to say that T is likely <6–9 years as this is the current knowledge of egg longevity in the environment.[12] Because parasite breeders within a given breeding year (N_t) are separated among hosts, N_t is function of the effective number of breeders within each host (N_b) of the subpopulation and the proportional offspring contributions of each N_b (Figure 8.1B).[13] I refer readers to Criscione and Blouin[13] for a detailed description of a model for subdivided parasite breeders that can be used to estimate N_t from measures of the N_b's. Here I draw attention to the fact that the single-sample estimators can be used to estimate the N_b of a given host. One simply would collect and genotype eggs/larvae from an individual person. Moreover, the N_b values themselves may be of epidemiological use especially if one does not have a means to directly count adult parasites in a person (e.g. schistosome parasites).[13] If N_b estimates correlate to actual intensities of infection (a relationship that still warrants testing), then N_b estimates could provide a more accurate depiction of infection intensities among hosts compared to other surrogate methods such as eggs per gram of feces. N_b estimates could also be important in helping determine the role an individual host has in contributing to a parasite's subpopulation N_e (or N_t in the case of *Ascaris*).[13]

In this chapter, I have illustrated the feasibility of using of single-sample estimators in estimating generational N_e estimates for subpopulations (households) of *A. lumbricoides*. In order to illustrate different concepts and applications that one could use with N_e estimates, I have made several assumptions and extrapolations with these data. Nonetheless, these estimates have shed additional light on the population and thus epidemiological dynamics of *Ascaris* in Jiri. Overall, the household N_e estimates were low (~100) and it appears that they were stable over time even with chemotherapy treatment (though a more formal test is needed). Comparison of metapopulation N_e (N_{eT}) between the island model and the single-sample estimators further elucidated transmission patterns in that subpopulations appear to be contributing fairly equally to overall dispersal of *Ascaris* across the metapopulation. Thus, among-subpopulation dynamics were relatively stable such that this comparison did not support extinction/recolonization dynamics. Because seed banks can slow the rate of genetic change,[65] it may be that the long-lived *Ascaris* eggs (a.k.a. parasite seed bank) are what contributed to the stability in N_e over time, the lack of genetic differentiation between time periods for a given household,[41] and prevention of subpopulation extinction.

Lastly, most of my discussion has focused on the short-term inference of N_e, which will be comparable across studies and species, as a means to monitor the impact of control programs on genetic diversity and population dynamics. N_e also provides long-term inference in relation to adaptive potential. For instance, a threefold reduction in N_e from 10^4 to 10 is likely to reduce adaptive potential. However, it is appreciated that drift will be mostly irrelevant in reducing adaptive potential if the threefold reduction is from 10^7 to 10^4. Also, there is likely no magic N_e below which all parasite species are likely to go extinct and additional demographic factors that may vary among parasitic species will also be important.[61] Clearly what is needed are more estimates of N_e from parasites before one can begin to conclude about the adaptive potential. For instance, if the small effective sizes in Jiri are reflective of *Ascaris* in other places, then it is interesting to speculate that the reason why drug resistance has not been reported for *Ascaris* is that the low effective sizes have be an impediment to the evolution of drug resistance. Indeed, even the N_{eT} and coalescent N_e were both low (~1300). In contrast, coalescent N_e on the order of 10^6–10^7 has been estimated for populations of trichostrongylid nematodes,[78,79] a group with several species that have evolved drug resistance.[4,5] The use of the single-sample estimators[56,57] will facilitate N_e comparisons among parasite species/populations that differ in life history/demographic attributes, thus allowing future studies to explore the relationship between parasite N_e and adaptive potential.

References

1. Frankham R. Genetics and extinction. *Biol Conserv* 2005;**126**:131–40.
2. Frankham R. Challenges and opportunities of genetic approaches to biological conservation. *Biol Conserv* 2010;**143**:1919–27.
3. Willi Y, Van Buskirk J, Hoffmann AA. Limits to the adaptive potential of small populations. *Annu Rev Ecol Evol Syst* 2006;**37**:433–58.
4. Wolstenholme AJ, Fairweather I, Prichard R, von Samson-Himmelstjerna G, Sangster NC. Drug resistance in veterinary helminths. *Trends Parasitol* 2004;**20**:469–76.
5. James CE, Hudson AL, Davey MW. Drug resistance mechanisms in helminths: is it survival of the fittest? *Trends Parasitol* 2009;**25**:328–35.
6. Gorton MJ, Kasl EL, Detwiler JT, Criscione CD. Testing local-scale panmixia provides insights into the cryptic ecology, evolution, and epidemiology of metazoan animal parasites. *Parasitology* 2012;**139**:981–97.
7. Peng W, Criscione CD. Ascariasis in people and pigs: New inferences from DNA analysis of worm populations. *Infect Genet Evol* 2012;**12**:227–35.

8. Roberts LS, Janovy Jr J. *Foundations of Parasitology*. 8th ed. New York, NY: McGraw-Hill; 2009.
9. Zhou CH, Yuan K, Tang X, Hu N, Peng W. Molecular genetic evidence for polyandry in *Ascaris suum*. *Parasitol Res* 2011;**108**:703–8.
10. Criscione CD, Anderson JD, Raby K, et al. Microsatellite markers for the human nematode parasite *Ascaris lumbricoides*: development and assessment of utility. *J Parasitol* 2007;**93**:704–8.
11. Olsen LS, Kelley GW, Sen HG. Longevity and egg-production of *Ascaris suum*. *Trans Am Microsc Soc* 1958;**77**:380–3.
12. Roepstorff A, Mejer H, Nejsum P, Thamsborg SM. Helminth parasites in pigs: new challenges in pig production and current research highlights. *Vet Parasitol* 2011;**180**: 72–81.
13. Criscione CD, Blouin MS. Effective sizes of macroparasite populations: a conceptual model. *Trends Parasitol* 2005;**21**:212–7.
14. Leles D, Gardner SL, Reinhard K, Iniguez A, Araujo A. Are *Ascaris lumbricoides* and *Ascaris suum* a single species? *Parasit Vectors* 2012;**5**:42.
15. Nejsum P, Betson M, Bendall RP, Thamsborg SM, Stothard JR. Assessing the zoonotic potential of *Ascaris suum* and *Trichuris suis*: looking to the future from an analysis of the past. *J Helminthol* 2012;**86**:148–55.
16. Anderson TJC, Romero-Abal ME, Jaenike J. Genetic structure and epidemiology of *Ascaris* populations: patterns of host affiliation in Guatemala. *Parasitology* 1993;**107**: 319–34.
17. Peng W, Anderson TJC, Zhou X, Kennedy MW. Genetic variation in sympatric *Ascaris* populations from humans and pigs in China. *Parasitology* 1998;**117**:355–61.
18. Peng WD, Yuan K, Hu M, Zhou XM, Gasser RB. Mutation scanning-coupled analysis of haplotypic variability in mitochondrial DNA regions reveals low gene flow between human and porcine *Ascaris* in endemic regions of China. *Electrophoresis* 2005;**26**:4317–26.
19. Anderson TJC. *Ascaris* infections in humans from North America: molecular evidence for cross infection. *Parasitology* 1995;**110**:215–9.
20. Nejsum P, Parker ED, Frydenberg J, et al. Ascariasis is a zoonosis in Denmark. *J Clin Microbiol* 2005;**43**:1142–8.
21. Arizono N, Yoshimura Y, Tohzaka N, et al. Ascariasis in Japan: is pig-derived *Ascaris* infecting humans? *Jpn J Infect Dis* 2010;**63**:447–8.
22. Criscione CD, Anderson JD, Sudimack D, et al. Disentangling hybridization and host colonization in parasitic roundworms of humans and pigs. *Proc R Soc B* 2007;**274**: 2669–77.
23. Falush D, Stephens M, Pritchard JK. Inference of population structure using multilocus genotype data: linked loci and correlated allele frequencies. *Genetics* 2003;**164**:1567–87.
24. Corander J, Marttinen P. Bayesian identification of admixture events using multilocus molecular markers. *Mol Ecol* 2006;**15**:2833–43.
25. Manel S, Gaggiotti OE, Waples RS. Assignment methods: matching biological questions with appropriate techniques. *Trends Ecol Evol* 2005;**20**:136–42.
26. Anderson EC, Thompson EA. A model-based method for identifying species hybrids using multilocus genetic data. *Genetics* 2002;**160**:1217–29.
27. Vaha JP, Primmer CR. Efficiency of model-based Bayesian methods for detecting hybrid individuals under different hybridization scenarios and with different numbers of loci. *Mol Ecol* 2006;**15**:63–72.
28. Zhou C, Li M, Yuan K, Deng S, Peng W. Pig *Ascaris*: an important source of human ascariasis in China. *Infect Genet Evol* 2012;**12**:1172–7.
29. Barton NH. The role of hybridization in evolution. *Mol Ecol* 2001;**10**:551–68.
30. Arnold ML. Natural hybridization and the evolution of domesticated, pest and disease organisms. *Mol Ecol* 2004;**13**:997–1007.

31. de Meeus T, McCoy KD, Prugnolle F, et al. Population genetics and molecular epidemiology or how to "debusquer la bête." *Infect Genet Evol* 2007;**7**:308–32.
32. Walker M, Hall A, Basanez MG. Individual predisposition, household clustering and risk factors for human infection with *Ascaris lumbricoides*: new epidemiological insights. *PLoS Negl Trop Dis* 2011;**5**:e1047.
33. Criscione CD, Poulin R, Blouin MS. Molecular ecology of parasites: elucidating ecological and microevolutionary processes. *Mol Ecol* 2005;**14**:2247–57.
34. Criscione CD, Blouin MS. Minimal selfing, few clones, and no among-host genetic structure in a hermaphroditic parasite with asexual larval propagation. *Evolution* 2006;**60**:553–62.
35. Criscione CD. Parasite co-structure: broad and local scale approaches. *Parasite* 2008;**15**:439–43.
36. Archie EA, Luikart G, Ezenwa VO. Infecting epidemiology with genetics: a new frontier in disease ecology. *Trends Ecol Evol* 2009;**24**:21–30.
37. Steinauer ML, Blouin MS, Criscione CD. Applying evolutionary genetics to schistosome epidemiology. *Infect Genet Evol* 2010;**10**:433–43.
38. Storfer A, Murphy MA, Evans JS, et al. Putting the "landscape" in landscape genetics. *Heredity* 2007;**98**:128–42.
39. Brooker S. Spatial epidemiology of human schistosomiasis in Africa: risk models, transmission dynamics and control. *Trans R Soc Trop Med Hyg* 2007;**101**:1–8.
40. Sork VL, Waits L. Contributions of landscape genetics – approaches, insights, and future potential. *Mol Ecol* 2010;**19**:3489–95.
41. Criscione CD, Anderson JD, Sudimack D, et al. Landscape genetics reveals focal transmission of a human macroparasite. *PLoS Negl Trop Dis* 2010;**4**:e665.
42. Legendre P, Anderson MJ. Distance-based redundancy analysis: testing multispecies responses in multifactorial ecological experiments. *Ecol Monogr* 1999;**69**:1–24.
43. McArdle BH, Anderson MJ. Fitting multivariate models to community data: a comment on distance-based redundancy analysis. *Ecology* 2001;**82**:290–7.
44. Woolhouse MEJ, Dye C, Etard JF, et al. Heterogeneities in the transmission of infectious agents: implications for the design of control programs. *Proc Natl Acad Sci USA* 1997;**94**:338–42.
45. Woolhouse MEJ, Etard JF, Dietz K, Ndhlovu PD, Chandiwana SK. Heterogeneities in schistosome transmission dynamics and control. *Parasitology* 1998;**117**:475–82.
46. Gurarie D, Seto EYW. Connectivity sustains disease transmission in environments with low potential for endemicity: modeling schistosomiasis with hydrologic and social connectivities. *J R Soc Interface* 2009;**6**:495–508.
47. Wright S. Evolution in Mendelian populations. *Genetics* 1931;**16**:97–159.
48. Fisher RA. *The Genetical Theory of Natural Selection*. Oxford: Oxford University Press; 1930.
49. Waples RS. Spatial-temporal stratifications in natural populations and how they affect understanding and estimation of effective population size. *Mol Ecol Resour* 2010;**10**:785–96.
50. Templeton AR. *Population Genetics and Microevolutionary Theory.* Hoboken, New Jersey: John Wiley & Sons, Inc.; 2006.
51. Hedrick PW. *Genetics of Populations*. 4th ed. Sudbury, MA: Jones and Bartlett Publishers; 2011.
52. Wang JL. Estimation of effective population sizes from data on genetic markers. *Philos Trans R Soc Lond B Biol Sci* 2005;**360**:1395–409.
53. Hare MP, Nunney L, Schwartz MK, et al. Understanding and estimating effective population size for practical application in marine species management. *Conserv Biol* 2011;**25**:438–49.
54. Hudson PJ, Rizzoli A, Grenfell BT, Heesterbeek H, Dobson AP, editors. *The Ecology of Wildlife Diseases*. Oxford: Oxford University Press; 2002.

55. Waples RS. A bias correction for estimates of effective population size based on linkage disequilibrium at unlinked gene loci. *Conserv Genet* 2006;**7**:167–84.
56. Waples RS, Do C. Linkage disequilibrium estimates of contemporary N_e using highly variable genetic markers: a largely untapped resource for applied conservation and evolution. *Evol Appl* 2010;**3**:244–62.
57. Wang JL. A new method for estimating effective population sizes from a single sample of multilocus genotypes. *Mol Ecol* 2009;**18**:2148–64.
58. Antao T, Perez-Figueroa A, Luikart G. Early detection of population declines: high power of genetic monitoring using effective population size estimators. *Evol Appl* 2011; **4**:144–54.
59. England PR, Luikart G, Waples RS. Early detection of population fragmentation using linkage disequilibrium estimation of effective population size. *Conserv Genet* 2010;**11**: 2425–30.
60. Norton AJ, Gower CM, Lamberton PHL, et al. Genetic consequences of mass human chemotherapy for *Schistosoma mansoni*: population structure pre and post-praziquantel treatment in Tanzania. *Am J Trop Med Hyg* 2010;**83**:951–7.
61. Nunney L. The limits to knowledge in conservation genetics: the value of the effective population size. *Evol Biol* 2000;**32**:179–94.
62. Williams-Blangero S, Subedi J, Upadhayay RP, et al. Genetic analysis of susceptibility to infection with *Ascaris lumbricoides*. *Am J Trop Med Hyg* 1999;**60**:921–6.
63. Williams-Blangero S, VandeBerg JL, Subedi J, et al. Genes on chromosomes 1 and 13 have significant effects on *Ascaris* infection. *Proc Natl Acad Sci USA* 2002;**99**: 5533–8.
64. Williams-Blangero S, Criscione CD, VandeBerg JL, et al. Host genetics and population structure effects on parasitic disease. *Philos Trans R Soc Lond B Biol Sci* 2012;**367**: 887–94.
65. Waples RS. Seed banks, salmon, and sleeping genes: Effective population size in semelparous, age-structured species with fluctuating abundance. *Am Nat* 2006;**167**: 118–35.
66. Waples RS, Do C. LDNE: a computer program for estimating effective population size from data on linkage disequilibrium. *Mol Ecol Resour* 2008;**8**:753–6.
67. Jones OR, Wang J. COLONY: a program for parentage and sibship inference from multilocus genotype data. *Mol Ecol Resour* 2010;**10**:551–5.
68. Waples RS, Waples RK. Inbreeding effective population size and parentage analysis without parents. *Mol Ecol Resour* 2011;**11**:162–71.
69. Frankham R. Effective population size/adult population size ratios in wildlife: a review. *Genet Res* 1995;**66**:95–107.
70. Pray LA, Goodnight CJ, Stevens L, Schwartz JM, Yan G. The effect of population size on effective population size: an empirical study in the red flour beetle *Tribolium castaneum*. *Genet Res* 1996;**68**:151–5.
71. Wright S. Isolation by distance. *Genetics* 1943;**28**:114–38.
72. Wang J, Caballero A. Developments in predicting the effective size of subdivided populations. *Heredity* 1999;**82**:212–26.
73. Whitlock MC, Barton NH. The effective size of a subdivided population. *Genetics* 1997; **146**:427–41.
74. Whitlock MC, McCauley DE. Indirect measures of gene flow and migration: Fst not= 1/(4Nm+1). *Heredity* 1999;**82**:117–25.
75. Kuhner MK. Coalescent genealogy samplers: windows into population history. *Trends Ecol Evol* 2009;**24**:86–93.
76. Seyfert AL, Cristescu MEA, Frisse L, Schaack S, Thomas WK, Lynch M. The rate and spectrum of microsatellite mutation in *Caenorhabditis elegans* and *Daphnia pulex*. *Genetics* 2008;**178**:2113–21.

77. Waples RS, England PR. Estimating contemporary effective population size on the basis of linkage disequilibrium in the face of migration. *Genetics* 2011;**189**:633–44.
78. Blouin MS, Dame JB, Tarrant CA, Courtney CH. Unusual population genetics of a parasitic nematode: mtDNA variation within and among populations. *Evolution* 1992; **46**:470–6.
79. Archie EA, Ezenwa VO. Population genetic structure and history of a generalist parasite infecting multiple sympatric host species. *Int J Parasitol* 2011;**41**:89–98.

CHAPTER

9

Transmission Dynamics of *Ascaris lumbricoides* – Theory and Observation

T. Déirdre Hollingsworth, James E. Truscott, Roy M. Anderson

Imperial College London, London, UK

OUTLINE

Ascaris: The Neglected Parasite
http://dx.doi.org/10.1016/B978-0-12-396978-1.00009-4

Copyright © 2013 Elsevier Inc. All rights reserved.

INTRODUCTION

Among the helminth parasites that infect humans, the roundworm *Ascaris lumbricoides* is the most widely distributed across the globe and one of the most difficult to control. This is largely due to its very high fecundity and, concomitantly, high transmission potential. However, the inability of the human host to mount an effective immune response to block parasite entry despite repeated exposure, and the robust nature of the egg infective stage excreted in the feces of infected people which have good tolerance to a wide range of environmental conditions, are also central to its wide distribution and often high abundance.

Its abundance in many regions of the world with poor sanitation facilities has resulted in much research on its biology and epidemiology (see Chapter 7), and on how best to control infection and spread (see Chapter 15). Some of the most detailed studies were carried out under the auspices of the Rockefeller program on the control of soil-transmitted helminths in the southern regions of the United States, Panama, Puerto Rico, China and West Africa in the early 1900s. Researchers such as Norman Stoll, who worked at the Rockefeller Institute for Medical Research at Princeton and the Rockefeller University in New York after graduating from Johns Hopkins University, and who coined the phrase "This Wormy World" in 1947, created the template for the epidemiological study of helminth infections and the methods employed to measure the prevalence and intensity of infection based on fecal egg output from infected individuals. These methods are still widely used today.[1] Many of the papers published by Stoll and his many collaborators such as William Cort from Johns Hopkins in the first three decades of the 1900s still stand today as some of the best and most detailed studies of the biology and epidemiology of intestinal nematodes.[1–4] Following on from this pioneering work, others contributed greatly from countries such as Japan and South Korea after the Second World War, where control programs based on improved sanitation and chemotherapy, concomitant with economic growth and development, resulted in parasite eradication in most urban areas during the period of the 1950s to the 1980s.[5–8] During the same period, an epidemiologist at the London School of Hygiene and Tropical Medicine – George Macdonald – first turned attention to the use of mathematical models to study the transmission of helminths in a pioneering paper on the schistosome parasites.[9] Reviews of the history of model development for helminth infections in both ecology and epidemiology are provided by Anderson and May[10] and Basáñez et al.[11]

Macdonald used a simple deterministic differential equation to denote changes over time in the mean worm burden per person and highlighted the concept of a basic reproductive number, R_0, to measure transmission success, following on from Ross's studies of malaria which had been expanded on by Macdonald.[12,13] The basic reproductive number for helminth macroparasites is essentially the number of successful offspring that a parasite is capable of producing.[14,15] More precisely, it is the average number of female offspring produced throughout the lifetime of a mature female worm, which themselves achieve reproductive maturity in the human host in the absence of any density-dependent constraints.[16] If R_0 is less than unity the parasite is unable to sustain transmission. If it is greater than or equal to unity the parasite persists in the defined human community.

Macdonald also turned to the question of mating success in a dioecious worm where the presence of female and male worms in an infected person is essential for the production of viable eggs to continue the life-cycle of the parasite. He defined the concept of a "breakpoint" in transmission, where transmission is halted due to mating success falling below the value required to sustain transmission. This concept has been elaborated on in more recent work to include various assumptions about both the distribution of worm numbers person and parasite reproductive biology.[10,17]

It was not until the early 1980s that attention returned to the transmission dynamics and population biology of helminth parasites of humans, largely stimulated by a rapid growth in the use of mathematical models in the study of ecology and population biology of free-living species.[15,16] The first papers published in this period of renewed interest introduced the concepts and terminologies of population ecology to the epidemiological study of infectious agents. These concepts included population regulation by density-dependent fecundity and mortality, frequency distributions of organisms per spatial location or, in the case of parasites, per host, non-linear processes in transmission and population stability, and the notion of multiple population steady states or equilibria. Mathematical formulation was essential in mapping out the impact of various biological and epidemiological processes (often highly nonlinear in form) on population abundance and population behavior following perturbation induced by activities such as control interventions.

This chapter describes the more recent developments in this field and their role in the study of the epidemiology and control of *A. lumbricoides*, with a focus on the various uses of mathematical models, the questions that can be addressed by their use, and future needs in model development and application. Throughout, our emphasis is on making the methods transparent to epidemiologists, parasitologists and public health professionals with a minimum of technical detail. For those who "get their kicks" from equations we provide the key references to formulations and derivations.

The chapter is organized as follows. The first section charts the uses of mathematical models and types of model described in the literature to date and key insights from these models. The second section outlines the basic structure of a deterministic model and the properties that arise from a simple hybrid structure in which the distribution of worm numbers per person is treated as a probability distribution within deterministic framework. This model is then extended to include age structure. The third section outlines a set of key questions that arise in furthering our understanding of the transmission dynamics of the parasite and its control at a population level by various forms of community-based treatment and how models can currently be used to address them. The concluding section summarizes future needs in model development, parameter estimation, comparison of prediction and observation, and data capture in the modern age of web-based databases.

WHAT ARE MATHEMATICAL MODELS USED FOR?

Mathematics provides a universal language which gives precision to the description of pattern and process in the world around us. It also provides a set of approaches or tools to help in analyzing pattern and process. This applies equally in both the physical and biological worlds. The use of mathematical methods has until recently been much more common in the physical and chemical sciences and engineering. Biology and medicine by contrast have remained disciplines where description and observation dominate without the use of formal mathematical tools to assess whether a hypothesis is indeed capable of generating the patterns and processes observed. In part this is to be expected given the very multivariate nature of many biological problems and the complex nonlinear systems that dominate the organization and functioning of living organisms. The arguments against using simplified mathematical representations of complex biological processes hinge on the following observation: how can a crude simplification capture the known complexity of real biological systems? The counterarguments are many and varied, but center on two key issues. First, it is often the case that a few processes dominate outcome or observed pattern, even in very complex systems. It is better to explore systematically how each individual process, when added to the model step by step, affects outcome. Second, in very complex systems, without some sort of formal representation and defined tools of analysis, rarely (if at all) will it be possible to understand how each process, variable or parameter influences observed pattern. The human immune system is a good example. With an ever-expanding list of cell types and chemical entities for communication between cells involved in even the simplest immune response, the range of possible outcomes is unlikely to be revealed by description and observation alone. Gaining an understanding of multivariate systems with complex dynamics requires the use of mathematics.

Recently, the reticence to use mathematics in biology and medicine has changed, and in some disciplines, such as the epidemiology of infectious agents, mathematics has become a tool of choice. It is used, for example, in defining control options and their potential impact for pandemics of directly transmitted pathogens such as those created by the HIV, SARS, and influenza A viruses.[18–25] Concomitantly, the subject of mathematical model formulation and analysis has entered most graduate programs on the epidemiology of infectious agents at research-intensive universities.

Understanding Observed Patterns of Infection

In the case of soil-transmitted helminths such as *Ascaris*, mathematical models have many uses. At the simplest level, they aim to explore how observed pattern is influenced by various biological and epidemiological processes. The simplest example is that of the relationship between two key epidemiological measures, the prevalence, p (measuring the fraction of the population infected), and the average intensity of infection (worm load or an indirect measure such as eggs per gram of feces), M. Both are summary statistics of the frequency distribution of parasite numbers per host (Figure 9.1). Statistical procedures for fitting well-understood probability models such as the Poisson distribution to observed data reveal that the Poisson model (which is based on the assumption that each event occurs at random, and independent of other events) is a poor mirror of observation. For the Poisson, the mean equals the variance in value. Observed distributions have variances greatly in excess of the mean, denoting high aggregation of worms in the human population. The negative binomial model which has two parameters, the mean M and a parameter k which varies inversely with the degree of parasite aggregation within the host population, fits much better for most studies where worm numbers are recorded by expulsion chemotherapy. This probability model reveals that the relationship between prevalence and intensity is highly nonlinear with definition:

$$p = 1 - (1 + M/k)^{-k} \tag{9.1}$$

As illustrated in Figure 9.2, large changes in the mean M may lead to only small changes in the prevalence p, when the degree of aggregation is high ($k < 1$). This simple observation, based on a well-known probability model, has important practical implications. Where community-based chemotherapy reduces average worm loads by a large degree, this may

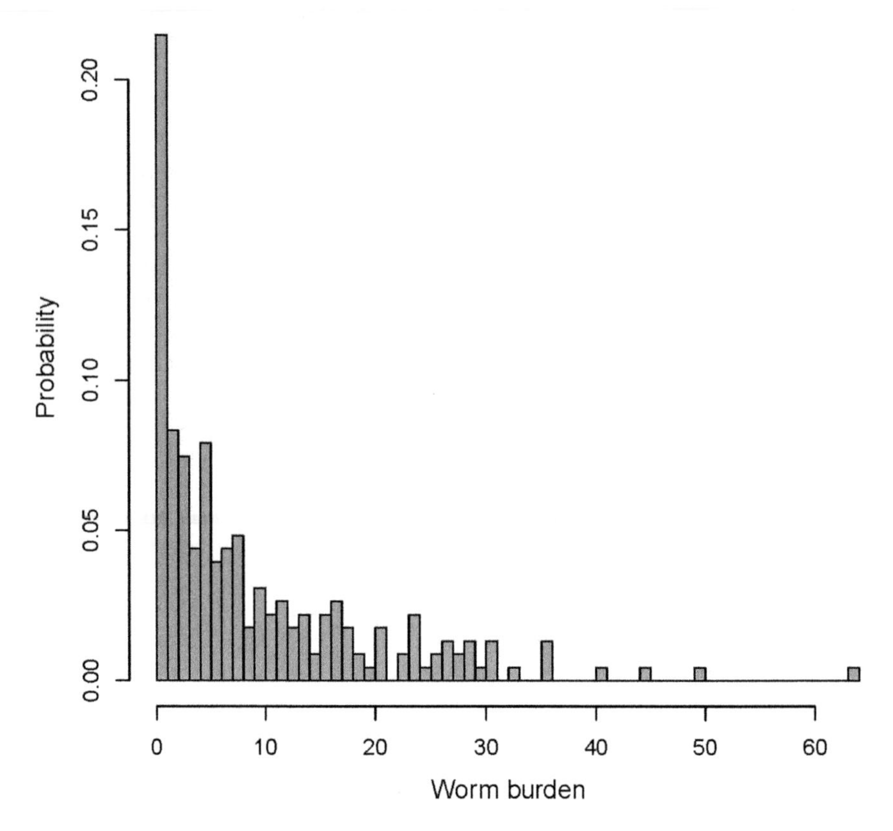

FIGURE 9.1 Distribution of *Ascaris* worm numbers per individual in a study in the Pulicat Lake region of Southern India.[30] Many individuals are uninfected while others are infected by large numbers of worms.

only result in very small – and perhaps undetectable – changes in prevalence. As such, prevalence is a poor measure of the impact of a control program (Figure 9.2).

The ecological or epidemiological process generating the negative binomial model of aggregated distributions of worm numbers per host, where most harbor few worms and a few harbor many, are many and varied.[26] One set of generative processes, and perhaps the most likely, is where the infection of each host is governed by random infection events with Poisson mean number of worms in an individual M, but where the mean differs in value for each host in a manner described by a flexible continuous distribution such as the gamma. This within and between host variability would generate a negative binomial at the population level.[27] Other processes, such as clumped acquisition of worms, will also affect the distribution of worms in an individual and at the population level.[28]

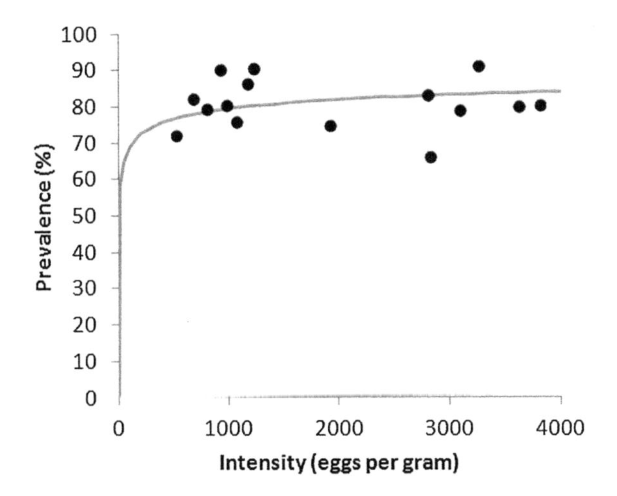

FIGURE 9.2 **The relationship between prevalence and mean egg output, from a study in Myanmar.**[60] The dots are the observed relationships and the solid line is the fitted curve, given in Eq. (9.1), with a k value of 0.194. For large changes in infection intensity there may be almost no change in prevalence.

The Biological and Ecological Determinants of Parasite Abundance

One of the benefits of using models are that, for simpler model structures in particular, larger scale behavior can be summarized by relatively simple expressions which can give powerful insights on the underlying processes. This type of analysis generates a template for understanding how each individual parameter influences the reproductive or transmission success of the parasite. This is encapsulated in the definition of R_0 by derivation from the basic model (which is described in the next section).[16] This model is illustrated by means of a flow chart of the life-cycle of *Ascaris*, definition of the key parameters determining flow through the life-cycle (Figure 9.3), a simple equation or equations to capture this flow by reference to changes in the mean worm burden per person M, and the derivation of an expression for the basic reproductive number R_0.

For the flow chart represented in this figure, and by ignoring for the time being mating probabilities, density dependence in fecundity and host age, the equation for R_0 is given by:

$$R_0 = (s\lambda\beta N d_1 d_2)/[(\mu + \mu_1)(\mu_2 + \beta N)] \tag{9.2}$$

The parameters (which are presented in the flow chart) are defined as follows. The parameter s represents the sex ratio of worms, λ denotes average egg production per female worm, β is the infection rate of hosts defined as the rate of contact between humans and infective stages times

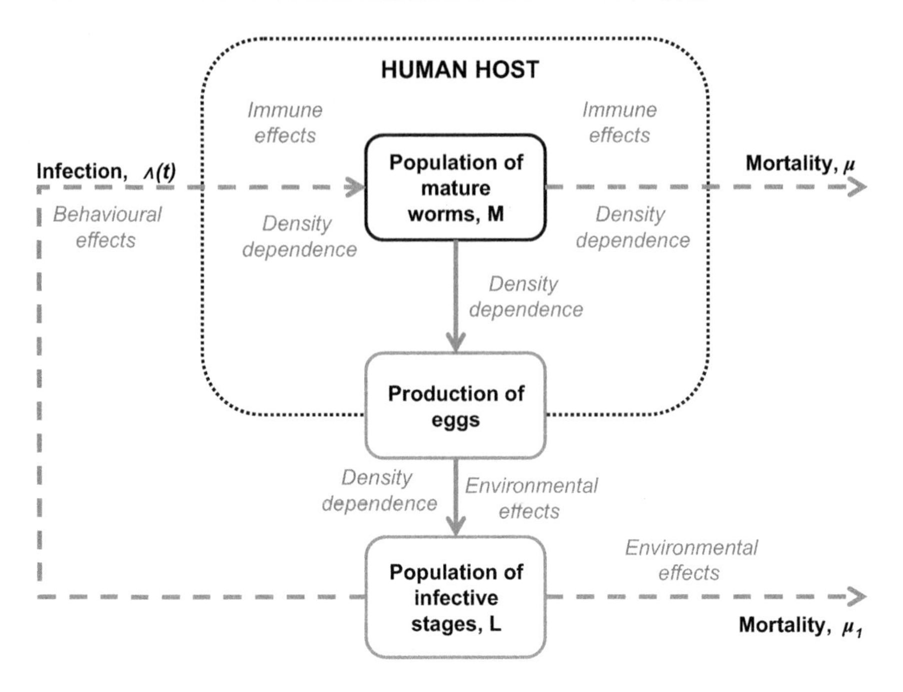

FIGURE 9.3 Schematic flow diagram of the populations of worms and eggs in the life-cycle of *Ascaris*.[10]

the probability that a contact results in parasite establishment, N is human population density, d_1 is the proportion of infective stages that gain entry to the host that survives to reach sexual maturity to contribute to fecundity, d_2 is the fraction of eggs excreted from the infected host that survive to become infective stages, μ is the human death rate ($1/\mu =$ life expectancy), μ_1 is adult worm death rate in the host, and μ_2 is the death rate of infective stages in the external environment.[16] Inspection of Eq. (9.2) shows that the numerator represents fecundity and transmission success while the denominator denotes losses due to death in various stages in the life-cycle. Or, put another way, the expression for R_0 is simply fecundity and transmission success times life expectancy of the various stages in the life-cycle. Thus, different soil-transmitted helminths (STH) achieve $R_0 > 1$ by different balances of longer life expectancy and low infectiousness or shorter life expectancy and increased infectiousness. In reality, this representation for R_0 is too simple in the face of known biological complexity. However, it is fairly straightforward to add density dependence in processes such as fecundity and survival, mating probabilities, seasonality in transmission, acquired immunity and host age and sex structure.[10,14] None of these really add much to the insights gained from the simple version denoted in Eq. (9.2).

Moving beyond transmission success, simple theory can provide insights into the dynamics of how a population of *Ascaris* is likely to respond to perturbation induced either by seasonal environmental or host behavioral factors, and – most importantly – the impact of control measures such as chemotherapy. This is illustrated by a very simple example, where we ignore age structure and mating probabilities on the assumption that aggregation of worms is high (k small) such that the breakpoint is just above a mean of zero worms per host, but include worm aggregation and density dependence in fecundity. In this case we can obtain a closed form solution for the equilibrium worm burden M^*:

$$M^* = k\left(R_0^{1/(k+1)} - 1\right)\Big/(1 - z) \tag{9.3}$$

Here R_0 and k are as defined above for Eqs (9.1) and (9.2), and z is a measure of the strength of density dependence in fecundity (the rapidity with which per capita fecundity declines as worm burden rises – see Figure 9.4).[16] Note how density dependence, transmission success, and the degree of worm aggregation are the sole determinants of average

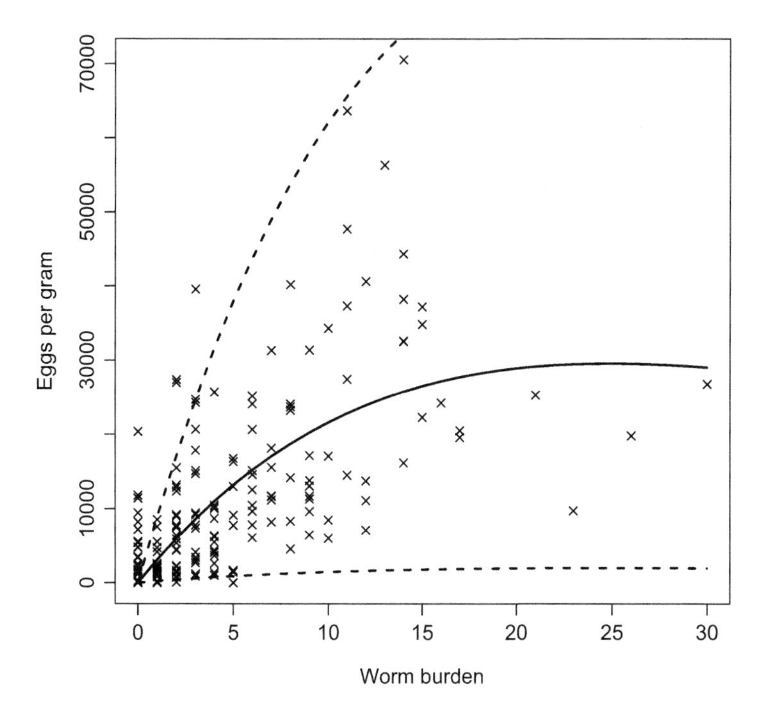

FIGURE 9.4 Density dependence in fecundity. The number of eggs output by individuals with different worm burdens in Pulicat, India.[30] Individual data are shown as crosses, and the black line is the fitted model (Eq. (9.8)), with light gray dotted lines giving the 95% confidence intervals.

worm load in this model. Recent analyses have suggested that the weight of female worms also affects this relationship, but this has yet to be captured in transmission models.[29]

The properties of the simple model are revealed using conventional methods to examine stability and dynamical behavior. The system has two stable equilibria. Either there are no parasites, an equilibrium mean worm burden of 0, or the system is at the equilibrium determined by R_0 and the other parasite characteristics, M^* (solid lines on Figure 9.5). In addition there is an unsteady state (dashed lines on Figure 9.5). If the system starts above this line, parasite loads will increase to the usual equilibrium. If, however, the system is perturbed to a state where the mean worm burden is below this line (e.g. by an extremely successful treatment program) then the population will crash to zero – i.e. the

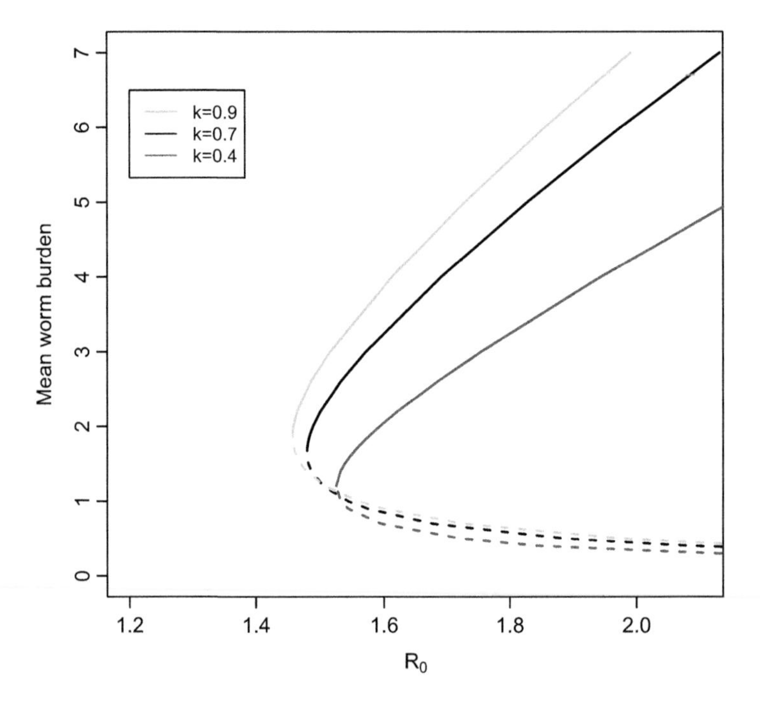

FIGURE 9.5 **Equilibrium worm burden as a function R_0 for simple model (Eq. 9.6) and different values of worm distribution shape parameter, *k*.** Solid lines represent stable worm burdens and broken lines unstable burdens. In a population with a given R_0, if worm burdens can be brought below the dotted line by interventions, then the breakpoint has been crossed and the cycle of transmission halted, leading to extinction. However, if worm burdens are not reduced to this level, the system will return to the stable equilibrium, increasing to baseline levels. The higher the level of aggregation (smaller *k*) the lower the breakpoint. These plots are generated for the model expressed in Eq. (9.6) with the parameters $1/(\mu + \mu_1)$, the life expectancy of the worm in the human host, of 1 year and fecundity parameter, $\gamma = 0.05$.

"breakpoint" has been crossed. The simple rebound behavior is exactly what is observed in reality, following the cessation of a community-based chemotherapy program or in the intervals between rounds of treatment (Figure 9.6). The existence of the breakpoint behavior is less readily observed since the breakpoint is predicted to lie at such low mean worm burdens (Figure 9.5).

Broadly speaking, by comparison with other infectious agents, such as viral and bacterial pathogens that can induce immunity and have short generation times of a few days, the dynamics of helminth parasites are fairly predictable and simple in form.[14] This is largely due to two factors; the inability of the human host to generate protective immunity to reinfection and the long life expectancy of helminth parasites in the human host which range from many months to many years.[10] The insights from these simple models have given many useful insights, and continue to inform planning of intervention programs (see below).

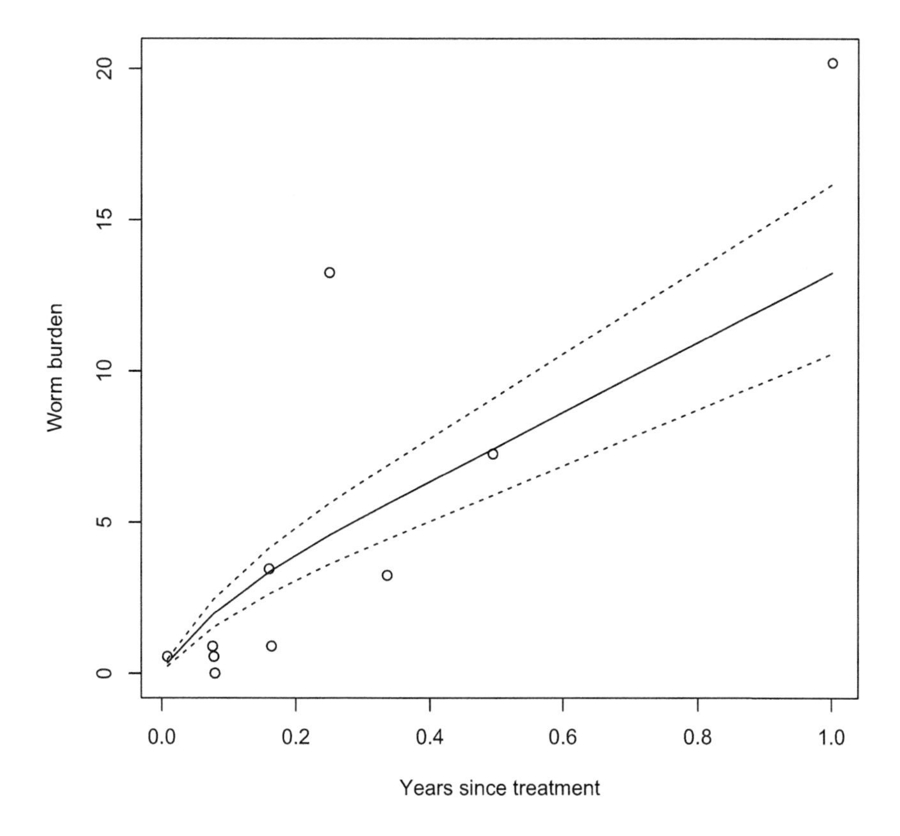

FIGURE 9.6 **Rebound of *Ascaris* worm burden following treatment; data and fitted model (Eqs. (9.4) and (9.5)).** Fitted parameters: $R_0 = 5.5$, treatment efficacy = 98%. Other parameter values: $\sigma = 1/\text{yr}$, $k = 0.57$, $\gamma = 0.04$. *Data and parameter values from.*[33]

Parameter Estimation

Mathematical models play an important role in parameter specification and estimation. The specification of a model structure involves the definition and inclusion of a number of key biological (e.g. parasite life expectancy in the host) and epidemiological (e.g. the force or rate of infection) parameters. Ideally, the typical values for these should be estimated independently from any data which a model is fitted to, such as temporal changes in mean intensity and the prevalence of infection, in order to test its accuracy in describing observed pattern.

A good example of such data is given in Figure 9.1, where, using worm expulsion techniques, the frequency distribution of *Ascaris* in a set of villages in India provided estimates of the aggregation parameter k.[30] A similar example is given in Figure 9.4 where fecal egg counts prior to worm expulsion provide estimates of the strength of density dependence in fecundity, z (see Chapter 7 for further examples).

It is often the case, however, that some parameters cannot be estimated without recourse to model fitting to observed changes in the key outcome variables of intensity and prevalence. Obvious examples are the rate of infection, β, and R_0, both of which may be dependent on host age and gender. To fill this need it is possible to draw on a range of statistical methods used in other areas of epidemiology and science in general. These statistical tools are being used to fit dynamic models in many areas of infectious disease epidemiology,[31,32] and are currently being developed in this field. This will allow models to be fitted to transmission dynamics from multiple settings, aiding their future development and applicability.

THE STRUCTURE OF MATHEMATICAL MODELS FOR THE TRANSMISSION DYNAMICS OF *ASCARIS*

We now outline the detail of some common model structures for the transmission dynamics of *Ascaris*. This section is useful, but not essential, for the understanding of the final section on insights from modeling of mass treatment interventions.

Basic Model

The simplest model structure represents the population biology of the mature adult worms (M) in the human host and the free-living infective egg stage (E), taking into account maturation delays in eggs (τ_2) and adult worms (τ_1) in progression to, respectively, infectivity and the egg-producing adult stage. As described by Anderson and May,[10] two simple equations are as follows:

$$\frac{dM}{dt} = \beta E(t-\tau_1)d_1 - \mu\sum_0^\infty ip(i) - \sum_0^\infty \mu_1(i)ip(i), \tag{9.4}$$

$$\frac{dE}{dt} = sd_2\varphi N\sum_0^\infty \lambda(i)ip(i,t-\tau_2) - \mu_2 E - \beta NE. \tag{9.5}$$

The parameters d_1 (fraction surviving to maturation in the human host), d_2 (fraction surviving to egg infectivity), μ (human death rate), μ_1 (death rate of adult worms), μ_2 (death rate of infective eggs), s (sex ratio), β (rate of infection per person per egg), and N (human population density) are all as defined for R_0 in Eq. (9.2). The term $\mu(i)$ represents density dependence in adult worm survival in the host (perhaps due to immunological responses) while $\lambda(i)$ represents density dependence in adult worm fecundity (Figure 9.4). The term $p(i)$ denotes the probability that a person harbors i worms and φ represents the mating probability in a person with i worms.

The model is deterministic in structure, but it contains probability elements in the mating probability φ and the worm distribution terms $p(i)$. If we make a phenomenological assumption concerning the distribution of worms to match observed patterns, the analysis is simpler. It is therefore assumed that the distribution is negative binomial in form with mean M and aggregation parameter k (Figure 9.1 is an example of the worm distribution). Further simplifications can be achieved by noting the different timescales of the two variables in the model. The rate of turnover of the eggs is much, much greater than that of adult worms and the developmental delays τ_2 is much shorter than τ_1, and hence little is lost in understanding the dynamics by collapsing the two equations into one for the mean worm burden M. If we retain the mating probability term and assume that density dependence acts solely on fecundity in the human host we arrive at the equation:

$$\frac{dM(t)}{dt} = (\mu+\mu_1)(R_0\Psi(M)-1)M(t). \tag{9.6}$$

Here $1/(\mu+\mu_1)$ is the life expectancy of the adult worm in the human host taking account of adult parasite and human mortality, R_0 is the basic reproductive number, and $\Psi(M)$ is a function that embeds density dependent fecundity and the mating probability.[14] Explicitly, if female and male worms are distributed jointly in a negative binomial pattern, and they are polygamous (as *Ascaris* is thought to be) where females are mated if at least one male is present, then[17]:

$$\Psi(M) = f(M,k)\left[1-\left(\frac{1+M(2-z)/2k}{1+M(1-z)/k}\right)^{-k-1}\right] \tag{9.7}$$

Here, $f(M,k)$ is the density-dependent fecundity function:

$$f(M,k) = [1 + M(1-z)/k]^{(-k-1)} \tag{9.8}$$

where $z = \exp(-\gamma)$ and γ is the exponent of the exponential decay function describing the decline in per capita fecundity as worm burden rises (see Figure 9.4).

This all looks rather complicated, but the reason for seeking this derivation from the basic model is to explore an important epidemiological question; namely, where does the breakpoint in transmission created by the mating probability effect lie? Is it close to zero worms per host, or is it somewhat higher?

As described by May[17] and Anderson and May,[10] the high degree of aggregation of *Ascaris* in the human population (where k values typically lie in the range of 0.1 to 0.5) results in the breakpoint lying close to zero worms per host. The model defined in Eq. (9.6) has two stable equilibria (zero and M^* worms per host) separated by an unstable equilibrium defined as the breakpoint (see Figure 9.5 and explanation above). A rough calculation based on a study of *Ascaris* in Iran[33] suggests that the breakpoint for a k value of 0.5, a population mean worm burden of 22 parasites per host, an R_0 of approximately 4–5, and a parasite life expectancy of around 1 year in the human host, gives a breakpoint of around 0.1 to 0.3 worms per host. In other words it is very low, and in practical terms with respect to control by community-based chemotherapy, it will only become of relevance if treatment is very intense and very frequent, suggesting that additional sanitation measures will be required for local elimination.

Age Structure in the Human Population

The prevalence and intensity of *Ascaris* infection vary with host age as illustrated in Figure 9.7 from a study in India.[30] Most epidemiological studies of *Ascaris* reveal a convex curve in intensity with respect to age, with the highest intensities in school-aged children (5–14 years of age). This pattern may be due to age-related exposure, the dynamics of new infections and the slow buildup of acquired immunity or some combination of both. It is straightforward to extend the model in the previous section to include age structure using a partial differential equation with the derivatives of time and age.[10] The hybrid age-structured model takes the form:

$$\frac{\delta M(a,t)}{\delta t} + \frac{\delta M(a,t)}{\delta a} = \Lambda(a,t) - f_3(M,k)M(a,t) \tag{9.9}$$

Here the transmission function, $\Lambda(a,t)$, is defined as:

$$\Lambda(a,t) = f_1(a,M,k)\left(\int_0^\infty l(a)M(a,t)f_2(M,k)da\right)\left(\int_0^\infty l(a)da\right)^{-1} \tag{9.10}$$

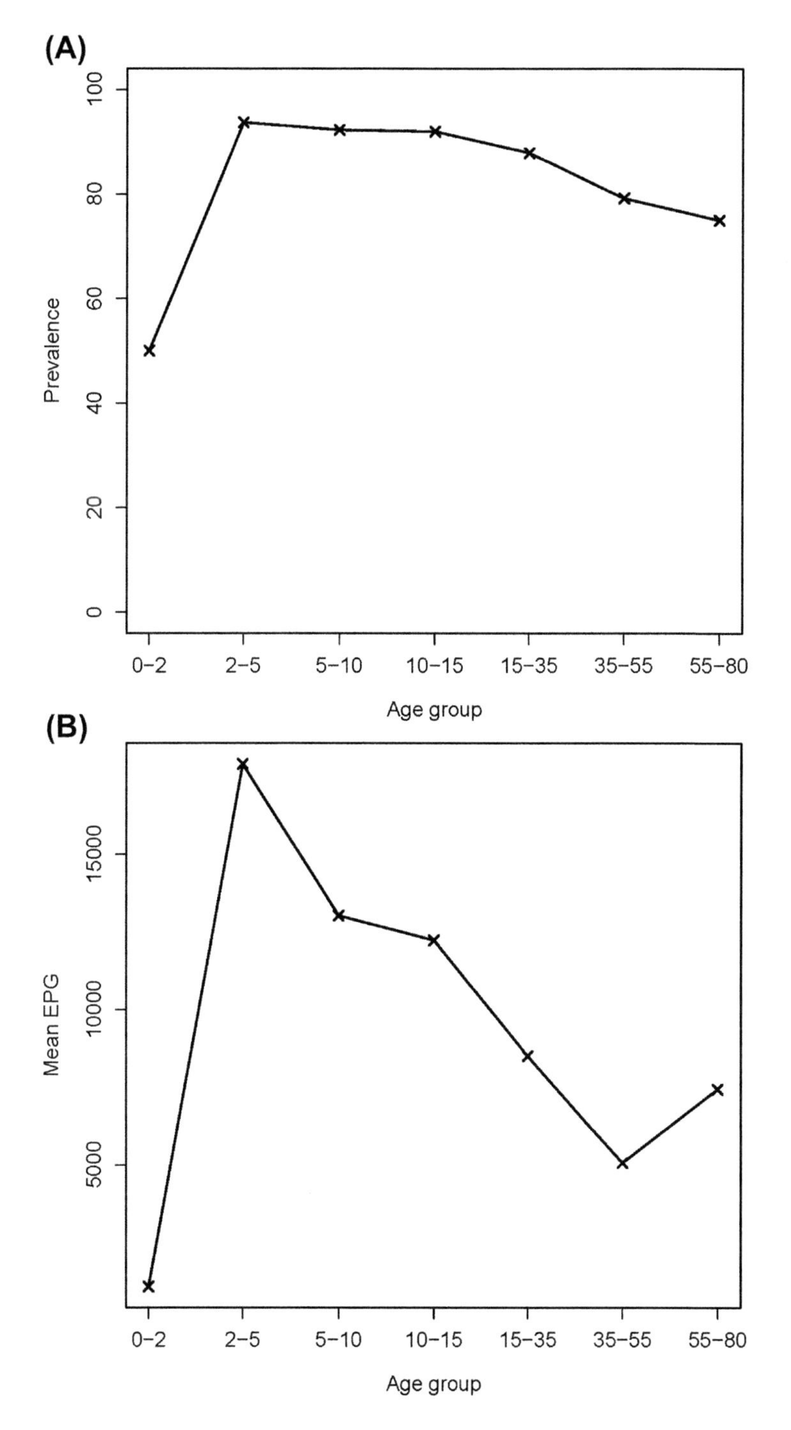

FIGURE 9.7 Horizontal patterns of (A) prevalence and (B) intensity from a study in Pulicat, India.[30]

The functions f_1, f_2, and f_3 represent, respectively: the collapsed details of parasite survival, reproduction, and transmission via segments of the life-cycle not involving the mature worms in humans; density-dependent constraints on parasite fecundity; and survival of adult parasites in the human host. The term $l(a)$ denotes the age-dependent survival function for the human host. In some countries in sub-Saharan Africa this is roughly an exponential decay term. In this formulation mating probabilities have been ignored based on the assumption of high degrees of worm aggregation. Analytical insights from this model are difficult to obtain, but details of equilibrium properties under various assumptions on human survival and age dependence in infection rates are given by Anderson and May.[14]

For most applications numerical work is required to solve Eq. (9.10) and this can form the basis for parameter estimation of the age-specific infection rates ($\Lambda(a,t)$). Parameterization of this model, however, is not straightforward, given that age dependence in parasite establishment in the human host may involve a degree of acquired immunity where parasite survival and fecundity depend on an integral of past exposure for each individual in the host population. A simpler approach to the inclusion of age structure in the human population is described in a subsequent section dealing with heterogeneities, where a crude three age class model is presented to mimic infection in pre-school children (0–4 years of age, school-aged children (5–14 years of age), and the rest (>14 years of age) (see later section). Some formal elaborations of the model to include the slow buildup of immunity in a manner dependent on accumulated past exposure are given in Anderson and May.[18] However, the relevance of these models remains in question at present in the absence of a consensus on the role of immunity in *Ascaris* infection in humans.[34]

The basic structure of the age-structured model has been used to construct a user friendly software interface (EpiWorm) to help policy makers in the design of control programs for intestinal worm infections and other helminths (http://www.schoolsandhealth.org). As is often the case with user friendly software used by those who are not specialists in mathematical model use in epidemiology, the assumptions underpinning the software and the uncertainties surrounding parameter estimation are not always clearly specified.

Stochastic Individual-based Simulation Models

There are many situations where a fully stochastic individual-based model is better able to capture the fine details of transmission and control impact at individual and community-based levels. An individual-based model is one in which each individual is included explicitly in the model, rather than a model of the average behavior of the population, as we have described above. This allows more explicit description of individual variability, but can be more challenging to analyze. One particular aspect of the epidemiology of helminths is the propensity for predisposition to heavy or light infection in the human host.[14,35–38] The reason for this is some poorly understood combination of genetic, social, and behavioral factors, with perhaps behavioral and social factors playing a greater role in many settings. The location of a home in a village, the social standing and degree of poverty of its family occupiers, the prevailing sanitation conditions plus the numbers of young children in the family could all play a role in exposure to infective stages. Predisposition is one factor contributing to the aggregated distributions of worm numbers per host and it often persists over many rounds of treatment where intensity measures enable the heavy and lightly infected individuals to be examined before and after each round of chemotherapy.

The description of predisposition ideally requires and individual-based model as first used by Anderson and Medley[39] to explore various aspects of community-based treatment and how predisposition influenced the choice of treatment program. The model used was age and gender stratified, with a gamma distribution for the mean exposure rate of individual hosts. Density dependence in fecundity, age dependence in exposure (the mean of the gamma distribution varied with age to give convex curves of intensity by age) and full human demography were also incorporated. This study examined two different assumptions concerning the generative mechanisms for worm aggregation – predisposition due to host genetics and environmental factors. One conclusion from this study was that worm aggregation was only significantly reduced from its baseline level (prior to the implementation of mass chemotherapy), if the causative factor of worm aggregation was predisposition not environmental factors.

It is somewhat surprising that individual-based stochastic models have not been used more widely to examine epidemiology, control-related questions, and policy issues since the early studies in 1985. This is especially the case for parameter estimation from data on intensity of infection stratified by age and gender, pre- and post-treatment, using Markov chain Monte Carlo (MCMC) methods. More is said about this later.

COMMUNITY-BASED CHEMOTHERAPY

With greatly increased focus on how best to control soil-transmitted helminths (STH) in poor regions of the world in recent years,[40–42] a great deal of attention is now focused on how models can best help public health workers to devise optimal and cost-effective control interventions. The ground work for current activity was laid in models published in the 1980s, but it is only recently for the STH group of parasites that these approaches have come back into focus[10,16,39] and questions raised on how best to improve upon these deterministic, hybrid, and stochastic frameworks. Most of current activity is centered on chemotherapeutic treatment, either targeted at children of school age, those most at risk (predisposed to heavy infection[35]), or the entire community. The donation of drugs in the 2012 London Declaration is targeted at school children, following the school-based de-worming targets published by the WHO.[40,43,44] Although effective and cheap drugs are available (often donated free by the pharmaceutical companies that manufacture them), it is widely recognized that, ultimately, treatment alone will not eliminate *Ascaris* infection. Effective sewage disposal is essential, as so clearly illustrated in South Korea, Japan, and the southern states in the US in their elimination campaigns.

Mathematical models can be used to investigate different strategies of treatment and other control options such as sanitation that reduces the pool of infective stages, but they reveal how difficult it is to reduce the population abundance of a parasite that evades creating protective immunity in the human host. After chemotherapeutic treatment to expel the worms, reinfection immediately starts to replenish the parasites in the human host via the pool of infective stages in the habitat of the community. As discussed earlier, bounce-back via reinfection to pre-treatment levels is the norm, unless treatment is both intense across all age classes in the population and takes place repeatedly.

The existence of a breakpoint in transmission (an unstable equilibrium, as discussed above), resulting from adult worm mating probabilities, does imply that repeated treatment could lower the average worm load to a point where it falls below the breakpoint and hence the parasite population moves to extinction at the stable equilibrium of $M = 0$. In practice, this breakpoint is close to an average of zero worms per host due to the high degrees of parasite aggregation observed in most if not all communities.

An illustrative example of repeated treatment using the basic model defined in Eqs. (9.4) and (9.5) is presented in Figure 9.8. The parameter assignments are set to mimic *Ascaris* in a setting with moderate to high levels of transmission. Note the jagged pattern of mean intensity over time as each round of treatment reduces worm burdens followed by a period of reinfection before the next treatment. In one example the frequency of treatment is sufficient to cross the breakpoint and the mean worm load moves to extinction. Elimination will persist provided no immigration of infected persons occurs to repopulate the habitat with infective stages.

Sanitation measures act to reduce R_0 due to the reduction in the pool of infective stages in the community and, concomitantly, the average force of

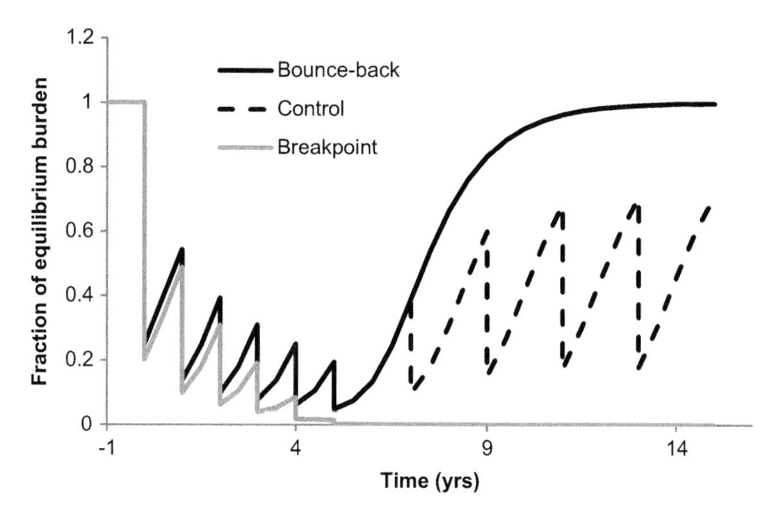

FIGURE 9.8 **The effect of treatment on the dynamics of infection intensity, simulated using Eq. (9.9).** In all scenarios yearly treatment is introduced for six rounds. In the bounce-back scenario (black solid line), the treatment program is halted. If treatment is continued at two-yearly intervals (black dashed line) then intensity bounces back, but to lower levels. If, however, coverage levels are high enough (gray line), the break-point may be reached after six years of annual treatment, and so intensity will not bounce back when the programme stops. Simulations for $k = 0.15$, $R_0 = 4.5$, $L_1 = 1$ year, treatment efficacy 75% (black lines) and 80% (gray line).

infection. Used in combination with community-based treatment, it is the only long-term solution for the control of STHs as well illustrated by the outcomes in Japan and South Korea post-World War II.

Perhaps the best way to summarize needs in this area is to list a set of key questions that surround *Ascaris* control by chemotherapeutic treatment, and these are presented in Table 9.1. The answers all hinge on four important issues identified by mathematical studies of the population biology and transmission dynamics of STHs. The first relates to the ability of the parasite population to bounce back to pre-control levels once treatment ceases. The second is what factors determine the speed of bounce-back and the third relates to the breakpoint in transmission and where it is located (how close to zero worms per host). The fourth concerns the expected lifespan of *Ascaris* eggs in the external environment, since this provides a reservoir of infection to repopulate the human host once treatment ceases.

Bounce-back

The first two issues are fairly straightforward to address. Bounce-back is inevitable if treatment stops in the absence of effective sewage disposal. *Ascaris* populations are very resilient to perturbations. Bounce-back times

TABLE 9.1 Key issues in the control of soil-transmitted helminths by chemotherapy[42]

1. For a given transmission level, how often should mass or targeted chemotherapy be administered to sustain infection prevalence and intensity below defined levels?
2. In terms of cost effectiveness, is it best to target school children, those predisposed to heavy infection, or the entire community?
3. As the prevalence and intensity fall after repeated rounds of treatment, can the interval between treatments increase, and by how much?
4. How is the interval between treatments influenced by the species mix of STHs in the community?
5. How do the demography of the population and the starting geographical distribution of infection influence the structure of the optimum treatment program where resources are finite?
6. What level of infection across a community should trigger mass chemotherapy to minimize morbidity?
7. Is elimination in a defined area possible by chemotherapy alone?
8. How might repeated mass treatment influence the evolution and spread of drug resistance – and how can this risk be minimized and how can it be best monitored?
9. What should be the target of STH control programs?
10. What are the best indicators of assessing the impact of STH control?
11. What is the optimal design and sampling requirements of programs for monitoring and evaluation?

are an interesting problem. For the simplest model structures of parasite infection – death, the time required to return to a fraction f of the pre-treatment equilibrium worm burden, t_f, from a zero worm burden in the treated individuals is given by:

$$t_f = -\ln(1-f)/\mu_1 \tag{9.11}$$

This result is revealing; it states that it is only the life expectancy $(1/\mu_1)$ of the adult worm that determines bounce-back time. In other words it is the generation time that determines this key aspect of intervention. Generation time is an ecological concept and it defines the average time for a parasite from birth to achieve sexual maturity. It is an essential aspect of the epidemiology of all infectious agents. It is not to be confused with the sum of the development times around the life-cycle from birth of an egg to maturation to an infective stage, and on to sexual maturity in the human host. Ecological generation time is this plus the weightings induced by mortality at all stages of the life-cycle. For the STH worms these generation times are in the domain of a few months to many years. *Ascaris* has one of the faster generation times for a helminth, and this ensures rapid bounce-back times. A key indicator of this is the need to treat every 3 months or so to bring *Ascaris* intensity to low levels in areas of high transmission intensity.[45]

The assumptions embedded in the model on which Eq. (9.11) is calculated are far too simple to reflect reality – but they provide a rough guideline. If more reality is added by way of a dynamic transmission term that reflects a reduction in the rate of infection as a consequence of treatment (the community wide effect of treatment), then Eq. (9.4) becomes more complex. One example based on no breakpoint and a Poisson distribution of worm numbers gives the following expression:

$$t_f \approx L_1 \left[\frac{1}{(R_0 - 1)} \ln\left(\frac{f}{f_1}\right) + \left(\frac{R_0}{(R_0 - 1)}\right) \ln\left(\frac{1 - f_1}{1 - f}\right) \right] \tag{9.12}$$

Here f is as defined in Eq. (9.11), the time taken for the worm burden to fall to a fraction f of the pre-treatment burden, e.g. 99%, L_1 is parasite life expectancy in the human host ($1/\mu_1$) and $1 - f_I$ is treatment coverage.[46] Note that in the dynamic model transmission success in the community, R_0, now enters as an important variable in determining bounce-back time. This is illustrated in Figure 9.9 for various parasite life expectancies and R_0 values. The importance of R_0 in determining the bounce-back time highlights the importance of understanding the local context when designing an intervention program. An obvious need is to extend these analyses to try to get approximations for the bounce-back time t_f, with aggregated distributions of worms and mating probabilities inserted.

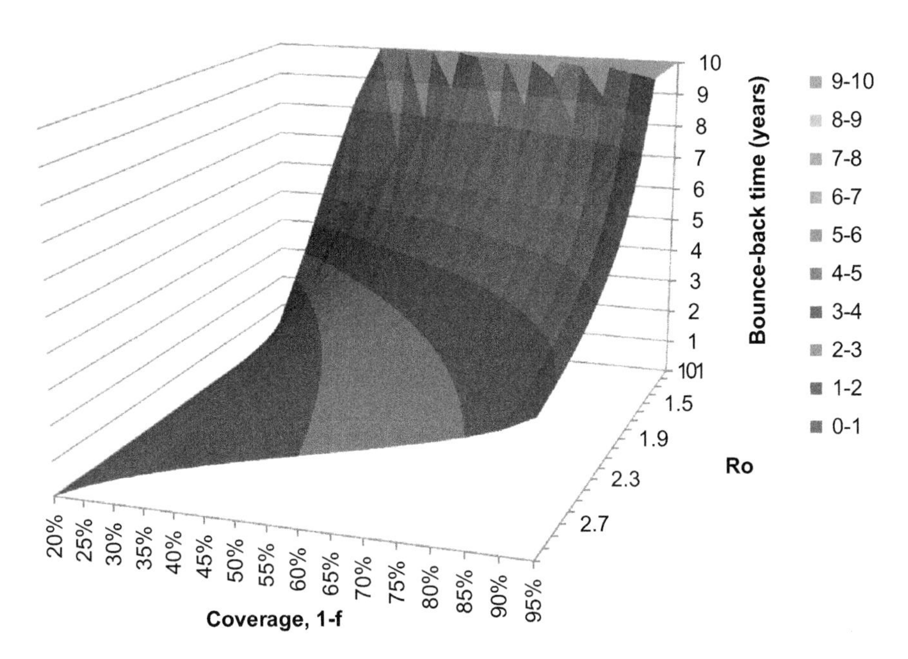

FIGURE 9.9 The average time taken for the average worm burden to bounce back to 80% of baseline, as in Eq. (9.12) for $L_1 = 1$ year.[46] For low treatment coverage, $1 - f_I$, and high R_0, bounce-back is very rapid (left), but for better treatment coverage, or lower R_0 it is slower.

Critical Treatment Coverage

One further application of models to the design of community-based chemotherapy intervention studies is to use them to get insights into what fraction of the host community needs to be treated per interval of time to lower R_0 to below unity in value. For simple hybrid models without age structure it is fairly straightforward to calculate the critical fraction of the population, g_c, that must be treated per unit of time to lower the effective reproductive number (transmission success under the influence of community-based chemotherapy), R, below unity in value[16]:

$$g_c = (1 - \exp[(1 - R_0)\tau/L_1])/h \qquad (9.13)$$

Here, h defines drug efficacy as a fraction of worm expelled by treatment, and τ is the time interval between treatments. For the commonly used drugs to treat *Ascaris*, h is usually around 0.9 to 0.95 in value.[47,48] The parameter g_c is defined in the same time units as used for parasite life expectancy (fraction treated per unit of time). Note that it is transmission success, R_0, and parasite life expectancy, L_1 (representing the rate of turnover of parasites in the human host in the absence of treatment), that define how many and how often treatment should take place. The influence of R_0 on the impact of the treatment program and the value of g_c are presented in Figure 9.10, which shows how changes in the fraction treated influence the mean intensity of infection and the prevalence at equilibrium. This figure again indicates how limited the measure of prevalence is of the success of community-based treatment.

Targeting by Age

Community-based treatment can take different forms ranging from the treatment of the entire population of all ages above 12 months of age, the treatment of school-aged children (5–15 years of age), and targeted treatment of only those with high worm loads. The last strategy benefits from the observation of predisposition to heavy infection in STH first recorded by Schad and Anderson[35] for hookworm, and subsequently observed for *Ascaris*,[38,49] *Trichuris*,[50] and schistosomes[36] infections. Targeting is rarely used in practice despite simulation studies recording its effectiveness.[39] The reasons relate to the perceived cost of identifying those with heavy infections. However, the evidence for predisposition suggests that you may only need to identify them once, but this depends on how predisposition varies over time for soil-transmitted helminths.[51,52] The drivers of predisposition may be multifactorial, including household factors and genetic factors.[49,53,54] Whatever the drivers, targeting by intensity, whether at an individual or household level, may be a method of getting maximum impact for lower numbers of treatments. More

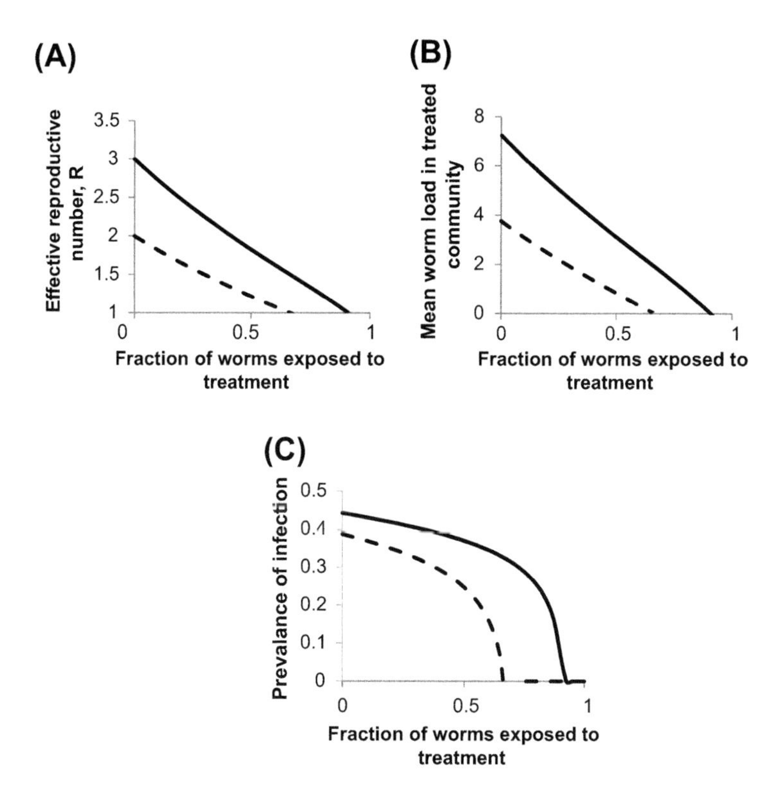

FIGURE 9.10 The influence of treatment coverage, g, in a yearly treatment campaign on (A) the reproductive number, (B) the equilibrium mean worm load in the population, and (C) equilibrium prevalence of infection. For a simulation in which $k = 0.15$, $R_0 = 3$ (solid line) and $R_0 = 2$ (dashed line).

simulation studies are need to further explore this issue, combining economic costs with impacts on transmission (i.e. the effective reproductive number).

Current practice is largely focused on school-age targeting, based on the ease of access to school children, and increasingly as more drug donations are made, on community-wide treatment. The age distribution of the human population where the parasites are endemic is often of importance in predicting the impact of community-based or school-aged children-based treatment programs.[42] The explicit representation of this distribution in transmission dynamic models is therefore highly desirable. A simple illustration of this point is provided by calculations of the fraction of worms treated by a school-age (5–14 years of age) targeted program. This fraction F is related to the age-specific death rate of humans, $l(a)$, which defines the age distribution, the age intensity of infection profile, $M(a)$, and the fraction of children in the population who attend school, θ, and who are therefore captured in the treatment program.

$$F = \theta\left(\int_{5}^{15} l(a)M(a,t)f_2(M,k)da \middle| \int_{0}^{\infty} l(a)M(a,t)f_2(M,k)da\right) \quad (9.14)$$

Here, $f_2(M,k)$ is the density-dependent fecundity term of Eq. (9.11). In many situations where cross-sectional surveys of intensity by age have been carried out the quantity F can be calculated. As illustrated in Table 9.2, *Ascaris* is a parasite for which the proportion of worms treated is around 30–50%, which is higher than for other soil-transmitted helminths,[55] due to the convex age-intensity profiles (Figure 9.7). This implies that only treating school-aged children may have some impact on the effective reproductive number of the parasite in the entire community. This reflects the results observed in two empirical studies, 3-monthly treatment of particular age groups of children has resulted in moderate changes in the intensity and prevalence of infections in adults.[45,56]

To gain a more accurate picture of how the treatment of just a segment of the population, such as school-aged children, influences the overall transmission dynamics of the parasite, the impact of not only age structure but also exposure to infective stages produced by all age groups in the population must be taken into account.[45,56,57] As formulated above, the age-structured models assume that exposure to infective stages across all age classes is random and independent of which age class produces the eggs or infective larvae. However, the spatial structure of egg deposition and infective stage development arising from one age group, plus their concomitant contact with other age groups, is in reality unlikely to be random in a defined community. Instead, it may be more likely that infective stages produced by school-aged children are deposited in areas closer to habitation and hence acquired by all age groups of the population,

TABLE 9.2 Fraction of worm population or egg output in 5–14-year-olds[55]

Parasite	Percentage of the total population of parasite or egg output in the 5–14-year-old age groups	Measure of parasite load	Data source	Country
Ascaris lumbricoides	49.4	Worm numbers	61	Burma
Ascaris lumbricoides	33.2	Worm numbers	30	India
Ascaris lumbricoides	27.6	Worm numbers	33	Iran

while those arising from adults are less likely to come into contact with children. Therefore the model should include heterogeneous mixing.[58]

A simple way to mimic non-random contact is to stratify the population into two age groups, namely: school-aged children (5–14) and the rest (0–4 and 15+ combined for simplicity, although patterns may also be different between these two groups), and assume different contact patterns with the infective stages within and between these larger age groupings. Such a stratification of hosts groups has the further advantage of facilitating the modeling of school-based treatment programs. We assume that the child and adult age groups have negative-binomially distributed worm distributions with the same shape parameter, k, but different means, M_c and M_a, respectively. The means evolve independently according to the degree of contact of each group with a common infectious "reservoir." The model equations are[55] (extending the approach by Chan et al.)[58]:

$$\begin{aligned} \frac{dM_c}{dt} &= \beta_c l - \sigma M_c, \\ \frac{dM_a}{dt} &= \beta_a l - \sigma M_a \end{aligned} \tag{9.15}$$

The quantity l is the per capita infectiousness of the shared reservoir. The parameters β_c and β_a determine the strength of contact with the reservoir for children and adults, respectively. The dynamics of the infectious reservoir are described by the following equation:

$$\frac{d}{dt}l = \frac{R_0\mu_2\sigma}{\lambda(\beta_c n_c p + \beta_a n_a q)}\left[f_2(M_c,k)n_c p + f_2(M_a,k)n_a p\right] - \mu_2 l \tag{9.16}$$

The parameters in Eqs. (9.15) and (9.16) are as defined earlier. The parameters n_c and n_a represent the proportion of the population in the two age classes and p and q the fraction of egg output that enters the reservoir from children aged 5–14 years and other age groups, respectively.

We investigate the effect of regular school-based treatment on the evolution of worm burdens in the community for three scenarios:

A. Homogeneous model. As defined in Eq. (9.6). Treatment is applied to a fraction g' of the population with efficacy h and at intervals of τ years.

B. Heterogeneous model. As described in Eqs. (9.15) and (9.16) assuming children and adults are identical epidemiologically. Treatment is applied to a fraction g of school-aged children with efficacy h and at intervals of τ years

C. Heterogeneous model with heterogeneous exposure. As described in Eqs. (9.15) and (9.16) assuming that children are both twice as potent a source of eggs and have twice the infectious contact rate. Treatment is applied to a fraction g of school-aged children with efficacy h and at intervals of τ years.

Treatment in the homogeneous model is made comparable with the heterogeneous model by setting coverage to $g' = g^{*}n_c$. Simulations of these different scenarios are presented in Figure 9.11, where the worm burdens in school-aged children and other age groups (where applicable) and averaged across the community are presented for different modeled scenarios, helminthes, and treatment intervals. The unbalanced mixing

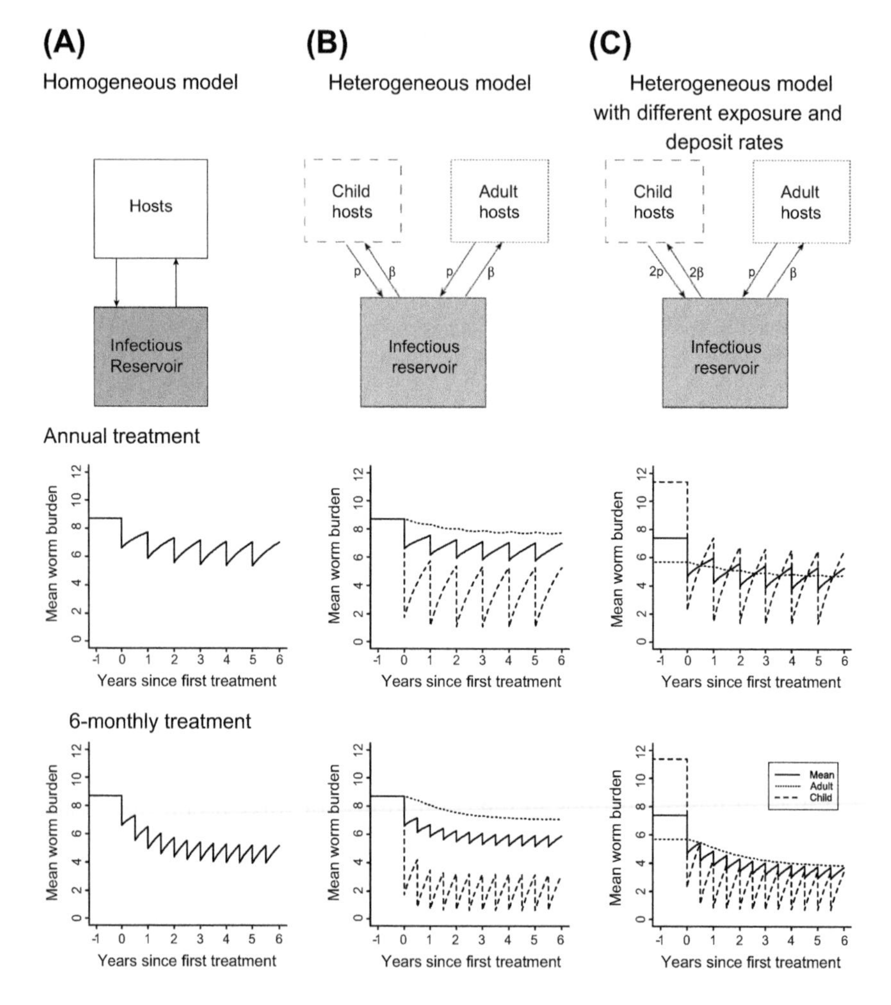

FIGURE 9.11 Effect of regular treatment on mean *A. lumbricoides* worm burden for different models. (A) Homogeneous population (left column), (B) heterogeneous population with uniform transmission dynamics (central column), and (C) heterogeneous population with greater contribution from children (right column) as in the text. The two rows represent annual and half-yearly treatment, respectively. For all runs, basic reproduction number is 3 and worm lifespan is 1 year. Other parameters (as defined for Eqs. (9.15) and (9.16)): $\mu_2 = 5/\text{yr}$, $k = 0.7$, $z = 0.93$. *Reproduced by license from.*[55]

(scenario C) results in a higher worm burden in the children than in the adults, as is seen in several settings (note this model does not include any immunity). All models have the same mid-range R_0 value of 3. We have simulated these scenarios for *A. lumbricoides*, with a life expectancy of 1 year (Figure 9.11).

Under these scenarios, and for these parameter settings, treatment of children has a modest impact on transmission, even at high levels of efficacy (95%) and coverage (85%). There are benefits for children in terms of period of low average worm burdens. The impact on the rest of the community is limited, due to the proportion of the worm population actually reached by treatment. The proportion of the population who are children being 30% is at the high end of school-attending fraction of the population.[55] Increasing the frequency of treatments to six monthly, there are some small additional benefits (bottom row in Figure 9.11).

When we divide the population into two groups who both interact equally with the reservoir (scenario B), the direct effect of school-based treatment on school-aged children is clearly shown, and is what would be measured in a monitoring and evaluation program among the children. The indirect effect on adults is much smaller. It should be noted that the rate of bounce-back after treatment is slightly lower in this model than in scenario A. This means that homogeneous mixing models, such as those earlier in this chapter, when used to describe non-uniform treatment regimens (targeted at some portion of the population) will always underestimate the time to recover to pre-treatment levels.

When we assume that children overcontribute to infection (scenario C), which is likely to be a more realistic scenario resulting in higher worm burdens in school-aged children, the effect of treatment on the school-aged group is quite pronounced. However, for these parameter values the impact at the community level is only marginally improved, due again to the small proportion of worms treated.

DISCUSSION AND FUTURE NEEDS

Much of the basic framework for the study of the transmission dynamics of helminth infections was laid down in the 1960s and 1980s. Rather little has been achieved since that time in model development and parameter estimation. Concomitantly, little use has been made of the insights gained from mathematical models in the design of public health policy for the control of STH and schistosome infections. A good illustration of the general ignorance of these insights among public health policy makers is well illustrated by the recommendations for STH control by the World Health Organization (WHO), which still advocates the use of prevalence as a marker of intervention success.[40,59] As illustrated in Figure 9.2, orders of magnitude change can occur in the mean intensity of infection with little change in prevalence. More generally, efforts to refine parameter estimates of key factors that control the rate at which infection bounces back to pre-control levels following an intervention has been very limited in the past two decades. This is despite the obvious importance of these parameters in the design of effective control programs based on mass chemotherapy.

There are a number of important needs if progress in epidemiological understanding is to be made and if the design of interventions is to be based on quantitative calculation. In epidemiological domains, the key unknowns are the importance of density-dependent constraints on parasite establishment in the human host and adult parasite survival, the patterns of mixing in infected communities with respect to exposure to infective stages contributed to the environment by individuals perhaps of different ages, and the importance of acquired immunity versus exposure to infection as determinants of convex intensity profiles by age.[10,11,29] All require more focused study. Mathematical model-based simulation studies can help in the absence of hard information, to understand their relative importance in shaping observed pattern and in parasite population responses post-treatment. More work of this nature needs to be carried out, given the obvious difficulties in designing epidemiological studies for measurement and the time scales over which such studies will have to be performed.

Perhaps the most urgent need is for better estimates of the key parameters. These include the severity of density dependence in fecundity (the parameter Y), estimates of parasite life expectancy in the human host, egg survival under different environmental conditions, age dependence in aggregation, sex ratios in adult worm populations, and age-specific forces of infection. Estimation of many of these parameters requires worm expulsion studies with eggs per gram data taken before expulsion and detailed follow-up during reinfection with appropriate stratification by host age and gender. Model fitting to observed patterns of reinfection is one approach using age and gender stratified models. Stochastic individual-based models should ideally be used, with MCMC methods employed in parameter estimation.

The final need relates to the design of intervention programs based on school age or community-based chemotherapeutic treatment. *Ascaris* is perhaps the most difficult STH to control by chemotherapy due to high rates of reinfection and the robust egg infective stage which can persist in the environment for many months under favorable environmental conditions. Currently, the much increased volume of drug donations to treat those infected with STH in developing countries combined with a much enhanced focus on the neglected tropical diseases demands that efforts to intervene on large scales in specific countries are based on designs that get the most benefit for the use of a defined quantity of treatment. At present, much treatment is dispensed with little quantitative analysis of who to treat and how often (see Table 9.1). Mathematical models, especially age-structured formulations and individual-based stochastic models could be used more frequently to design optimal community-based treatment programs given the availability of a known drug availability resource. Such models could be put in user friendly software packages (with appropriate health warnings on assumptions and uncertainties about parameter estimates) to help policy makers in the design of control programs. The new generation of young epidemiologist and public health workers is much more familiar with the use of mathematical and statistical methods in biology and medicine, and may have been exposed to courses with mathematical modeling content. It is to be hoped that they will lead a move to base control policy design on calculation as well as verbal discussion.

Acknowledgments

RMA, TDH, and JET acknowledge the Bill and Melinda Gates Foundation for research grant support.

References

1. Stoll NR. This wormy world. *J Parasitol* 1947;**33**:1–18.
2. Cort W, Riley W, Sweet WA, Schapiro L, Stoll NR. Studies on hookworm, *Ascaris* and *Trichuris* in Panama. V. An analysis of hookworm infestation in areas in Panama uninfluenced by control measures. *Am J Hyg* 1929;**9**:62–97.
3. Cort W, Schapiro L, Riley W, Stoll NR. A study of the influence of the rainy season on the level of helminth infestations in a Panama village. *Am J Hyg* 1929;**10**:62–97.
4. Cort WW, Schapiro L, Stoll NR. A study of reinfection after treatment with hookworm and *Ascaris* in two villages in Panama. *Am J Epidemiol* 1929;**19**:614–25.
5. Seo BS, Cho SY, Chai JY. Frequency distribution of *Ascaris lumbricoides* in rural Koreans with special reference on the effect of changing endemicity. *Kisaengchunghak Chapchi* 1979;**17**:105–13.
6. Seo BS, Lee SH, Chai JY. *An evaluation of the student directed mass control programme against ascariasis in Korea. Collected papers on the control of soil-transmitted helminthiases.* 2nd ed. Tokyo: Japan: Asian Parasite Control Organisation; 1983.
7. Seo BS, Rim HJ, Cho SY, Ahn JH, Kwak JW, Lee JW, et al. The prevalence of intestinal helminthes in inhabitants of Cheju Do. *Kisaengchunghak Chapchi* 1972;**10**:100–8.
8. Kim JH, Yoon JJ, Lee SH, Seo BS. 1970. Parasitologial studies of Korean forces in South Vietnam: II. A comparative study on the incidences of intestinal parasites. *Kisaengchunghak Chapchi* 1970;**8**:30–5.
9. Macdonald G. The dynamics of helminth infections, with special reference to schistosomes. *Trans R Soc Trop Med Hyg* 1965;**59**:489–506.
10. Anderson RM, May RM. Helminth infections of humans: mathematical models, population dynamics, and control. *Adv Parasitol* 1985;**24**:1–101.

11. Basáñez MG, McCarthy JS, French MD, Yang GJ, Walker M, Gambhir M, et al. A research agenda for helminth diseases of humans: modelling for control and elimination. *PLoS Negl Trop Dis* 2012;**6**:e1548.
12. Macdonald G. The analysis of equilibrium in malaria. *Trop Dis Bull* 1952;**49**:813–29.
13. Ross R. Some *a priori* pathometric equations. *Br Med J* 1915;**1**:546–7.
14. Anderson RM, May RM. *Infectious Diseases of Humans: Dynamics and Control*. Oxford, UK: Oxford University Press; 1991.
15. Anderson RM, May RM. Prevalence of schistosome infections within molluscan populations: observed patterns and theoretical predictions. *Parasitology* 1979;**79**:3–94.
16. Anderson RM. Depression of host population abundance by direct life cycle macroparasites. *J Theor Biol* 1980;**82**:283–311.
17. May RM. Togetherness among schistosomes – effects on dynamics of infection. *Math Biosci* 1977;**35**:301–43.
18. Anderson RM, May RM. Herd immunity to helminth infection and implications for parasite control. *Nature* 1985;**315**:493–6.
19. Anderson RM, May RM. Vaccination and herd immunity to infectious diseases. *Nature* 1985;**318**:323–9.
20. Anderson RM, Donnelly CA, Ferguson NM, Woolhouse ME, Watt CJ, Udy HJ, et al. Transmission dynamics and epidemiology of BSE in British cattle. *Nature* 1966;**382**:779–88.
21. Ferguson NM, Cummings DA, Cauchemez S, Fraser C, Riley S, Meeyai A, et al. Strategies for containing an emerging influenza pandemic in Southeast Asia. *Nature* 2005;**437**:209–14.
22. Riley S, Fraser C, Donnelly CA, Ghani AC, Abu-Raddad LJ, Hedley AJ, et al. Transmission dynamics of the etiological agent of SARS in Hong Kong: impact of public health interventions. *Science* 2003;**300**:1961–6.
23. Fraser C, Donnelly CA, Cauchemez S, Hanage WP, Van Kerkhove MD, Hollingsworth TD, et al. Pandemic potential of a strain of influenza A (H1N1): early findings. *Science* 2009;**324**:1557–61.
24. Fraser C, Hollingsworth TD, Chapman R, De Wolf F, Hanage WP. Variation in HIV–1 set-point viral load: epidemiological analysis and an evolutionary hypothesis. *Proc Nat Acad Sci USA* 2007;**104**:17441–6.
25. Hollingsworth TD, Ferguson NM, Anderson RM. Will travel restrictions control the international spread of pandemic influenza? *Nat Med* 2006;**12**:497–9.
26. Anderson RM, Gordon DM. Processes influencing the distribution of parasite numbers within host populations with special emphasis on parasite-induced host mortalities. *Parasitology* 1982;**85**(Pt 2):373–98.
27. Bartlett MS. *Stochastic Population Models in Ecology and Epidemiology*. London: Methuen; 1960.
28. Walker M, Hall A, Basanez MG. Trickle or clumped infection process? An analysis of aggregation in the weights of the parasitic roundworm of humans, *Ascaris lumbricoides*. *Int J Parasitol* 2010;**40**:1373–80.
29. Walker M, Hall A, Anderson RM, Basanez MG. Density-dependent effects on the weight of female *Ascaris lumbricoides* infections of humans and its impact on patterns of egg production. *Parasit Vectors* 2009;**2**:11.
30. Elkins DB, Haswell-Elkins M, Anderson RM. The epidemiology and control of intestinal helminths in the Pulicat Lake region of Southern India. I. Study design and pre- and post-treatment observations on *Ascaris lumbricoides* infection. *Trans R Soc Trop Med Hyg* 1986;**80**:74–92.
31. Cauchemez S, Valleron AJ, Boelle PY, Flahault A, Ferguson NM. Estimating the impact of school closure on influenza transmission from sentinel data. *Nature* 2008;**452**:50–4.

32. Griffin JT, Hollingsworth TD, Okell LC, Churcher TS, White M, Hinsley W, et al. Reducing *Plasmodium falciparum* malaria transmission in Africa: a model-based evaluation of intervention strategies. *PLoS Med* 2010;**7**.
33. Croll NA, Anderson RM, Gyorkos TW, Ghadirian E. The population biology and control of *Ascaris lumbricoides* in a rural community in Iran. *Trans R Soc Trop Med Hyg* 1982;**76**:187–97.
34. Dold C, Holland CV. Investigating the underlying mechanism of resistance to *Ascaris* infection. *Microbes Infect* 2011;**13**:624–31.
35. Schad GA, Anderson RM. Predisposition to hookworm infection in humans. *Science* 1985;**228**:537–40.
36. Bensted-Smith R, Anderson RM, Butterworth AE, Dalton PR, Kariuki HC, Koech D, et al. Evidence for predisposition of individual patients to reinfection with *Schistosoma mansoni* after treatment. *Trans R Soc Trop Med Hyg* 1987;**81**:651–4.
37. Haswell-Elkins M, Elkins D, Anderson RM. The influence of individual, social group and household factors on the distribution of *Ascaris lumbricoides* within a community and implications for control strategies. *Parasitology* 1989;**98**(Pt 1):125–34.
38. Haswell-Elkins M, Elkins D, Anderson RM. Evidence for predisposition in humans to infection with *Ascaris*, hookworm, *Enterobius* and *Trichuris* in a South Indian fishing community. *Parasitology* 1987;**95**(Pt 2):323–37.
39. Anderson RM, Medley GF. Community control of helminth infections of man by mass and selective chemotherapy. *Parasitology* 1985;**90**(Pt 4):629–60.
40. *WHO. Helminth control in school-age children.* Geneva, Switzerland: World Health Organization; 2011.
41. WHO. Soil-transmitted helminthiases: estimates of the number of children needing preventive chemotherapy and number treated. *Wkly Epidemiol Rec* 2009;**2011**(86):257–66.
42. Anderson R, Hollingsworth TD, Truscott J, Brooker S. Optimisation of mass chemotherapy to control soil-transmitted helminth infection. *Lancet* 2012;**379**:289–90.
43. *London Declaration. London declaration on neglected tropical diseases*; 2012.
44. *WHO. Accelerating work to overcome the global impact of neglected tropical diseases – a roadmap for implementation.* Geneva: World Health Organization; 2012.
45. Thein-Hlaing, Than-Saw, Myat-Lay-Kyin. The impact of three-monthly age-targeted chemotherapy on *Ascaris lumbricoides* infection. *Trans R Soc Trop Med Hyg* 1991;**85**: 519–22.
46. Fowler AC, Hollingsworth TD, Anderson RM. The dynamics of Ascaris lumbricoides infections; 2013. In preparation.
47. Marriner SE, Morris DL, Dickson B, Bogan JA. Pharmacokinetics of albendazole in man. *Eur J Clin Pharmacol* 1986;**30**:705–8.
48. Vercruysse J, Behnke JM, Albonico M, Ame SM, Angebault C, Bethony JM, et al. Assessment of the anthelmintic efficacy of albendazole in school children in seven countries where soil-transmitted helminths are endemic. *PLoS Negl Trop Dis* 2011; **5**:e948.
49. Walker M, Hall A, Basanez MG. Individual predisposition, household clustering and risk factors for human infection with *Ascaris lumbricoides*: new epidemiological insights. *PLoS Negl Trop Dis* 2011;**5**:e1047.
50. Bundy DA, Cooper ES, Thompson DE, Didier JM, Anderson RM, Simmons I. Predisposition to *Trichuris trichiura* infection in humans. *Epidemiol Infect* 1987;**98**:65–71.
51. Holland CV, Asaolu SO, Crompton DWP, Stoddart RC, Macdonald R, Torimiro SE. The epidemiology of *Ascaris lumbricoides* and other soil-transmitted helminths in primary school children from Ile-Ife, Nigeria. *Parasitology* 1989;**99**(Pt 2):75–85.
52. Quinnell RJ, Slater AF, Tighe P, Walsh EA, Keymer AE, Pritchard DI. Reinfection with hookworm after chemotherapy in Papua New Guinea. *Parasitology* 1993;**106**: 79–85.

53. Shapiro AE, Tukahebwa EM, Kasten J, Clarke SE, Magnussen P, Olsen A, et al. Epidemiology of helminth infections and their relationship to clinical malaria in southwest Uganda. *Trans R Soc Trop Med Hyg* 2005;**99**:18–24.
54. Quinnell RJ. Genetics of susceptibility to human helminth infection. *Int J Parasitol* 2003;**33**:1219–31.
55. Anderson RM, Truscott JE, Pullan RL, Brooker SJ, Hollingsworth TD. How effective is school-based deworming for the community-wide control of soil-transmitted helminths? *PLoS Negl Trop Dis* 2013;**7**(2):e1001076.
56. Asaolu SO, Holland CV, Crompton DWP. Community control of *Ascaris lumbricoides* in rural Oyo State, Nigeria: mass, targeted and selective treatment with levamisole. *Parasitology* 1991;**103**:291–8.
57. Bundy DAP, Wong MS, Lewis LL, Horton J. Control of geohelminths by delivery of targeted chemotherapy through schools. *Trans R Soc Trop Med Hyg* 1990;**84**:115–20.
58. Chan MS, Guyatt HL, Bundy DAP, Medley GF. The development and validation of an age-structured model for the evaluation of disease control strategies for intestinal helminths. *Parasitology* 1994;**109**:389–96.
59. *WHO. Preventive Chemotheraphy in Human Helminthiasis.* Geneva, Switzerland: World Health Organization; 2006.
60. Thein Hlaing, Than S, Htay Htay A, Myint L, Thein Maung M. Epidemiology and transmission dynamics of *Ascaris lumbricoides* in Okpo village, rural Burma. *Trans R Soc Trop Med Hyg* 1984;**78**:497–504.
61. Thein Hlaing, Saw T, Lwin M. Reinfection of people with *Ascaris lumbricoides* following single, 6-month and 12-month interval mass chemotherapy in Okpo village, rural Burma. *Trans R Soc Trop Med Hyg* 1987;**81**:140–6.

PART IV

HOST AND PARASITE GENETICS

CHAPTER

10

From the Twig Tips to the Deeper Branches: New Insights into Evolutionary History and Phylogeography of *Ascaris*

Martha Betson[‡], *Peter Nejsum*[†], *J. Russell Stothard*[*]

[*] Liverpool School of Tropical Medicine, Liverpool, UK
[†] University of Copenhagen, Denmark
[‡] The Royal Veterinary College, London, UK

OUTLINE

Ascaris: The Neglected Parasite
http://dx.doi.org/10.1016/B978-0-12-396978-1.00010-0

Copyright © 2013 Elsevier Inc. All rights reserved.

INTRODUCTION

Infections with the parasitic large roundworm give rise to ascariasis, which is a particularly common disease throughout the world.[1,2] Initially described by Linnaeus in 1758, *Ascaris lumbricoides* has a cosmopolitan distribution being found in both temperate and tropical regions, especially where there is adequate moisture matched with inadequate sanitation and hygiene. The closely-related *A. suum* is present in pigs worldwide, though its distribution is strongly influenced by local farming practices and animal husbandry.[3] As *Ascaris* worms do not directly multiply within their host, to complete its life-cycle the parasite requires the ingestion of fertile eggs produced by adult females within the intestinal lumen and passed out in the host feces. In the environment the parasite eggs mature and become infective. Upon entry into the host by the oral route, eggs hatch within the duodenum releasing the third stage rhabditiform larvae which penetrate across the intestinal mucosa eventually reaching the liver and then the lungs. Larvae then ascend the trachea to be swallowed and pass down the esophagus into the ileum where they molt twice and further mature and mate, typically residing there for 1–2 years unless expelled by host immune response or de-worming medications.[1]

EVOLUTIONARY HISTORY OF ASCARIDOIDEA AND HOST POTENTIAL

With more than 50 described genera, nematode worms within the superfamily Ascaridoidea are ubiquitous helminth parasites being found in a variety of mammals inclusive of farmed livestock (e.g. *Ascaris suum* in pigs and *Toxocara vitulorum* in cattle), a variety of domesticated and wild mammals (e.g. *Parascaris equorum* in horses and *Baylisascaris procyonis* in raccoons), as well as in man (e.g. *Ascaris lumbricoides*), notwithstanding occasional infections in non-human primates such as chimpanzees (see Table 10.1). Attempts to bring taxonomic and evolutionary systematic order to classification of these nematodes has not always been successful, as morphologically distinctive characters are often lacking and those used are frequently judged to be homoplastic (i.e. misleading) by character congruence testing, in which the distributions of shared character states are compared among taxa.[4] The use of molecular phylogenetics, targeting nuclear ribosomal 18S and 28S regions augmented with the mitochondrial ribosomal 12S or *cox*2, has attempted to provide systematic order[5] and is summarized in schematic form in Figure 10.1.[4,6,7] The use of a combination of nuclear and mitochondrial markers yielded trees with higher

TABLE 10.1 Summary of definitive hosts for selected species belonging to the family Ascarididae.

Species	Definitive host(s)
Ascaris lumbricoides	Humans[1] Non-human primates[100,101]
Ascaris suum	Pigs[102] (Chimpanzees)[40] (Humans)[103]
Baylisascaris ailuri	Red pandas[104]
Baylisascaris columnaris	Skunks[102]
Baylisascaris procyonis	Racoons[10]
Baylisascaris schroederi	Giant pandas[105]
Baylisascaris transfuga	Bears Pandas[102]
Parascaris equorum	Equids[102]
Toxascaris leonina	Dogs Wolves Foxes Cats[102]
Toxocara canis	Dogs[102] Wolves[106,107]
Toxocara cati	Cats[102]
Toxocara malaysiensis	Cats[108]
Toxocara pteropodis	Fruit bats[102]
Toxocara vitulorum	Cattle[102] Buffaloes[109] Bisons[110]

(Humans, Chimpanzees) indicates that while it is clear that *A. suum* can infect these hosts, it appears to be better adapted to the pig host.[103]

support for each clade than when nuclear or mitochondrial markers were analyzed on their own.[6] It is evident that the inter-generic relationships of these groups within Ascaridoidea are open to debate; however, relationships within the family Ascarididae are perhaps less contentious. The former problem is likely due to the poor phylogenetic performance of these loci across "deep" evolutionary time spanning hundreds of millions of years[5] and is also confounded by poor taxonomic sampling, meaning that samples available for analysis are not evenly distributed across genera. Thus some genera are very well represented (e.g. *Toxocara* spp.) while others are not (e.g. *Parascaris* spp.),[8] leading to sampling biases.

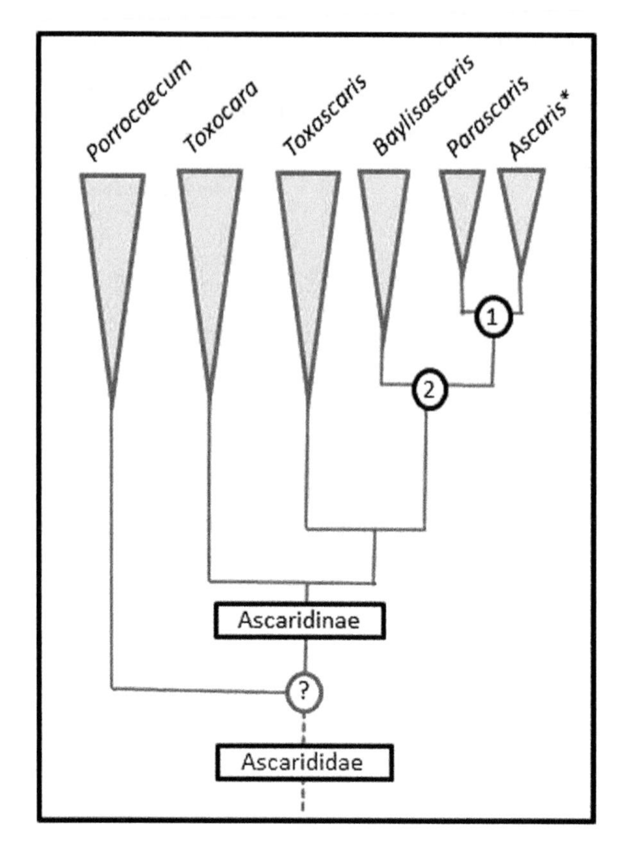

FIGURE 10.1 Schematic representation of phylogenetic trees based upon Li et al.,[6] Nadler and Hudspeth,[4] and Zhu et al.[7] The *Ascaris* genus is indicated with an asterisk. More cryptic genera such as *Lagochilascaris* are omitted as presently there is no molecular information on these taxa. A key question is to place the nodes 1 and 2 in a precise evolutionary time which might better reveal the ancestry and host potential of proto-*Ascaris*.

In addition, for some genera there has been very limited sampling of populations within a species, resulting in failure to detect intraspecific variation and its significance.

It is therefore unclear at present when the evolutionary split, as depicted in nodes 1 and 2 of the schematic tree (Figure 10.1), took place but this would appear to impact upon the parasitic host range of these worms,[9] which might have some bearing on the human–pig epidemiology debate (see below). It is notable that only *Ascaris* can complete its direct life-cycle in full within humans and non-human primates whereas for worms of other genera, humans are paratenic hosts. For example, the definitive hosts for *B. procyonis* are racoons, which become infected through direct ingestion of eggs in the environment (in the case of juveniles) or ingestion of larvae through predation on intermediate hosts (small mammals and birds).[10] Accidental ingestion of *B. procyonis* eggs by humans in areas where racoons and humans live in close proximity can cause a form of visceral, neural or ocular larva migrans.[10–14] Nonetheless, it is likely that the host range of many of these worms is catholic, open to evolutionary opportunity when given sufficient epidemiological potential.

An evolutionary quirk of *Parascaris*, shared with *Ascaris*, is that nuclear DNA is greatly reduced during early development in all somatic cells but not in germ-line cells. This developmentally-regulated DNA rearrangement, known as chromatin diminution, is highly specific with respect to timing and chromosomal location.[15,16] These non-Mendelian processes have frustrated the application of genomic fingerprinting techniques such as RAPDs (randomly amplified polymorphic DNA) and therefore one must always remain aware of unusual molecular drive processes that confound molecular systematic approaches.[17] Interestingly, sequence analysis of chromosomal breakage regions (specific sites where elimination of DNA takes place) has shown that these regions are conserved between *Ascaris* and *Parascaris*, suggesting that these sites were present in a common ancestor of these parasites.[18]

SPECIES CONCEPTS

The distinction between *A. lumbricoides* and *A. suum* and whether they should be regarded as one or two species still engenders debate today.[19,20] We will discuss this issue below but for simplicity and convenience we will use *A. lumbricoides* and *A. suum* to refer to human and pig *Ascaris*, respectively. We now briefly review the more common species concepts which are broadly applicable. The conclusion one reaches on whether there are one or two *Ascaris* species is, of course, highly influenced by the species definition used and this issue is often ignored in the debate. Species is the lowest taxonomic unit used in biology to classify organisms into taxonomic ranks but a range of different species concepts have been proposed largely tailored to a generalized concept of "organism" which does, of course, differ in application across the Metazoa.[21]

The major classes of contemporary species definitions include the *phenetic species concept* which groups organisms that are phenotypically similar and can be distinguished from other organisms into a coherent phenetic cluster.[22] Thus if no morphological differences can be found between *Ascaris* obtained from humans and pigs they must be considered a single operational taxonomic unit (OTU) or "one" species. The *phylogenetic species concept* typically takes a cladistic approach where a species can be defined as an irreducible clade of organisms that is diagnostically distinguishable from others by analysis of synapomorphic characters (traits shared by two or more taxa and their most recent common ancestor but not by more distant ancestors) and consideration of a parental pattern of ancestry and descent. In seeking the most parsimonious explanation for depicting character changes, recourse to taxonomic "outgroups" is often needed for investigation of the polarity of characters and their stepwise changes. This, of course, can be problematic, especially if there is limited taxonomic material for inspection and "key" linking groups are missing, i.e. so-called gaps in the evolutionary record. Identification of an evolutionary informative character or "diagnostic" marker will therefore in this case suggest that *A. suum* and *A. lumbricoides* still should be designated two different species names. The *evolutionary species concept* defines a species as a single lineage which maintains its identity from other such lineages (evolving separately as inferred from tree-building methods often using probabilistic approaches such as maximum likelihood), and has a unitary evolutionary role, tendencies and historical fate through time. So here focus is on evolutionary independence due to unique features that can be ascribed to the organisms.

The most commonly used concept for a sexually reproducing organism, however, is the *biological species concept* proposed by Mayr: "Species are groups of actually or potentially interbreeding populations, which are reproductively isolated from other such groups."[23] As some organisms can mate but not produce fertile offspring, Bock advocated a reformulation: "a species is a group of actually or potentially interbreeding populations of organisms which are genetically isolated in nature from other such groups,"[24] emphasizing the lack of gene flow between populations as central for evolution of species and definition hereof. Barriers to gene flow, and therefore the evolution of species, can be pre-zygotic and post-zygotic.[22] Pre-zygotic barriers can be physical or temporal, i.e. if *A. suum* and *A. lumbricoides* are not found in the same host at the same time, mating cannot take place. They could also be found in the same host at the same time but be unable to recognize each other, or, if they do, transfer of male gametes may not take place or eggs may not be fertilized (gametic incompatibility). The nature of asynchronous mating populations in *Ascaris* should also not be forgotten, especially given that parasites' eggs can remain dormant but viable within the environment for several years. Another restriction in gene flow can be post-zygotic. If *A. suum* and *A. lumbricoides* can mate, fertilized eggs may die, the produced off-spring (F_1 hybrid) may be sterile or have reduced viability (fitness) or lastly the worms may have reduced viability or fertility in F_2 or backcross generations. According to the biological species concept, cases of cross-infection or identification of hybrids does therefore not imply that *Ascaris* in humans and pigs are one species. In addition, where two populations are attributed to the same species based on the phenetic or phylogenetic concepts, which is often the case in "newly developed species," the biological species concept will define them as two if there is no gene flow between the populations. In this case morphological differences and diagnostic markers, which can be used to distinguish between species, will arise over time. The use of molecular markers to "tag" and "track" worms is therefore an invaluable tool to explore whether there is gene flow between *Ascaris* populations in pig and human and thereby settle the taxonomic status according to the biological species concept. This would aim to bring some better taxonomic stability to the long and often historically confusing list of named species within *Ascaris*.

APPLICATION OF BIOCHEMICAL AND MOLECULAR TOOLS

Early studies investigating the genetic diversity of *Ascaris*, and in particular differences between worms from human and pig hosts, used a variety of biochemical and immunological methods including one-dimensional and two-dimensional electrophoresis of proteins and isoenzyme analysis, which investigates polymorphism at isoenzyme loci such as mannose phosphate isomerase and lactate dehydrogenase.[25–29] Molecular techniques were first applied to investigate the genetic diversity of *Ascaris* and transmission between pigs and humans in the early 1990s. The RAPD (random amplified polymorphic DNA) method, which involves PCR amplification of genomic DNA using random primers to produce a profile of amplified fragments that can be compared between individuals/populations, was employed in a handful of studies but was soon superseded by other methods.[29,30] PCR amplification of mitochondrial and nuclear markers followed by restriction enzyme digestion to identify polymorphic restriction sites has been widely used.[31–40] Alternative methods employed to detect polymorphisms in PCR-amplified mitochondrial (e.g. *cox*1 and *nad*1) and nuclear markers (e.g. ITS-1) include direct sequencing[39–46] and single-strand conformation polymorphism (SSCP),[38,47–49] which is based on the fact that single-stranded DNA fragments of different sequence will adopt different conformations that can be detected by gel electrophoresis. Direct sequencing has become increasingly popular with dramatic reductions in sequencing costs. In contrast to these methodologies targeting specific markers, amplified fragment length polymorphism (AFLP), a technique for whole genome fingerprinting, has been used to detect genetic diversity in *Ascaris* from humans, pigs, and captive chimpanzees in Denmark.[35,36,50] Recently, more than 30 microsatellite markers, representing autosomal and sex-linked loci, have been described in *Ascaris*.[51] Microsatellites are repeating sequences of 2–6 base pairs in the nuclear genome, which can vary in length between individuals and populations. The *Ascaris* microsatellites have since been applied in a number of studies investigating population genetics.[42,43,46,52,53] At the present time, PCR amplification of nuclear and mitochondrial markers followed by direct sequencing and microsatellite analysis are the most common methods used for detecting genetic variation in *Ascaris*. In addition, complete mitochondrial genome sequencing is now becoming a realistic means of comparing the genetic diversity of different *Ascaris* isolates.[54,55]

EXPERIMENTAL STUDIES

A classical way to explore host affiliation and define species is to conduct experimental infection studies and these have been used to illuminate the relationship between *A. suum* and *A. lumbricoides* and their host specificity. In this way Takata infected 17 volunteers with *A. suum* and found egg excretion among one-third of the participants suggesting that pig *Ascaris* can establish and mature in the human host.[56] However egg excretion was found in all of seven other humans infected with *A. lumbricoides* suggesting differences in host preference. Differences related to quality and infectivity of eggs can, however, not be ruled out. Likewise, Galvin infected pigs with *A. suum* and *A. lumbricoides* and was also able to demonstrate that cross-infection can occur,[57] but there was also clear evidence of host preference both with respect to number of egg-excreting individuals and number of worms in the intestine of the pigs. In this study, more larvae of *A. lumbricoides* were isolated from the lungs and small intestine of rabbits whereas the opposite was the case for *A. suum* in pigs suggesting that differences in establishment rate in pigs cannot purely be attributed to egg quality.[57] Numbers of larvae isolated from the lungs of mice 7 days post-infection were also higher when infected with genotype G1 (predominantly found in humans) compared with genotype G3 (predominantly found in pigs).[38] In contrast, while genotype G3 established well in pigs, none of the pigs infected with eggs of genotype G1 were found to excrete eggs and only one immature worm was found among 47 infected pigs. However, one caveat to this study is that as eggs from only two female worms of each genotype were included in the study, the differences may purely reflect worm to worm variations in establishment rate.[58]

Several case reports on accidental infection of humans after handling infective *A. suum* eggs or after contact with pig manure also suggest that cross-infections can take place.[59–62] The broader host spectrum is further supported by the finding of zoo chimpanzees infected with *A. suum*[36] and the evidence would seem to suggest that a permanent transmission cycle has been established.[40] However, the fact that the viability of excreted eggs seems to be reduced suggests that chimpanzees are not an ideal host for *A. suum*. Mature *Ascaris* thought to be of pig origin have also been reported in the intestine and bile duct of sheep and cattle,[63,64] despite the large differences in the alimentary tract between monogastrics and ruminants. Even though *A. lumbricoides* and *A. suum* can be found in heterologous hosts and cross-infections can occur between them and despite concerns related to the design of the above-mentioned studies, it seems safe to assume that *Ascaris* is most adapted to its "appropriate" host suggesting that certain aspects of speciation have occurred between them.

INSIGHTS INTO A. *LUMBRICOIDES/SUUM* COMPLEX

The gross morphology of *Ascaris* worms from pigs and humans is indistinguishable without exception, but subtle differences in denticle morphology and lip shape between worms from the two hosts have been described.[65–67] Biochemical studies revealed contrasting profiles in lysates obtained from human and pig worms (both adults and larvae).[25,26,68] Differences in the properties of trypsin inhibitors between *A. lumbricoides* and *A. suum* were also identified.[27] However, the worms compared in these studies were not sympatric, meaning that differences could be due to geographical and/or intrinsic variability in the *Ascaris* populations rather than due to host specificity. In addition, the fact that worms were generally collected by different methods from humans and pigs (chemo-expulsion versus collection in abattoirs) may also have influenced protein profiles. Attempts to apply isoenzyme analysis to determine variation between pig and human *Ascaris* from a variety of locations were not very informative due to these same issues and/or the low level of polymorphism at the loci examined.[28–31]

Molecular analysis of *Ascaris* populations in areas where pigs and humans live in close proximity (sympatric) has been carried out in Guatemala, China, and Brazil. Anderson et al. concluded that *Ascaris* in humans and pigs in Guatemala represented two different reproductive populations with little gene flow between them,[31,34] although no single diagnostic marker was found which could distinguish between pig and human worms. Researchers in China came to an analogous conclusion, based on application of similar molecular techniques to human and pig worms collected from six provinces.[47,48,69] However, a recent reanalysis of the Chinese mitochondrial data provided evidence for a high level of gene flow between human and pig-derived *Ascaris*.[70] In addition, based on fecal samples collected from humans and pigs in Brazil, shared mitochondrial haplotypes were found[71] and human worms from East Africa had some haplotypes in common with pig worms from China,[42,43] suggesting recent or contemporary gene flow between worms from the two hosts or retention of ancestral polymorphisms.

Detailed genetic analysis based on microsatellite loci identified 4 and 7% of worms ($n = 129$) from China and Guatemala, respectively, as being hybrids.[52] A similar study of 258 worms from China found evidence of hybrids and also 19 examples of "pig" worms infecting humans and one example of a "human" worm infecting a pig.[46] Together these results suggest that cross-infections can occur between the pig and human worm populations and that fertile offspring are produced (hybrids). In addition, cross-infection and cases of hybrids seem to be a more frequent event in humans than in pigs. A possible explanation for the difference between these data and previous findings is that earlier studies may have lacked the power to detect hybridization and transmission between hosts due to the use of only a few markers.

In countries where *Ascaris* infections are rare among humans, molecular evidence (together with epidemiological findings) suggests that the vast majority of infections result from cross-infections from pigs. This has been demonstrated for 10 worms from nine patients in North America, for 32 worms found in humans in Denmark and for 11 worms from patients in the UK.[32,35,72] Interestingly, in Japan it was found that humans were infected with both "pig" and "human" worms.[41,45] Two out of nine patients infected with "human-like" worms had a history of travel to *Ascaris*-endemic areas but the others did not, suggesting that a human–human transmission cycle may still exist for *Ascaris* in Japan, or that ingestion of contaminated imported foods is taking place.

GLOBAL AND LOCAL GENETIC DIVERSITY

Increased sampling of *Ascaris* worms from a variety of global locations has uncovered greater levels of genetic diversity in this parasite. For example, at least 50 *cox*1 and 35 *nad*1 haplotypes have been identified to date.[42,43,45,47,71] Recent work from Brazil has shown that ITS-1, which is present in around 40 copies in the *Ascaris* nuclear genome,[73] shows intra-individual variability and 21 genotypes/haplotypes have been identified.[44]

As well as partitioning between host species, a number of studies have investigated whether there is evidence for geographical genetic structuring on a macro- and micro-scale. Anderson et al. compared samples from pigs in Guatemala, the Philippines, Switzerland, Scotland, and Peru and humans in Guatemala, Bangladesh, Madagascar, and Peru and found that geographical differences explained 17% of genetic variation observed.[34] Differences in mitochondrial haplotype frequencies were observed in *Ascaris* from China and worms from Guatemala,[69] and *Ascaris* from humans in Bangladesh and Nepal fell into separate genetic clusters to worms from Guatemala and Denmark.[35] Although analysis of a mitochondrial marker provided little evidence for geographical clustering of worm populations in East Africa and China,[42,43,70] based on microsatellite markers, strong differentiation between worm populations from China, Guatemala, and Nepal and between populations from Zanzibar and Uganda was observed.[42,43,52] Thus, overall there is a strong consensus supporting structuring of *Ascaris* at a macro-scale (between countries) but why this is not reflected at the mitochondrial DNA level remains to be explored.

The situation is more complex when considering differentiation of *Ascaris* populations at a smaller scale. There was evidence for genetic structuring of *Ascaris* populations between villages in Guatemala,[33] but no obvious restriction to gene flow between pig worms from different parts of Denmark or between human worms within Zanzibar or Uganda.[42,43,50] In addition, identical *Ascaris* haplotypes were found circulating in north and south-east Brazil.[71] Results from China are somewhat conflicting but the most recent detailed analyses suggest that the genetic diversity of *Ascaris* in humans varies between provinces, although there is no distinct geographical distribution pattern.[47,48,69,70] On the other hand, extremely detailed analysis from Nepal found substantial genetic structuring of *Ascaris* within one village (sampling area of 14 km^2) and evidence for the existence of multiple foci of transmission which were stable over time (see Chapter 8).[53] There was also evidence of genetic substructuring at the level of the individual host in Guatemala and Denmark, but not in China.[33,37,50] The varying results found in different regions may reflect differences in sampling methods, the resolution of the genetic markers used and local farming practices and migration patterns, among other factors.

EVOLUTIONARY HISTORY AND SPREAD OF ASCARIS

Can the archeological record and genetic analysis of contemporary *Ascaris* worms from pigs and humans from a variety of locations provide us with clues as to the evolutionary history and global spread of *Ascaris* in these two host species? A number of hypotheses have been proposed with reference to the origins of *Ascaris* in humans and pigs according to key points within known human history: (1) *A. lumbricoides* (in humans) and *A. suum* (in pigs) originated through speciation from a common ancestor when pigs were domesticated by humans, or (2) *A. lumbricoides* derived from *A. suum* with *A. suum* surviving as a persistent ancestor, or (3) *A. suum* derived from *A. lumbricoides* with the persistent ancestor being *A. lumbricoides*, or (4) *A. lumbricoides* and *A. suum* are in fact the same species, or (5) multiple host colonization events occurred after the ancestral population was geographically subdivided.[19,52] Some of these postulates are more easily testable than others against current evidence.

From the archaeological record, it is clear that human ascariasis has been present for a long time in the Old and New Worlds.[70,74-81] The oldest find in Europe dates from ~30,000–25,000 years before present (BP), in Africa from 2050–1770 BC, and in the Americas from 2277 BC.[75,78] It is disappointing that this archeological record is strongly geographically biased and that Africa is especially underinvestigated. The oldest *Ascaris* eggs discovered so far (~30,000–25,000 BP) were found in coprolites from a cave in Arcy-sur-Cure, France.[78] It has been argued that this finding suggests that humans were parasitized before pigs as the eggs predate the first pig domestication around 9000 years ago.[76,78,82] However, it has not been confirmed whether the coprolites came from humans or bears, as both were co-inhabiting the Arcy-sur-Cure cave. Thus, an alternative explanation is that this finding represents an ancestral Ascarid in bears which could also infect humans and subsequently diverged to produced *Ascaris* and *Baylisascaris*.

One way to address this would be the application of molecular techniques to archeological samples. The extraction of ancient DNA and amplification of *Ascaris*-specific markers has been carried out on a subset of more recent samples, which included isolated eggs and coprolites. It has been possible to amplify and sequence (directly or after subcloning) nuclear and mitochondrial markers including 18S rDNA and cytochrome b.[80,81,83,84] Sequence alignment has revealed exact or near exact matches with sequences obtained from modern *A. lumbricoides* or *A. suum* but in most cases it has not been possible to unambiguously assign the archaeological samples to either species, partly due to the lack of a definitive diagnostic marker to distinguish between the two. Nonetheless linking such genetic information to older archeological samples is very appealing especially if future laboratory methods show improved ability to retrieve parasite DNA.

Anderson et al. suggest that one explanation of their genetic data from human and pig *Ascaris* from a number of geographical locations is that current parasite populations infecting humans and pigs resulted from a single host shift.[34] However, an alternative explanation is that multiple host shifts have occurred with subsequent merging of worm populations in the two hosts. The fact that pig and human *Ascaris* are closely related at a genetic level may reflect a complex evolutionary history and so could be due to multiple host colonization events (hypothesis 5 above). This was suggested by Criscione et al. who recorded that worms from Nepal, China, and Guatemala assorted first by geography and then by host with, however, low bootstrap support.[52] Supplementary analysis using more samples from additional locations is required to provide further support for this hypothesis. Transfer of *Ascaris* worms from humans to pigs (or vice versa) at multiple locations across the globe would be consistent with genetic analysis of wild boar (*Sus scrofa*, the wild ancestor of domestic pigs), which revealed multiple centers of pig domestication across Eurasia.[85] Based on sequence analysis of mitochondrial markers, *Ascaris* does not fall into clusters dependent on host or geographical origin, but rather three or more clades containing pig and human worms from different geographic locations.[40,45,70,71] This likewise suggests a complex evolutionary history with multiple host switches and may explain why a single diagnostic marker to distinguish between pig and human *Ascaris* has not been identified to date.

The difference in genetic diversity in human worms between southern and central/northern provinces of China has been linked to human migration patterns and differences in seasonal transmission patterns between the regions.[70] The greater relative diversity of mitochondrial markers in *Ascaris* sampled from Africa than elsewhere may indicate that this parasite originated in this region and then spread to other parts of the world.[42,43,47,71] This may not be so surprising given that the likely evolution of early hominids is known to have taken place in Africa with several later diaspora events.[86] If indeed these early migrating humans were parasitized it would have facilitated the subsequent spread of a proto-*Ascaris* throughout their dispersal range in successive waves of emigration. It is without doubt that in recent times many Africans with ascariasis were transported to the New World as part of the slave trade.[87]

In light of the findings summarized above, we would argue that *A. lumbricoides/suum* is one species based on the phenetic, phylogenetic, and evolutionary species concepts, as worms are difficult, if not impossible, to distinguish morphologically and to date no diagnostic marker has been discovered which is able to distinguish between them. As discussed, the evolutionary history of *A. lumbricoides* and *A. suum* is complex and it is difficult to come to conclusions as to their evolutionary independence given the historical origins are contentious and that near "outgroup" taxa are lacking. In contrast, however, there is more support for *A. lumbricoides* and *A. suum* to be considered as two separate species based on the biological species concept. Although there is evidence for cross-transmission of *Ascaris* between pigs and humans and formation of hybrids, there are high levels of genetic differentiation between worms from pigs and humans in *Ascaris*-endemic areas based on nuclear markers, suggesting that these populations are reproductively isolated. It is likely that the hybrids have reduced fitness/viability (leading to a postzygotic barrier), otherwise there would be many more shared haplotypes and no population structuring between the two hosts. The presence of hybrids perhaps represents a dynamic phase of population differentiation at the tips of the evolutionary tree.

IMPLICATIONS FOR CONTROL: A FOCUS ON ZANZIBAR

Despite these subtle ambiguities and conjectures on the taxonomy of *Ascaris*, what relevance does this have for disease control? In Africa, for example, there are now many interventions delivering anthelmintic drugs to people across large geographical swathes within the endemic area, in a general attempt to control neglected tropical diseases. Monitoring and surveillance of these programs is often challenging and has been perhaps best recorded on the islands off East Africa, in particular Zanzibar (Pemba and Unguja). Here there is an interesting history of surveillance and control of ascariasis;[88,89] over the last decade there have been several campaigns distributing albendazole and/or mebendazole, often alongside praziquantel (for schistosomiasis) or ivermectin (for lymphatic filariasis).[88] During this time there have been some clear declines in the prevalence of ascariasis in children with the spatial structure of infections by village revealing "hot spots" of transmission. However, rather surprisingly there was a general increase in the prevalence of ascariasis in school children, albeit small, in the 2004–2006 period.[90–93] Cure rates with albendazole and mebendazole (and in combination with ivermectin[94]) have been documented, confirming concerns on the diminishing performance of these drugs.[95]

In an attempt to explain this putative "drug resistance," adult *Ascaris* retrieved from Zanzibar by chemo-expulsion have been subjected to molecular DNA analysis. Inspection of the genetic variation in beta-tubulin genes associated with resistance did not provide evidence to support a genetic basis of diminished drug efficacy with these loci,[96] although a broader screen of the genetic diversity of *Ascaris* at mitochondrial and microsatellite loci[42] pointed towards some interesting alternatives. Foremost was that the *Ascaris* population on Zanzibar was heterogeneous with some spatial structuring across villages as revealed by mitochondrial DNA (Figure 10.2). This might therefore point towards the supposition that the known heterogeneities in parasitological cure by village (JRS, unpublished data) may have an underlying heritable component, the extent and mechanism of which is presently unknown. Present hot spots of infection at Kandwi village, for example, may also suggest different behavioral practices (e.g. geophagy) and local environmental heterogeneities (e.g. soil type) which themselves might encourage the transmission of certain genotypes of *Ascaris*. It is enigmatic why "pig-like" worms on Zanzibar are still found despite the cessation of

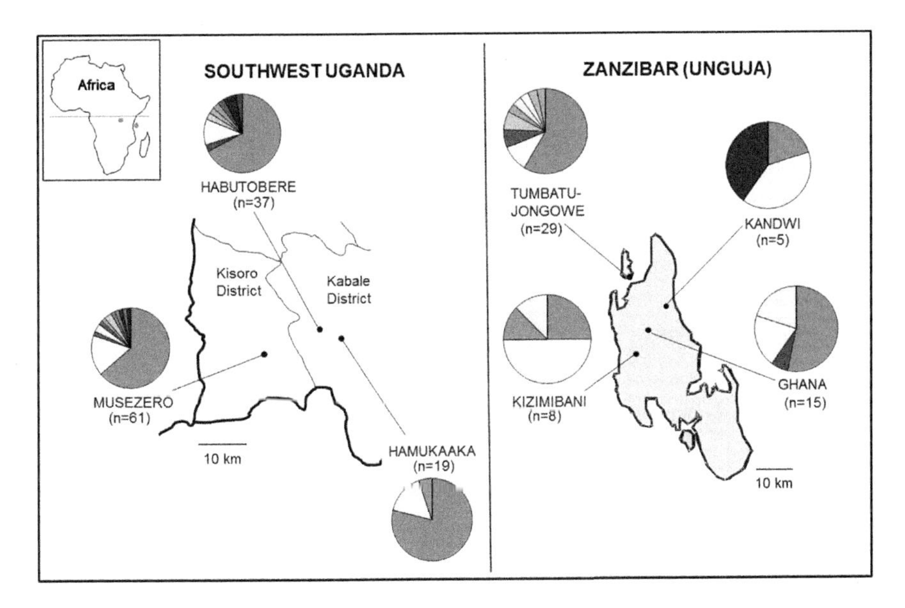

FIGURE 10.2 ***Cox*1 haplotype diversity in *Ascaris* sampled from humans living in three Ugandan and four Zanzibari villages.** Pie charts show relative proportions of different haplotypes at each location. It is apparent that some locations have greater diversity than others, despite a similar number of worms sampled.

farmed pig production several decades ago, so greater attention on these historical environmental heterogeneities would be well placed,[42] which might reveal a much longer-term asynchronicity of worm populations. At the very least, documenting present genetic information relating to *Ascaris* on Zanzibar will become ever more important when set against future molecular investigations and control goals, especially if an archive collection of worms (preserved in ethanol) or genomic DNA is maintained for future investigations as more advanced genetic profiling techniques come online.

CONCLUDING REMARKS AND FUTURE STUDIES

It is heartening to see the growing interest in the molecular scrutiny of *Ascaris* given its global importance as a longstanding disease across the globe. A key question still remains – what is the present genomic diversity of *Ascaris* across its endemic range and are there genomic islands of variation which manifest themselves at the phenotypic level in terms of host specificity, pathogenicity, and drug sensitivity? Since the genome and transcriptome of *A. suum* is now available[97–99] (see Chapter 11) and that of *A. lumbricoides* is on its way, an important step will be to compare the genomes and launch a more detailed transcriptomic exploration of *Ascaris* in the two hosts which might reveal a more complicated set of heritable factors and epigenetic traits that are required to maintain parasite transmission across different definitive hosts. From this molecular perspective, the key challenge is to scale up the relevance of such genetic and genomic studies and quickly apply them to ongoing experimental studies and on-the-ground control interventions.

Acknowledgments

We warmly thank Professor Celia Holland for inviting us to write this chapter and feature our joint work on the biology of *Ascaris*. We also gratefully acknowledge the long list of colleagues we have worked with in Africa and Europe who have generously given their time to the study of neglected tropical diseases.

References

1. Crompton DW. *Ascaris* and ascariasis. *Adv Parasitol* 2001;**48**:285–375.
2. Crompton DWT. Aspects of human helminthiasis in sub-Saharan Africa. *Helminthologia* 2003;**40**(2):109–15.
3. Nansen P, Roepstorff A. Parasitic helminths of the pig: factors influencing transmission and infection levels. *Int J Parasitol* 1999;**29**(6):877–91.
4. Nadler SA, Hudspeth DSS. Phylogeny of the ascaridoidea (Nematoda: Ascaridida) based on three genes and morphology: hypotheses of structural and sequence evolution. *J Parasitol* 2000;**86**(2):380–93.
5. Wasmuth J, Schmid R, Hedley A, Blaxter M. On the extent and origins of genic novelty in the phylum Nematoda. *PLoS Negl Trop Dis* 2008;**2**(7):e258.
6. Li Y, Niu L, Wang Q, Zhang Z, Chen Z, Gu X, et al. Molecular characterization and phylogenetic analysis of ascarid nematodes from twenty-one species of captive wild mammals based on mitochondrial and nuclear sequences. *Parasitology* 2012;**139**(10): 1329–38.
7. Zhu X, Gasser RB, Jacobs DE, Hung GC, Chilton NB. Relationships among some ascaridoid nematodes based on ribosomal DNA sequence data. *Parasitol Res* 2000;**86**(9): 738–44.
8. Muller R. *Worms and Human Disease*. Wallingford: CABI Publishing; 2002.
9. Okulewicz A, Lonc E, Borgsteede FHM. Ascarid nematodes in domestic and wild terrestrial mammals. *Pol J Vet Sci* 2002;**5**(4):277–81.
10. Gavin PJ, Kazacos KR, Shulman ST. *Baylisascarias*. *Clin Microbiol Rev* 2005;**18**(4):703–18.
11. Hung T, Neafie RC, Mackenzie IR. *Baylisascaris procyonis* infection in elderly person, British Columbia, Canada. *Emerg Infect Dis* 2012;**18**(2):341–2.
12. Kelly TG, Madhavan VL, Peters JM, Kazacos KR, Silvera VM. Spinal cord involvement in a child with raccoon roundworm (*Baylisascaris procyonis*) meningoencephalitis. *Pediatr Radiol* 2011;**42**(3):369–73.
13. Mes TH. Technical variability and required sample size of helminth egg isolation procedures. *Vet Parasitol* 2003;**115**(4):311–20.
14. Saffra NA, Perlman JE, Desai RU, Kazacos KR, Coyle CM, Machado FS, et al. *Baylisascaris procyonis* induced diffuse unilateral subacute neuroretinitis in New York City. *J Neuroparasitol* 2010;**1**:N100401.
15. Muller F, Tobler H. Chromatin diminution in the parasitic nematodes *Ascaris suum* and *Parascaris univalens*. *Int J Parasitol* 2000;**30**(4):391–9.

16. Tobler H, Etter A, Muller F. Chromatin diminution in nematode development. *Trends Genet* 1992;**8**(12):427–32.
17. Rollinson D, Stothard JR. Identification of pests and pathogens by Random Amplification of Polymorphic DNA (RAPDs). In: Hawkworth DL, editor. *Identification and Charaterisation of Pest Organisms*. Wallingford: CAB International; 1994. p. 447–59.
18. Bachmann-Waldmann C, Jentsch S, Tobler H, Muller F. Chromatin diminution leads to rapid evolutionary changes in the organization of the germ line genomes of the parasitic nematodes *A. suum* and *P. univalens*. *Mol Biochem Parasitol* 2004;**134**(1): 53–64.
19. Leles D, Gardner SL, Reinhard K, Iniguez A, Araujo A. Are *Ascaris lumbricoides* and *Ascaris suum* a single species? *Parasit Vectors* 2012;**5**:42.
20. Peng W, Criscione CD. Ascariasis in people and pigs: new inferences from DNA analysis of worm populations. *Infect Genet Evol* 2012;**12**(2):227–35.
21. De Queiroz K. Species concepts and species delimitation. *Syst Biol* 2007;**56**(6):879–86.
22. Futuyma DJ. *Evolutionary Biology*. 3rd ed. Sunderland, MA. Sinauer Associates; 1998.
23. Mayr E. *Systematics and the Origin of Species*. New York: Columbia University Press; 1942.
24. Bock WJ. Species concepts, speciation, and macroevolution. In: Iwatsuki K, Raven PH, Bock WJ, editors. *Modern Aspects of Species*. Tokyo: University of Tokyo Press; 1986. p. 31–57.
25. Abebe W, Tsuji N, Kasuga-Aoki H, Miyoshi T, Isobe T, Arakawa T, et al. Lung-stage protein profile and antigenic relationship between *Ascaris lumbricoides* and *Ascaris suum*. *J Parasitol* 2002;**88**(4):826–8.
26. Abebe W, Tsuji N, Kasuga-Aoki H, Miyoshi T, Isobe T, Arakawa T, et al. Species-specific proteins identified in *Ascaris lumbricoides* and *Ascaris suum* using two-dimensional electrophoresis. *Parasitol Res* 2002;**88**(9):868–71.
27. Hawley JH, Peanasky RJ. *Ascaris suum*: are trypsin inhibitors involved in species specificity of Ascarid nematodes? *Exp Parasitol* 1992;**75**(1):112–8.
28. Ibrahim AP, Conway DJ, Hall A, Bundy DA. Enzyme polymorphisms in *Ascaris lumbricoides* in Bangladesh. *Trans R Soc Trop Med Hyg* 1994;**88**(5):600–3.
29. Nadler SA, Lindquist RL, Near TJ. Genetic structure of midwestern *Ascaris suum* populations: a comparison of isoenzyme and RAPD markers. *J Parasitol* 1995;**81**(3): 385–94.
30. Nadler SA. Microevolutionary patterns and molecular markers: the genetics of geographic variation in *Ascaris suum*. *J Nematol* 1996;**28**(3):277–85.
31. Anderson TJC, Romero-Abal ME, Jaenike J. Genetic structure and epidemiology of *Ascaris* populations: patterns of host affiliation in Guatemala. *Parasitology* 1993;**107** (Pt 3):319–34.
32. Anderson TJC. *Ascaris* infections in humans from North America: molecular evidence for cross-infection. *Parasitology* 1995;**110**(Pt 2):215–9.
33. Anderson TJC, Romero-Abal ME, Jaenike J. Mitochondrial DNA and *Ascaris* microepidemiology: the composition of parasite populations from individual hosts, families and villages. *Parasitology* 1995;**110**(Pt 2):221–9.
34. Anderson TJC, Jaenike J. Host specificity, evolutionary relationships and macrogeographic differentiation among *Ascaris* populations from humans and pigs. *Parasitology* 1997;**115**(Pt 3):325–42.
35. Nejsum P, Parker ED, Frydenberg J, Roepstorff A, Boes J, Haque R, et al. Ascariasis is a zoonosis in Denmark. *J Clin Microb* 2005;**43**(3):1142–8.
36. Nejsum P, Grondahl C, Murrell KD. Molecular evidence for the infection of zoo chimpanzees by pig *Ascaris*. *Vet Parasitol* 2006;**139**(1–3):203–10.
37. Peng W, Anderson TJ, Zhou X, Kennedy MW. Genetic variation in sympatric *Ascaris* populations from humans and pigs in China. *Parasitology* 1998;**117**(Pt 4):355–61.

38. Peng W, Yuan K, Hu M, Peng G, Zhou X, Hu N, et al. Experimental infections of pigs and mice with selected genotypes of *Ascaris Parasitology* 2006;**133**:651–7.
39. Leles D, Araujo A, Vicente AC, Iniguez AM. Molecular diagnosis of ascariasis from human feces and description of a new *Ascaris* sp. genotype in Brazil. *Vet Parasitol* 2009;**163**(1–2):167–70.
40. Nejsum P, Bertelsen MF, Betson M, Stothard JR, Murrell KD. Molecular evidence for sustained transmission of zoonotic *Ascaris suum* among zoo chimpanzees (*Pan troglodytes*). *Vet Parasitol* 2010;**171**(3–4):273–6.
41. Arizono N, Yoshimura Y, Tohzaka N, Yamada M, Tegoshi T, Onishi K, et al. Ascariasis in Japan: is pig-derived *Ascaris* infecting humans? *Jpn J Infect Dis* 2010;**63**(6):447–8.
42. Betson M, Halstead FD, Nejsum P, Imison E, Khamis IS, Sousa-Figueiredo JC, et al. A molecular epidemiological investigation of *Ascaris* on Unguja, Zanzibar using isoenyzme analysis, DNA barcoding and microsatellite DNA profiling. *Trans R Soc Trop Med Hyg* 2011;**105**(7):370–9.
43. Betson M, Nejsum P, Llewellyn-Hughes J, Griffin C, Atuhaire A, Arinaitwe M, et al. Genetic diversity of *Ascaris* in southwestern Uganda. *Trans R Soc Trop Med Hyg* 2012;**106**(2):75–83.
44. Leles D, Araujo A, Vicente AC, Iniguez AM. ITS1 intra-individual variability of *Ascaris* isolates from Brazil. *Parasitol Int* 2010;**59**(1):93–6.
45. Snabel V, Taira K, Cavallero S, D'Amelio S, Rudohradska P, Saitoh Y. Genetic structure of *Ascaris* roundworm in Japan and patterns of its geographical variation. *Jpn J Infect Dis* 2012;**65**(2):179–83.
46. Zhou C, Li M, Yuan K, Deng S, Peng W. Pig *Ascaris*: an important source of human ascariasis in China. *Infect Genet Evol* 2012;**12**(6):1172–7.
47. Peng WD, Yuan K, Hu M, Zhou XM, Gasser RB. Mutation scanning-coupled analysis of haplotypic variability in mitochondrial DNA regions reveals low gene flow between human and porcine *Ascaris* in endemic regions of China. *Electrophoresis* 2005;**26**(22): 4317–26.
48. Peng WD, Yuan K, Zhou XM, Hu M, El Osta YGA, Gasser RB. Molecular epidemiological investigation of *Ascaris* genotypes in China based on single-strand conformation polymorphism analysis of ribosomal DNA. *Electrophoresis* 2003;**24**(14): 2308–15.
49. Zhu XQ, Gasser RB. Single-strand conformation polymorphism (SSCP)-based mutation scanning approaches to fingerprint sequence variation in ribosomal DNA of ascaridoid nematodes. *Electrophoresis* 1998;**19**(8–9):1366–73.
50. Nejsum P, Frydenberg J, Roepstorff A, Parker ED. Population structure in *Ascaris suum* (Nematoda) among domestic swine in Denmark as measured by whole genome DNA fingerprinting. *Hereditas* 2005;**2005**(142):7–14.
51. Criscione CD, Anderson JD, Raby K, Sudimack D, Subedi J, Rai DR, et al. Microsatellite markers for the human nematode parasite *Ascaris lumbricoides*: development and assessment of utility. *J Parasitol* 2007;**93**(3):704–8.
52. Criscione CD, Anderson JD, Sudimack D, Peng W, Jha B, Williams-Blangero S, et al. Disentangling hybridization and host colonization in parasitic roundworms of humans and pigs. *Proc Biol Sci* 2007;**274**(1626):2669–77.
53. Criscione CD, Anderson JD, Sudimack D, Subedi J, Upadhayay RP, Jha B, et al. Landscape genetics reveals focal transmission of a human macroparasite. *PLoS Negl Trop Dis* 2010;**4**(4):e665.
54. Liu GH, Wu CY, Song HQ, Wei SJ, Xu MJ, Lin RQ, et al. Comparative analyses of the complete mitochondrial genomes of *Ascaris lumbricoides* and *Ascaris suum* from humans and pigs. *Gene* 2012;**492**(1):110–16.
55. Okimoto R, Macfarlane JL, Clary DO, Wolstenholme DR. The mitochondrial genomes of two nematodes, *Caenorhabditis elegans* and *Ascaris suum*. *Genetics* 1992;**130**(3):471–98.

56. Takata I. Experimental infection of man with *Ascaris* of man and the pig. *Kitasato Arch Exp Med* 1951;**23**(4):49–59.
57. Galvin TJ. Development of human and pig *Ascaris* in the pig and rabbit. *J Parasitol* 1968;**54**(6):1085–91.
58. Nejsum P, Roepstorff A, Anderson TJ, Jorgensen C, Fredholm M, Thamsborg SM. The dynamics of genetically marked *Ascaris suum* infections in pigs. *Parasitology* 2009;**136**(2): 193–201.
59. Crewe W, Smith DH. Human infection with pig *Ascaris (A. suum)*. *Ann Trop Med Parasitol* 1971;**65**(1):85.
60. Jaskoski BJ. An apparent swine *Ascaris* infection of man. *J Am Vet Ass* 1961;**138**(9): 504–5.
61. Lord WD, Bullock WL. Swine *Ascaris* in humans. *N Eng J Med* 1982;**306**(18):1113.
62. Phillipson RF, Race JW. Human infection with porcine *Ascaris*. *BMJ* 1967;**3**(5569):865.
63. McDonald FE, Chevis RA. *Ascaris lumbricoides* in lambs. *N Z Vet J* 1965;**13**(2):41.
64. Roneus O, Christensson D. Mature *Ascaris suum* in naturally infected calves. *Vet Parasitol* 1977;**3**(4):371–5.
65. Ansel M, Thibaut M. Value of the specific distinction between *Ascaris lumbricoides* Linne 1758 and *Ascaris suum* Goeze 1782. *Int J Parasitol* 1973;**3**(3):317–9.
66. Maung M. *Ascaris lumbricoides* Linne, 1758 and *Ascaris suum* Goeze, 1782: morphological differences between specimens obtained from man and pig. *Southeast Asian J Trop Med Public Health* 1973;**4**(1):41–5.
67. Sprent JFA. Anatomical distinction between human and pig strains of *Ascaris*. *Nature* 1952;**170**(4328):627–8.
68. Kennedy MW, Qureshi F, Haswell-Elkins M, Elkins DB. Homology and heterology between the secreted antigens of the parasitic larval stages of *Ascaris lumbricoides* and *Ascaris suum*. *Clin Exp Immunol* 1987;**67**(1):20–30.
69. Peng W, Zhou X, Cui X, Crompton DW, Whitehead RR, Xiong J, et al. Transmission and natural regulation of infection with *Ascaris lumbricoides* in a rural community in China. *J Parasitol* 1998;**84**(2):252–8.
70. Zhou C, Li M, Yuan K, Hu N, Peng W. Phylogeography of *Ascaris lumbricoides* and *A. suum* from China. *Parasitol Res* 2011;**109**(2):329–38.
71. Iniguez AM, Leles D, Jaeger LH, Carvalho-Costa FA, Araujo A. Genetic characterisation and molecular epidemiology of *Ascaris* spp. from humans and pigs in Brazil. *Trans R Soc Trop Med Hyg* 2012;**106**(10):604–12.
72. Bendall RP, Barlow M, Betson M, Stothard JR, Nejsum P. Zoonotic ascariasis, United Kingdom. *Emerg Infect Dis* 2011;**17**(10):1964–6.
73. Pecson BM, Barrios JA, Johnson DR, Nelson KL. A real-time PCR method for quantifying viable *Ascaris* eggs using the first internally transcribed spacer region of ribosomal DNA. *Appl Environ Microbiol* 2006;**72**(12):7864–72.
74. Cockburn A, Barraco RA, Reyman TA, Peck WH. Autopsy of an Egyptian mummy. *Science* 1975;**187**(4182):1155–60.
75. Patrucco R, Tello R, Bonavia D. Parasitological studies of coprolites of pre-Hispanic Peruvian populations. *Curr Anthropol* 1983;**24**(3):393–4.
76. Goncalves ML, Araujo A, Ferreira LF. Human intestinal parasites in the past: new findings and a review. *Mem Inst Oswaldo Cruz* 2003;**98**(Suppl. 1):103–18.
77. Harter S, Le Bailly M, Janot F, Bouchet F. First paleoparasitological study of an embalming rejects jar found in Saqqara, Egypt. *Mem Inst Oswaldo Cruz* 2003;**98** (Suppl. 1):119–21.
78. Loreille O, Bouchet F. Evolution of ascariasis in humans and pigs: a multi-disciplinary approach. *Mem Inst Oswaldo Cruz* 2003;**98**(Suppl. 1):39–46.
79. Seo M, Guk SM, Kim J, Chai JY, Bok GD, Park SS, et al. Paleoparasitological report on the stool from a Medieval child mummy in Yangju, Korea. *J Parasitol* 2007;**93**(3):589–92.

80. Leles D, Araujo A, Ferreira LF, Vicente AC, Iniguez AM. Molecular paleoparasitological diagnosis of *Ascaris* sp. from coprolites: new scenery of ascariasis in pre-Colombian South America times. *Mem Inst Oswaldo Cruz* 2008;**103**(1):106–8.
81. Oh CS, Seo M, Lim NJ, Lee SJ, Lee EJ, Lee SD, et al. Paleoparasitological report on *Ascaris* aDNA from an ancient East Asian sample. *Mem Inst Oswaldo Cruz* 2010;**105**(2): 225–8.
82. Mason IL, editor. *Evolution of Domesticated Animals*. London: Longman; 1984.
83. Botella HG, Vargas JAA, de la Rosa MA, Leles D, Reimers EG, Vicente ACP, et al. Paleoparasitologic, paleogenetic and paleobotanic analysis of XVIII century coprolites from the church La Concepcion in Santa Cruz de Tenerife, Canary Islands, Spain. *Mem Inst Oswaldo Cruz* 2010;**105**(8):1054–6.
84. Loreille O, Roumat E, Verneau O, Bouchet F, Hanni C. Ancient DNA from *Ascaris*: extraction amplification and sequences from eggs collected in coprolites. *Int J Parasitol* 2001;**31**(10):1101–6.
85. Larson G, Dobney K, Albarella U, Fang M, Matisoo-Smith E, Robins J, et al. Worldwide phylogeography of wild boar reveals multiple centers of pig domestication. *Science* 2005;**307**(5715):1618–21.
86. Willoughy PR. *The Evolution of Modern Humans in Africa: A Comprehensive Guide*. Lanham, MD: AltaMira Press; 2007.
87. Hoeppli R. Parasitic diseases in Africa and the western hemisphere: early documentation and transmission by the slave trade. *Acta Tropica* 1969;**10**(Suppl.):1–240.
88. Knopp S, Mohammed KA, Rollinson D, Stothard JR, Khamis IS, Utzinger J, et al. Changing patterns of soil-transmitted helminthiases in Zanzibar in the context of national helminth control programs. *Am J Trop Med Hyg* 2009;**81**(6):1071–8.
89. Stothard JR, French MD, Khamis IS, Basanez M-G, Rollinson D. The epidemiology and control of urinary schistosomiasis and soil-transmitted helminthiasis in schoolchildren on Unguja Island, Zanzibar. *Trans R Soc Trop Med Hyg* 2009;**103**(10):1031–44.
90. Knopp S, Mohammed KA, Khamis IS, Mgeni AF, Stothard JR, Rollinson D, et al. Spatial distribution of soil-transmitted helminths, including *Strongyloides stercoralis*, among children in Zanzibar. *Geospatial Health* 2008;**3**(1):47–56.
91. Knopp S, Mohammed KA, Stothard JR, Khamis IS, Rollinson D, Marti H, et al. Patterns and risk factors of helminthiasis and anemia in a rural and a peri-urban community in Zanzibar, in the context of helminth control programs. *PLoS Negl Trop Dis* 2010;**4**(5): e681.
92. Sousa-Figueiredo JC, Basanez MG, Mgeni AF, Khamis IS, Rollinson D, Stothard JR. A parasitological survey, in rural Zanzibar, of pre-school children and their mothers for urinary schistosomiasis, soil-transmitted helminthiases and malaria, with observations on the prevalence of anaemia. *Ann Trop Med Parasitol* 2008;**102**(8):679–92.
93. Stothard JR, Imison E, French MD, Sousa-Figueiredo JC, Khamis IS, Rollinson D. Soil-transmitted helminthiasis among mothers and their pre-school children on Unguja Island, Zanzibar with emphasis upon ascariasis. *Parasitology* 2008;**135**(12): 1447–55.
94. Knopp S, Mohammed KA, Speich B, Hattendorf J, Khamis IS, Khamis AN, et al. Albendazole and Mebendazole administered alone or in combination with Ivermectin against *Trichuris trichiura*: a randomized controlled trial. *Clin Infect Dis* 2010;**51**(12): 1420–8.
95. Stothard JR, Rollinson D, Imison E, Khamis IS. A spot-check of the efficacies of albendazole or levamisole, against soil-transmitted helminthiases in young Ungujan children, reveals low frequencies of cure. *Ann Trop Med Parasitol* 2009;**103**(4):357–60.
96. Diawara A, Drake LJ, Suswillo RR, Kihara J, Bundy DAP, Scott ME, et al. Assays to detect beta-tubulin codon 200 polymorphism in *Trichuris trichiura* and *Ascaris lumbricoides*. *PLoS Negl Trop Dis* 2009;**3**(3):e397.

97. Huang CQ, Gasser RB, Cantacessi C, Nisbet AJ, Zhong W, Sternberg PW, et al. Genomic-bioinformatic analysis of transcripts enriched in the third-stage larva of the parasitic nematode *Ascaris suum*. *PLoS Negl Trop Dis* 2008;**2**(6):e246.
98. Jex AR, Liu S, Li B, Young ND, Hall RS, Li Y, et al. *Ascaris suum* draft genome *Nature* 2011;**479**:529–33.
99. Ma X, Zhu Y, Li C, Shang Y, Meng F, Chen S, et al. Comparative transcriptome sequencing of germline and somatic tissues of the *Ascaris suum* gonad. *BMC Genomics* 2011;**12**:481.
100. Kalema-Zikusoka G, Rothman JM, Fox MT. Intestinal parasites and bacteria of mountain gorillas (*Gorilla beringei beringei*) in Bwindi Impenetrable National Park, Uganda. *Primates* 2005;**46**(1):59–63.
101. Mbaya AW, Udendeye UJ. Gastrointestinal parasites of captive and free-roaming primates at the Afi Mountain Primate Conservation Area in Calabar, Nigeria and their zoonotic implications. *Pak J Biol Sci* 2011;**14**(13):709–14.
102. Anderson RC. *Nematode parasites of vertebrates. Their development and transmission.* Wallingford: CAB International; 1992.
103. Nejsum P, Betson M, Bendall RP, Thamsborg SM, Stothard JR. Assessing the zoonotic potential of *Ascaris suum* and *Trichuris suis*: looking to the future from an analysis of the past. *J Helminthol* 2012;**86**(2):148–55.
104. Yang GY, Wang CD. Advances on parasites and parasitology of *Ailurus fulgens*. *Chin J Vet Sci* 2000;**18**:206–8.
105. Zhang JS, Daszak P, Huang HL, Yang GY, Kilpatrick AM, Zhang S. Parasite threat to panda conservation. *Ecohealth* 2008;**5**(1):6–9.
106. Bryan HM, Darimont CT, Hill JE, Paquet PC, Thompson RC, Wagner B, et al. Seasonal and biogeographical patterns of gastrointestinal parasites in large carnivores: wolves in a coastal archipelago. *Parasitology* 2012;**139**(6):781–90.
107. Popiolek M, Szczesnaa J, Nowaka S, Myslajeka RW. Helminth infections in faecal samples of wolves *Canis lupus* L. from the western Beskidy Mountains in southern Poland. *J Helminthol* 2007;**81**(4):339–44.
108. Gibbons LM, Jacobs DE, Sani RA. *Toxocara malaysiensis* n. sp. (Nematoda: Ascaridoidea) from the domestic cat (*Felis catus* Linnaeus, 1758). *J Parasitol* 2001;**87**(3):660–5.
109. Roberts JA. The life cycle of *Toxocara vitulorum* in Asian buffalo (*Bubalus bubalis*). *Int J Parasitol* 1990;**20**(7):833–40.
110. Goossens E, Dorny P, Vervaecke H, Roden C, Vercammen F, Vercruysse J. *Toxocara vitulorum* in American bison (*Bison bison*) calves. *Vet Rec* 2007;**160**(16):556–7.

CHAPTER

11

Decoding the *Ascaris suum* Genome using Massively Parallel Sequencing and Advanced Bioinformatic Methods – Unprecedented Prospects for Fundamental and Applied Research

*Aaron R. Jex**, *Shiping Liu*†, *Bo Li*†, *Neil D. Young**, *Ross S. Hall**, *Yingrui Li*†, *Peter Geldhof*‡, *Peter Nejsum*§, *Paul W. Sternberg*||, *Jun Wang*†, *Huanming Yang*†, *Robin B. Gasser**

* The University of Melbourne, Parkville, Victoria, Australia
† BGI-Shenzhen, Shenzhen, PR China
‡ University of Ghent, Belgium
§ University of Copenhagen, Denmark
|| California Institute of Technology, Pasadena, CA, USA

Ascaris: The Neglected Parasite
http://dx.doi.org/10.1016/B978-0-12-396978-1.00011-2

Copyright © 2013 Elsevier Inc. All rights reserved.

OUTLINE

INTRODUCTION

Ascaris is among the commonest and most important parasites of humans globally, infecting ~1.2 billion people and causing significant nutritional deficiency, impaired physical and cognitive development, and, in severe cases, death.[1] Ascariasis has the greatest impact in people of 5–15 years of age, particularly in impoverished populations in developing countries (see Chapter 13). This disease also has a substantial impact on the growth and productivity of one of the world's major food animals, pigs, resulting in reduced growth, failure to thrive and mortalities[1,2] (see Chapter 14). Despite the impact of ascariasis and the increased attention in recent years through the Millennium Development Goals[3] and the Neglected Diseases Initiative,[4] treatment is confined to a handful of drugs, and the prevalence of the disease remains high in many areas of the world.[5]

Ascaris is transmitted directly by the fecal–oral route. The human and porcine hosts become infected by ingesting larvated *Ascaris* eggs from the environment (e.g. in contaminated water or food); third-stage larvae hatch from the eggs migrate through the intestinal wall into the bloodstream, and then through the liver and to the lungs (=hepatopulmonary migration), after which the larvae ascend the trachea, are swallowed and then develop as dieocious adults in the small intestine. The damage caused by the migrating larvae can be significant, in addition to the effects of the large worms (10–20 cm long) in the small intestine, relating to occlusion of the lumen, perforation of the intestinal wall or occlusion of bile/pancreatic ducts and death in hosts with high burdens.[6] Morphologically, *Ascaris* of humans (*Ascaris lumbricoides*) and pigs (*Ascaris suum*) are indistinguishable, and their status as separate species remains controversial,[7] with reported cases of *A. suum* infections in humans[8] (see Chapters 8 and 11). Given the close genetic relationships between *A. lumbricoides* and *A. suum*[9] as well as between the human and porcine hosts, the *Ascaris*-swine model[10] provides an excellent system to explore many aspects of *Ascaris* and ascariasis at the molecular level (see Chapters 14 and 16).

The development of high-throughput nucleic acid sequencing and bioinformatic technologies are having far-reaching implications for all areas of the biological sciences, including the study of geohelminths of major socioeconomic importance. These technologies provide the prospect of exploring these helminths on a scale that, even in the last few years, was unachievable for all but the largest of sequencing centers, providing unprecedented opportunities to investigate key aspects of the biology of these critically important parasites. Such advances also provide a window to revitalizing the urgently needed development of new interventions through the identification and characterization of novel drug and vaccine targets as well as the prospects of defining genetic and biological markers for improved diagnostic applications. To facilitate such efforts, we recently sequenced the 273 Mb draft genome of *A. suum* utilizing massively parallel sequencing and advanced bioinformatics and also undertook extensive comparative analyses.[11] In this chapter, we (1) describe the advanced methodologies used to sequence, assemble, and annotate the *A. suum* genome, (2) review the salient features of this genome and associated transcriptomes, and (3) emphasize the unique prospects that knowledge of this genome provides for investigations into the genetics, evolution, immunobiology, epidemiology, ecology of *Ascaris* from both pig and human hosts, disease processes in *Ascaris* as well as for the design of new interventions against ascariasis.

ADVANCED METHODOLOGIES ESTABLISHED TO SEQUENCE, ASSEMBLE, AND ANNOTATE THE GENOME AND TRANSCRIPTOMES OF *A. SUUM*

Here, we provide a summary of the methodologies used for the sequencing, assembly, and annotation of the *A. suum* genome and transcriptome. For this genome, significant technical hurdles needed to be overcome to enable its sequencing from a limited amount of genomic DNA from the reproductive tract from a single female worm. In general terms, Illumina-based technology[12] was used for the sequencing of the genome and transcriptome of *A. suum*. Bioinformatic data analyses were conducted in a Unix environment or Microsoft Excel 2007 using standard commands; scripts required to facilitate data analysis were designed using Perl, BioPerl, Java, and Python and are available via http://gasserlab.org/.

Sequencing of the Genome from the DNA from the Reproductive Tract from a Single Adult Female of *A. suum*

Although the advent of massively parallel sequencing technologies has vastly reduced the cost of large-scale -omics research, there has been a number of technical and analytical challenges associated with the *de novo* sequencing of large eukaryotic genomes, including, for example, that of *A. suum*. Irrespective of the technology chosen,[12] these platforms achieve massively parallelized sequencing by (1) fragmenting nucleic acids, (2) standardizing the ends of these fragments through the ligation of adaptors of a known sequence, and (3) immobilizing each fragment, either on beads in an oil emulsion matrix (e.g. 454 technology or SOLID) or on a glass slide (e.g. Illumina technology), prior to generating reads of ~100–600 bp (depending on the technology used). The first versions of these platforms achieved sequencing from one end of each fragment, with read quality decreasing significantly with increasing read length. Improvements to the sequencing chemistry have now enhanced read quality and length. In addition, most platforms now achieve sequencing from both ends of each fragment (called paired-end sequencing),[13] thus maximizing data for subsequent assembly. Therefore, it is now possible to sequence the ends of fragments much larger than twice the achievable read length, leaving an unsequenced gap of known size between each pair of reads of each fragment. This gap provides physical/spatial information that assists in the assembly process and, in particular, in traversing long, repetitive regions of eukaryotic genomes. The effectiveness can be improved further through the construction of large-insert, mate-paired libraries,[12] whereby the physical gap or insert between the paired reads can be up to 10 kb or more. Although this mate-paired approach is a major advance, the shearing process used to generate DNA fragments prior to library construction becomes increasingly inefficient as the insert size increases, thus requiring substantially more starting template for an effective build. Because of this inefficiency, the amount of sequence generated from a library tends to decrease with increasing insert size as well. Therefore, a common strategy is to generate sequence data from libraries representing a range of insert sizes using the large amounts of data from small-insert libraries (e.g. 200, 500, and 800 bp) to achieve a high level of genome coverage (i.e. to ensure that all regions of the genome are sequenced to an acceptable depth of usually ≥30 fold). Sequence data from large-insert libraries (2, 5, and 10 kb) are then used to bridge repetitive or unsequenced regions and to enhance the final assembly. We adopted this approach to sequence the nuclear genome of *Ascaris suum* using the Illumina Hi-seq platform (Figure 11.1).

To do this, we purified high molecular weight genomic DNA from the reproductive tract from a single adult female of *A. suum* by sodium-dodecyl sulfate/proteinase K digestion,[14] followed by phenol-chloroform extraction and ethanol precipitation.[15] We specifically selected reproductive tissues rather than muscle because, in our experience, this tissue yields a larger amount of high molecular weight DNA (unpublished finding). Particularly for larger fragment library construction, it is critical that genomic DNA is of high quality and high molecular weight, and available in a sufficient amount. We assessed DNA quality and quantity using a Qubit fluorometer dsDNA HS Kit (Invitrogen), standard agarose gel electrophoresis and a 2100 Bioanalyzer (Agilent, USA). The genomic DNA isolated from the reproductive tissue of a single female was sufficient to allow the construction of two short-insert libraries representing mean fragment sizes of 170 and 500 bp, respectively. However, it was not enough to allow the construction of large-insert libraries (i.e. mate-paired libraries) without further processing. Although genomic DNA from multiple individuals could be pooled, there is potential that genetic polymorphism among individuals could complicate or prevent an effective assembly of sequence data. Therefore, we elected to synthesize a substantial amount of template from an individual by whole genome amplification (WGA) using a multi-strand displacement method (REPLI-g Midi Kit, Qiagen). Using this technique, we generated ~40 μg of high molecular weight DNA from <200 ng of template, allowing the construction of 800 bp, and 2, 5, and 10 kb mate-paired libraries. By

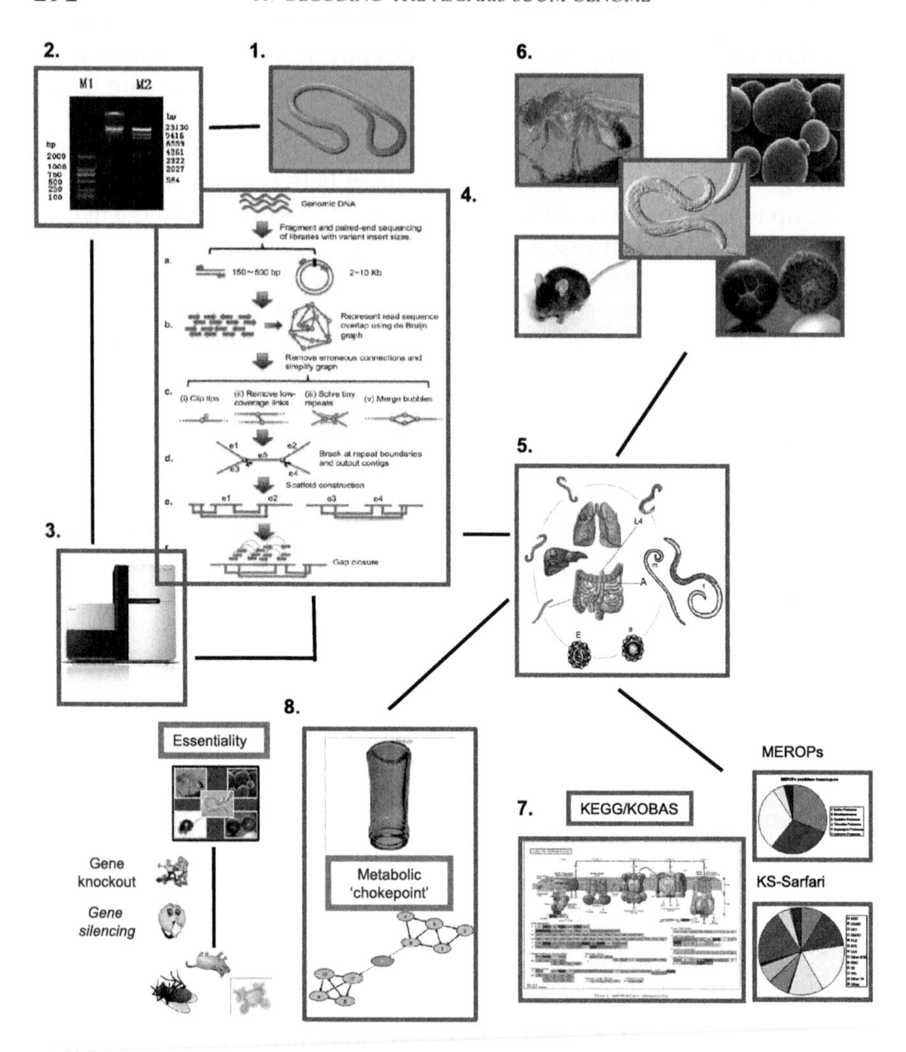

FIGURE 11.1 Flow diagram summarizing the sequencing and annotation of the *Ascaris suum* genome, consisting of (1) procurement of adult female *Ascaris suum*, (2) extraction of genomic DNA from adult female uterine tissue, (3) deep-sequencing on the Illumina Hi-Seq Platform, (4) scaffolded assembly using SOAPdenovo, (5) RNA-sequencing of larval and adult male and female transcripts to support gene predictions, (6) functional annotation based on homology with key model organisms including *Caenorhabditis elegans* and *Drosophila melanogaster*, (7) further functional annotation based on generalist (i.e. KEGG) and specialist [e.g. MEROPs (peptidases) and KS-Sarfari (kinases)] protein databases, and (8) prediction of druggability based on assessment of lethal phenotypes in homologous genes (i.e. essentiality) in model organisms and/or identification of metabolic chokepoints.

sequencing all short and mate-paired libraries, we generated ~39 Gb of usable paired-end sequence data, from which adaptors were trimmed and low-quality sequences and read-duplicates removed.[11]

Because both template heterozygosity and coverage affect an assembly,[16] we estimated the heterozygosity in the *A. suum* dataset by calculating the frequency of occurrence of each 17 bp *k*-mer (i.e. each unique combination of 17 nucleotides occurring in the dataset) within the genomic sequence dataset (from the 170 bp library) and, based on this calculation, estimated[17] the genome size to be ~300 Mb, suggesting a mean coverage of the genome of >80-fold, which we deemed to be more than adequate for assembly.

Genome Assembly

Although short-read platforms provide unprecedented capacity to sequence large genomes in a cost- and time-effective manner, the assembly of short-read data presents significant technical challenges. Therefore, a wide-array of advanced and complex mathematical algorithms has been developed for the assembly process. The principles of these mathematical approaches have been reviewed thoroughly[18] and are, thus, not repeated here. For the assembly of the genome of *A. suum*, we used the program SOAPdenovo[16] to join overlapping, single-end read data from the short-insert library datasets into contigs using a de Bruijn graph approach,[18] and then connected, through an iterative process, contigs into scaffolds using paired-end data from the large-insert mate-paired libraries.[11] Between each iteration of the scaffold assembly phase, we conducted local assemblies in gap regions between the contigs using sequence data from the short-insert libraries. Due to the technical constraints of the de Bruijn graph assembly algorithm, this scaffolding approach requires that a heavy "weighting" be applied in favor of data generated from the mate-paired libraries. However, because the WGA process has the potential to introduce substitution errors, we needed to account for this weighting by remapping all raw reads to the final assembly using the program Maq[19] to produce a final, unbiased consensus sequence. This approach yielded a high-quality assembly of ~273 Mb represented by ~1600 contigs of >2 kb, with an N50 of 408 kb (50% of all nucleotides in the assembly are in contigs of ≥408 kb in length) and an N90 of 80 kb (90% of all nucleotides in the assembly are in contigs of ≥80 kb in length) (Table 11.1).

Gene Prediction and Annotation

Following the assembly, a multi-step process is required to predict the coding regions of the genome and their function based on comparisons

TABLE 11.1 Some salient features of the *Ascaris suum* genome

Estimated genome size in megabases	309
Total number of base pairs (bp) within assembled scaffolds	272,782,664
N50 length in bp; total number >2 kb in length	407,899; 1618
N90 length in bp; total number >N90 length	80,017; 748
GC content (%) of whole genome	37.9
Repetitive sequences (%)	4.4
Proportion (%) of genome that is coding (exonic; including introns)	5.9; 44.2
Number of putative coding genes	18,542
Gene size (mean bp)	6536
Average coding domain length (mean bp)	983
Average exon number per gene (mean)	6
Gene exon length (mean bp)	153
Gene intron length (mean bp)	1081
GC content (%) in coding regions	45
Number of transfer RNAs	255

with experimental data available for homologous genes in model organisms, such as *Caenorhabditis elegans* (www.wormbase.org) and *Drosophila melanogaster* (www.flybase.org). In a prokaryotic organism identifying coding genes, which lack introns, is often based on the prediction of long open reading frames (ORFs), followed by a homology-based inference of function. This process can be enhanced based on codon usage and/or AT-richness using pattern predictive modeling (e.g. Hidden Markov Modeling, HMM).[20] However, generally, a simple ORF prediction will, at least, provide an acceptable draft annotation. For eukaryotic genomes, in which intronic regions can be tens of kilobases or longer in length, such an approach is not possible. Predictive modeling based on codon usage, AT-richness, and intron-exon boundary "splice" signal sequences is required using various algorithms.[21] Most programs use a training set of well-defined transcripts to initiate and then drive the modeling process. Therefore, for *A. suum*, we isolated total RNAs from third-stage larvae (L3s) from eggs ($n = 500{,}000$), L3s migrating through liver parenchyma ($n = 60{,}000$) or lungs ($n = 80{,}000$) or fourth-stage larvae (L4s) from the small intestine ($n = 30{,}000$) as well as from the somatic musculature or reproductive tracts from two individual adult males and two adult females. Messenger RNAs were purified using polyadenylated

Sera-mag oligo(dT) beads, fragmented (300–500 nt) and reverse-transcribed using random hexamers. Short-insert libraries were constructed for paired-end RNA sequencing using the Illumina Hi-seq platform. Using this approach, we generated 15–25 million paired-end reads from each of the libraries representing each tissue, stage, and sex. Again, low-quality reads and potential contaminants were removed, and adaptors trimmed. Following this process, all reads were pooled into one large dataset and assembled using the program Oases.[22]

Because there is currently no consensus approach for the prediction of genes in eukaryotes, we used an integrated strategy, involving *de novo*-, homology-, and evidence-based methods. *De novo* gene prediction was performed on a repeat-masked genome using three programs (Augustus, GlimmerHMM and SNAP)[16]; training models were generated from a subset of the transcriptomic dataset representing 1355 distinct genes. The homology-based prediction was conducted by comparison with complete genomic data for *C. elegans*,[23] *Pristionchus pacificus*,[24] and *Brugia malayi*.[25] Evidence-based gene prediction was conducted by aligning all RNA-Seq data generated during the study[11] against the assembled genome using TopHat,[26] with cDNAs predicted from the resultant data using Cufflinks.[27] Following the prediction of genes, a non-redundant gene set representing homology-based, *de novo*-predicted and RNA-Seq-supported genes, was generated using Glean (http://sourceforge.net/projects/glean-gene). All Glean-predicted genes were retained, as were all genes supported by RNA-Seq data and those predicted using two or more *de novo* methods (i.e. Augustus, GlimmerHMM and/or SNAP). The protein coding sequences encoded by these predicted genes were then inferred using BestORF (www.softberry.com).

Once a consensus gene set had been predicted from the *A. suum* genome, the next major step was to annotate the sequences and predict their function(s). Although extensive functional data have been amassed for model organisms, such as *C. elegans*, through gene knockout, knock-down, and protein localization studies (see www.wormbase.org), such data are not available for most parasites because of the complexities and challenges in conducting experimental studies of most metazoan parasites *in vitro*. Although gene silencing appears to be possible in *A. suum*,[28,29] this research is in its infancy for this parasite. Therefore, we inferred the function of the *A. suum* predicted gene set using homology-based comparisons with a wide range of datasets. These comparisons included assessing the predicted peptides for conserved protein domains classified in several databases (e.g. SProt, Pfam, and ProDom) using the program InterProScan[30] and, on the basis of these data, classifying each transcript according to functional hierarchies using the gene ontology (GO) database,[31] providing additional information on the location of activity within the cell ("cellular component"), basic molecular role/s ("molecular function"), and specific biological role/s ("biological process") of each transcript for which sufficient data were available. More specific inferences of biological function were achieved using homology-based comparisons among functional orthologues and/or conserved biological pathways (e.g. oxidative phosphorylation) defined in the Kyoto Encyclopedia of Genes and Genomes (KEGG) (www.kegg.com). However, although comparison using the KEGG database allows the identification of the functions of thousands of orthologous sequences, this approach is only applicable to conserved genes involved in the same biological function in a broad range of organisms (e.g. metabolic, DNA replication or cell signaling pathways). Using this approach, function could be predicted for ~12.5% (2500) of the genes in *A. suum*.

To increase the number of functionally annotated genes in the *A. suum* genome, the amino acid sequence inferred from each coding domain was compared by BLASTp with protein sequences available in a wide range of databases. Numerous databases can be utilized for this purpose. For *A. suum*, substantial functional information was inferred based on homology with genes encoded by *C. elegans* and linked to extensive functional/phenotypic data archived in WormBase (www.wormbase.org) representing decades of experimental investigations of *C. elegans*. Given the relatively close biological relationship between *A. suum* and *C. elegans*, this database clearly represented the richest source of data to infer the function of each predicted gene. In addition to this resource, functional information could also be collected based on similar comparisons to other model organisms, such as *D. melonagaster*[32] and *Mus musculus*,[33] and/or homology to sequences contained in curated general or specific protein databases, including UniProt,[34] SwissProt or TrEMBL[35] as well as MEROPS[36] (peptidases), KS-Sarfari (kinases) and GPCR-Sarfari (G-protein-coupled receptors) (http://www.sarfari.org), and the Transporter Classification DataBase (TCDB).[37]

In addition to these classifications, because such proteins are known to play key roles in the host–parasite interaction, including in evasion or modulation of the host immune responses and degradation of host tissues, we predicted excretory/secretory (ES) proteins for *A. suum*. First, using the program Phobius,[38] signal peptides, inferred to allow trafficking across the cell membrane, were predicted by neural network and Hidden Markov Modeling. These molecules were then compared by BLASTp analysis to curated proteins in the signal peptide database (SPD)[39] and in an ES database containing published proteomic data for *B. malayi*,[40,41] *Meloidogyne incognita*,[42] and *Schistosoma mansoni*.[43] A peptide was classified as being a representative of the *A. suum* secretome if it encoded a signal peptide and was supported by having an homologue in the SPD database and/or the ES proteomes predicted for other parasitic helminths.

SALIENT CHARACTERISTICS OF THE GENOME AND TRANSCRIPTOMES OF *A. SUUM*: TOWARD UNDERSTANDING THE MOLECULAR LANDSCAPE OF THE PARASITE AND THE DESIGN OF NEW INTERVENTIONS

Characterizing the Genome

The *A. suum* genome was sequenced at 82-fold coverage, producing a final draft assembly of 272,782,664 bp (N50 = 407 kb; N90 = 80 kb; 1618 contigs of >2 kb) (Table 11.1) with a mean GC-content of 37.9%. Notably, repetitive sequence (which was identified using the program Tandem Repeats Finder [TRF][44]) in the *A. suum* assembly is remarkably low (~4.4% of the total assembly) relative to that reported for other metazoan genomes sequenced to date,[23,32,45] including those sequenced employing the same approach as used for *A. suum*.[16] There were various possible explanations for this low repeat content. First, it was possible that the assembly of the repeat content was poor and, thus, the genomic assembly was not reflective of the true repetitive content of the genome. To assess this possibility, we mapped all of our raw genomic sequence data (i.e. reads) to the final assembly (using the program BWA[46]) and assessed the depth of coverage achieved for repetitive regions of the genome relative to non-repetitive flanking regions (500 bp regions on either side of each repeat). If significantly more repetitive content was present in the raw data relative to the assembly, we would anticipate the repetitive regions to be covered much more deeply than their flanking regions, as has been demonstrated for genes of variable copy number in other studies.[47] For the current *A. suum* assembly, no such difference in coverage was detected; indeed, the mean coverage of the repetitive regions in the genome (~68-fold) was slightly lower than the non-repetitive flanking regions (76-fold coverage). A second possibility was that the repetitive sequence data were present in the assembly but not identified using Tandem Repeat Finder. To assess this possibility, we explored the repeat content of the genome using the programs RepeatMasker,[48] LTR_FINDER,[49] PILER,[50] and RepeatScout.[51] Although no additional repeat content was detected, this approach did allow us to identify transposable elements encoded within the repetitive sequence data found. Indeed, ~75% (i.e. 3.2% of the total assembly) of these repetitive sequences represented at least 22 families (8 LTR, 12 LINE, and 2 SINE) of retro-transposons and eight families of DNA transposons (91 distinct sequences in total). This richness of transposable element families is comparable with that predicted for other genomes of parasitic helminths,[25,52,53] suggesting a third explanation for the low repeat content in the *A. suum* genome.

Clearly, one of the major, interesting aspects of the genomic structure of some ascaridoids, including *A. suum*, is chromatin diminution,[54] wherein a proportion of the genomic content is lost from somatic tissue relative to the germ-line. The precise mechanism governing diminution is not understood, nor has the genomic content lost during this process been fully characterized. However, evidence suggests a significant bias towards the loss of repetitive sequences.[54] When we explored the repeat content of the *A. suum* genome, we found no evidence of Tas2 transposons,[11] predicted to be within the content lost during chromatin dimunition in *A. suum*,[55] which appears to support the hypothesis that our genome assembly reflects the somatic genome of this species, thus explaining its low repeat content. Intriguingly, the present gene set inferred for *A. suum* includes *fert-1* and *rpS19G*, which, although originally proposed to be germ-line specific (i.e. lost from the somatic genome during diminution),[54] were found to be highly transcribed both in muscle and reproductive tissues in adult worms sequenced during our study.[11] It is worthy to note that previous information on diminution for this species is based on comparisons between germ-line and intestinal tissues. Our findings suggest that, if the low repeat content in our assembly is the consequence of somatic chromatin diminution, the genomic content lost during diminution might vary among individuals and/or tissues. We emphasize, as we proposed previously,[11] that investigations of chromatin diminution between and among individual cells (i.e. sperm or eggs), stages and tissues of *A. suum* would be timely, given the availability of the current assembly, and achievable, considering that high-quality genomic DNA can be generated using a WGA-based sequencing approach.

The Gene Set

Using transcriptomic data from adult and larval stages of *Ascaris*, *de novo* predictions and homology-based searching (see "Gene prediction and annotation," above), we identified 18,542 genes, with a mean length of 6.5 kb (see Table 11.2). In comparison with complete genome sequence data for other nematodes, such as *B. malayi*,[25] *C. elegans*,[23] *Meloidogyne hapla*[56] or *Pristionchus pacificus*,[24] the *A. suum* genes have much longer introns than predicted previously for other nematodes, which relates to the larger genome size of this parasite. Further analyses suggested that our integrated approach to gene prediction was indeed robust. Almost 80% of the genes predicted in the current gene set of *A. suum* are supported by transcriptomic data. In addition, 80% have a homologue in *B. malayi* ($n = 12{,}853$), *C. elegans* ($n = 12{,}779$), *M. hapla* ($n = 10{,}482$), and/or *P. pacificus* ($n = 11{,}865$), with nearly 9000 being common among all species

TABLE 11.2 Summary of the major protein classes represented by the *Ascaris suum* gene set

Family	BLASTp (10^{-5})
Peptidases	456
Metallo-protease	184
Serine protease	132
Cysteine protease	90
Threonine protease	22
Aspartic protease	18
Kinases	609
TK	94
CK1	83
CMGC	67
CAMK	54
AGC	34
STE	34
RGC	15
TKL	20
Other	72
Atypical	9
Phosphatases	257
Protein-tyrosine	68
Receptor protein tyrosine	17
Serine-threonine	64
Dual specificity	39
GTPases	169
Receptors	649
GPCRs	279
Transporters	1797
Channels/pores	477
Porters	462
P-P-bond-hydrolysis-driven transporters	382
Auxiliary transport proteins	104

examined. Of the entire *A. suum* gene set, 2370 genes (representing 279 known biological pathways) had an orthologue in the KEGG database, with close agreement (95%) to the representation of such orthologues in *C. elegans*, suggesting that the gene set is complete. Using homology data for model organisms, including *C. elegans*, *Drosophila melanogaster*, *Saccharomyces cerivisiae*, and *Mus musculus*, and information available in all accessible protein and/or conserved protein domain databases, we were able to assign possible functions to >70% of the genes predicted for *A. suum*.

Molecular Groups Involved in Key Biological Functions

Transporters and Channels

Receptors and transporters play central functional roles in the cell. Examples of key groups among these proteins are G-Protein Coupled Receptors (GPCRs), which interact with GTPases and are involved in a large range of signal transduction pathways. A range of transport proteins (i.e. channels, pores, and porters) are also important, which enable the active or passive transport of molecules and/or ions across phospholipid membranes. Based on homology to curated GPCR databases (e.g. GPCR-SARFARI and IUPHAR, accessible via www.ebi.ac.uk and www.iuphar-dg.org, respectively), we predicted ~280 GPCRs for *A. suum*, including 166 class A rhodopsin-like, 97 class B secretin-like, 9 class C metabotropic glutamate/pheromone family, and 6 frizzled receptors. In addition, of >1800 transporters predicted to be encoded in the *A. suum* genome,[11] we identified 477 channel or pore proteins, including 272 voltage-gated (VICs) and 98 ligand-gated (LICs) ion channels. VICs are activated by the electro-potential gradient across cellular membranes, play key roles in the cell, including the transmission of nerve impulses and the excitation of muscle fibers, and thus represent potent targets for new nematocides. LICs are activated through the specific binding of a ligand (e.g. acetylcholine) and are known targets for nematocidal drugs, such as macrocyclic lactones (e.g. ivermectin),[57] levamisole,[58] and aminoacetonitrile derivatives (AAD).[59] Given the history of LICs as drug targets, these proteins remain a major resource for novel anti-parasite compounds. In addition, we detected 462 transporters (e.g. small molecule porter proteins), of which the major facilitators ($n = 155$), cation symporters ($n = 71$), and resistance-nodulation-cell division (RND; $n = 56$) superfamilies were most abundant. These proteins are usually associated with the transport, via diffusion or pumping, of small molecules and/or ions (e.g. potassium, hydrogen, glucose or sodium ions) across cellular membranes. Porters are essential for maintaining osmotic balance between the cell and the external milieu, generating electro-chemical potentials across membranes, and the trafficking of important small molecules and/or ions linked to a variety of metabolic pathways.

Key Classes of Enzymes

Enzymes with essential functions include peptidases, kinases, phosphatases, and GTPases. All of them, in parasites, represent attractive targets for novel drugs or vaccines. The *A. suum* peptidome is represented by 456 sequences in the five major classes of peptidases (aspartic, cysteine, metallo-, serine, and threonine), with the metallo- ($n = 184$: 41.0%) and serine proteases ($n = 132$: 30.0%) predominating. Major, abundant families within these classes included secreted peptidases representing the M12A "astacins," S09X cholinesterases and acetylcholinesterases, S33 prolyl aminopeptidases, and the C1A "papain" (e.g. cathespins) and C2A "calpain-like" cysteine proteases. Such secreted peptidases are of major interest, given their presence in the excretory/secretory (ES) products of many parasitic helminths and central roles in tissue invasion and degradation (e.g. during migration and/or feeding) and/or immune evasion/modulation.[60,61] ES peptidases (including aminopeptidases and/or cysteine proteases) are likely to play key roles in these processes in *Ascaris* and represent important drug and/or vaccine targets.[62–66]

The kinases and phosphatases, which function in cell-cycle progression, metabolism, transcription, and DNA replication, are key enzymes often associated with signal transduction pathways. Such molecules are known to be highly transcribed in *C. elegans* sperm[67] and are male enriched in some parasitic nematodes,[68,69] suggesting key functional roles in major development and/or reproduction pathways. For example, receptor kinases, such as *daf-1* and *daf-4*, are critical in dauer formation[70]; homologues exist in parasitic nematodes which are thought to play a major role in the establishment of larvae in the host animal.[68,69] Based on homology with *C. elegans* genes and/or sequences in the KS-Sarfari database, we defined the kinome and phosphatome of *A. suum* as ~600 kinases and 250 phosphatases, respectively. Among the *A. suum* kinome, most abundant are the tyrosine, casein, CMGC, and CAMK kinases. The phosphatome, although smaller, was represented by at least 17 receptor and 68 conventional tyrosine, 64 serine/threonine, and 39 dual-specificity phosphatases. Given their involvement in numerous essential signaling (cell-cycle) pathways linked to developmental and/or reproductive processes, these molecules are attractive as novel targets for anti-parasite drugs.[71–74] Recent evidence suggests that norcantharidin analogues, which appear to have specific inhibitory activity against PP1 and PP2A phosphatases, are promising candidate nematocides.[75]

Considering their relative conservation and crucial functions in membrane and vesicle transport, signal transduction, cell division, and differentiation as well as the regulation of transcription and translation, GTPases are another key group of enzymes suitable as targets for anti-parasitic drugs.[74] Based on homology with molecules in *C. elegans*, 169 GTPases are encoded in the *A. suum* genome, including 135 small GTPases (Ras superfamily) representing the Rab, Ras, Rho, and Ran subfamilies. Examples of these homologues include *eft-1*, *fzo-1*, *glo-1*, and *rho-1*, which play essential roles in embryonic, larval, and/or reproductive development (see www.wormbase.org). Given the important roles that GTPases play in a range of species,[76] such molecules are likely to be essential in *A. suum* and other parasitic nematodes and, thus, also represent an attractive class of enzymes worthy of in-depth functional exploration toward drug discovery.

Key Molecules Involved in Parasite–Host Interactions – ES Peptides

ES peptides are central to understanding the immunological relationship between a parasite and its host. We predicted the secretome of *A. suum* to comprise at least 775 proteins with diverse functions. Notable among them are ~70 secreted proteases (including families S9, S3, M12, and C1) which have known roles in host-tissue degradation, required for feeding, tissue penetration, and/or larval migration for a range of helminths,[60] including *Ascaris*[1] and in inducing and modulating host immune defences.[61,77] Indeed nearly half of the proteins representing the predicted *A. suum* secretome are homologues of peptides with immunomodulatory or immunogenic activity in other parasites[61] (see Table 11.3). Most abundant among them are O-linked glycosylated proteins ($n = 300$), which are targeted by various pattern recognition receptors associated with host dendritic cells and by IgM antibodies, and linked to a Th2 immune response.[61] Other members of the *A. suum* secretome are predicted to direct or evade immune responses, including an ES-62 leucyl aminopeptidase homologue, which has been shown to inhibit B-cell, T-cell, and mast cell proliferation/response, promote an alternative activation of the host macrophages, through the inhibition of the Toll-like receptor signaling pathway, and induce a Th2 response through the inhibition of IL-12p70 production by dendritic cells.[61] Additional, immunomodulatory molecules predicted from the *A. suum* secretome include homologues of the *B. malayi* cystatin CPI-2 (i.e. another B-cell inhibitor), several TGF-β and macrophage initiation factor mimics, neutrophil inhibitors, various oxidoreductases, and close homologues of platelet anti-inflammatory factor α.[61] The *A. suum* ES peptides which appear to have a role in immune evasion include homologues of various galectins, thought to mimic host proteins

TABLE 11.3 Summary of the *Ascaris suum* excretory/secretory proteins, predicted to play a key role in immuno-modulation, -regulation and/or -evasion in the host animal

Predicted role	Number of genes (%)
Immunogen	301 (78.4)
O-linked glycan	300
Omega-1 ribonuclease	1
Immunogen/inhibitor	17 (4.4)
SCP/TAPs (VAL)[a]	17
Inhibitor/immunomodulator	26 (6.8)
Aminopeptidase	2
Cystatin	2
Serpin	11
SmSP1[b]	1
Galectin	8
Calreticulin	2
Mimicry	18 (4.7)
C-type lectin	7
Lectin	6
MIF[c]	2
TGF-β like	2
Oxidoreductase	14 (3.6)
Glutathione peroxidase	5
Peroxidase	1
Superoxide dismutase	3
Thioredoxin peroxidase	5
Anti-inflammatory	5 (1.3)
PAFA[d]	5
Others	3 (0.8)
ALT1,2[e]	3

[a] *SCP/TAPS (VAL), Sperm Coating Protein/Tpx-1/Ag5/PR-1/Sc7 (venom allergen-like) proteins.*
[b] *SmSP1, serine protease inhibitor (serpin) 1 encode in* Schistosoma mansoni *(human blood fluke).*
[c] *MIF, macrophage initiation factor 4 mimic.*
[d] *PAFA, platelet anti-inflammatory α.*
[e] *ALT1, 2, abundant larval transcripts 1 and 2, encoded in* Brugia malayi.

(e.g. vertebrate macrophage mannose or CD23 [low affinity IgE] receptors).[61] Taken together, these data suggest that *A. suum* has complex strategies for manipulating, blocking, and/or evading immune responses in the host. Understanding these strategies, particularly in the early phases of the infection process, is likely to be central to developing vaccination approaches.

Elucidating the Biology of A. *Suum* and Parasite–Host Interactions from Transcriptomic Datasets

In addition to utilizing the transcriptomic data to assist in annotating the *A. suum* genome, the RNA-Seq data generated[11] allowed transcription profiles in different developmental stages and tissues to be explored. To do this, we aligned all paired-end reads for each transcript library to the predicted *A. suum* gene set using the program TopHat,[26] and calculated quantitative levels of transcription (i.e. reads per kilobase per million reads [RPKM][47]) using the program Cufflinks.[78]

Reproduction and Development in the Adult Stage

Based on read alignments, we determined organ-specific transcription profiles for *A. suum*. Because of the large size of the adult worm (10–20 cm), we were able to explore transcription in the musculature and the reproductive tracts of adult male and female *A. suum* individuals. Using a predictive networking approach,[79] we defined large clusters of genes whose transcripts are significantly enriched in reproductive tissues of the male or female. The male-enriched reproductive tissue cluster has three times more genes ($n > 1771$) than its female counterpart ($n = 596$). We inferred functions based on Gene Ontology (GO) terms, with both male and female gene sets being highly enriched for terms, such as embryonic, larval, and genital development as well as reproduction and growth. Among the male-enriched transcripts is a range of genes associated specifically with sperm and/or spermatogenesis, including *fer-1*, *spe-4*, *-6*, *-9*, *-10*, *-15*, and *-41*, *alg-4*, and *msp-57* (see www.wormbase.org). Notable among the female-enriched genes is a large variety of genes linked to oogenesis/egg laying (e.g. *cat-1*, *unc-54*, *cbd-1*, and *pqn-74*), vulva development (e.g. *noah-1*, *nhr-25*, *cog-1*, and *pax-3*) and/or embryogenesis (e.g. *cam-1* and *unc-6*; see www.wormbase.org). Although the functions of these genes have been explored in *C. elegans* (primarily a hermaphroditic nematode), we discovered organ-specific genes associated with reproduction and embryogenesis for a dioecious, parasitic nematode.

Hepatopulmonary Migration of Larval Stages

Ascaris larvae undertake an extensive migration through their host's body before they establish as adults in the small intestine. Following the ingestion of infective eggs and their gastric passage, third-stage larvae (L3s)[80] emerge from the hatched eggs in the intestine and penetrate the intestinal wall; they then undergo, via the bloodstream, an arduous hepatopulmonary migration. The complexity of this migration coincides with important developmental changes in the nematode.[1] Clearly, this migration requires tightly and rapidly regulated transcriptional changes in the parasite. We explored this aspect by characterizing the transcription profiles of infective L3s (from eggs), L3s from the liver or lungs of the host, and L4s from the small intestine. Peptidases ($n = 87$) are significantly enriched in either or both of the migrating L3s (from liver or lungs). Conspicuous among them are transcripts encoding secreted peptidases of various families (e.g. C1/C2, M1, M12, S9, and S33), which have known roles in tissue penetration and degradation during feeding and/or migration in various helminths,[60] implying a significant role in *Ascaris* larvae. Considering the complex nature of migration to and through specific tissues and organs, with transitions being critical for parasite development, a key role for molecules associated with chemosensory pathways is highly likely. Such molecules have been studied extensively in *C. elegans*,[81] with numerous homologues being identified in larval transcripts in our study.[11] With few exceptions, all of these homologues relate to olfactory chemosensation linked to volatile compounds (e.g. alcohols, aldehydes or ketones), suggesting that the olfactory detection of molecular gradients is central to the navigation of *A. suum* larvae during migration. Lastly, considering the substantial host attack against migrating *Ascaris* larvae, ES proteins likely play crucial roles in immune modulation and/or evasion during hepatopulmonary migration. This hypothesis is supported by the abundant transcription linked to orthologues of *Bm-alt-1*, *Bm-cpi-2*, and *mif-4* in the larval stages of *A. suum*, particularly migrating L3s. Collectively, these data shed significant light on the molecular biology of a crucial component of the *Ascaris* life-cycle and will provide an important foundation for future research into the immunobiology and developmental processes of *A. suum*.

Prediction and Prioritization of Drug Targets

A major challenge in the treatment and control of the world's major neglected helminths is the limited arsenal of drugs available for their control. Presently, the control of ascariasis in humans is almost exclusively dependent upon oral dosage with albendanzole/mebendazole.[82] The main advantage of using these compounds, in addition to their current efficacy, is their low cost (~US$0.05 per dose[83]). Although there are major efforts to expand the use of these compounds for the treatment of *Ascaris* and other major human soil-transmitted helminths,[84] e.g. widespread usage of these anthelmitics has the potential to lead to the emergence of drug resistance in *Ascaris*, as has been observed for some strongylid nematodes of livestock.[83] Given the reliance on a small number of drugs (i.e. piperazine and pyrantel, albendazole and mebendazole), and that few new anthelmintics (i.e. aminoacetylnitriles[59] and cyclooctodepsipeptides[85]) have been discovered in the past two decades using traditional screening methods, alternative means of drug discovery are needed.[86,87] Genomic-guided drug target or drug discovery has major potential to provide such an alternative. The goal of genome-guided analysis is to identify genes or molecules whose inactivation by one or more drugs will selectively kill parasites but not harm their hosts. However, most parasitic nematodes are difficult to produce or maintain outside of their host(s), or to subject to gene-specific inactivation by RNAi or morpholinos.[88] Therefore, it is not practical to assess gene function or essentiality (i.e. linked to lethality upon perturbation) in these nematodes directly. However, essentiality can be inferred by functional genomic information from model organisms (e.g. lethality in *C. elegans* and *D. melanogaster*),[89] and this approach has indeed yielded effective targets for nematocides.[75] In *Ascaris*, we predicted >600 proteins with close homology to essential peptides in *C. elegans* and *D. melanogaster*. These molecules represented a diverse range of functional classes, including various transporters or channels (e.g. 44 voltage-gated ion channels), GTPases, phosphatases (including PP1 and PP2A homologues),[75] kinases, and peptidases. We expanded these predictions by employing an alternative strategy[51] of identifying molecules inferred to be linked to "chokepoints" (i.e. enzymatic reactions that uniquely produce or consume a substrate) in key metabolic pathways. It is hypothesized that the disruption of such enzymes would lead to the toxic accumulation (i.e. for unique substrates) or starvation (i.e. for unique products) of metabolites within cells. Using this combined approach, we predicted >200 metabolic chokepoints associated with genes known to have lethal knockout/knockdown phenotypes in *C. elegans* and *D. melanogaster*. Five key molecules were encoded by single-copy genes, which might restrict the emergence of drug resistance. One of these five candidate molecules, IMP dehydrogenase (GMP reductase), has been reported to be a target for anti-cancer and anti-viral therapies.[90,91] Therefore, analogues of known inhibitors of IMP dehydrogenase might be tested for selective and specific anti-nematode activity. Clearly, the "druggable" genome of *Ascaris* provides a solid basis for rational drug design, aimed at controlling parasitic nematodes of major socioeconomic impact worldwide.

CONCLUSIONS, IMPLICATIONS, AND MAJOR PROSPECTS FOR FUNDAMENTAL AND APPLIED RESEARCH

Using massively parallel (Illumina) sequencing technology, we sequenced the genome of *A. suum* from the reproductive tract of a single adult female worm.[11] From six paired-end sequencing libraries (insert sizes: 0.17 kb to 10 kb), we generated 39 Gb of usable short-read sequence data, equating to $\geq$80-fold coverage (following filtering) of the 273 Mb genome. We assembled the short-reads, constructed scaffolds, in a step-by-step manner, and then closed intra-scaffold gaps.[16] Transposable elements, non-coding RNAs, and the protein coding gene set were inferred using a combined, predictive modeling and homology-based approach. To enable gene predictions and explore key molecules associated with larval migration, reproduction, and development, we sequenced messenger RNA from infective L3s (from eggs), migrating L3s from the liver or lungs of the host, and L4s from the small intestine, as well as muscle and reproductive tissues from adult male and female worms. All proteins predicted from the gene set were annotated using databases for conserved protein domains, GO annotations, and model organisms (i.e. *C. elegans*, *D. melanogaster*, and *M. musculus*). Essentiality and drug target predictions were conducted using established methods.

The *A. suum* genome determined contains at least 18,500 protein-coding genes, whose predicted products include ~1800 channels and transporters, ~600 kinases, >600 receptors, and ~450 peptidases. Notably, the *A. suum* secretome (~750 molecules) is rich in peptidases, which are likely linked to the penetration and degradation of host tissues, and an assemblage of molecules likely to misdirect or weaken host immune responses. This genome provides a comprehensive resource to the scientific community for a range of future genomic, genetic, evolutionary, biological, ecological, and epidemiological investigations and to underpin the development of new interventions (drugs, vaccines and diagnostic tests) against ascariasis and other helminthiases.

Although various studies have given improved insights into the systematics, phylogeography, population genetics, biology, immunobiology, and epidemiology of *Ascaris* using molecular tools[6] (see also Chapters 1, 7, 8, and 10), there have been limited global studies of the molecular biology, biochemistry, and physiology of *Ascaris* spp., parasite–host relationships and ascariasis using -omic technologies. Characterizing the genome and transcriptomes of *A. suum* using the advanced technologies described here provides a solid foundation to explore the systems biology of this parasite and design entirely new treatment, diagnostic, and control strategies.

From an epidemiological viewpoint, changes in the temporal and spatial distribution of *Ascaris* spp. and/or their hosts might also be monitored using population genetic or metagenomic approaches.[92] Another question of potential significance is whether *Ascaris* species are developing genetic resistance against current anthelmintics, which have been routinely used for many years. Although resistance against such compounds had not been reported in the past for ascaridoids, there is now evidence of emerging anthelmintic resistance in *Parascaris equorum* populations in various countries.[93,94] With a reference genome available for *A. suum*, it would now be possible to genetically compare multiple populations of adult worms predicted to be resistant or susceptible to one or more compounds, and then to assess links between genotype and a drug-resistant phenotype on a genome-wide scale. Transcriptomic analyses might also be used to underpin comparisons between susceptible and resistant worms. Such analyses might allow the definition of "resistance markers," which could then be used in a molecular test for the direct and specific molecular detection of drug resistance. These examples suggest that there are numerous exciting fundamental areas and questions to tackle using modern genomic tools.

Comparative genome-wide sequencing of well-defined *A. suum* and *A. lumbricoides* samples could finally address the key question as to the specific status of these two parasites.[9] In addition, the definition of a wide range of genetic markers for use in specific and sensitive diagnostic tools could provide a foundation to address questions regarding the complex network of ecological and biological factors involved in parasite–host–environment interactions and the immunological idiosyncrasies of porcine and human hosts in endemic and non-endemic regions. It would also be particularly interesting to explore the resistance and susceptibility of genotypes of human and porcine hosts to *Ascaris* infection. For instance, elucidating the relationship between host genotype and phenotype in response to *Ascaris* and/or intervention strategy (e.g. treatment) would be informative and could provide an understanding of the genetic basis of disease.

The deep sequencing and bioinformatic approaches now established provide the throughput and depth-of-coverage required to rapidly define *de novo* the nuclear genomes of *A. lumbricoides* and other ascarioids of human and animal health importance. Repertoires of drug targets can now be inferred on a genome-wide scale. Such genomic-guided drug target or drug discovery could have significant advantages over traditional screening methods. In *Ascaris*, at least 629 proteins with essential homologues (associated with lethal phenotypes following gene perturbation) in *C. elegans* and *D. melanogaster* were identified. Among these are

87 channels or transporters, which represent protein classes already recognized as anthelmintic targets for compounds, including macrocyclic lactones, levamisoles, and amino-acetonitrile derivatives (AADs). Moreover, at least 225 genes predicted to be essential in *A. suum* are linked to specific metabolic chokepoints, with five molecules (including IMP dehydrogenase) representing high priority drug targets, which warrant future evaluation. Although we have focused specifically on essentiality[95] (namely lethality) as an indicator of "druggability," traditionally, many anthelmintic compounds have caused paralysis/paresis, rather than directly killed the parasite, usually through the inhibition/excitation of signaling pathways associated with contractility or locomotion. Such pathways are likely highly conserved among nematodes and, thus, inhibitors of components of these pathways might have broad applicability as drug targets. For example, a recent, post-genomic study of receptors associated with monoaminergic signaling (linked to decision-making processes during locomotion) in nematodes[96] found significant conservation and proposed parallel evolution for the complement of monoamine receptors encoded by *Caenorhabditis* spp., *B. malayi*, and *A. suum*. Additional "core" signaling and/or functional nodes are almost certainly present in these species. Therefore, *A. suum* is uniquely placed for investigations into the structural and functional conservation of neuronal signaling pathways in nematodes, particularly considering that the large size of this parasite would allow direct investigations of neuron-specific gene transcription using RNA-Seq. Such investigations could have major implications for anti-nematode drug discovery.

In conclusion, characterizing the *A. suum* genome has identified a broad range of key classes of molecules, with major relevance to understanding the molecular biology of *A. suum*, shedding new light on the exquisite complexities of the host–parasite interplay on an immunobiological level, and paving the way for future fundamental molecular explorations, with unique prospects of designing new methods to control one of the world's most important parasitic nematodes. Clearly, an integrated use of -omic technologies will now underpin future investigations of the systems biology of *A. suum* and ascariasis on a scale not possible previously, and will provide unprecedented prospects for developing new diagnostic and intervention strategies. This focus is now crucial, given the major impact of *Ascaris* and other soil-transmitted helminths, which affect billions of people and animals worldwide.[97] Although these parasites are seriously neglected, particularly in terms of funding for fundamental research and the development of new drugs, vaccines, and diagnostics, genomic and other -omic technologies provide new hope for the discovery of new and improved interventions against ascariasis.

Acknowledgments

This project was funded by the Australian Research Council (ARC; LP110100018). Other support from the Australian Academy of Science, the Australian-American Fulbright Commission, Melbourne Water Corporation, the Victorian Life Sciences Computation Initiative (VLSCI) and the IBM Collaboratory is gratefully acknowledged. PWS is an investigator with the Howard Hughes Medical Institute (HHMI). ARJ held a CDA1 (Industry) from the National Health and Medical Research Council of Australia.

References

1. Crompton DW. *Ascaris* and ascariasis. *Adv Parasitol* 2001;**48**:285–375.
2. Kipper M, Andretta I, Monteiro SG, Lovatto PA, Lehnen CR. Meta-analysis of the effects of endoparasites on pig performance. *Vet Parasitol* 2011;**181**:316–20.
3. Anon. *Investing in Development: A Practical Plan to Achieve the Millenium Development Goals*. Washington, DC: United Nations Development Program; 2005.
4. Frankish H. Initiative launched to develop drugs for neglected diseases. *Lancet* 2003;**362**:135.
5. Hotez P, Molyneux DH, Fenwick A, Kumaresan J, Sachs SE, Sachs JD, et al. Control of neglected tropical diseases. *N Engl J Med* 2007;**357**:1018–27.
6. Dold C, Holland CV. *Ascaris* and ascariasis. *Microbes Infect* 2011;**13**:632–7.
7. Anderson TJ. The dangers of using single locus markers in parasite epidemiology: *Ascaris* as a case study. *Trends Parasitol* 2001;**17**(4):183–8.
8. Nejsum P, Parker Jr ED, Frydenberg J, Roepstorff A, Boes J, Haque R, et al. Ascariasis is a zoonosis in Denmark. *J Clin Microbiol* 2005;**43**:1142–8.
9. Peng W, Yuan K, Hu M, Gasser RB. Recent insights into the epidemiology and genetics of *Ascaris* in China using molecular tools. *Parasitology* 2007;**134**:325–30.
10. Boes J, Helwigh AB. Animal models of intestinal nematode infections of humans. *Parasitology* 2000;**121**:S97–111.
11. Jex AR, Liu S, Li B, Young ND, Hall RS, Li Y, et al. *Ascaris suum* draft genome. *Nature* 2011;**479**:529–33.
12. Metzker ML. Sequencing technologies – the next generation. *Nat Rev Genet* 2010;**11**:31–46.
13. Fullwood MJ, Wei CL, Liu ET, Ruan Y. Next-generation DNA sequencing of paired-end tags (PET) for transcriptome and genome analyses. *Genome Res* 2009;**19**:521–32.
14. Gasser RB, Hu M, Chilton NB, Campbell BE, Jex AR, Otranto D, et al. Single-strand conformation polymorphism (SSCP) for the analysis of genetic variation. *Nat Protoc* 2006;**1**:3121–8.
15. Sambrook J, Russell DW. *Molecular Cloning: A Laboratory Manual*. Cold Spring Harbor, NY: Cold Spring Harbor Laboratory; 2001.
16. Li R, Fan W, Tian G, Zhu H, He L, Cai J, et al. The sequence and *de novo* assembly of the giant panda genome. *Nature* 2010;**463**:311–7.
17. Lander ES, Waterman MS. Genomic mapping by fingerprinting random clones: a mathematical analysis. *Genomics* 1988;**2**:231–9.
18. Miller JR, Koren S, Sutton G. Assembly algorithms for next-generation sequencing data. *Genomics* 2010;**95**:315–27.
19. Li H, Ruan J, Durbin R. Mapping short DNA sequencing reads and calling variants using mapping quality scores. *Genome Res* 2008;**18**:1851–8.
20. Besemer J, Lomsadze A, Borodovsky M. GeneMarkS: a self-training method for prediction of gene starts in microbial genomes. Implications for finding sequence motifs in regulatory regions. *Nucleic Acids Res* 2001;**29**:2607–18.

21. Brent MR. How does eukaryotic gene prediction work? *Nat Biotechnol* 2007;**25**:883–5.
22. Zerbino DR, Birney E. Velvet: algorithms for *de novo* short read assembly using de Bruijn graphs. *Genome Res* 2008;**18**:821–9.
23. CSC. Genome sequence of the nematode *C. elegans*: a platform for investigating biology. *Science* 1998;**282**:2012–8.
24. Dieterich C, Clifton SW, Schuster LN, Chinwalla A, Delehaunty K, Dinkelacker I, et al. The *Pristionchus pacificus* genome provides a unique perspective on nematode lifestyle and parasitism. *Nat Genet* 2008;**40**:1193–8.
25. Ghedin E, Wang S, Spiro D, Caler E, Zhao Q, Crabtree J, et al. Draft genome of the filarial nematode parasite *Brugia malayi*. *Science* 2007;**317**:1756–60.
26. Trapnell C, Pachter L, Salzberg SL. TopHat: discovering splice junctions with RNA-Seq. *Bioinformatics* 2009;**25**:1105–11.
27. Trapnell C, Williams BA, Pertea G, Mortazavi A, Kwan G, van Baren MJ, et al. Transcript assembly and quantification by RNA-Seq reveals unannotated transcripts and isoform switching during cell differentiation. *Nat Biotechnol* 2010;**28**:511–5.
28. Xu MJ, Chen N, Song HQ, Lin RQ, Huang CQ, Yuan ZG, et al. RNAi-mediated silencing of a novel *Ascaris suum* gene expression in infective larvae. *Parasitol Res* 2010;**107**:1499–503.
29. Chen N, Xu MJ, Nisbet AJ, Huang CQ, Lin RQ, Yuan ZG, et al. *Ascaris suum*: RNAi mediated silencing of enolase gene expression in infective larvae. *Exp Parasitol* 2011;**127**:142–6.
30. Quevillon E, Silventoinen V, Pillai S, Harte N, Mulder N, Apweiler R, et al. InterProScan: protein domains identifier. *Nucleic Acids Res* 2005;**33**:W116–20.
31. Harris MA, Clark J, Ireland A, Lomax J, Ashburner M, Foulger R, et al. The Gene Ontology (GO) database and informatics resource. *Nucleic Acids Res* 2004;**32**:D258–61.
32. Adams MD, Celniker SE, Holt RA, Evans CA, Gocayne JD, Amanatides PG, et al. The genome sequence of *Drosophila melanogaster*. *Science* 2000;**287**:2185–95.
33. Waterston RH, Lindblad-Toh K, Birney E, Rogers J, Abril JF, Agarwal P, et al. Initial sequencing and comparative analysis of the mouse genome. *Nature* 2002;**420**:520–62.
34. Wu CH, Apweiler R, Bairoch A, Natale DA, Barker WC, Boeckmann B, et al. The Universal Protein Resource (UniProt): an expanding universe of protein information. *Nucleic Acids Res* 2006;**34**:D187–91.
35. Boeckmann B, Bairoch A, Apweiler R, Blatter MC, Estreicher A, Gasteiger E, et al. The SWISS-PROT protein knowledgebase and its supplement TrEMBL in 2003. *Nucleic Acids Res* 2003;**31**:365–70.
36. Rawlings ND, Barrett AJ, Bateman A. MEROPS: the peptidase database. *Nucleic Acids Res* 2010;**38**:D227–33.
37. Saier Jr MH, Yen MR, Noto K, Tamang DG, Elkan C. The Transporter Classification Database: recent advances. *Nucleic Acids Res* 2009;**37**:D274–8.
38. Kall L, Krogh A, Sonnhammer EL. Advantages of combined transmembrane topology and signal peptide prediction – the Phobius web server. *Nucleic Acids Res* 2007;**35**:W429–32.
39. Chen Y, Zhang Y, Yin Y, Gao G, Li S, Jiang Y, et al. SPD – a web-based secreted protein database. *Nucleic Acids Res* 2005;**33**:D169–73.
40. Bennuru S, Semnani R, Meng Z, Ribeiro JM, Veenstra TD, Nutman TB. *Brugia malayi* excreted/secreted proteins at the host/parasite interface: stage- and gender-specific proteomic profiling. *PLoS Negl Trop Dis* 2009;**3**:e410.
41. Hewitson JP, Harcus YM, Curwen RS, Dowle AA, Atmadja AK, Ashton PD, et al. The secretome of the filarial parasite, *Brugia malayi*: proteomic profile of adult excretory-secretory products. *Mol Biochem Parasitol* 2008;**160**:8–21.
42. Bellafiore S, Shen Z, Rosso MN, Abad P, Shih P, Briggs SP. Direct identification of the *Meloidogyne incognita* secretome reveals proteins with host cell reprogramming potential. *PLoS Pathog* 2008;**4**:e1000192.

43. Cass CL, Johnson JR, Califf LL, Xu T, Hernandez HJ, Stadecker MJ, et al. Proteomic analysis of *Schistosoma mansoni* egg secretions. *Mol Biochem Parasitol* 2007;**155**:84–93.
44. Benson G. Tandem repeats finder: a program to analyze DNA sequences. *Nucleic Acids Res* 1999;**27**:573–80.
45. IHGSC. Initial sequencing and analysis of the human genome. *Nature* 2001;**409**:860–921.
46. Li H, Durbin R. Fast and accurate long-read alignment with Burrows-Wheeler transform. *Bioinformatics* 2010;**26**:589–95.
47. Mortazavi A, Williams BA, McCue K, Schaeffer L, Wold B. Mapping and quantifying mammalian transcriptomes by RNA-Seq. *Nat Methods* 2008;**5**:621–8.
48. Tarailo-Graovac M, Chen N. Using RepeatMasker to identify repetitive elements in genomic sequences. *Curr Protoc Bioinformatics* 2009. Chapter 4:Unit 4 10.
49. Xu Z, Wang H. LTR_FINDER: an efficient tool for the prediction of full-length LTR retrotransposons. *Nucleic Acids Res* 2007;**35**:W265–8.
50. Edgar RC, Myers EW. PILER: identification and classification of genomic repeats. *Bioinformatics* 2005;**21**(Suppl. 1):i152–8.
51. Price AL, Jones NC, Pevzner PA. *De novo* identification of repeat families in large genomes. *Bioinformatics* 2005;**21**(Suppl. 1):i351–8.
52. Berriman M, Haas BJ, LoVerde PT, Wilson RA, Dillon GP, Cerqueira GC, et al. The genome of the blood fluke *Schistosoma mansoni*. *Nature* 2009;**460**:352–8.
53. Mitreva M, Jasmer DP, Zarlenga DS, Wang Z, Abubucker S, Martin J, et al. The draft genome of the parasitic nematode *Trichinella spiralis*. *Nat Genet* 2011;**43**:228–35.
54. Muller F, Tobler H. Chromatin diminution in the parasitic nematodes *Ascaris suum* and *Parascaris univalens*. *Int J Parasitol* 2000;**30**:391–9.
55. Aeby P, Spicher A, de Chastonay Y, Muller F, Tobler H. Structure and genomic organization of proretrovirus-like elements partially eliminated from the somatic genome of *Ascaris lumbricoides*. *EMBO J* 1986;**5**:3353–60.
56. Opperman CH, Bird DM, Williamson VM, Rokhsar DS, Burke M, Cohn J, et al. Sequence and genetic map of *Meloidogyne hapla*: A compact nematode genome for plant parasitism. *Proc Natl Acad Sci USA* 2008;**105**:14802–7.
57. Campbell WC, Fisher MH, Stapley EO, Albers-Schonberg G, Jacob TA. Ivermectin: a potent new antiparasitic agent. *Science* 1983;**221**:823–8.
58. Qian H, Robertson AP, Powell-Coffman JA, Martin RJ. Levamisole resistance resolved at the single-channel level in *Caenorhabditis elegans*. *Faseb J* 2008;**22**:3247–54.
59. Kaminsky R, Ducray P, Jung M, Clover R, Rufener L, Bouvier J, et al. A new class of anthelmintics effective against drug-resistant nematodes. *Nature* 2008;**452**:176–80.
60. McKerrow JH, Caffrey C, Kelly B, Loke P, Sajid M. Proteases in parasitic diseases. *Annu Rev Pathol* 2006;**1**:497–536.
61. Hewitson JP, Grainger JR, Maizels RM. Helminth immunoregulation: the role of parasite secreted proteins in modulating host immunity. *Mol Biochem Parasitol* 2009;**167**:1–11.
62. Robinson MW, Dalton JP, Donnelly S. Helminth pathogen cathepsin proteases: it's a family affair. *Trends Biochem Sci* 2008;**33**:601–8.
63. Sajid M, McKerrow JH. Cysteine proteases of parasitic organisms. *Mol Biochem Parasitol* 2002;**120**:1–21.
64. Dalton JP, Brindley PJ, Knox DP, Brady CP, Hotez PJ, Donnelly S, et al. Helminth vaccines: from mining genomic information for vaccine targets to systems used for protein expression. *Int J Parasitol* 2003;**33**:621–40.
65. Smooker PM, Jayaraj R, Pike RN, Spithill TW. Cathepsin B proteases of flukes: the key to facilitating parasite control? *Trends Parasitol* 2010;**26**:506–14.
66. Redmond DL, Smith SK, Halliday A, Smith WD, Jackson F, Knox DP, et al. An immunogenic cathepsin F secreted by the parasitic stages of *Teladorsagia circumcincta*. *Int J Parasitol* 2006;**36**:277–86.

67. Jiang M, Ryu J, Kiraly M, Duke K, Reinke V, Kim SK. Genome-wide analysis of developmental and sex-regulated gene expression profiles in *Caenorhabditis elegans. Proc Natl Acad Sci USA* 2001;**98**:218–23.
68. Cantacessi C, Jex AR, Hall RS, Young ND, Campbell BE, Joachim A, et al. A practical, bioinformatic workflow system for large data sets generated by next-generation sequencing. *Nucleic Acids Res* 2010;**38**. e171.
69. Campbell BE, Nagaraj SH, Hu M, Zhong W, Sternberg PW, Ong EK, et al. Gender-enriched transcripts in *Haemonchus contortus* – predicted functions and genetic interactions based on comparative analyses with *Caenorhabditis elegans. Int J Parasitol* 2008;**38**:65–83.
70. Hu PJ. Dauer. *WormBook* 2007:1–19.
71. Campbell BE, Boag PR, Hofmann A, Cantacessi C, Wang CK, Taylor P, et al. Atypical (RIO) protein kinases from *Haemonchus contortus* – promise as new targets for nematocidal drugs. *Biotechnol Adv* 2011;**29**:338–50.
72. Cohen P. Protein kinases – the major drug targets of the twenty-first century? *Nat Rev Drug Discov* 2002;**1**:309–15.
73. Campbell BE, Hofmann A, McCluskey A, Gasser RB. Serine/threonine phosphatases in socioeconomically important parasitic nematodes – prospects as novel drug targets? *Biotechnol Adv* 2011;**29**:28–39.
74. Renslo AR, McKerrow JH. Drug discovery and development for neglected parasitic diseases. *Nat Chem Biol* 2006;**2**:701–10.
75. Campbell BE, Tarleton M, Gordon CP, Sakoff JA, Gilbert J, McClusky A, et al. Norcantharidin analogues with nematocidal activity in *Haemonchus contortus. Bioorg Med Chem Lett* 2011;**21**:3277–81.
76. Bourne HR, Sanders DA, McCormick F. The GTPase superfamily: conserved structure and molecular mechanism. *Nature* 1991;**349**:117–27.
77. Maizels RM, Yazdanbakhsh M. Immune regulation by helminth parasites: cellular and molecular mechanisms. *Nat Rev Immunol* 2003;**3**:733–44.
78. Roberts A, Pimentel H, Trapnell C, Pachter L. Identification of novel transcripts in annotated genomes using RNA-Seq. *Bioinformatics* 2011;**27**:2325–9.
79. Zhong W, Sternberg PW. Genome-wide prediction of *C. elegans* genetic interactions. *Science* 2006;**311**:1481–4.
80. Geenen PL, Bresciani J, Boes J, Pedersen A, Eriksen L, Fagerholm HP, et al. The morphogenesis of *Ascaris suum* to the infective third-stage larvae within the egg. *J Parasitol* 1999;**85**:616–22.
81. Bargmann CI. Chemosensation in C. *elegans*. In: Community TCeR, editor. *Wormbook: Wormbook*; October 25, 2006.
82. WHO. *Deworming for Health and Development*. Geneva: World Health Organization; 2005.
83. Montresor A, Zin TT, Padmasiri E, Allen H, Savioli L. Soil-transmitted helminthiasis in Myanmar and approximate costs for countrywide control. *Trop Med Int Health* 2004; **9**:1012–5.
84. Jex AR, Lim YA, Bethony JM, Hotez PJ, Young ND, Gasser RB. Soil-transmitted helminths of humans in Southeast Asia-towards integrated control. *Adv Parasitol* 2011;**74**: 231–65.
85. Harder A, Schmitt-Wrede HP, Krucken J, Marinovski P, Wunderlich F, Willson J, et al. Cyclooctadepsipeptides – an anthelmintically active class of compounds exhibiting a novel mode of action. *Int J Antimicrob Agents* 2003;**22**:318–31.
86. Geary TG, Woo K, McCarthy JS, Mackenzie CD, Horton J, Prichard RK, et al. Unresolved issues in anthelmintic pharmacology for helminthiases of humans. *Int J Parasitol* 2010;**40**:1–13.
87. Keiser J, Utzinger J. The drugs we have and the drugs we need against major helminth infections. *Adv Parasitol* 2010;**73**:197–230.

88. Geldhof P, Visser A, Clark D, Saunders G, Britton C, Gilleard J, et al. RNA interference in parasitic helminths: current situation, potential pitfalls and future prospects. *Parasitology* 2007;**134**:609–19.
89. Lee I, Lehner B, Crombie C, Wong W, Fraser AG, Marcotte EM. A single gene network accurately predicts phenotypic effects of gene perturbation in *Caenorhabditis elegans*. *Nat Genet* 2008;**40**:181–8.
90. Chen L, Pankiewicz KW. Recent development of IMP dehydrogenase inhibitors for the treatment of cancer. *Curr Opin Drug Discov Devel* 2007;**10**:403–12.
91. Graci JD, Cameron CE. Mechanisms of action of ribavirin against distinct viruses. *Rev Med Virol* 2006;**16**:37–48.
92. Gilbert JA, Laverock B, Temperton B, Thomas S, Muhling M, Hughes M. Metagenomics. *Methods Mol Biol* 2011;**733**:173–83.
93. von Samson-Himmelstjerna G. Anthelmintic resistance in equine parasites – detection, potential clinical relevance and implications for control. *Vet Parasitol* 2012;**185**:2–8.
94. Reinemeyer CR. Anthelmintic resistance in non-strongylid parasites of horses. *Vet Parasitol* 2012;**185**:9–15.
95. Chen WH, Minguez P, Lercher MJ, Bork POGEE. an online gene essentiality database. *Nucleic Acids Res* 2012;**40**:D901–6.
96. Komuniecki R, Law WJ, Jex A, Geldhof P, Gray J, Bamber B, et al. Monoaminergic signaling as a target for anthelmintic drug discovery: receptor conservation among the free-living and parasitic nematodes. *Mol Biochem Parasitol* 2012;**183**:1–7.
97. Hotez PJ, Fenwick A, Savioli L, Molyneux DH. Rescuing the bottom billion through control of neglected tropical diseases. *Lancet* 2009;**373**:1570–5.

CHAPTER

12

Genetics of Human Host Susceptibility to Ascariasis

Sarah Williams-Blangero *, *Mona H. Fenstad* †, *Satish Kumar* *, *John Blangero* *

* Texas Biomedical Research Institute, San Antonio, TX, USA
† Norwegian University of Science and Technology, Trondheim, Norway

OUTLINE

Ascaris: The Neglected Parasite
http://dx.doi.org/10.1016/B978-0-12-396978-1.00012-4

Copyright © 2013 Elsevier Inc. All rights reserved.

INTRODUCTION

Infections with roundworms (*Ascaris lumbricoides*) remain major public health concerns in many areas of the world. Epidemiologically, *A. lumbricoides* is the most common of the soil-transmitted helminth infections in humans, currently affecting over a quarter of the world's population[1,2] (see Chapter 13). The global impact of *Ascaris* infection has remained remarkably constant even though effective anthelmintics are available. Despite its worldwide distribution and long-term consequences for human health, ascariasis is considered to be a neglected disease.[3] Even among the soil-transmitted helminths themselves, *Ascaris* has been relatively poorly studied.[2,4]

The long-term health impact of *Ascaris* infection tends to be underappreciated, perhaps because the disease occurs in areas of the world also affected by more severe infectious diseases with very apparent, immediate effects. However, infection with *A. lumbricoides* can have both major short-term and subtle long-term effects on health. Surgical intervention may be required to treat severe cases associated with very high worm loads that can cause intestinal blockage.[5] The more common, subtle consequences of infection for health are associated with deficits in cognitive and physical growth and development.[2,6–8] For example, levels of *Ascaris* infection have been associated with poor performance in school and on some types of cognitive tests.[8–12] The association of *Ascaris* infection with deficits in physical growth and development has been demonstrated repeatedly.[7,13–16]

In addition to the direct health impact of *Ascaris* infection, the disease also may have consequences for responses to other infectious diseases. For example, there is evidence that helminth infections can negatively impact the immune response to infection with HIV and tuberculosis.[15,16] The consequences of *Ascaris* infection are complex, and must be considered carefully. Recent studies have suggested that while *Ascaris* infections have clear negative effects on risk for some infectious diseases, they may be associated with decreased risk for malaria[19,20] (see Chapter 4).

EPIDEMIOLOGY OF ASCARIASIS: THE ROLE OF NONRANDOM CLUSTERING

It has long been recognized that *Ascaris* infections are typically overdispersed, meaning that a small proportion of the available human host population harbors the majority of the parasitic worm population.[21,22] As reviewed by Holland,[2] this epidemiological pattern has been observed repeatedly in populations throughout the world (see also Chapter 7).

In addition to overdispersion, clustering of infections also has been observed frequently. This clustering is often seen at the familial or household level. For example, Forrester and colleagues[23] documented household clustering of heavy *Ascaris* infections in a population in Mexico. They went on to suggest a familial or genetic basis to this clustering.[24] In a study of *Ascaris* infections in the Poyang lake region of China, clustering of infections by household was observed.[25] Similarly, Walker and colleagues[26] examined *Ascaris* infections in approximately 3000 individuals belonging to about 500 families living in Dhaka, Bangladesh. They found significant evidence for clustering by household, although they fit no formal genetic models that might account for such clustering.[26]

Studies examining reinfection with *A. lumbricoides* have repeatedly provided evidence that individuals are predisposed to infection. This predisposition has been documented in numerous studies and numerous populations[27–29] (see review in[2]) (see also Chapter 7). Recently, Mehta and colleagues[30] documented maternal effects on risk for *Ascaris* infection, with children of mothers with *Ascaris* infections having increased risk of infection.

The epidemiological literature on *A. lumbricoides* provides significant evidence for nonrandom clustering of infection that is due to shared exposure, worm genetic factors, host genetic factors or, as is more likely, some combination of all of these causal players. As part of our long-running research program, we have focused on the potential role of host genetic factors in *Ascaris* infection. Differentiating between household effects (or shared exposure) and effects associated with individual risk requires an advanced genetic approach. Quantitative genetic approaches can be used to distinguish the effects of individual predisposition from household effects. However, in order to utilize such quantitative genetic approaches, some kind of family information is required. Even in the absence of genetic marker information (which itself can be used to infer pedigree relationships), pedigree information for large extended families crossing multiple households does provide you with power for genetic analysis. Pedigrees of the type required for such analyses can be reconstructed using information on relationships among individuals within households as outlined by Williams-Blangero and Blangero.[31] This standardized interview process allows efficient collection of pedigree information, and facilitates subsequent reconstruction of extended pedigrees crossing multiple households. The process has been used effectively in populations with varying family structures throughout the world,[31] and was used to gather family information in a population endemic for *Ascaris* infection in eastern Nepal.

The rest of this chapter will focus specifically on how we make inferences about the potential role of host genetics in *Ascaris* infection and what our current state of knowledge is regarding genetic determinants of *Ascaris* burden in humans.

MEASURING THE EXTENT OF HOST GENETIC VARIATION IN INFECTION

How do we begin to assess the importance of host genetic variation in a phenotype as complex as *Ascaris* variation? Fortunately, there is a large body of study designs and statistical genetic methods that now allow us to largely disentangle the roles of shared environment and host genetic factors. Most of the needed framework comes from quantitative genetics. To start with, we will ignore the major problem of shared environment and focus purely on host genetic factors. It is clear that human quantitative genetics is primarily concerned with characterizing the range of genetic variation in populations that, given variation in environmental exposures and population substructure, results in an observable range of phenotype that in our specific case is a measure of *Ascaris* burden. An overriding challenge, as compared to analyses of monogenic traits, is that multiple loci may be contributing to the phenotype, so the effect size of any locus is likely to be relatively small. In addition, there may be multiple types of sequence variation in play, ranging from substitution of single nucleotides to large chromosomal rearrangements. Even if the majority of genetic effects were confined to a single locus, this could be due either to a single variable site or to multiple rare alleles segregating in the population(s) under study.

Heritability is the proportion of the total variance of a phenotype that is attributable to the additive effects of alleles. It represents an estimate of the relative extent of genetic variation in a given phenotype. Thus, heritability provides us with a single measure of how important a role genes are likely to play in the causal determination of a variable human trait such as susceptibility to *Ascaris* infection or burden. In a classical variance-components-based approach to quantitative genetic analysis, heritability is readily estimated by decomposing the phenotypic covariance between pairs of individuals based on their kinship:

$$\begin{aligned} Cov(i,j) &= 2\phi_{ij}\sigma_g^2 \\ Cov(i,i) &= \sigma_g^2 + \sigma_e^2 \\ h^2 &= \sigma_g^2/\sigma_P^2 \end{aligned} \tag{12.1}$$

where Cov(i, j) is the covariance between different individuals i and j, and Cov(i, i) represents the variance for the ith individual, ϕ_{ij} is the kinship coefficient between i and j, σ_g^2 is the additive genetic variance, σ_e^2 is the error (sometimes termed environmental) variance, σ_P^2 is the total phenotypic variance of the trait given by $\sigma_P^2 = \sigma_g^2 + \sigma_e^2$, and h^2 is the additive genetic heritability which measures the relative contribution of additive genetic factors to the overall observed phenotypic variance. Twice the "kinship" coefficient in this context refers to the expected proportion of alleles shared *identical by descent* (IBD) by two individuals given their degree of relatedness: siblings share half their alleles IBD, half-siblings one-fourth of their alleles, and so on. Alleles shared IBD are necessarily also *identical by state* (IBS), a distinction that we discuss in more detail below. The genetic variance is a cumulative variance; it represents the summation over all additive genetic factors for the phenotype. Hence, depending upon the phenotype, it may represent the influence of a single genetic variant or that of many hundreds of genetic variants.

If variation in a phenotype were due entirely to genetic causes (and these could be clearly discerned) the heritability of the trait would be 1. Due to multifactorial causation and measurement error, a typical range of heritability for many quantitative traits is 30–80% of total variance, and estimates may differ widely from one study to another due to sampling error. The heritability is a critical measure of the importance of within-population genetic variation. This single metric conveys whether or not the search for the individual contributing genes is merited for a given phenotype.

FUNCTIONAL GENETIC VARIATION IS THE SOURCE OF HERITABILITY

What is the biological source of heritability in humans? Ultimately, it comes from observable functional genetic variation at the sequence level. A functional variant is one that influences the focal phenotype via some molecular mechanism. Thus, functional variants can be considered to be phenotype specific in this context. If the variant influences a quantitative trait (such as measured *Ascaris* burden), we term it a *quantitative trait nucleotide* (QTN) variant. The effect of a functional variant on the phenotype can be quantified by the QTN-specific variance that is given by $\sigma_q^2 = 2p(1-p)\alpha^2$, where p is the minor allele frequency of the QTN and α is one-half the difference between phenotypic means of the two homozygotes. Biologically, we expect α to be determined by biophysical molecular properties of the QTL and to be relatively constant across populations. The term $2p(1-p)$ is also known as the expected heterozygosity of the underlying genotype and measures the variance of a trait that is scored as the number of minor alleles in the diploid genotype. The relative genetic signal intensity for this QTN is given by the QTN-specific heritability $h_q^2 = \sigma_q^2/\sigma_P^2$, where σ_P^2 is the total variance of the phenotype. The relative genetic signal for a quantitative trait locus (QTL) is determined by the sum of the QTN-specific heritabilities (although these must be corrected for possible linkage disequilibrium among variant sites) in the immediate region of the QTL and thus will be influenced by all of the relevant functional variants in the region. In algebraic form, the QTL-specific heritability is

$$h_Q^2 = \frac{\sum 2p_i(1-p_i)\alpha_i^2}{\sigma_P^2} = \sum h_{qi}^2$$

where the summation is over the functional variants in the regions of the QTL. Similarly, the total heritability of the phenotype is given by the sum of all of the QTL-specific heritabilities over the whole genome or $h^2 = \sum h_{Qi}^2$.

ASSESSING HERITABILITY IN HUMAN PEDIGREES

At this point, we can employ variance decomposition theory to derive a model that allows for the infection-related phenotypic variance to be decomposed into components that are due to host genetic factors and random environmental factors. For n host individuals, the $n \times n$ phenotypic covariance matrix (Ω) is given by the following variance component model:

$$\mathbf{\Omega} = (2\mathbf{\Phi}h^2 + \mathbf{I}e^2)\sigma_p^2 \tag{12.2}$$

where 2Φ is twice the kinship matrix (i.e. the coefficient of relationship matrix) among individuals which is derived from the host pedigree structure, and **I** is the identity matrix (with ones on its diagonal and zeros elsewhere). The identity matrix simply conveys the assumed structure of the remaining environmental (which also includes all model error components) component in which everyone experiences a unique person-specific environment. The phenotypic variance of the focal trait is given by σ_p^2 while the relative variance components associated with each of the structuring matrices are the additive genetic heritability in the host (h^2), and the proportion of phenotypic variance due to random environmental factors ($e^2 = 1 - h^2 - w^2$). The phenotypic covariance model essentially captures all of the information necessary for making gross genetic inferences. It predicts how phenotypically correlated pairs of individuals will be as a function of their genetic relatedness.

Using a standard maximum likelihood variance estimation method as implemented in the statistical genetics computer package, SOLAR,[32] we can estimate the parameters of this model along with any mean effect parameters (such as covariate effects). In order to estimate this model, we assume an underlying multivariate density has generated our data which is obligately violated for traits like worm and egg counts. We therefore typically perform direct inverse Gaussian transformations on all phenotypes prior to analysis. Even after such transformation, these traits are not exactly normal. However, prior analyses have shown that as long as kurtosis is not excessive, the maximum likelihood analyses provide valid results.[33]

Using this approach, we can test to see whether a host genetic component is necessary to explain the observed phenotypic covariances among individuals. Likelihood ratio tests (LRT) are formed as twice the difference in ln likelihoods. Because of boundary conditions (variances must be greater than or equal to zero), the resultant test statistics are distributed as mixtures of chi-square distributions. For example, when comparing a model that includes host genetic factors and random environmental factors with a baseline model only including random environmental factors (i.e. $h^2 = 0$), the LRT is distributed as a 50:50 mixture of a point mass at zero and a chi-square distribution with 1 degree of freedom.

IDENTIFYING SPECIFIC GENES RESPONSIBLE FOR HERITABILITY OF *ASCARIS* BURDEN

The basic knowledge that host genetic factors are involved in the observed distribution of *Ascaris* burden is of limited immediate translational value. The most important reason for doing host genetic analysis is to identify the actual causal genes underlying human variation in disease risk. Knowledge of causal genes (and their causal sequence variants) can provide new windows into understanding disease and can directly provide novel drug targets. How does one go about localizing and identifying these causal genetic factors that ultimately determine heritability? Many recent advances in analysis of human quantitative traits have been made in the context of genetically complex diseases.[34,35]

Generally, prior to identification of causal genes, we must first localize a QTL to a restricted genomic region. In the absence of complete sequence information for all study participants (which still lies in the future in terms of technical and economic feasibility), gene localization depends either on the random effect of known genetic markers assessed via linkage, or the main effect of the markers via association.

In linkage analysis, we expand the standard variance component model for a pedigree to also include an additional component that reflects the chromosomal location-specific relatedness among individuals. This new structuring matrix, the so-called IBD matrix (for identity by descent probability matrix) must be calculated from genome-wide genetic marker information such as highly polymorphic short tandem repeat (STR) markers or, now more commonly, from high density SNP marker sets. Allowing for location-specific QTL effects at location l in the genome, we now write the covariance model as:

$$\boldsymbol{\Omega} = (2\boldsymbol{\Phi}h_r^2 + \widehat{\boldsymbol{\Pi}}_l h_{Ql}^2 + \mathbf{I}e^2)\sigma_p^2 \tag{12.3}$$

where $\widehat{\boldsymbol{\Pi}}_l$ refers to the estimate of the IBD probability matrix at position l in the genome, h_{Ql}^2 is the heritability due to the QTL, and h_r^2 refers to the remaining residual heritability after controlling for the QTL (and any covariates). If h_{Ql}^2 is significantly greater than zero (using the same likelihood ratio test for a single variance component as described above), then there is evidence for a QTL at genomic position l that influences worm burden. One quirk of the field must be noted here. Instead of using a standard LRT statistic for tests of linkage hypotheses, we utilize the statistic LOD = LRT/(2 ln(10)). The localization of a human QTL is typically imprecise. This precision is noted by the 1-LOD support interval which is the region which captures a one-unit LOD score drop around a given QTL peak. This area is typically 10–15 cM in studies of humans. Thus, linkage-derived QTL regions typically involve approximately 10–15 Mb of human genome sequence.

Association methods can also be used to identify QTL regions using genome-wide high density SNP arrays. This approach depends upon alleles shared, not necessarily IBD, but *identical by state* (IBS). If a marker is physically close enough to a functional variant, linkage between the functional and marker alleles will not be disrupted by meiotic recombination: that is, the alleles are in *linkage disequilibrium* (LD[36]). Association is conceptually and mathematically simpler than linkage analysis: association analysis is typically based on *measured genotypes* (MG) either for individual markers (often diallelic single nucleotide polymorphisms, SNPs, which are closely spaced throughout the human genome), or on *haplotype blocks* of marker alleles in high LD. In family-based studies, association may be measured as the main effect of the measured genotype in a mixed model along with the random effects of allele-sharing.[37] In principle, the main effect may be tested in unrelated individuals as well, although the test is then susceptible to confounding effects of population stratification.[38]

If one is so lucky as to have an actual functional variant in hand – or a marker in perfect LD with a functional variant – measured genotype (MG) analysis is necessarily the most powerful test of genetic effect. However, LD falls off quite sharply with physical distance (by approximately 250 kb), which is both a blessing and a curse. It is why association is capable of providing high-resolution positional data, but it also places a severe limitation on detectable effect. The observed effect of a marker locus $h_m^2 = h_q^2 \rho^2$ where h_q^2 and ρ^2 – both unobserved – are the effect size of the functional variant and the squared correlation between the marker and functional genotypes, respectively. Thus, genome-wide studies of association using 500,000 to 2M SNPs typically localize QTLs that have effective regions of support of approximately 500 kb of sequence to search through for functional variants. This is much smaller ($\sim 20\times$) than that typically observed for a linkage-derived QTL.

QUANTITATIVE GENETICS STUDIES OF ASCARIASIS

All genetic studies of *Ascaris* infection have employed the methods (or closely related ones) briefly outlined above. The first quantitative genetic study of susceptibility to *Ascaris* infection was conducted in the Jirel population of eastern Nepal.[39] This study clearly demonstrated genetic effects on risk for infection with *A. lumbricoides* using a variance components approach for discriminating between genetic effects and household effects on worm loads. Analyses of *Ascaris* burden were conducted using data collected from 1261 members of a single complex pedigree which provided outstanding statistical power. The single pedigree included over 26,000 pairs of relatives which were informative for the genetic analyses. Additionally, because the entire population ultimately forms a single pedigree, the problem of shared environment is greatly diminished since the pedigree is composed of many (primarily) nuclear families living in separate households. Thus, the confounding of shared exposure is minimized when using very large pedigrees with individuals living in separate households. The heritability of each measure of *Ascaris* worm burden assessed in the pedigree members was determined using the variance components approach as described above and implemented in SOLAR.[32,40] As mentioned previously, the heritability defines the proportion of the variation in a trait which is attributable to additive genetic factors.

Ascaris burden was assessed in a variety of ways in this study. Egg counts were determined from fecal samples, direct worm counts were determined in stools collected for a 96-hour period following treatment with albendazole, and total weight of the worms was determined.[39]

Our initial genetic results demonstrated significant heritabilities for each of the measures of *Ascaris* burden in the Jirel population. Egg counts were significantly heritable with approximately 30% of the variation in egg counts being attributable to genetic factors. Worm counts showed a significant heritability, with approximately 36% of the variation in total worm count being attributable to genetic factors. And finally, total worm weight was significantly heritable with about 34% of the variation in total worm weight attributable to genetic factors.

The results of these initial quantitative genetic studies of *Ascaris* in the Jirel population strongly support the hypothesis that the individual predisposition to infection repeatedly observed in earlier studies by other investigators[27–29,41,42] was in fact due to genetic effects.

These results also appear to be consistent across host species. A heritability of *Ascaris* egg counts of approximately 30% has since been found in the pig model.[43]

GENOME SCANS FOR GENES INFLUENCING *ASCARIS* INFECTION

To date, only two genome-wide scans for genes influencing traits associated with *Ascaris* infection have been published. Of all the techniques used to assess genetic influences on *Ascaris* infection, genome-wide approaches are the most powerful (strongest) genetic approaches because they are unbiased. All possible genetic factors are considered with no *a priori* hypotheses governing the range of genes considered.

Work in the Jirel population of eastern Nepal involved a genome scan of a total of 1258 individuals belonging to a single extended pedigree.[44,45] *Ascaris* egg counts were determined for each individual and each person also was characterized for approximately 370 STR markers evenly spaced across the autosomes. The first report generated from this scan localized two genes influencing risk for *Ascaris* infection in the original core of the Jirel pedigree which included 444 individuals, 27.2% of whom were infected.[44] Because of the complexity of the pedigree, over 6200 pairs of relatives informative for genetic analysis were present in the sample.

Two loci were found to exert significant effects on *Ascaris* burden in the branch of the Jirel pedigree examined.[44] The strongest linkage signal was found on chromosome 13 near the q terminus (13q32–q34). An excellent candidate gene was located in the region *TNFSF13B*. The gene is an important regulator of B cell activation and Ig secretion.[46–48] The other significant quantitative trait locus influencing *Ascaris* worm burden was located on chromosome 1 (1p32).

The Jirel data were further analyzed upon completion of the genotyping for all 1258 individuals.[45] This new analysis resulted in the localization of three significant quantitative trait loci. The linkage to chromosomal region 13q33 remained in the larger sample, confirming that *TNFSF13B* is likely involved in determining risk for *Ascaris* infection. The LOD score plot for this QTL is shown in Figure 12.1. The peak observed LOD was 3.37 occurring at 113 cM with a 1-LOD support interval extending from 104 to 117 cM. The genome-wide (corrected for multiple testing) p-value was 0.013. There is an additional significant linkage to chromosomal region 11p14 (LOD = 3.19, genome-wide $p = 0.020$) which encompasses 54 known genes in the QTL support interval. A potential candidate in this area is *CD59* which is a membrane-bound inhibitor of the cytolytic membrane attack complex (MAC) of complement.[49] The third significant QTL linked to *Ascaris* egg counts was found at chromosomal region 8q23 in a region encompassing over 300 known or predicted genes. Two possible candidate genes were suggested for this region: first, a gene thought to be involved in mediation of viral infection, *MAL2*[50]; and second, a member of the collectin family which is involved in innate immunity, *COLEC10*.[51]

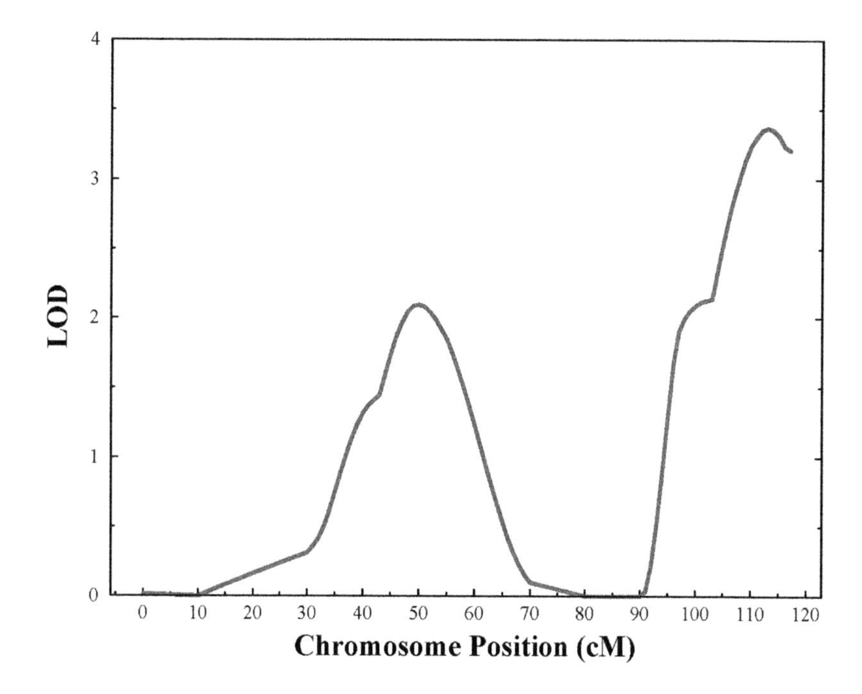

FIGURE 12.1 Quantitative trait linkage analysis results showing a major QTL influencing *Ascaris* burden on chromosome 13 in the Jirel population.

Besides our studies, one genome-wide linkage study considered IgE levels specific to *Ascaris* as part of an overall assessment of the genetics of allergen-specific IgE levels in relation to asthma.[52] A total of 653 family members from eight large pedigrees from Costa Rica were characterized for both IgE specific to *Ascaris* and 387 STR markers spaced across the autosomes. The heritability of IgE specific to *Ascaris* was 0.33 indicating that approximately one-third of the variation was attributable to genetic factors. The linkage analysis revealed a suggestive linkage of IgE to *Ascaris* on chromosome 7q (LOD 2.72) at a position of 153 cM.

As outlined above, to date, six QTLs have been unambiguously linked to risk for *Ascaris* infection.[44,45] In addition, four suggestive linkages for genes influencing traits related to *Ascaris* infection have been found.[45,52]

FURTHER ASSOCIATION MAPPING OF THE CHROMOSOME 13 QTL USING HIGH DENSITY SNPs

Recent technological advances have dramatically changed the search for genes influencing complex traits such as risk for *Ascaris* infection. Rather than having only about 400 markers spaced across the genome, it is now possible to type hundreds of thousands of SNPs in large numbers of samples very efficiently and cost effectively. Following up on our linkage signal on chromosome 13, we typed 912 people for a total of 500,000 SNPs using the Illumina 660W-Quad DNA analysis kit. For the purposes of this chapter, we focus only on the chromosome 13 QTL region in an attempt to fine-map this QTL in hopes of identifying the underlying causal gene(s). We employed a standard measured genotype analysis for each SNP allowing for residual heritability as described above. Figure 12.2 shows the results of our QTL region-specific association analyses in the approximately 10.25 Mb corresponding to the 13 cM QTL linkage support interval. In this region, we typed a total of 2387 SNP markers that passed quality control procedures. In Figure 12.2, the solid horizontal line shows the cut-off for QTL region-wide significance accounting for the number of effectively independent tests (a procedure which also takes into account the LD between SNPs). One significant association result stands out involving the SNP rs12866842. The minor allele of this SNP was associated with increased worm burden ($p = 1.5 \times 10^{-6}$). This SNP is located in an intergenic region between the *TNFSF13B* and *IRS2* genes. Both of these genes are viable candidates for a gene influencing susceptibility to *Ascaris* infection. We have long noted the potential role of *TNFSF13B* and its potential relevance for helminthic infection has been described above. *IRS2* is important in nitric oxide (NO) production, an important mediator of inflammation. Our localization results from linkage analysis and QTL region-specific association

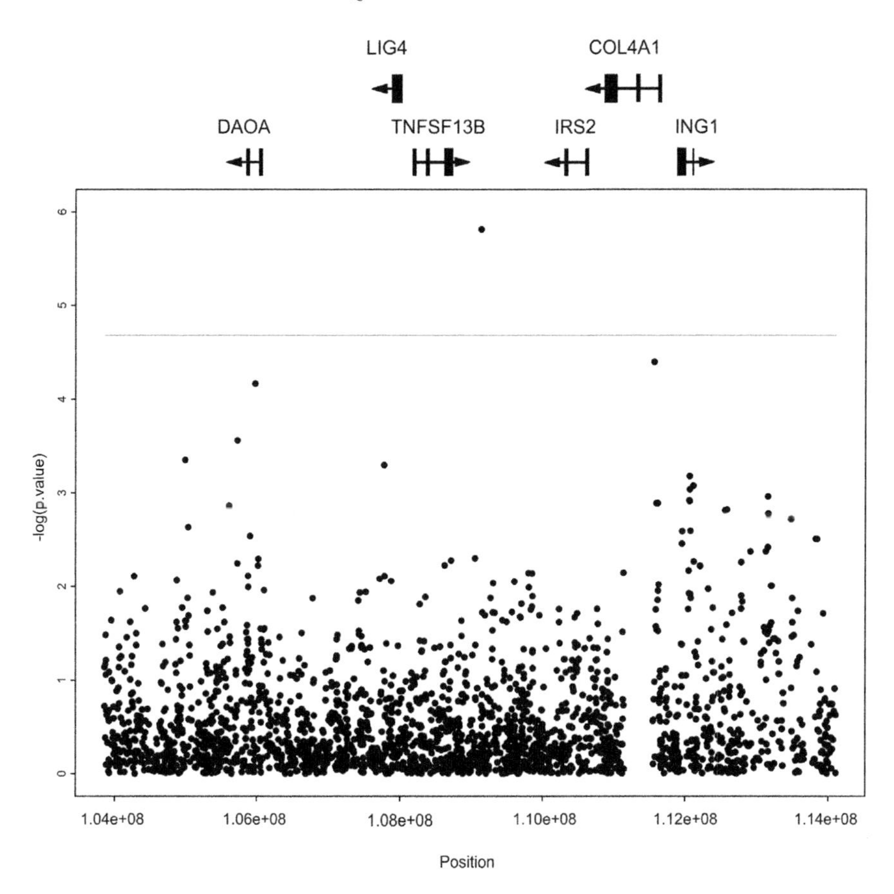

FIGURE 12.2 Fine mapping association analysis results of the chromosome 13 *Ascaris* burden QTL linkage region using high density common SNPs.

analysis have now reduced our focus to a smaller region on chromosome 13 around the *TNFSF13B* and *IRS2* genes.

SEQUENCING OF *TNFSF13B* AS A GENE UNDERLYING THE QTL FOR *ASCARIS* INFECTION

TNFSF13B has been suggested as a gene influencing susceptibility to *Acaris* infection in both our linkage analyses of STR marker data and in our family-based association analyses of high density SNP data. To further explore this candidate gene, we sequenced the exons and putative promoter of *TNFSF13B* in 200 unrelated members that best capture the founding genomes of the Jirel pedigree who had *Ascaris* egg count data.

We used the exact same standard Sanger-style sequencing procedures and primer designs as detailed in Fenstad et al.[53] Two kb of the proximal promoter (upstream of the translation start site) and all six exons (translated or untranslated) of the 13q QTL candidate gene, *TNFSF13B* (NM_006573.3) were sequenced. We found 13 variants as a result of this sequencing. The linkage disequilibrium plot detailing the squared correlations among these variants is shown in Figure 12.3. In this plot, dark red blocks document variants that are in high linkage disequilibrium with one another and are, therefore, redundant. Variants that show little correlation with other variants (such as rs60972017) are effectively independent. When we performed association analysis using these 13 variants, only the G17226A variant in intron 3 was significantly correlated with *Ascaris* egg count ($p = 0.0357$).

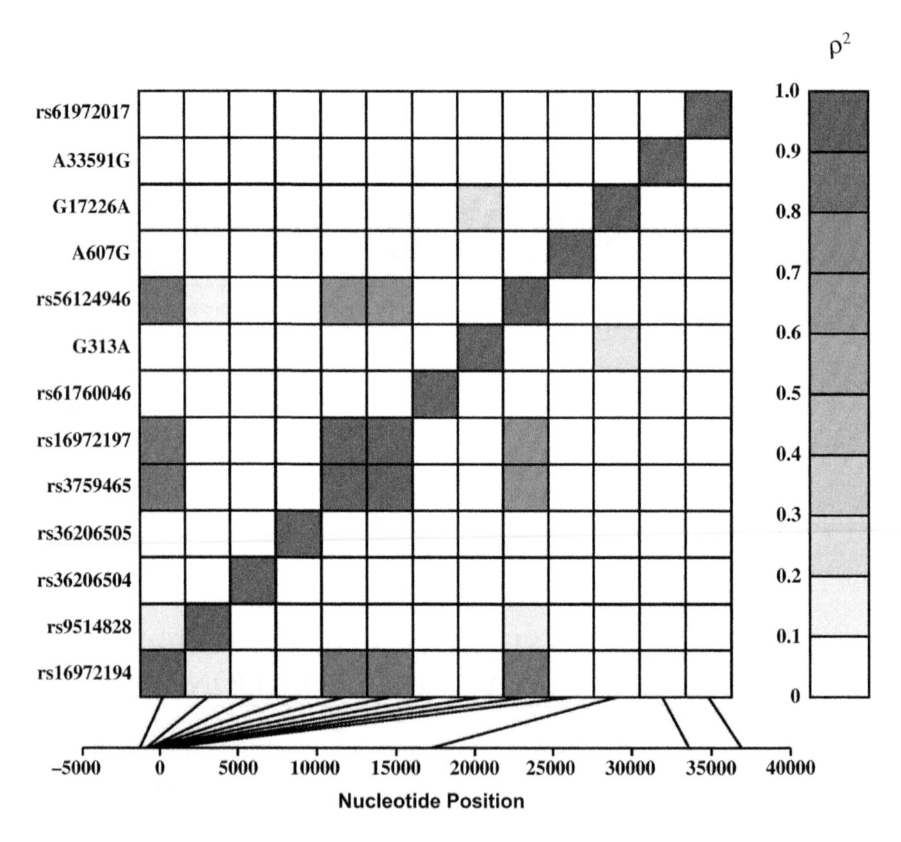

FIGURE 12.3 Observed linkage disequilibrium (measured by the squared correlation coefficient between genotypes) in the chromosome 13 *Ascaris* burden positional candidate gene, TNFSF13B, in 200 unrelated individuals of the Jirel ethnic group.

In order to better understand any potential underlying functional correlation, we further explored the G17226A variant in another data set that we had available to us. This data set includes genome-wide transcriptional profiles for participants in the San Antonio Family Heart Study.[54] In this data set, the G17226A variant in intron 3 was significantly associated with increased gene expression of *TNFSF13B* in peripheral blood mononuclear cells (PBMCs) from Mexican Americans ($p = 0.014$). This finding suggests that the observed variant may be reflecting a functional signal relating to gene regulation. However, because we have not exhaustively sequenced this gene relative to potential regulatory sites, our result could still be due to linkage disequilibrium with another untyped variant.

While our results continue to be consistent for a potential role of *TNFSF13B*, they are not comprehensive. Ideally, we should assess all possible sequence variation within the chromosome 13 QTL region. For example, the *IRS2* gene and its surrounding area should also be deeply sequenced. Additionally, in an ideal situation, all individuals should be directly sequenced or at least exactly imputed by utilizing the high density SNP framework and given that appropriately chosen pedigree members (i.e. those that capture the most of the originating founder genomes and who are appropriately positioned to maximally allow for exact Mendelian imputation of their descendants) are sequenced. Regardless, *TNFSF13B* continues to be our primary positional candidate gene for this QTL.

OTHER CANDIDATE GENE STUDIES OF ASCARIASIS

As reviewed by Quinnell,[55] there have been few candidate gene studies of risk for *Ascaris* infection. The candidate gene studies that have been conducted generally base their gene selection on the hypothesized relationship between asthma and helminthic infection.[56,57] Numerous observations from epidemiological studies have suggested that resistance to *Ascaris* infection may be associated with risk for asthma. As reviewed by Hopkin,[58] it has been proposed that over evolutionary time, parasitic worm infections may have selected for genes in humans that confer resistance to *Ascaris* and susceptibility to asthma. Studies have demonstrated a relationship between current *Ascaris* infection and decreased risk for asthma[59] (see Chapter 2).

Based upon our original nomination of the 13q33-34 *Ascaris* infection influencing QTL, Acevedo and colleagues[60] assessed potential genetic effects at three positional candidate genes (*LIG4*, *TNFSF13B*, and *IRS2*) related to the immune system on quantitative levels of immunoglobulins specific to *Ascaris*. They found that a missense coding variant Thr9Ile (T9I, rs1805388) in the *LIG4* was significantly associated ($p = 0.024$) with specific IgE levels to *Ascaris* in a population of 1064 individuals from Cartagena, Colombia. Bioinformatic analysis using the PolyPhen-2 program[61] to assess the likely functional effect expected of this non-synonymous variant predicts that this variant is possibly damaging (PolyPhen-2 score = 0.520). However, in the Jirel sample, we find that there is no relationship between this variant and *Ascaris* burden ($p = 0.818$) which suggests that it is unlikely to be a cause of the original QTL. Acevedo et al.[60] also identified an intronic *TNFSF13B* variant (rs10508198) that was not associated with *Ascaris*-specific IgE, but was strongly associated ($p = 0.003$) with IgG levels specific to *Ascaris*. We currently have no information on this variant in our study of the Jirels. Additionally, this group of investigators found no significant association with a 3′ UTR variant near the *IRS2* gene.[60]

With regard to other genomic regions, Ramsay and colleagues[62] assessed the potential association of the β_2-adrenergic receptor (*ADRB2*) gene with *Ascaris* infection as part of their studies of asthma in a Venezuelan population. They examined two coding variants, Gly16Arg (G16R, rs1042713) and Glu27Gln (E27Q, rs1042714), in this gene and genotyped 126 children from Coche Island who were also characterized for levels of *Ascaris* infection. The results of this study indicated that children who were homozygous for Arg16 had significantly higher *Ascaris* egg counts ($p < 0.001$) than children who were not homozygous for the allele. The gene was also associated with significantly higher levels of *Ascaris* specific IgE ($p = 0.002$). PolyPhen-2 analysis predicts that both the R16G variant (PolyPhen-2 score = 0.043) and the E27Q variant (PolyPhen-2 score = 0.009) are most likely to be benign (suggesting little alteration of the functional capacity of the protein). Consistent with this bioinformatic prediction, both of these variants are available in our Jirel sample and neither show any relationship to *Ascaris* egg counts (R16G, $p = 0.904$; E27Q, $p = 0.071$). These results suggest that it is unlikely these variants are directly influencing infection levels.

Finally, a study in China has assessed the association between variants in a number of genes related to the Th-2 and Th-1 immune cytokine signaling pathway that have been previously associated with asthma and *Ascaris* infection in children. Utilizing data from 614 children, they determined that a variant (G4219A, rs324015) of the *STAT6* gene was significantly associated ($p = 0.001$) with low *Ascaris* egg counts.[63] However, when we examine our Jirel-derived data for this gene, we do not observe any significant association with *Ascaris* egg counts for this marker ($p = 0.309$) or any other marker in or near this gene.

MORE COMPLEX MODELS: HOST/WORM GENETIC EFFECTS AND SPATIAL FACTORS INFLUENCING HOST *ASCARIS* BURDEN

We have so far only considered simple models of host genetic variation influencing observed *Ascaris* burden phenotypes in humans. However, greater model complexity may be required. Genetic variation in the worms (see Chapter 8) is likely to influence these phenotypes, as is shared environmental determinants. Using our variance component framework, it is possible to test more complicated (and biologically reasonable) models regarding the potential determinants of host worm burden. We have recently published the first simultaneous examination of host and worm genomic effects on host burden phenotypes[64] that we summarize here. We initially wanted to determine if both host genes and worm genes are important for determining the distribution of *Ascaris* worm burden in the Jirel population. For this analysis we utilized data available from 320 individuals (including 141 males and 179 females ranging in age from 3 years to 79 years) for whom we had collected all of the worms expelled post-albendazole treatment. For this analysis, our measure of worm burden was direct count of worms obtained over a 4-day fecal collection period. The 1094 *Ascaris* worms collected from these individuals were then characterized for a total of 23 genetic markers.[65,66]

Using variance decomposition theory, we derived a model that allowed the infection-related phenotypic variance to be decomposed into components that are due to host genetic factors, genetic variation among the worms, and random environmental factors. Expanding upon our previous quantitative genetic model (Eq. (12.2)), the resulting phenotypic covariance matrix is then given by:

$$\mathbf{\Omega} = (2\mathbf{\Phi}h^2 + \mathbf{R}_w w^2 + \mathbf{I}e^2)\sigma_p^2 \tag{12.4}$$

$\mathbf{R}_w$ represents the estimated coefficient of relationship[67] between parasites across hosts (and averaged within host) as measured (see[65,66]) from highly polymorphic short tandem repeat polymorphism in the *Ascaris* worms collected from the Jirel human host population, and $\mathbf{I}$ is the identity matrix. The additive genetic heritability in the worms is denoted by w,^{2} and the proportion of phenotypic variance due to random environmental factors is now $e^2 = 1 - h^2 - w$.^{2}

Using this approach, we tested whether nested models that eliminate specific variance components adequately captured the variation accounted for in this general model. Likelihood ratio tests (LRT) were formed as twice the difference in ln likelihoods. See Williams-Blangero et al.[64] for technical details incurred by testing of boundary conditions.

All worm burden traits were normalized using an inverse Gaussian transformation prior to analysis. Covariates for worm burden as assessed by *Ascaris* total count included sex, age, and number of days of sampling post-albendazole treatment. Only age was significant (showing a decrease in total worm count with age, $p = 0.0160$). Formal testing of the variance component models revealed that for *Ascaris* total worm count, the model which included only host genes fits the data nearly as well as the most general model which included both host genetic and worm genetic effects. Thus, the host genetic model represents a parsimonious model that fits as well as a more general model that included both host and worm genetic factors. Additive genetic host factors accounted for approximately 54% of the total variation in total worm counts. Models that only included worm genetic factors or no genetic factors were both rejected as being significantly different from the general model. *A priori*, one might expect host genetic effects to be the primary influence on total worm count since each worm represents an independent infection. We would expect strong selection for infectivity in the worms since infection must occur in order for the worm to reproduce. Therefore, worm counts are also a function of the survival of individual worms and represent a life history or fitness-related traits that would be expected to have low levels of genetic variation.

The above test of worm genetic variation suffers from the relatively small sample size (320 hosts) that was incurred by only considering hosts for which we had also genetically assayed the collected worms. Therefore, we also decided to examine potential spatial effects influencing host worm burden. The inclusion of a random spatial process may also allow inference on worm population genetic structure effects, such as inference of isolation by distance or spatial autocorrelation effects on worm kinship and, therefore, may be an indirect proxy for worm genetic variation. For this analysis, we utilized data on 1108 individuals (544 males and 564 females ranging in age from 3 to 85 years).

We employed a standard exponential decay model to parameterize the expected correlations between hosts as a function of geographic distance.[64] Because host genetic factors are explicitly accounted for by the pedigree information, any resulting spatial component may represent either: (1) a geographic component of exposure patterns due to local environmental conditions such as temperature or moisture, or (2) a component representing worm population genetic differentiation resulting from an isolation-by-distance process. In order to account for such a potential spatial process, we modified our variance component model as follows:

$$\mathbf{\Omega} - (2\mathbf{\Phi}h^2 + \exp(-\lambda\mathbf{D})s^2 + \mathbf{I}e^2)\sigma_p^2 \qquad (12.5)$$

where $\mathbf{D}$ is a matrix of geographic distances (in kilometers) between individuals as assessed from geographic coordinates of the houses that they live in, λ is the exponential decay parameter, and s^2 is a new relative variance component that measures the relative proportion of variance accounted for by the random spatial process. The term $\exp(-\lambda\mathbf{D})$ generates a valid spatially decaying positive correlation matrix that structures the spatial variance component. As $\lambda \to \infty$, this model becomes a standard shared household model in which individuals within a household exhibit a correlation of 1 while across households the resulting correlation is zero. This model can also be expanded to allow for host genetic $\times$ spatial interactions by adding an additional variance component that is structured by the Hadamard product (i.e. the matrix operator that involves element by element multiplication) of the two relevant covariance kernels (i.e. covariance structuring matrices derived from either the pedigree information or the spatial information).

Using this spatial model, we found that both a host genetic factor and a random spatial component are required to best fit the data on *Ascaris* worm burden.[64] We observed no evidence of host genetic $\times$ spatial interaction. The observed estimate of the exponential decay parameter in the general model was 3.92 which suggests that the expected spatially varying correlation is halved by a distance of 0.177 km. In the general model, host genetics accounts for approximately 27.2 ($\pm$9.6%) of the total variation in *Ascaris* worm count while spatial factors account for an additional 9.6% ($\pm$4.1%). Thus, host genetic factors still appear to be the single largest determinant of phenotypic variation. The observed spatial component may reflect the worm genetic component or some other environmental spatial process but this still remains to be resolved.

Our interpretations for these more complex models are still equivocal. Direct assay of between-worm genetic variation appears to show that genetic variation within the parasite is not important for explaining host variation in total worm burden. However, our much larger approximate study using spatial variation strongly suggests that some type of spatial decay process is important. Such decay may still be due to underlying worm genetics or may simply be due to spatial variation in exposure that would represent a case where environmental sharing also influences the phenotypic correlations among individuals.

FUTURE OF GENETIC RESEARCH ON *ASCARIS* INFECTION: DEEP SEQUENCING TO IDENTIFY RARE FUNCTIONAL VARIANTS

We have outlined our long-term approach to studying the genetic basis of *Ascaris* infection in an isolated human population from eastern Nepal. The statistical genetic models that we have developed are dependent upon the availability of large pedigrees that cut across multiple

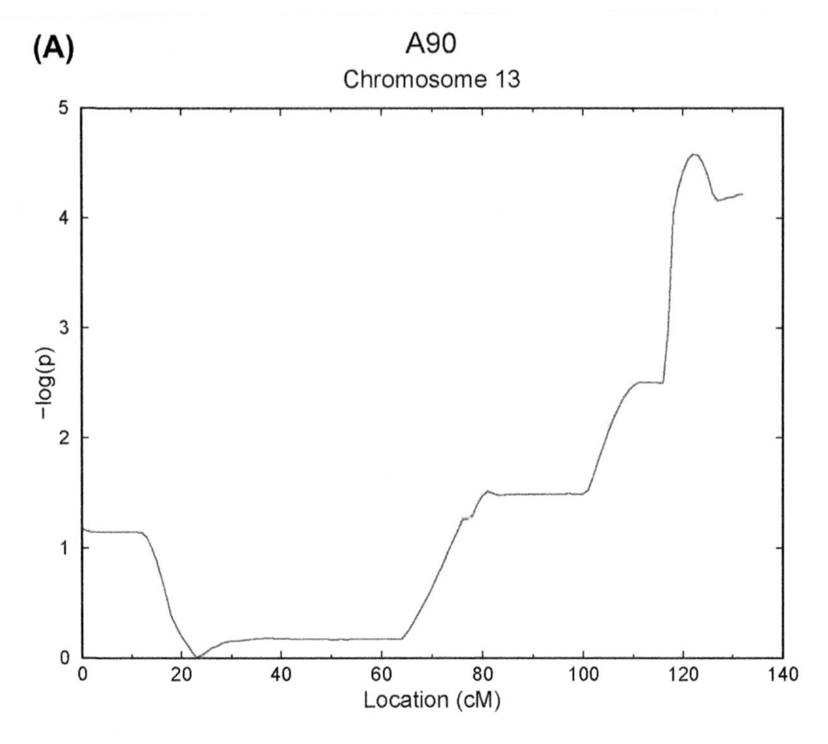
(A)
A90
Chromosome 13
−log(p)
Location (cM)
0
1
2
3
4
5
0
20
40
60
80
100
120
140

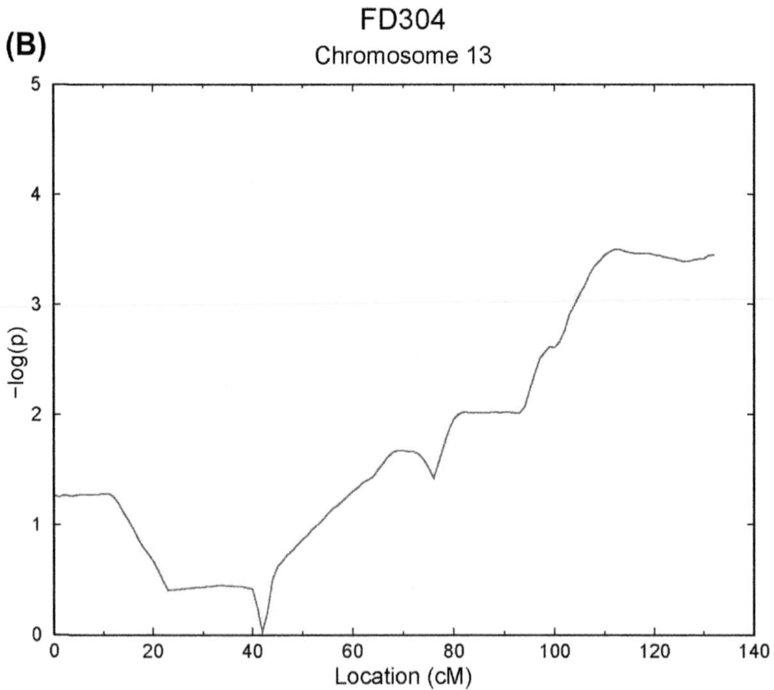
(B)
FD304
Chromosome 13
−log(p)
Location (cM)
0
1
2
3
4
5
0
20
40
60
80
100
120
140

households (environments). Using genetic markers, we have localized several QTL influencing *Ascaris* burden in this population including a major one on chromosome 13. Using common SNP variants, we further localize potential functional variants to the region including the *TNFSF13B* and *IRS2* genes, both of which represent reasonable biological candidate genes. Initial classical sequencing of the *TNFSF13B* coding (and promoter) regions has identified a rare variant that is correlated with both *Ascaris* burden and *TNFSF13B* gene expression levels (in PBMCs). Thus, *TNFSF13B* continues to be an excellent candidate for partial explanation of our original chromosome 13 QTL. However, this region (and other genome regions harboring additional QTLs) still requires exhaustive evaluation. Next-generation sequencing is dramatically changing our approach to sequence variant detection and typing. Our current goal is to perform whole genome sequencing in this population to capture all existing genetic variation, both common and rare. It is becoming clear in these early days of large-scale sequencing that there is a vast amount of rare functional variation segregating in the human genome.[32]

What is the best way to study rare functional variation influencing host susceptibility? Studies of very large pedigrees, such as the Jirel study, represent the most powerful design for the identification of rare functional variants. In fact, pedigrees are the only appropriate design for studying the many private (i.e. occurring in only one lineage) functional variants that occur in humans. For such rare variants, the extended pedigree design is powerful because Mendelian transmission leads to numbers of potential allelic copies greatly in excess of those expected from random selection of unrelated individuals. Alternatively, conventional epidemiological studies of unrelated individuals can only capture a single copy of such private variants by definition, a result that is not amenable to statistical inference. We already have preliminary evidence that such rare variants may be the underlying source of our chromosome 13 *Ascaris* burden QTL.

Figure 12.4 shows the results of lineage-specific quantitative trait linkage analysis in the Jirel pedigree for chromosome 13. Such analyses (using a standard pedigree-based association model in which the founder-specific IBD probability vector is the focal covariate) examine the potential for a given founder to have transmitted a unique rare variant to his descendants that correlates with *Ascaris* burden. Examination of approximately 100 of the largest male founder lineage reveals that our original linkage signal is driven by two such lineages. (The results for females are similar, as

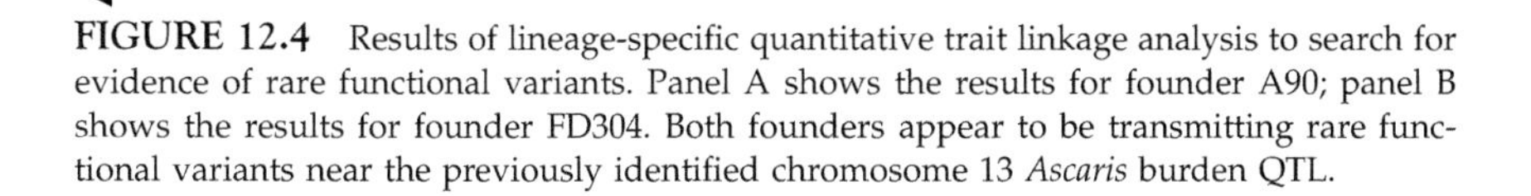

FIGURE 12.4 Results of lineage-specific quantitative trait linkage analysis to search for evidence of rare functional variants. Panel A shows the results for founder A90; panel B shows the results for founder FD304. Both founders appear to be transmitting rare functional variants near the previously identified chromosome 13 *Ascaris* burden QTL.

is expected in a largely monogamous population.) These two lineages, A90 and FD304, are the only ones that show genome-wide significant linkage to this region. Founder A90 has nine measured descendants that when transmitted a hypothetical variant in this genome region exhibit substantially higher *Ascaris* burdens ($p = 2.8 \times 10^{-5}$). Similarly, founder FD304 has 21 measured descendants that also exhibit significantly higher *Ascaris* burdens ($p = 3.7 \times 10^{-4}$) when transmitted such a variant in this region. This result suggests that at least two rare functional variants are segregating through the pedigree. These variants may or may not be in the same gene. Such rare variants are generally not in linkage disequilibrium with other sequence variants and therefore need to be directly identified and genotyped. Only through whole genome sequencing will we be able to identify such functional variants contributing to our existing (and still to be found) QTLs influencing *Ascaris* burden.

The field of human complex disease genetics is rapidly moving towards the measurement of comprehensive genetic information that is only possible with whole genome sequencing. The coming availability of such important data will mean that we no longer have to rely upon signals resulting from correlated surrogate non-functional genetic variation to find regions likely to harbor genes influencing disease susceptibility. Our future challenge will be to separate the functional wheat from the chaff in as efficient a manner as possible. For diseases as complex as *Ascaris* infection which involves host genetic, worm genetic, and environmental factors, we believe that our focus on very large human pedigrees will offer a good natural study design for identifying both common and rare functional variants playing a role in disease burden.

Acknowledgments

We gratefully acknowledge the Jirel people for their participation in our studies of helminthic infection. The new data reported in this chapter were generated under protocols approved by the University of Texas Health Science Center at San Antonio Institutional Review Board and the Nepal Health Research Council. This work was supported by National Institutes of Health grants AI44406, AI37091, HL045522, and MH59490. The lymphocyte expression profiling was supported by a generous donation from the Azar/Shepperd families with additional funds from ChemGenex Pharmaceuticals, Ltd. The work presented in this chapter was conducted in facilities that were constructed with partial support from NIH grants C06 RR017515 and C06 RR013556.

References

1. Holland CV. *Ascaris, Trichuris,* hookworm *and Enterobius*. In: *Topley and Wilson's Parasitology.* 10th ed. UK: Holder Arnold; 2005. p. 712–36.
2. Holland CV. Predisposition to ascariasis: patterns, mechanisms, and implications. *Parasitology* 2009;**136**:1537–47.

3. Hotez PJ, Brindley PJ, Bethony JM, King CH, Pearce EJ, Jacobson J. Helminth infections: the great neglected tropical disease. *J Clin Invest* 2008;**118**:1311–21.
4. Lustigman S, Geldhof P, Grant WN, Osei-Atweneboana MR, Sripa B. A research agenda for helminth disease of humans: basic research and enabling technologies to support control and elimination of helminthiases. *PLoS Neglected Trop Dis* 2012;**6**(4):e1445.
5. Crompton DW. Ascaris and ascariasis. *Adv Parasitol* 2001;**48**:285–375.
6. Dold C, Holland CV. Investigating the underlying mechanism of resistance to *Ascaris* infection. *Microbes Infect* 2011;**13**:624–31.
7. Hall A, Hewitt G, Tuffrey V, de Silva N. A review and meta-analysis of the impact of intestinal worms on child growth and nutrition. *Maternal Child Nutr* 2008;**4**:118–236.
8. O'Lorcain P, Holland CV. The public health importance of *Ascaris lumbricoides*. *Parasitology* 2000;**121**:S51–71.
9. Jardim-Botelho A, Raff S, Rodrigues Rde A, Hoffman HJ, Diemert DJ, Correa-Oliveira R, et al. Hookworm, *Ascaris lumbricoides* infection and polyparasitism associated with poor cognitive performance in Brazilian school children. *Trop Med Int Health* 2008;**13**: 994–1004.
10. Ezeamama AE, Firedman JF, Acosta LP, Bellinger DC, Langdon GC, Manalo DL, et al. Helminth infection and cognitive impairment among Filipino children. *Am J Trop Med Hyg* 2005;**72**:540–8.
11. Hadidjaja P, Bonang E, Suyardi MA, Abidin SA, Ismid IS, Margono SS. The effect of intervention methods on nutritional status and cognitive function of primary school children infected with *Ascaris lumbricoides*. *Am J Trop Med Hyg* 1998;**59**:791–5.
12. Oninla SI, Onayade AA, Owa JA. Impact of intestinal helminthiases on the nutritional status of primary school children in Osun state, south-western Nigeria. *Ann Trop Med Parasitol* 2010;**104**:583–94.
13. Oberhelman RA, Guerrero ES, Fernandez ML, Silio M, Mercado D, Comiskey N, et al. Correlations between intestinal parasitosis, physical growth, and psychomotor development among infants and children from rural Nicaragua. *Am J Trop Med Hyg* 1998;**58**: 470–5.
14. Ulukanligil M, Seyrek A. Anthropometric status, anaemia, and intestinal helminthic infections in shantytown and apartment schoolchildren in the Sanliurfa province of Turkey. *Eur J Clin Nutr* 2004;**58**:1056–61.
15. Phathammavong O, Moazzam A, Xaysomphoo D, Phengsavanh A, Kuroiwa C. Parasitic infestation and nutritional status among schoolchildren in Vientiane, Lao PDR. *J Paediatr Child Health* 2007;**43**:689–94.
16. Mbuh JV, Nembu NE. Malnutrition and intestinal helminth infections in schoolchildren from Dibanda, Cameroon. *J Helminthol* 2013;**87**:46–51.
17. Bentwich Z, Kalinkovish A, Weisman Z. Immune activation is a dominant factor in the pathogenesis of African AIDS. *Immunol Today* 1995;**16**:187–91.
18. Bentwich Z, Kalinkovish A, Weisman Z, Borkow G, Beyers N, Beyers AD. Can eradication of helminthic infections change the face of AIDS and tuberculosis? *Immunol Today* 1999;**20**:485–7.
19. Nacher M. Interactions between worms and malaria: good worms or bad worms? *Malaria J* 2011;**10**:259.
20. Boel M, Carrara VI, Rijken M, Proux S, Nacher M, Pimanpanarak M, et al. Complex interactions between soil-transmitted helminths and malaria in pregnant women on the Thai-Burmese border. *PLoS Neglected Trop Dis* 2010;**4**:e887.
21. Croll NA, Ghadirian F. Wormy persons: contributions to the nature and patterns of overdispersion with *Ascaris lumbricoides, Ancylostoma duodenale, Necator americanus*, and Trichuris trichiura. *Trop Geogr Med* 1981;**33**:241–8.
22. Anderson RM, May RM. Helminth infections of humans: mathematical models, population dynamics, and control. *Adv Parasitol* 1985;**24**:1–10.

23. Forrester JE, Scott ME, Bundy DA, Golden MN. Clustering of *Ascaris lumbricoides* and *Trichuris trichiura* infections within households. *Trans R Soc Trop Med Hyg* 1988;**82**: 282–8.
24. Forrester JE, Scott ME, Bundy DAP, Golden MN. Pre-disposition of individuals and families in Mexico to heavy infection with *Ascaris lumbricoides* and *Trichuris trichiura* infection. *Trans R Soc Trop Med Hyg* 1990;**84**:272–6.
25. Ellis MK, Raso G, Li Y-S, Rong Z, Chen H-G, McManus DP. Familial aggregation of human susceptibility to co- and multiple helminth infections in a population from Poyang Lake region, China. *Int J Parasitol* 2007;**37**:1153–61.
26. Walker M, Hall A, Basanez MG. Individual predisposition, household clustering, and risk factors for human infection with *Ascaris lumbricoides*: new epidemiological insights. *PLoS Negl Trop Dis* 2011;**5**:e1047.
27. Elkins D, Haswell-Elkins MR, Anderson RM. The epidemiology and control of intestinal helminthes in the Pulicat region of Southern India. I. Study design and pre- and post-treatment observations on *Ascaris lumbricoides* infection. *Trans R Soc Trop Med Hyg* 1986;**80**:774–92.
28. Holland CV, Asaolu SO. Crompton DWT, Stoddart RC, MacDonald R, Torimiro SEA. The epidemiology of *Ascaris lumbricoides* and other soil-transmitted helminthes in primary school children from Ile-Ife, Nigeria. *Parasitology* 1989;**99**:275–85.
29. Hall A, Anwar KS, Tomkins A. Intensity of reinfection with *Ascaris lumbricoides* and its implications for parasite control. *Lancet* 1992;**339**:1253–7.
30. Mehta RS, Rodriguez A, Chico M, Guadalupe I, Broncano N, Sandoval C, et al. Maternal geohelminth infections are associated with an increased susceptibility to geohelminth infection in children: a case–control study. *PLoS Negl Trop Dis* 2012;**6**(7):e1753.
31. Williams-Blangero S, Blangero J. Collection of pedigree data for genetic analysis in isolate populations. *Hum Biol* 2006;**78**:89–107.
32. Almasy L, Blangero J. Multipoint quantitative trait linkage analysis in general pedigrees. *Am J Hum Gen* 1998;**62**:1198–211.
33. Blangero J, Williams JT, Almasy L, Williams-Blangero S. Mapping genes influencing human quantitative trait variation. In: Crawford M, editor. *Anthropological Genetics*. Cambridge: Cambridge University Press; 2007. p. 306–33.
34. Blangero J. Localization and identification of human quantitative trait loci: King Harvest has surely come. *Curr Opin Gen Devel* 2004;**14**:233–40.
35. Almasy L, Blangero J. Human QTL linkage mapping. *Genetica* 2009;**136**:333–40.
36. Lewontin RC, Kojima K. The evolutionary dynamics of complex polymorphisms. *Evolution* 1960;**14**:458–72.
37. Boerwinkle E, Chakraborty R, Sing CF. The use of measured genotype information in the analysis of quantitative phenotypes in man. I. Models and analytical methods. *Ann Hum Gen* 1986;**50**:181–94.
38. Devlin B, Roeder K. Genomic control for association studies. *Biometrics* 1999;**55**:997–1004.
39. Williams-Blangero S, Subedi J, Updahayay RP, Manral DB, Rai DR, Jha B, et al. Genetic analysis of susceptibility to infection with *Ascaris lumbricoides*. *Am J Trop Med Hyg* 1999;**60**:921–6.
40. Blangero J, Almasy L. Multipoint oligogenic linkage analysis of quantitative traits. *Gen Epidemiol* 1997;**14**:959–64.
41. Chan L, Kan DP, Bundy DAP. The effect of repeated chemotherapy on age-related predisposition to *Ascaris lumbricoides* and *Trichuris trichiura*. *Parasitology* 1992;**104**:371–7.
42. Thein-Hlaing Than Saw, Myat Lay Kyin. The impact of three-monthly age-targeted chemotherapy on *Ascaris lumbricoides*. *Trans R Soc Trop Med Hyg* 1991;**81**:519–22.
43. Nejsum P, Roepstorff A, Jørgensen CB, Fredholm M, Göring HH, Anderson TJ, et al. High heritability for *Ascaris* and *Trichuris* infection levels in pigs. *Heredity (Edinb)* 2009;**102**(4):357–64.

44. Williams-Blangero D, VandeBerg JL, Subedi J, Aivaliotis MJ, Rai DR, Upadhayay RP, et al. Genes on chromosomes 1 and 13 have significant effects on *Ascaris* infection. *Proc Nat Acad Sci USA* 2002;**2002**(99):5533–8.
45. Williams-Blangero S, VandeBerg JL, Subedi J, Jha B, Corrêa-Oliveira R, Blangero J. Localization of multiple quantitative trait loci influencing susceptibility to infection with *Ascaris lumbricoides*. *J Infect Dis* 2008;**197**:66–71.
46. Moore PA, Belvedere O, Orr A, Pieri K, LaFleur DW, Feng P, et al. BLyS: member of the tumor necrosis factor family and B lymphocyte stimulator. *Science* 1999;**285**:260–3.
47. Schneider P, MacKay F, Steiner V, Hofmann K, Bodmer JL, Holler N, et al. BAFF, a novel ligand of the tumor necrosis factor family, stimulates B cell growth. *J Exp Med* 1999;**189**: 1747–56.
48. Yan M, Marsters SA, Grewal IS, Wang H, Ashekenazi A, Dixit VM. Identification of a receptor for BLyS demonstrates a crucial role in humoral immunity. *Nat Immunol* 2000;**1**:37–41.
49. Kimberly FC, Sivasankar B, Morgan BP. Alternative roles for *CD59*. *Mol Immunol* 2007;**44**:73–81.
50. Bello-Morales R, Fedetz M, Alcina A, Tabarés E, López-Guerrero JA. High susceptibility of a human oligodendroglial cell line to herpes simplex type 1 infection. *J Neurovirol* 2005;**11**:190–8.
51. Gupta G, Surolia A. Collectins: sentinels of innate immunity. *Bioessays* 2007;**29**:452–64.
52. Hunninghake GM, Lasky-Su J, Soto-Quirós ME, Avila L, Liang C, Lake SL, et al. Sex-stratified linkage analysis identifies a female-specific locus for IgE to cockroach in Costa Ricans. *Am J Respir Crit Care Med* 2008;**177**:830–6.
53. Fenstad MH, Johnson MP, Roten LT, Aas PA, Forsmo S, Klepper K, et al. Genetic and molecular functional characterization of variants within TNFSF13B, a positional candidate preeclampsia susceptibility gene on 13q. *PLoS One* 2010;**5**(9).
54. Göring HH, Curran JE, Johnson MP, Dyer TD, Charlesworth J, Cole SA, et al. Discovery of expression QTLs using large-scale transcriptional profiling in human lymphocytes. *Nat Gen* 2007;**39**:1208–16.
55. Quinnell RJ. Genetics of susceptibility to human helminth infection. *Int J Parasitol* 2003;**33**:1219–31.
56. Smits HH, Everts B, Hartgers FC, Yazdanbakhsh M. Chronic helminth infections protect against allergic diseases by active regulatory processes. *Current Allergy and Asthma Report* 2010;**10**:3–12.
57. Flohr C, Quinnell RJ, Britton J. Do helminth parasites protect against atopy and allergic disease? *Clin Exper Allergy* 2009;**39**:20–32.
58. Hopkin J. Immune and genetic aspects of asthma, allergy, and parasitic worm infections: evolutionary links. *Parasite Immunol* 2009;**31**:267–73.
59. Cardosa LS, Costa DM, Almeida MC, Souza RP, Carvalho EM, Araujo MI, et al. Risk factors for asthma in a helminth endemic area in Bahia. *J Parasitol Res* 2012;796–820.
60. Acevedo N, Mercado D, Vergara C, Sanchez J, Kennedy MW, Jimenez S, et al. Association between total immunoglobulin E and antibody responses to naturally acquired *Ascaris lumbricoides* infection and polymorphisms of immune system related *LIG4, TNFSF13B,* and *IRS2* genes. *Clin Exp Immunol* 2009;**157**:282–90.
61. Adzhubei IA, Schmidt S, Peshkin L, Ramensky VE, Gerasimova A, Bork P, et al. A method and server for predicting damaging missense mutations. *Nat Methods* 2010;**7**:248–9.
62. Ramsay CE, Hayden CM, Tiller KJ, Burton PR, Palenque IHM, Lynch NR, et al. Association of polymorphisms in the β_2-adrenoreceptor gene with higher levels of parasitic infection. *Hum. Gen* 1999;**104**:269–74.
63. Peisong G, Mao X-Q, Enomoto T, Feng Z, Gloria-Bottini F, Bottini E, et al. An asthma-associated genetic variant of *STAT6* predicts low burden of ascaris worm infestation. *Genes Immun* 2004;**5**:58–62.

64. Williams-Blangero S, Criscione CD, VandeBerg JL, Correa-Oliveira R, Williams KD, Subedi J, et al. Host genetics and population structure effects on parasitic disease. *Philos Trans R Soc Lond B Biol Sci* 2012;**367**(1590):887–94.
65. Criscione CD, Anderson JD, Rabyt K, Sudimack D, Subedi J, Rai DR, et al. Microsatellite markers for the human nematode parasite *Ascaris lumbricoides*: development and assessment of utility. *J Parasitol* 2007;**93**:704–8.
66. Criscione CD, Anderson JD, Sudimack D, Subedi J, Upadhayay RP, Jha B, et al. Landscape genetics reveals focal transmission of a human macroparasite. *PLoS Negl Trop Dis* 2010;**4**:e665.
67. Queller DC, Goodnight KF. Estimating relatedness using genetic markers. *Evolution* 1989;**43**:258–75.

PART V

CLINICAL ASPECTS AND PUBLIC HEALTH

CHAPTER

13

Ascaris lumbricoides and Ascariasis: Estimating Numbers Infected and Burden of Disease

Simon J. Brooker, Rachel L. Pullan
London School of Hygiene & Tropical Medicine, London, UK

OUTLINE

INTRODUCTION

Ascaris lumbricoides is an intestinal nematode (or soil-transmitted helminth, STH) that is among the most common of all chronic infections of humans. It occurs and flourishes where climatic conditions permit the survival and development of its free-living stages and where poverty,

Ascaris: The Neglected Parasite
http://dx.doi.org/10.1016/B978-0-12-396978-1.00013-6

Copyright © 2013 Elsevier Inc. All rights reserved.

inadequate water and sanitation facilities, and poor hygienic practices promote fecal contamination of the environment and fecal–oral contacts. Unsurprisingly, *Ascaris* infection is prevalent in many less developed countries today and historically was endemic in some developed countries.

The majority of individuals with *A. lumbricoides* will exhibit no signs or symptoms. This is because pathology is strongly related to the number of worms present (the intensity of infection),[1–2] and most individuals harbor only a few worms.[3–4] In the minority (5%) of individuals who harbor large worm burdens, infection can result in clinical disease, including intestinal, biliary, and pancreatic obstructions (see Box 13.1). If left untreated, these complications can prove fatal.[5] In addition to these acute manifestations, intervention studies have demonstrated that moderate worm burdens are associated with reversible growth deficits in children.[6–7] Chronic ascariasis has also been implicated in some studies with reduced cognitive performance and school peformance.[7–10]

Reliably estimating the extent of the problem of *A. lumbricoides* and ascariasis is difficult for a number of reasons. First, *A. lumbricoides* infection is typically diagnosed by the microscopic detection of the parasite's eggs in stool samples, a procedure that misses light infection.[11] Second, microscopy does not detect non-fecund infections (single worm or single sex) which may represent an important minority of infections.[12] Thus, any estimates of infection prevalence are likely to be conservative and an underestimate. Third, much of the acute clinical morbidity caused by ascariasis is non-specific, making attribution problematic. Fourth, the chronic effects of infection on growth and cognition are subtle, making their impact underappreciated and revealed only by well-designed intervention studies.[7,13] These challenges have contributed to the debate as to the true impact of *A. lumbricoides* and ascariasis. They have also meant that estimates of the global distribution and disease burden of *A. lumbricoides* and ascariasis have inevitably been based on informed approximations, using the best available information.[14–21]

This chapter reviews efforts to define the global distribution and disease burden of *A. lumbricoides* and ascariasis. It begins by evaluating the biological limits imposed by climatic factors to *A. lumbricoides* transmission and the influence of urbanization and control measures on contemporary distributions. Next, past and current estimates of the global population infected with *A. lumbricoides* are reviewed. The Global Burden of Disease (GBD) study is then introduced, detailing how estimates of the disease burden caused by ascariasis have been calculated, both in the original 1990 GBD study and the recent 2010 study. Finally, the limitations of the GBD approach to estimating the full extent of the problem of ascariasis are discussed.

BOX 13.1

IMPACT OF *ASCARIS LUMBRICOIDES* AND ASCARIASIS

Ascariasis refers to the spectrum of disease manifestations caused by infection with *Ascaris lumbricoides*. The pathology caused by infection corresponds to the life-cycle stage and the intensity of infection. Acute disease can arise due to larval migrations or adult worms causing intestinal obstruction, whereas chronic disease is due to the insidious effects of infection on nutrition and cognition and is most common among children, due to age-related intensity patterns. The main features of ascariasis include:

- Pulmonary migration induces hypersensitivity, which can manifest as asthma (see Chapter 2). The reaction to migration can be severe resulting in cough, hypersecretion of mucus, and bronchiolar inflammation, which is usually subclinical.
- The intestinal phase of infection is generally asymptomatic, but heavy infection can cause physiological abnormalities in the small intestine resulting in malabsorption of nutrients, vitamin A and other micronutrients, nutritional deficiency and growth failure, especially in children.[57]
- Moderate and heavy infection in children may also adversely affect cognitive development.[7–9]
- Heavy infection can cause serious complications, the most common of which is small bowel obstruction by a bolus of worm, leading to gastrointestinal discomfort, vomiting, and occasionally intussusception and death.[5] *Ascaris*-related intestinal obstruction most commonly occurs in the ileum of young children.
- Female *A. lumbricoides* worms can migrate up the common bile duct into the liver where they can cause bile duct obstruction leading to cholangiitis or pancreatitis.[58,59] The worms can also die, releasing eggs. Granulomatous reactions around the dead worm can result in liver abscess and acute upper abdominal pain, sometimes with fever and jaundice.
- Liver abscesses can be caused by female *A. lumbricoides* worms migrating up the common bile duct into the liver where they die, releasing eggs. Histologically there is a granulomatous reaction round the dead worm with release of the eggs.

THE GLOBAL LIMITS OF TRANSMISSION

In common with most parasitic infections, the transmission of *A. lumbricoides* is constrained by temperature and humidity. This is because the eggs excreted by female worms require a period of embryonation in the external environment prior to being picked up and ingested by humans. Experimental studies suggest that maximum rates of embryonation of *Ascaris* ova occur at temperatures between 28 and 32°C, with embryonation arresting below 5 and above 38°C.[22] Similarly, at low humidity (atmospheres less than 80% saturation) *Ascaris* ova do not embryonate.[23] These differing rates of development and survival will influence parasite establishment in the human host and hence observed limits of transmission, which can be observed at global and country levels. Figure 13.1A presents the global limits of *A. lumbricoides*, based on spatial analysis of available data and a range of environmental factors.[19] This analysis shows that high and low land surface temperature and extremely arid environments limit *A. lumbricoides* transmission. In particular, the prevalence of *A. lumbricoides* is generally <4% in areas where maximum land surface temperature exceeds 35°C, and drops to <1% by 40°C. Prevalence is also <2% in areas classified as "arid" (as defined by an aridity index which combines data on annual total rainfall and the ability of the atmosphere to remove water through evapotranspiration) and <0.1% in "hyper-arid" areas.

Within these biological limits local endemicity will be mediated by settlement patterns and urbanization. Interestingly, the prevalence of *A. lumbricoides* is higher in urban and peri-urban than in rural settings, especially in Africa.[19] Such differences may reflect variations in inadequate water and sanitation or higher population density in peri-urban and urban areas compared to rural areas.

A SHRINKING GLOBAL DISTRIBUTION

A further factor influencing the contemporary distribution of *A. lumbricoides* is local control efforts. Historically, *A. lumbricoides*, as well as *Trichuris trichiura* and hookworm, was prevalent in parts of Europe, Japan, South Korea, Taiwan, the Caribbean, and the southern states of America, but sustained control efforts and economic development helped to eliminate intestinal nematodes from these countries.[17,24–28] An example of such sustained control is illustrated for South Korea in Figure 13.2. In the early 1960s, *A. lumbricoides* and other intestinal nematodes were highly prevalent and in response to this situation, the Korean government began a national parasite control program, supported by the *Parasitic Diseases*

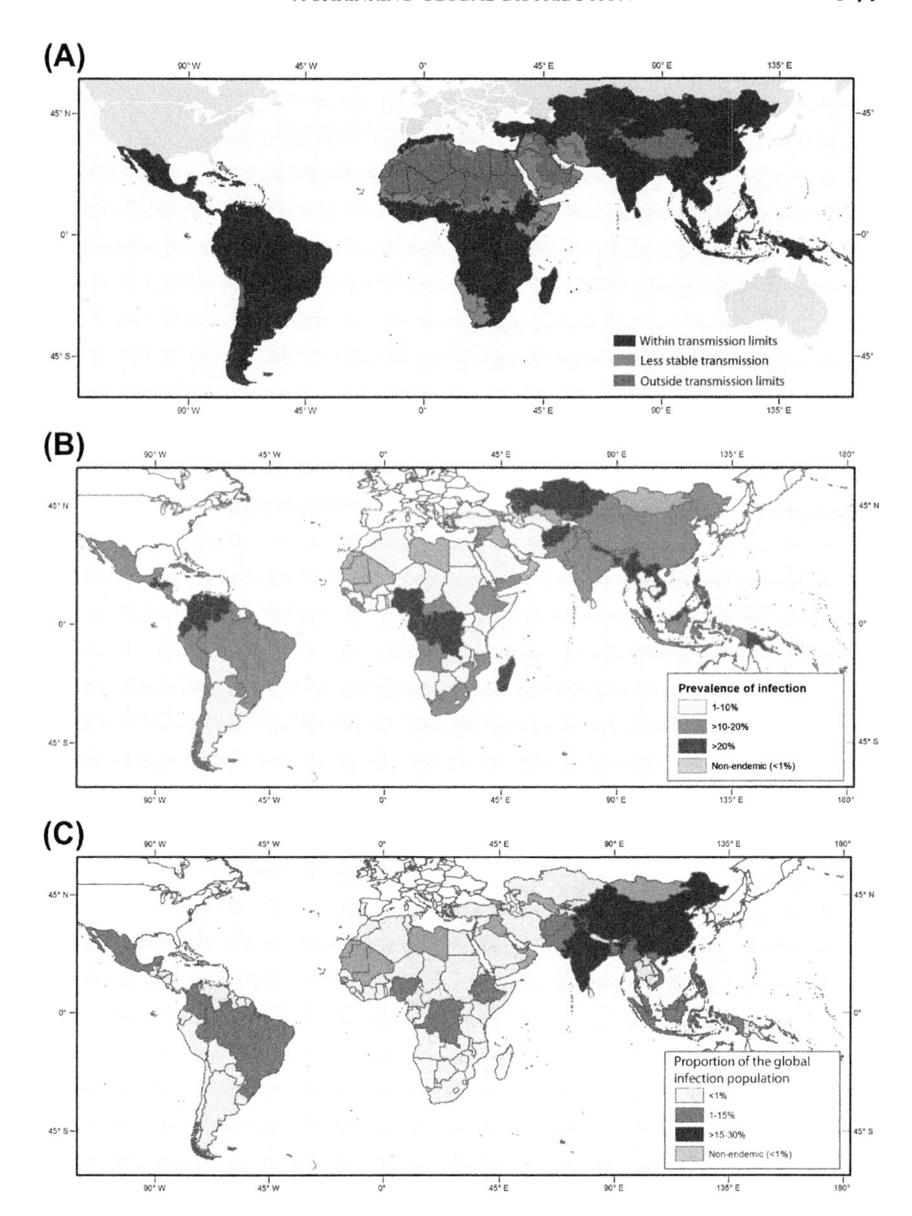

FIGURE 13.1 (A) Global limits of *Ascaris lumbricoides* transmission. Areas are defined as stable, less stable or no risk, on the basis of temperature and aridity. Adapted from ref.[19] (B) The prevalence of *A. lumbricoides* infection, based on available empirical information. (C) The proportion of the global population infected (0.762 billion) by country (unpublished estimates).

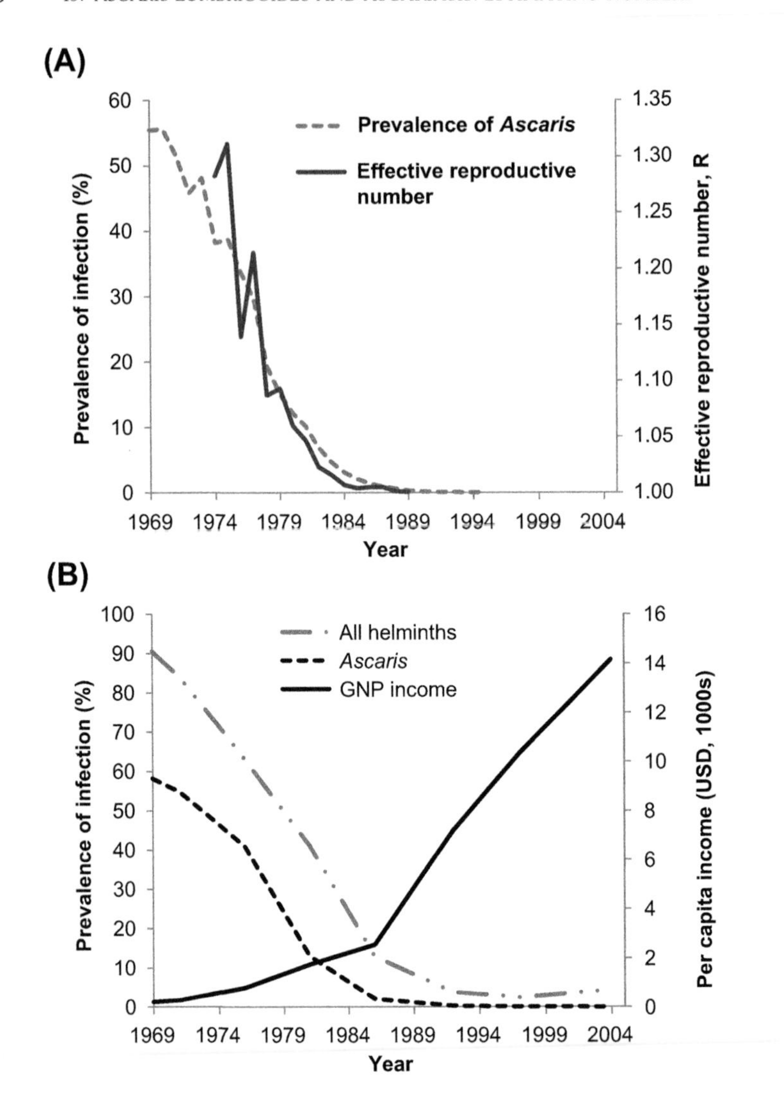

FIGURE 13.2 Changing patterns of (A) the prevalence and transmission (as estimated by the effective reproductive number, R) of *Ascaris lumbricoides* and (B) other intestinal nematodes and nation gross domestic product (GDP) in South Korea, 1969–2004. *Data adapted from ref.*[25]

Prevention Act. This law stipulated that all school children should be screened by fecal examination periodically and all infected children treated with anthelmintics.[25] Having instituted these control activities, the prevalence and transmission of *A. lumbricoides* began to decrease rapidly (Figure 13.2A). The implementation of large-scale control coincided with dramatic economic development that saw large-scale social infrastructure projects, including water and sewage systems, which undoubtedly facilitated reductions in the prevalence of infection (Figure 13.2B).

Today, *A. lumbricoides* has been almost completely eliminated from South Korea as well as Japan and Taiwan.[25–27] The prevalence of *A. lumbricoides* is also rapidly decreasing in China,[29] especially in urban areas where urbanization has been accompanied by improvements in water and sanitation. Globally, country level analysis suggests that the prevalence of *A. lumbricoides* is negligible in those countries with a gross domestic product >US$20,000 in 2010.[19]

Since 2000, a number of other countries have implemented large-scale campaigns using anthelmintic drugs, including albendazole or mebendazole, through either school-based de-worming programs or community-based lymphatic filariasis elimination, and such campaigns will, no doubt, contribute to lower levels of infection in the future (see Chapter 15).

GLOBAL NUMBERS AT RISK AND INFECTED

Possibly the first global estimates of *A. lumbricoides* were provided by Normal Stoll in his seminal *This Wormy World* study.[20] He suggested that 0.644 billion people out of a global 1947 population of the 2.167 billion were infected with *A. lumbricoides*, which was equivalent to 30% of the world's population. Some 40 years later, David Crompton[17] estimated a global prevalence of 22%, including some 1.008 billion people in 1988, an estimate derived independently from previous estimates and without extrapolation from Stoll's values. Subsequently, Don Bundy and colleagues estimated the total number of *A. lumbricoides* infection to be 1.382 billion in 1990, representing 34% of the world's population.[15] Inevitably, the number of prevalence studies included in these different estimates varies, hence the reliability of estimates at global and country levels also varies.

Based on the global limits of transmission (Figure 13.1A), Pullan and Brooker[19] estimated that 5.23 billion people (0.99 billion of school-going age) were living in areas of stable *A. lumbricoides* transmission worldwide in 2010, and hence at risk of infection. The vast majority (71%) of these individuals were living in Asia and Oceania, with only 18% living in Africa and the Middle East, and 11% in Latin America and the Caribbean (LAC) (Table 13.1 and Figure 13.1A).

Being at risk of infection is not the same as being infected, however. Using data from the *Global Atlas of Helminth Infection*,[30–32] Table 13.2 provides estimates of the global number individuals infected with *A. lumbricoides* in 2010, by region. These estimates are derived through

TABLE 13.1 Population at risk of *Ascaris lumbricoides* infection in 2010, by region

Region	Total population (millions)	Population at risk (stable transmission)		Population at risk (unstable transmission)	
		N (million)	(%)	*N* (million)	(%)
Asia	**3487.0**	**3477.1**	**(99.7%)**	–	–
Central Asia	76.8	76.2	(99.3%)	–	–
East Asia	1336.9	1331.5	(99.6%)	–	–
South Asia	1498.6	1494.7	(99.7%)	–	–
Southeast Asia	574.7	574.7	(100.0%)	–	–
Latin America (LA) and the Caribbean	**550.5**	**539.7**	**(98.0%)**	**0.4**	**(0.1%)**
Caribbean	38.0	38.0	(100.0%)	–	–
Andean LA	49.5	39.6	(79.9%)	0.4	(0.8%)
Central LA	215.2	215.2	(100.0%)	–	–
Southern LA	55.0	54.1	(98.3%)	–	–
Tropical LA	192.8	192.8	(100.0%)	–	–
Sub-Saharan Africa (SSA)	**763.1**	**632.2**	**(82.8%)**	**87.0**	**(11.4%)**
Central SSA	84.4	84.2	(99.7%)	0.2	(0.3%)
East SSA	313.4	270.3	(86.2%)	23.1	(7.4%)
Southern SSA	64.9	63.6	(97.8%)	1.3	(2.0%)
West SSA	300.4	214.2	(71.3%)	62.2	(20.7%)
North Africa and Middle East	**410.8**	**235.6**	**(57.3%)**	**96.7**	**(23.5%)**
Oceania	**9.7**	**9.7**	**(100.0%)**	–	–
GLOBAL	**5221.2**	**4894.3**	**(94.0%)**	**184.1**	**(3.5%)**

Adapted from ref. 19.

a number of steps. First, areas within endemic countries that are biologically unsuitable for the transmission of *A. lumbricoides* are excluded on this basis.[19] On this basis, 223 districts were excluded and assumed to have zero prevalence. Second, for countries outside Africa, empirical prevalence estimates at a district level were generated. For districts without suitable data, provincial or national estimates were applied.

TABLE 13.2 Estimates of global numbers infected with *Ascaris lumbricoides* infection in 2010, by region

Region	Total population (millions)	Infected population (95% BCI[a])		Prevalence (95% BCI)	
Asia	**3487.0**	**552.4**	**(326.2–842.7)**	**15.9%**	**(9.4–24.2%)**
Central Asia	76.8	5.8	(3.1–9.5)	7.5%	(4.0–12.4%)
East Asia	1336.9	144.2	(71.1–246.1)	10.8%	(5.3–18.5%)
South Asia	1498.6	284.2	(179.8–411.4)	19.0%	(12.0–27.5%)
Southeast Asia	574.7	118.2	(72.2–175.7)	20.6%	(12.6–30.6%)
Latin America (LA) and the Caribbean	**550.5**	**80.8**	**(50.0–121.9)**	**14.7%**	**(9.1–9.1%)**
Caribbean	38.0	3.1	(2.2–4.4	8.2%	(5.8–11.6%)
Andean LA	49.5	10.1	5.5–16.1)	20.3%	(11.1–32.5%)
Central LA	215.2	38.9	(25.9–55.3)	18.1%	(12.1–25.7%)
Southern LA	55.0	5.7	(4.1–7.9)	10.4%	(7.4–14.3%)
Tropical LA	192.8	23.0	(12.3–38.2)	11.9%	(6.4–19.8%)
Sub-Saharan Africa (SSA)	**763.1**	**104.9**	**(63.7–159.2)**	**13.8%**	**(8.4–20.9%)**
Central SSA	84.4	18.1	(10.7–27.9)	21.5%	(12.7–33.0%)
East SSA	313.4	30.2	(18.1–46.6)	9.6%	(5.8–14.9%)
Southern SSA	64.9	8.3	(4.6–13.3)	12.9%	(7.1–20.5%)
West SSA	300.4	48.2	(30.2–71.4)	16.1%	(10.1–23.8%)
North Africa and Middle East	**410.8**	**23.0**	**(15.7–32.8)**	**5.6%**	**(3.8–8.0%)**
Oceania	**9.7**	**1.8**	**(1.0–2.8)**	**18.2%**	**(10.5–28.4%)**
GLOBAL	**5221.2**	**762.9**	**(456.5–1159.3)**	**14.6%**	**(8.8–22.3%)**

[a]*BCI = Bayesian credible interval, based on geo-statistical models and logit-normal distributions.*

Heterogeneity in prevalence within districts was estimated empirically for 64 districts, representing a total of 969 prevalence surveys ranging from 10 to 99 surveys per district. These data showed that within-district distributions were highly skewed and best described by a logit-normal distribution, and standard deviation increased with increasing prevalence. On this basis, a normal distribution was applied around the logit-transformed mean prevalence and subsequently back-transformed to estimate the proportion of the total district population in each five-percentile prevalence class, and summation of these estimates provide an overall estimate of numbers infected.

Third, for countries within Africa where detailed data were lacking, Bayesian geo-statistical modeling was used to predict the prevalence of infection, using available data and environmental information.[33–34] This approach is predicated on the role of environmental factors in influencing the large-scale geographic distributions of *A. lumbricoides*, in the absence of substantive control measures.[33,35–37] Fourth, prevalence estimates were adjusted for those countries* which have recently implemented large-scale treatment campaigns, through either school-based de-worming programs or community-based lymphatic filariasis elimination programs. Information about the coverage of these campaigns was assembled from relevant sources and adjustments were made that reflected treatment coverage levels, using a mathematical model of transmission dynamics.[38]

Based on these analyses, it is estimated that 0.762 billion people are infected with *A. lumbricoides* in 2010 (Table 13.2), representing 14.6% of the world's population. This estimate is markedly lower than previous estimates provided by Crompton[17] and Bundy et al.,[15] and reflect the dramatic reductions in prevalence in China and parts of Latin America and the Caribbean. Prevalence remains around 20% for countries in south and Southeast Asia, central and western Africa, and Andean Latin America (Figure 13.1B). In numerical terms, however, the greatest numbers of *A. lumbricoides* infection occur in China and India (Figure 13.1C).

GLOBAL DISEASE BURDEN

As noted above, quantifying the disease burden caused by ascariasis is difficult and there are therefore no direct estimates of morbidity. Instead,

*Bangladesh, Belize, Burkina Faso, Burundi, Cambodia, Cameroon, Cape Verde, Côte d'Ivoire, Democratic People's Republic of Korea, Dominican Republic, Ecuador, El Salvador, Guatemala, Guinea Bissau, Haiti, Honduras, Kenya, Lao People's Democratic Republic, Madagascar, Malawi, Mali, Myanmar, Nepal, Nicaragua, Niger, Peru, Philippines, Sierra Leone, Uganda, Venezuela, Vietnam.

estimates of burden have been extrapolated from data on the prevalence of infection, where is it assumed that only a fraction of infections, those which are intense, are associated with ascariasis.[16,39]

An essential prerequisite to estimating the global burden of ascariasis is a framework for assessing its burden relative to all other health conditions. The first comprehensive approach to estimating the global health burden, including ascariasis, was provided by the Global Burden of Disease (GBD) study.[40–41] For a summary measure of population health this study uses the Disability-Adjusted Life Years (DALYs), which incorporates both years of life lost from premature death (YLLs) and years of life lived with disability (YLDs) into a composite estimate. YLLs are computed by multiplying the number of deaths at each age x by a standard life expectancy at age x. YLDs are computed as the prevalence of different disease and injury sequelae multiplied by a disability weight for each sequela. Disability weights are a quantification of the severity of health loss associated with each sequel and range from 0 to 1, where 0 is commensurate with perfect health and 1 is commensurate with death.

The original GBD study – referred to here as the 1990 study – produced results for 1990 which were initially published in the World Development Report 1993: Investing in Health[42] and separately in 1994 and 1997.[40,43–44] Updated GBD estimates were subsequently produced for each year from 1999 to 2004 by the World Health Organization, but these estimates were based on updated epidemiology for only a subset of conditions. A major update to the GBD was initiated in 2007,[45] led by a core team of researchers who also developed the methodology, while expert groups conducted systematic reviews of incidence and prevalence of disease and disabling sequelae, reporting their figures to core team members. This initiative – referred to here as the 2010 GBD study – provided estimates for 2010, 2005, and 1990, with key findings published in 2012.[46,47]

Estimates for ascariasis in the 1990 study were based on analytical work by Chan et al.,[15–16] whereas the 2010 study estimates are based on a central modeling approach, utilizing empirical estimates of infection prevalence provided by an intestinal nematode expert group (led by the current authors). There are a number of important methodological differences between the 1990 study and the 2010 study, both in terms of overall study design and in how ascariasis estimates were calculated, including the following issues.

Project scope. The 1990 study provided estimates for 107 diseases and injuries and for these conditions 483 sequelae were identified. The 2010 study identified 291 diseases and injuries which have 1160 sequelae and provided estimates for 2010, 2005, and 1990. The 2010 study also provides estimates for 20 age groups instead of eight; and 21 regions instead of the 14 used in the 1990 study.

Age weighting and discounting. In the 1990 study, results were computed using DALYs that included discounting and age-weighting. This approach valued years lived as a young adult more highly than years spent as a child or older adult, a reflection of the societal emphasis on productivity. It also discounted future years at 3%, where each year in the future is valued 3% more than the current year. Critics of these assumptions argued that years lived by people of different ages and generations should be valued equitably[48–50] and consequently, based on a broad consensus, the 2010 study omitted both discounting and age-weighting.

Disease sequelae and disability weights. The 1990 study defined six broad disability classes and then mapped each sequela into the appropriate class, as judged by panels of healthcare professionals. Whereas the 2010 study derived disability weights from judgments of the general public about the health loss associated with the health state related to each disability sequela.

The disabling sequelae and disability weights assigned to ascariasis in the two GBD studies are presented in Table 13.3. An important change in the 2010 study from 1990 study was the removal of cognitive impairment as a sequela. This was justified by the perceived paucity of evidence on the cognitive impact of intestinal nematodes. Many of the trials examining the impact of *A. lumbricoides* and de-worming have small sample sizes or are only quasi-randomized. In their 2012 Cochrane review of the available randomized controlled trials of de-worming, Taylor-Robinson et al.[7] found contrasting evidence of nutritional benefits, and little support for cognitive or educational benefits. There was, however, a difference in studies that screened for intestinal nematode infections and mass de-worming studies, with significant nutritional gains in infected individuals, but little impact in unscreened population studies where any effect may be diluted. It should be noted, however, that the Cochrane review excluded studies that treated both intestinal nematodes and schistosomiasis. Furthermore, critics of the Cochrane review[13] have argued that many of the underlying trials of de-worming suffer from a number of methodological challenges, including non-assessment of treatment externalities, inadequate measurement of cognitive outcomes and school attendance, and sample attrition, and that further, better-designed studies are required. It is probable that, because of the insidious nature of intestinal nematode infections and the complexity of factors influencing cognition and education, it may be difficult to reach a definitive conclusion on the impact of ascariasis on cognition.

Estimating prevalence of infection. In both GBD studies, DALY estimates were based on extrapolating the population at risk of morbidity (intensely infected individuals) from empirical observations of the proportion of the population infected.

Chan et al.[15–16] based their prevalence estimates on an extensive search of the published literature for community-based studies (i.e. not hospital

TABLE 13.3 Disabling sequelae and disability weights for ascariasis included in the 1990 GBD study and 2010 study, stratified by intensity of infection

Study/intensity of infection	Sequelae and case definition	Disability weight
1990 STUDY		
Lower worm burden threshold	Contemporaneous cognitive deficit: reduction in cognitive ability in children aged 5–14 years, which occurs only while infection persists	0.006
Lower worm burden threshold	Cognitive impairment: delayed psychomotor development and impaired performance on language, motor skills, equivalent to 5–10 point deficit in IQ	0.024
Higher worm burden threshold	Intestinal obstruction: blockage of the intestines due to worm mass	0.463
2010 STUDY		
Light intensity infection	None	
Medium intensity infection	Mild abdominopelvic problems, including obstruction of the terminal ileum by a bolus of worms, which is common in children under 10 years	0.0108128
High intensity infection	Mild infectious disease, acute episode	0.0296199
	Wasting, underweight and marasmus	0.1244958

records or institutional data). In order to represent the geographical variation in prevalence within countries, they assumed that infection prevalences between communities in the same country were normally distributed and that the standard deviations of these normal distributions increase linearly with the mean prevalence. However, analysis of data available in the *Global Atlas of Helminth Infection* suggests that while standard deviations increase with prevalence, within-country distributions are more skewed than assumed and a logit-normal distribution, in fact, provides a better fit of the data (Pullan and Brooker, unpublished). Finally, prevalence estimates were calculated for four age groups, weighted according to the well-established age pattern of *A. lumbricoides* infection[3–4]: an age weight of 0.75 for 0–4 years, 1.2 for 5–14 years, and 1 for 15 and above years.

For the 2010 GBD study, prevalence estimates were provided by the intestinal nematodes expert group (Pullan and Brooker, unpublished) – see above and Table 13.2. Estimates were provided separately for the years 1990, 2005, and 2010, with age-adjusted estimates calculated using the same age weights as used by Chan et al.

Estimating disability. For many diseases and injuries, age–country–year estimates of incidence exist and can be used to estimate disability in terms of YLDs. This is not the case for intestinal nematodes and therefore estimates of disability need to be extrapolated from data on the prevalence of infection. To do this, Chan et al.[15–16] quantified relationships between prevalence of infection and worm burden and between worm burden and potential disability, based on the negative binomial distribution.[51] They further assumed that there is some age-specific worm burden threshold above which morbidity is likely to occur. Developmental and cognitive effects of infection in childhood were assumed to occur at lower worm burdens, and therefore lower worm burden thresholds were assumed to correspond to disability arising from cognitive and growth deficits.

Whereas Chan et al.[15–16] used worm burden to define the thresholds for morbidity, the core team in the 2010 study related prevalence of infection to the intensity of infection, as expressed by quantitative egg counts. This was because most literature on the health impacts of intestinal nematodes expresses results in these terms. Estimates of infection prevalence provided by the expert group were related to the intensity of infection using a negative binomial distribution and stratified by intensity: light = 1–1999 eggs per gram (epg); medium = 2000–3999 epg; and heavy = >4000 epg. Subsequently, the core modeling team assumed that mild abdominopelvic problems were associated with medium intensities and mild infectious disease with heavy intensities (Table 13.3).

To estimate the prevalence of wasting attributable to heavy *A. lumbricoides* infection, the core team adopted a two-stage approach. First, the prevalence of wasting among children under five years old was independently estimated using available data. Second, this wasting was apportioned to two underlying causes: (1) heavy intestinal nematode infection and (2) Protein Energy Malnutrition (PEM). The prevalence of wasting due to the former was calculated by shifting the 2006 WHO reference population weight-for-height distribution according to the product of (1) the prevalence of heavy intestinal nematode infections and (2) the average shift in weight-for-height per case of heavy intestinal nematode infection, based on a meta-analysis of randomized controlled trials of mass de-worming.[6] The pooled effect across identified studies was a change in weight-for-height *z*-score per individual with heavy intestinal nematode infection of 0.4938. Finally, the overall prevalence of wasting attributable to heavy intestinal nematode infections due to ascariasis versus trichuriasis versus hookworm was calculated by the relative prevalence of heavy intestinal infection due to each individual species. The prevalence of wasting due to PEM was calculated by simply subtracting the prevalence of wasting due to heavy intestinal nematode infections from the estimated overall prevalence of wasting.

Estimating mortality. For ascarisiasis, deaths are mainly due to intestinal obstruction, and to a lesser extent biliary or pancreatic disease.[5] Such acute complications are seen largely in children less than 10 years of age.[5] Each of the GBD studies employed a statistical model to estimate deaths by broad cause group, with a number of methodological refinements included in the 2010 study. The 1990 GBD study attributed 3000 global deaths due to ascariasis, whereas in the 2010 study 2824 deaths were attributed to ascariasis, with most deaths in populations from East Asia and South Asia (Table 13.4).

TABLE 13.4 Estimates of global deaths, cases of childhood wasting and years lived with disability (YLDs) due to ascariasis in 2010, by region

Region	Deaths	Wasting	YLDs
Asia	**1829**	**4,042,671**	**801,369**
Central Asia	21	112,823	5845
East Asia	389	521,579	80,506
South Asia	1028	2,832,863	503,191
Southeast Asia	391	575,406	211,826
Latin America (LA) and the Caribbean	**294**	**342,554**	**82,619**
Caribbean	11	16,833	1819
Andean LA	40	33,293	12,654
Central LA	150	184,563	43,489
Southern LA	18	10,625	2635
Tropical LA	75	97,240	22,022
Sub-Saharan Africa (SSA)	**582**	**101,1854**	**169,864**
Central SSA	118	158,960	27,710
East SSA	165	226,371	38,541
Southern SSA	39	11,630	4035
West SSA	260	614,893	99,579
North Africa and Middle East	**112**	**619,766**	**54,858**
Oceania	**8**	**4020**	**1889**
GLOBAL	**2825**	**6,020,865**	**1,110,600**

Taken from Vos et al.[47]

Both sets of GDB estimates are lower than those provided by de Silva et al. who independently estimated the incidence of *Ascaris*-induced intestinal obstruction to be in the range of 0–0.25 cases per year per 1000 in endemic areas and the case fatality rate to be up to 5%, with 10,500 deaths each year directly attributed to serious complications of ascariasis.[5,52] These estimates were based on a systematic review of the available literature, whereas the data sources by which the GBD estimates are derived are unclear.

Inclusion of uncertainty. No explicit estimate of uncertainty was included in the 1990 study. Whereas the 2010 study used statistical models that generated estimates of uncertainty when estimating YLLs and YLDs. Uncertainty was also included in the definition of disability weights and propagated through the modeling process.

BURDEN OF ASCARIASIS IN 2010

Table 13.5 presents the global estimates of DALYs due to intestinal nematodes, including ascariasis from the 1990 and 2010 studies. The differing methodology between the 1990 and 2010 studies makes it impossible to directly compare estimates between these studies. However, the 2010 study provides estimates for both 1990 and 2010, and these findings are assumed to supersede all previously published GBD results. In 2010, there were a total of 2.490 billion DALYs, with 54% of all DALYs due to non-communicable causes, compared to 35% due to communicable, maternal, nutritional, and neonatal causes, and 11% due to injuries. Globally, intestinal nematodes contributed

TABLE 13.5 Global DALYs (billions) due to ascariasis as estimated by the 1990 study and the 2010 study (both 1990 and 2010)

		2010 study	
Cause	**1990 study**	**1990 estimates**	**2010 estimates**
Intestinal nematodes	3.793	9.0084 (4.9932–15.3911)[a]	5.184 (2.979–8.811)
Ascariasis	1.817	4.2173 (2.2906–7.1483)	1.3148 (0.7126–2.3494)
Trichuriasis	1.006	0.8571 (0.465–1.4204)	0.6382 (0.3493–1.0614)
Hookworm	0.97	3.934 (2.0557–6.9832)	3.2311 (1.6951–5.732)

[a]*95% uncertainty interval.*
Taken from Bundy et al.[15] and Murray et al.[46]

5.184 million DALYs in 2010, with ascariasis contributing 1.3148 million, trichuriasis 0.6382 million, and hookworm 3.2311 million. The estimate for ascariasis is much lower than that for 1990, where ascariasis contributed 4.2173 billion DALYs – a decline that mirrors the precipitous reduction in the prevalence of *A. lumbricoides* in China and part of Asia.

LIMITATIONS OF THE DALY APPROACH AND A BROADER VIEW OF IMPACT

The use of DALYs as a measure of disease burden has its advantages and disadvantages. The main advantage is that DALYs provide a composite, internally consistent measure of population health which can be used to evaluate the relative burden of different diseases and injuries and compare population health by geographic region and over time. Combined with information on the effectiveness and cost of different interventions, such estimates can guide priority setting.[53] The main disadvantage of DALYs is that they focus solely on health and do not capture the broader societal impact of diseases. This is especially true for ascariasis and other intestinal nematodes which have subtle, lasting impacts on child development and education. For example, recent studies in Africa, as well as reanalysis of the extensive Rockefeller Foundation control programs in the southern United States at the beginning of the 20th century, have shown remarkable long-run effects on productivity and employment and wages of treating children at school age.[54–56] There is also an important equity issue that goes beyond health: ascariasis and other neglected tropical diseases (NTDs) affect the poorest communities and are, as Caroline Anstey of the World Bank has said, diseases of neglected populations. Thus, tackling ascariasis and other NTDs should be seen as part of broader efforts to reduce global poverty. This argument provides perhaps the most compelling case for de-worming, rather than some scientific quantification of health impacts.

Acknowledgments

SJB is supported by a Wellcome Trust Senior Fellowship in Basic Biomedical Science (098045) and RLP is supported by a grant from the Bill & Melinda Gates Foundation.

References

1. de Silva NR, Guyatt HL, Bundy DA. Worm burden in intestinal obstruction caused by *Ascaris lumbricoides*. *Trop Med Int Health* 1997;**2**:189–90.

2. Thein-Hlaing, Thane-Toe, Than-Saw, Myat Lay-Kyin, Myint-Lwin. A controlled chemotherapeutic intervention trial on the relationship between *Ascaris lumbricoides* infection and malnutrition in children. *Trans R Soc Trop Med Hyg* 1991;**85**:523–8.
3. Croll NA, Anderson RM, Gyorkos TW, Ghadirian E. The population biology and control of *Ascaris lumbricoides* in a rural community in Iran. *Trans R Soc Trop Med Hyg* 1982;**1982**(76):187–97.
4. Thein-Hlaing, Than-Saw, Htay-Htay-Aye, Myint-Lwin. Thein-Maung-Myint. Epidemiology and transmission dynamics of *Ascaris lumbricoides* in Okpo village, rural Burma. *Trans R Soc Trop Med Hyg* 1984;**78**:497–504.
5. de Silva NR, Guyatt HL, Bundy DA. Morbidity and mortality due to *Ascaris*-induced intestinal obstruction. *Trans R Soc Trop Med Hyg* 1997;**91**:31–6.
6. Hall A, Hewitt G, Tuffrey V, de Silva N. A review and meta-analysis of the impact of intestinal worms on child growth and nutrition. *Matern Child Nutr* 2008;**4**(Suppl. 1): 118–236.
7. Taylor-Robinson DC, Maayan N, Soares-Weiser K, Donegan S, Garner P. Deworming drugs for soil-transmitted intestinal worms in children: effects on nutritional indicators, haemoglobin and school performance. *Cochrane Database Syst Rev* 2012;**7**:CD000371.
8. Hadidjaja P, Bonang E, Suyardi MA, Abidin SA, Ismid IS, Margono SS. The effect of intervention methods on nutritional status and cognitive function of primary school children infected with *Ascaris lumbricoides*. *Am J Trop Med Hyg* 1998;**59**:791–5.
9. Watkins WE, Cruz JR, Pollitt E. The effects of deworming on indicators of school performance in Guatemala. *Trans R Soc Trop Med Hyg* 1996;**90**:156–61.
10. Watkins WE, Pollitt E. Effect of removing *Ascaris* on the growth of Guatemalan schoolchildren. *Pediatrics* 1996;**97**:871–6.
11. Glinz D, Silue KD, Knopp S, Lohourignon LK, Yao KP, Steinmann P, et al. Comparing diagnostic accuracy of Kato-Katz, Koga agar plate, ether-concentration, and FLOTAC for *Schistosoma mansoni* and soil-transmitted helminths. *PLoS Negl Trop Dis* 2010;**4**:e754.
12. Guyatt HL, Bundy DA. Estimation of intestinal nematode prevalence: influence of parasite mating patterns. *Parasitology* 1993;**107**(Pt 1):99–105.
13. Bundy DA, Kremer M, Bleakley H, Jukes MC, Miguel E. Deworming and development: asking the right questions, asking the questions right. *PLoS Negl Trop Dis* 2009;**3**:e362.
14. Brooker S, Clements AC, Bundy DA. Global epidemiology, ecology and control of soil-transmitted helminth infections. *Adv Parasitol* 2006;**62**:221–61.
15. Bundy DAP, Chan MS, Medley GF, Jamison D, Savioli L. Intestinal Nematode Infections. In: Murray CJL, Lopez AD, Mathers CD, editors. *Global Epidemiology of Infectious Disease*. Geneva: World Health Organization; 2004. p. 243–300. 2004.
16. Chan MS, Medley GF, Jamison D, Bundy DA. The evaluation of potential global morbidity attributable to intestinal nematode infections. *Parasitology* 1994;**109**(Pt 3): 373–87.
17. Crompton DW. The prevalence of Ascariasis. *Parasitol Today* 1988;**4**:162–9.
18. Crompton DW, Tulley JJ. How much Ascariasis is there in Africa? *Parasitol Today* 1987;**3**:123–7.
19. Pullan RL, Brooker SJ. The global limits and population at risk of soil-transmitted helminth infections in 2010. *Parasit Vectors* 2012;**5**:81.
20. Stoll NR. This Wormy World. *J Parasitol* 1947;**33**:1–18.
21. Warren KS, Bundy DAP, Anderson RM, Davis AR, Henderson DA, Jamison DT. Helminth infection. In: Jamison DT, Mosley WH, Measham AR, Bobadilla JL, editors. *Disease Control Priorities in Developing Countries*. New York: Oxford University Press; 1993. p. 131–60. 1993.
22. Seamster AP. Developmental studies concerning the eggs of *Ascaris lumbricoides* var. *suum*. *The American Midland Naturalist* 1950;**43**:450–68.

23. Otto GF. A study of the moisture requirements of the eggs of the horse, the dog, human and pig ascarids. *Am J Hyg* 1929;**10**:497–520.
24. Crompton DW. Ascaris and ascariasis. *Adv Parasitol* 2001;**48**:285–375.
25. Hong ST, Chai JY, Choi MH, Huh S, Rim HJ, Lee SH. A successful experience of soil-transmitted helminth control in the Republic of Korea. *Korean J Parasitol* 2006;**44**: 177–85.
26. Kobayashi A, Hara T, Kajima J. Historical aspects for the control of soil-transmitted helminthiases. *Parasitol Int* 2006;**55**(Suppl.):S289–91.
27. Komiya Y, Kunii C. The epidemiology of *Ascaris* infection in relation to its control program in Japan. *Jpn J Med Sci Biol* 1964;**17**:23–31.
28. Starr MC, Montgomery SP. Soil-transmitted helminthiasis in the United States: a systematic review – 1940–2010. *Am J Trop Med Hyg* 2011;**85**:680–4.
29. Chen YD, Tang LH, Xu LQ. Current status of soil-transmitted nematode infection in China. *Biomed Environ Sci* 2008;**21**:173–9.
30. Brooker S, Hotez PJ, Bundy DA. The global atlas of helminth infection: mapping the way forward in neglected tropical disease control. *PLoS Negl Trop Dis* 2010;**4**:e779.
31. Brooker S, Kabatereine NB, Smith JL, Mupfasoni D, Mwanje MT, Ndayishimiye O, et al. An updated atlas of human helminth infections: the example of East Africa. *Int J Health Geogr* 2009;**8**:42.
32. Brooker S, Rowlands M, Haller L, Savioli L, Bundy DA. Towards an atlas of human helminth infection in sub-Saharan Africa: the use of geographical information systems (GIS). *Parasitol Today* 2000;**16**:303–7.
33. Pullan RL, Gething PW, Smith JL, Mwandawiro CS, Sturrock HJ, Gitonga CW, et al. Spatial modelling of soil-transmitted helminth infections in Kenya: a disease control planning tool. *PLoS Negl Trop Dis* 2011;**5**:e958.
34. Global Atlas of Helminth Infection. *Technical Information – Development of Predictive Risk Models*. London: London School of Hygiene and Tropical Medicine; 2010.
35. Brooker S, Kabatereine NB, Tukahebwa EM, Kazibwe F. Spatial analysis of the distribution of intestinal nematode infections in Uganda. *Epidemiol Infect* 2004;**132**: 1065–71.
36. Ratard RC, Kouemeni LE, Ekani Bessala MM, Ndamkou CN, Sama MT, Cline BL. Ascariasis and trichuriasis in Cameroon. *Trans R Soc Trop Med Hyg* 1991;**85**:84–8.
37. Saathoff E, Olsen A, Kvalsvig JD, Appleton CC, Sharp B, Kleinschmidt I. Ecological covariates of Ascaris lumbricoides infection in schoolchildren from rural KwaZulu-Natal, South Africa. *Trop Med Int Health* 2005;**10**:412–22.
38. Chan MS, Guyatt HL, Bundy DA, Medley GF. The development and validation of an age-structured model for the evaluation of disease control strategies for intestinal helminths. *Parasitology* 1994;**109**(Pt 3):389–96.
39. Brooker S. Estimating the global distribution and disease burden of intestinal nematode infections: adding up the numbers – a review. *Int J Parasitol* 2010;**40**:137–44.
40. Murray CJ, Lopez AD, Jamison DT. The global burden of disease in 1990: summary results, sensitivity analysis and future directions. *Bull World Health Organ* 1994;**72**:495–509.
41. Murray CJL, Lopez AD. *The Global Burden of Disease*, vol. 1. Cambridge: Harvard University Press; 1996.
42. World Bank. *World development report 1993: investing in health*. Washington, DC: The World Bank; 1993.
43. Murray CJ, Lopez AD. Global mortality, disability, and the contribution of risk factors: Global Burden of Disease Study. *Lancet* 1997;**349**:1436–42.
44. Murray CJ, Lopez AD. Regional patterns of disability-free life expectancy and disability-adjusted life expectancy: global Burden of Disease Study. *Lancet* 1997;**349**:1347–52.
45. Murray CJ, Lopez AD, Black R, Mathers CD, Shibuya K, Ezzati M, et al. Global burden of disease 2005: call for collaborators. *Lancet* 2007;**370**:109–10.

46. Murray CJ, Vos T, Lozano R, et al. Disability-adjusted life years (DALYs) for 291 diseases and injuries in 21 regions, 1990-2010: a systematic analysis for the Global Burden of Disease Study. *Lancet* 2012;**380**(9859):2197–223.
47. Vos T, Flaxman AD, et al. The global burden of non-fatal health outcomes for 1,160 sequelae of 291 diseases and injuries 1990-2010: a systematic analysis. *Lancet* 2012;**380**:2163–96.
48. Anand S, Hanson K. Disability-adjusted life years: a critical review. *J Health Econ* 1997;**16**:685–702.
49. Arnesen T, Kapiriri L. Can the value choices in DALYs influence global priority-setting? *Health Policy* 2004;**70**:137–49.
50. Murray CJ, Acharya AK. Understanding DALYs (disability-adjusted life years). *J Health Econ* 1997;**16**:703–30.
51. Guyatt HL, Bundy DA, Medley GF, Grenfell BT. The relationship between the frequency distribution of *Ascaris lumbricoides* and the prevalence and intensity of infection in human communities. *Parasitology* 1990;**101**(Pt 1):139–43.
52. de Silva NR, Chan MS, Bundy DA. Morbidity and mortality due to ascariasis: re-estimation and sensitivity analysis of global numbers at risk. *Trop Med Int Health* 1997;**2**:519–28.
53. Laxminarayan R, Chow J, Shahid-Salles SA. Intervention cost-effectiveness: overview of main messages. In: Jamison DT, Breman JG, Measham AR, Alleyne C, Claeson M, Evans D, et al., editors. *Disease Control Priorities in Developing Countries*. 2nd ed. New York: Oxford University Press; 2006. p. 35–86. 2011/01/21.
54. Baird S, Hicks JH, Miguel E, Kremer M. *Worms at Work: Long-run Impacts of Child Health Gains*. Cambridge, MA: National Bureau of Economic Research; 2011.
55. Bleakley H. Disease and development: evidence from hookworm eradication in the American south. *Q J Econ* 2007;**122**:73–117.
56. Guyatt H. Do intestinal nematodes affect productivity in adulthood? *Parasitol Today* 2000;**16**:153–8.
57. Stephenson LS, Crompton DW, Latham MC, Schulpen TW, Nesheim MC, Jansen AA. Relationships between *Ascaris* infection and growth of malnourished preschool children in Kenya. *Am J Clin Nutr* 1980;**33**:1165–72.
58. Koumanidou C, Manoli E, Anagnostara A, Polyviou P, Vakaki M. Sonographic features of intestinal and biliary ascariasis in childhood: case report and review of the literature. *Ann Trop Paediatr* 2004;**24**:329–35.
59. Shah OJ, Zargar SA, Robbani I. Biliary ascariasis: a review. *World J Surg* 2006;**30**:1500–6.

CHAPTER

14

Impact of *Ascaris suum* in Livestock

Stig Milan Thamsborg, Peter Nejsum, Helena Mejer
University of Copenhagen, Denmark

OUTLINE

INTRODUCTION

The distribution of *Ascaris suum* in pigs is cosmopolitan and remarkably extensive but exact estimates do not exist. Although the prevalence and intensity vary with geographical region and production system from intensive herds to backyard pigs, a large proportion of swine herds are infected worldwide. Among poor pig farmers in developing countries

Ascaris: The Neglected Parasite
http://dx.doi.org/10.1016/B978-0-12-396978-1.00014-8

Copyright © 2013 Elsevier Inc. All rights reserved.

helminthosis in general ranks within the 10 most important pathogens worldwide.[1] And even though industralized in-door production with high levels of hygiene may have lower prevalence than more traditional systems,[2] recently implemented changes in management to accommodate for better welfare, i.e. group housing of dry sows (=non-lactating, most often pregnant sows) and sprinklers as a surrogate for wallowing, may increase the prevalence within these herds.[3]

This chapter will try to describe the impact of *A. suum* by revisiting some of the older controlled experiments and review some more recent studies under farming conditions, and examine if there is a quantitative relationship between worm load and pathogenicity. Nodular worms (*Oesophagostomum* spp.) and numerous other worms are commonly encountered co-infections to *A. suum* in pigs, particularly in natural settings and outdoor systems.[2] Understandably, effects of *A. suum* are difficult or impossible to single out from combined effects of these multiple parasite infections but data will nevertheless be used when relevant.

This chapter will examine the direct effects on health (clinical and subclinical effects) and performance (subclinical effects). It is evident that the underlying pathophysiological changes and thus impact will depend on level of exposure (dose level in experimental infections), stage of infection, build-up or maintenance of immunity, nutritional and physiological status, and possible other factors. Indirect effects on susceptibility to bacterial and viral infections or enhanced pathogenicity, related to, for example, the lung migratory phase or immunomodulatory capacity of *A. suum*,[4] will be summarized although the mechanisms behind the latter phenomenon are thoroughly addressed in Chapter 4. Focus will be on the natural host, the pig, but the chapter will also examine implications of *A. suum* in other livestock.

ASCARIS SUUM AND OTHER HELMINTHS IN PIG PRODUCTION SYSTEMS

The most important intestinal helminth fauna of pigs in Europe and other temperate regions includes nematodes like *Strongyloides ransomi* (mainly in suckling piglets = piglets in the suckling phase (4−10 weeks of age depending on production type)),[2] *A. suum* (fatteners = growing pigs after weaning up to time of slaughter (large fattener = finisher)),[2] *Trichuris suis* (fatteners),[2] *Oesophagostomum* spp. (adult pigs: sows and boars),[2] *Hyostrongylus rubidus* (sows and boars),[2] and *Globocephalus urosubulatus* (rarely found)[5,6] while spiruroid nematodes (e.g. *Ascarops strongylina, Physocephalus sexalatus*)[5,7] are also commonly found in warmer parts of the world. *Oesophagostomum dentatum* is overall the predominant *Oesophagostomum* species and is most often found as a single species infection on farms while mixed infections with *O. quadrispinulatum* are seen occasionally.[8]

A. suum is worldwide the most or second most (surpassed by *Oesophagostomum* spp. in older stock) prevalent intestinal species in domestic pigs, although the prevalence varies considerably with climate, production system (indoor/outdoor/scavenging), farm practices, herd, age, and breed, as evident from Chapter 16. High prevalences ranging from 17 to 35% have been reported from the Scandinavian countries by fecal analysis in fatteners and gilts (=pregnant pig before first farrowing[9]) and more recently 22% in Danish sows/gilts.[10] Adult worms were detected in 17% of finishers at slaughter in Canada.[11] In Japan, 15% slaughter pigs ($n = 129$) were found positive by coproscopy.[12] African studies, also based on fecal egg counts, revealed prevalences of 2% in growers on smallholder farms in Nigeria,[13] 13% in growers in Ghana,[14] and 40% with only small variation across age groups in Burkina Faso[15] and Uganda.[16] More details on prevalence in age groups and herds can be found in Chapter 16. Some of these coprological studies undoubtedly overestimate the prevalence[14] as low or moderate egg counts may be due to coprophagia and therefore represent false positive findings, especially if stocking rate is high.[17,18] In contrast, infections with only one worm or all worms of the same sex will result in underestimation of prevalence as compared to post-mortem-based studies. A higher prevalence in females than in males (33% vs. 14%) was found in Burkina Faso[15] which confirmed an earlier observation from Sweden that female finishers generally have higher egg excretion than castrates.[19] A range of other studies have not reported such differences. One has to bear in mind that in most modern intensive systems male piglets are castrated early in life whereas castration may take place later or not at all in other production systems. This may influence potential differences between sexes.

In the early 1990s 60% of Danish farms were infected with *A. suum*[20] but farming practices have now changed to loose housing of sows in dynamic groups, where sows in different stages of gestation move in and out on a weekly basis, and recently 76% of the surveyed Danish farms were found by coproscopy to be infected.[10] On 84 smaller organic farms with traditional husbandry in Austria, *A. suum* was present (fecal samples) in 30% of sow units and 59% of finishing units.[21] Corresponding figures for *Oesophagostomum* were 66 and 43%. More than 50% of pigs at slaughter had "milk spots" on the liver. In Germany, 7% of 144 breeding farms (sows only) had *A. suum* infections.[22] In a small Dutch survey including 36 farms, the prevalence of *A. suum*-positive farms (across age groups) was 50% for free-range, 73% for organic, and 11% for conventional farms (in-door with straw bedding).[23] The highest prevalences were seen among finishing pigs with about half of the examined groups being infected on free-range and organic farms, while none were found to be positive on conventional farms. A marked difference between production systems was also confirmed with regard to prevalence of milk spots: 8% vs. 1% in pigs slaughtered from Danish organic and conventional farms, respectively, based on standard recordings of condemnations from the abattoir.[24] In Denmark, all pigs from the organic farms would have stayed outside until at least weaning. A recent English study confirmed a higher prevalence of milk spot livers in outdoor as compared to indoor reared pigs.[25] Although it is tempting, based on these findings, to conclude that endoparasites are among the main disease problems in organic/free-range pig production, it may not be the case. Most studies measure prevalence or occurrence of parasites (see Chapter 9) while neglecting intensity of infections. Intensity of infections in a herd is probably a major determinant of impact in affected pigs, as will be clear from later paragraphs.

IMPACT ON CLINICAL HEALTH IN PIGS

The majority of *A. suum* infections in pigs are subclinical. The penetration of cecum and proximal colon by newly hatched third-stage larvae (L3) results in petechial bleedings of the mucosa only, whereas the following liver migration is accompanied by severe pathology, thoroughly described by Roepstorff.[26] We have found no clinical reports of this early intestinal and liver migratory phase (days 0–6 post infection (p.i.)). During the lung penetration phase 6–8 days p.i.[27] pigs are subject to respiratory stress (pneumonitis) reflected in increased breathing rate, dyspnea and dry coughing.[28–30] The condition resembles Loeffler's syndrome in humans infected with *A. lumbricoides*[26] and is associated with an increased number of blood eosinophils. Considering the large number of experimental infections performed in our and other laboratories, it is remarkable that barely any of these have reported such symptoms, which may indicate low incidence or only vague symptoms.[30] From a herd health viewpoint, it is difficult to define this pathological stage in an endemic situation, and presumably in most infected herds it will remain unconfirmed, although cases have been reported.[31] Earlier studies have also commented on this, e.g. Spindler[32] stating "the clinical manifestations of an invasion of the lungs by large numbers of ascarid larvae are sometimes spectacular and easily recognized" (see also Chapter 5).

The third-stage larvae (L3) molt to L4 when the nematodes return to the small intestine around day 10 p.i.,[26] and a further molt takes place 2 weeks later to reach L5, the final stage that will ultimately become sexually mature by days 42–49 p.i. About 40–50% of a single infection dose can be isolated from the small intestine at day 12 p.i.[26] but the majority will rapidly be expelled, and by day 25 p.i. only a small number of worms will remain and mature into egg-producing (patent) adults. These may stay in the intestine for 1 year but some may be expelled earlier.[33] If this happens, pigs may become reinfected.[34] The intestinal phase is associated with pathological changes in the mucosa.[35]

While the influence of *A. lumbricoides* on cognitive abilities has been recognized for years,[36] there is no experimental data to support this finding for *A. suum* in pigs.

IMPACT ON GROWTH AND PRODUCTIVITY IN PIGS – EXPERIMENTAL STUDIES

Though based on very few observations, early studies describing severe clinical disease in *A. suum*-infected pigs also reported dramatic reductions in weight gain (e.g. Spindler[32]). A later study showed beneficial effects (a 9% increase in weight gain and 10% better feed utilization) of pyrantel treatment in growing pigs naturally exposed to *A. suum* on contaminated soil.[37] The pigs were, however, exposed to mycoplasms at the same time. The first detailed nutritional balance studies with *A. suum* in experimentally infected pigs were performed in the 1980s. Young pigs, fed low or high protein diets, and infected with 200 2-week-old *A. suum*-larvae, showed reduced nitrogen absorption and retention, and reduced weight gains as compared to uninfected controls,[35] but none of these often cited differences were statistically significant, despite a remarkably high establishment of adult worms. However, the study clearly showed a significant influence of these luminal stages on the pathology of the small intestine at both dietary protein levels: increased wet and dry weight correlated with worm burden, hypertrophy of *tunica muscularis*, increased crypt length, and a trend of decreased villus height. In pigs on a (very) low protein diet (7.8% protein), Forsum et al.[38] found reduced weight gains in experimentally infected pigs (600 eggs/pig) during patency but not in the early stages of infection. The infection was associated with a 28% reduction in feed intake and heavier small intestinal tract (thickened muscular layer as in the previous study). In general, these trends were positively correlated to worm burden, but in pigs fed a normal protein-level diet statistical significance was not obtained. A Swedish study on naturally infected growers showed that female pigs but not castrates with a high parasite load (>10,000 epg at start of fattening and/or >20 worms at slaughter) had a 6% lower daily weight gain as compared to pigs with low (or nil) load.[19] The pigs were co-infected with *Oesophagostomum* spp. at low levels.

A study with single infections of 600, 6000, and 60,000 eggs per pig did not affect feed intake.[39] However, uninfected pigs grew 10% faster than the high infection group (not significant) and the same animals had 13%

more efficient utilization of feed. It has to be noted that the uninfected group unintentionally became infected (mean of 3.2 vs. 18.5 worms). Pigs each infected with 20,000 eggs and uninfected controls were compared in metabolic cages. Infected pigs had significantly reduced digestibility of dry matter (5.1%) and crude protein (6.8%) as well as lowered nitrogen retention (11.0%) 33–37 days p.i.[39] However, these authors failed to demonstrate any effect at earlier stages of infection. Urban et al.[40] reported 21% lower weight gain in untreated pigs in the finishing phase (days 53–110) after natural exposure to *A. suum* in a contaminated outside yard (days 0–52) as compared to pigs reared under similar conditions but treated twice with either fenbendazole or ivermectin during the exposure phase. The yard was also contaminated with *T. suis* (low numbers) which may have added to the poorer performance. To support this, a group trickle infected with (only) *A. suum* for the same period but kept in a clean, concrete pen did not exhibit reduced performance in the finishing phase. The study concluded that stringent control of environmental contamination, which is difficult to achieve in outdoor production systems, and strategic application of anthelmintics is needed to avoid long-term effects of infection. A replicated study[41] failed to find any beneficial effects of intensive anthelmintic treatment (fenbendazole or pyrantel) applied early in life on weight gains or feed utilization in growing pigs naturally exposed/infected with *A. suum* on a dirtlot. Bernardo et al.[42] reported a very limited but significant effect (<1%) of *A. suum* life-time burden (a composite measure based on repeated fecal egg counts during fattening) on average daily weight gain of 352 pigs, produced in 15 Canadian farms.

Two studies from the southern USA have investigated the combined effects of experimental *A. suum*, *Oesophagostomum* spp., and *S. ransomi* infections on performance of growing pigs.[43,44] The first three-factorial study examined the combined effects of pyrantel given continuously for 6 weeks, two protein levels, and order of infection and reported no significant effect of anthelmintic treatment on overall weight gains.[43] The pyrantel treatment resulted in only moderately reduced worm burdens. The other study compared uninfected controls, infected controls, and infected pigs treated with ivermectin after 34 days (exp. 1) or 37 days (exp. 2)[44]. In exp. 1, infected pigs had 12–14% reduced weight gains and 10% reduced carcass weight as compared to uninfected controls but no differences were found in exp. 2.

Based on this range of controlled, mostly experimental studies reported here, it is evident that it has been difficult to demonstrate – in a statistically meaningful way – consistent and major changes in parameters commonly associated with subclinical gastrointestinal parasitism (e.g. reduced weight gain, feed intake, digestibility, nitrogen retention), even when larger groups are used. In most cases, any observed differences were found in the late intestinal phase. At the commonly used levels of protein nutrition, several studies failed to show consistent changes in nitrogen absorption, digestibility or retention. One may speculate on the explanation for this. An obvious reason is the nature of *A. suum* infection, characterized by pronounced overdispersion (see Chapter 7), rapid expulsion of early stages and strong immune responses. One may control the infection levels, but the resulting worm burdens are highly unpredictable and it is often difficult to get any dose dependency on the performance parameters, in contrast to, for example, the gastrointestinal nematode infections of ruminants. For the same reasons, the method of inoculation, dose level, timing of measurements in relation to infection stage, stage of immunity, and age of animals become very important factors that are difficult to align between studies. Further, feed protein levels seem to affect the disease impact, and only a few proper two-factorial designs with appropriate statistical analysis have been performed. Also effects of feed components other than protein, e.g. minerals, cannot be ruled out. A logical consequence of this is that studies examining effects of *A. suum* in commercial farms (on-farm studies) are likely to face even more difficulties. This will be addressed in the next paragraph.

PRODUCTION LOSSES ASSOCIATED WITH ASCARIASIS IN PIGS – ON-FARM STUDIES

The liver migratory activity of *A. suum* leads to a marked focal inflammatory response and later whitish fibrotic lesions known as milk spots or white spots, which are easily identified at slaughter. The lesions are characterized as either superficial granulation tissue-type milk spots (small or large), which are visible gray–white interlobular septa, or lymphonodular milk spots which are elevated and rounded[26] (Figure 14.1). In practice at slaughter, milk spots are not differentiated. While other migratory nematodes may cause similar liver lesions in pigs, e.g. *Stephanurus dentatus* (kidney worm) in warmer regions and occasionally *Toxocara canis*, there is a general consensus that milk spots are caused by *A. suum* infections in commercial production systems in temperate climates.

The lesions are self-healing and disappear with time and milk spots thus represent a recent *A. suum* infection, in most cases within 5–6 weeks.[26] Despite the disappearance of milk spots, the general texture of livers after exposure is clearly hardened as a response to migration, and the livers are less suited for human consumption.[45] It should be noted that pigs without liver lesions may have adult worms in the intestine and vice versa. However, other investigators found a good correlation between herds with high levels of *A. suum* infection and a high proportion of livers with milk spots at slaughter.[46] Wagner and Polley[11] calculated a negative predictive value of 0.90 for milk spots as an indicator of adult

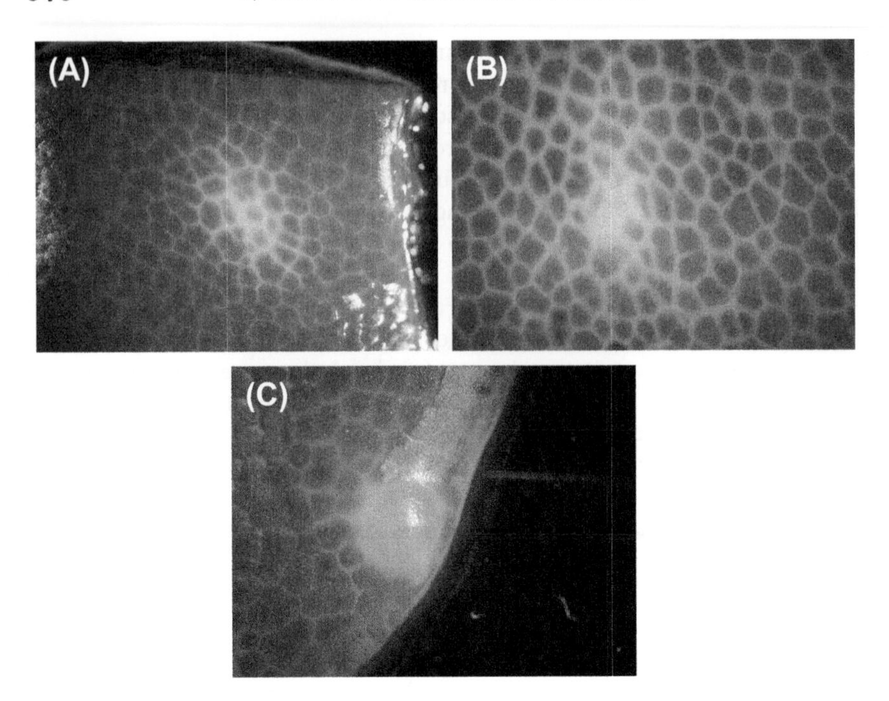

FIGURE 14.1 Liver white spots or milk spots caused by migrating *Ascaris suum* larvae: (A) small granulation-tissue type; (B) large granulation-tissue type; (C) lymphonodular type. *Photos courtesy of N.P.K. Hansen and A. Roepstorff.*[26]

worms, i.e. the absence of milk spots means that it is unlikely that adult worms will be found in a pig. There is a further caveat of using milk spots as a diagnostic indicator: in solidly immune pigs there is presumably a functional pre-hepatic barrier which may result in no (or very limited) liver migration even if animals are heavily exposed, e.g. 12 weeks' immunization followed by challenge with 10,000 eggs lead to no liver pathology,[47] and 14 weeks' continuous infection resulted in very few milk spots.[48] The noticeable decrease in number (or lack) of white spots suggests that the level of exposure is not reflected in the number of white spots in immune animals. If adult worms, established before the onset of the barrier, are eliminated by treatment, such animals are unlikely to perpetuate the infection, unless the immunological memory wanes.

Abattoirs incur losses due to downgrading or total condemnation of livers due to milk spots. Data from abattoirs are available from a range of countries and are in some countries used to monitor prevalence levels.[49] However, in many studies it is unclear which case definition has been applied, whether it be milk spots (present or not; quantified), hepatic lesions or condemnations, and in the case of condemnations, is there an upper acceptable limit of number of milk spots which does not lead to condemnation? A recent study from England reported a prevalence of 4.2% milk spot livers based on post-mortem examination performed on the slaughter line in an abattoir on more than 12,000 pigs monthly for 5 years (2005–2010).[49] The prevalence showed a clear seasonal pattern: increasing in summer/early autumn, peaking in September and declining in winter/spring. This marked seasonality, also reported from Denmark,[50,51] is probably related to lack of development of eggs to infectivity under temperatures of 15°C and increasing rate of development with increasing temperature, e.g. during the late spring and summer both indoor and outdoor in northern temperate climates.[19,52] An abattoir study in Northern Ireland in 1969–1989 supports this explanation by describing a strong positive correlation between annual prevalences of liver condemnations due to cirrhotic lesions with mean air temperatures in May.[53] Any year was strongly correlated with the previous year, probably indicating that high contamination levels were carried over to the following year. The same study reported an increase in prevalence from 4 to 9% during the period investigated and this was attributed to increased transmission. However, as pointed out by Roepstorff[26] intensification and increased use of anthelmintics during this period are more likely to cause reduced transmission, and the increase observed is thus perhaps due to acquisition of infections later in life as white spots only reflect recent exposure, as discussed above. From tropical regions, 8% of livers were reported condemned at slaughter slabs (basically equipped abattoirs) in Dar es Salaam, Tanzania ($N = 731$).[54]

Surveys using extended examination at slaughter have, for obvious reasons, in many cases reported higher prevalences of milk spot livers as compared to abattoir condemnation data after routine meat inspection: 44% in Saskatchewan ($N = 2500$),[11] 24% in Denmark ($N = 9186$),[50] 36% in Belgium during expected high season ($N = 20{,}758$),[55] and 31% in Poland ($N = 1000$).[56] The Danish study reported a markedly higher prevalence in females than entire males (26% vs. 11%) while females and castrated males (29%) were not significantly different. Production of entire males is a niche production in Denmark, and these animals are usually kept separately from the rest. The Belgian study reported that livers with few (number not mentioned) milk spots were trimmed while the Polish study (range: 1–30 milk spots per liver) mentioned that livers with 1–7 milk spots were trimmed and the remaining part approved for consumption while 8 or more spots lead to total condemnation.

A number of recent on-farm studies have estimated the production losses associated with *A. suum* by comparing treated with untreated groups within infected herds, whereas comparison of infected and uninfected herds have not been attempted, probably because it is very difficult, if not impossible, to declare any herd totally free of *A. suum*.[26] In a Croatian farm, three groups of piglets ($n = 90$) – (1) indoor, remaining untreated and worm free; (2) outdoor and untreated; (3) outdoor piglets treated with ivermectin at age 10, 30, and 50 days – were compared up to the time of slaughter at age 60 days (!).[57] No fecal examinations were performed but presence of *A. suum* eggs on the paddock and uptake of infection were confirmed. Parasites reduced live weight gain by 12% and slaughter weight by 9% (Group 3 vs. 2). At slaughter, there were no milk spots in livers of the indoor group, confirming absence of *A. suum* transmission, while fewer milk spots were found in the treated as compared to the untreated group.[57]

A randomized, controlled study (>22,000 pigs) investigating 5 days of treatment with flubendazole three times in a finishing round for fatteners, i.e. at day 0, 40, and 79 after entry, showed reduced *A. suum* load and an increased daily weight gain of 2.0–6.3% in treated pigs on three farms, whereas one farm had unaltered weight gains.[58] None of these differences were statistically significant and although there tended to be fewer rejected livers, better carcass conformation scores and lower mortality, these findings were not consistent across farms. A modified treatment regime with only two flubendazole treatments (days 2–6 and 36–40 in the fattening unit) versus placebo were tested in a randomized control study on two *A. suum*-only infected farms in Denmark (>4000 pigs).[59] No eggs were found in any pens with treated pigs (floor samples) while 18% untreated pens (low levels) were found to be infected. In both groups, 24–27% of livers had milk spots. No overall differences were found for weight gains which were ascribed to low levels of infection and overriding bacterial, intestinal infections (*Lawsonia intracellularis*).

As mentioned earlier, the presence of a mixture of intestinal helminth species is undoubtedly a common finding on many farms and from a practical point of view, it makes sense also to consider how a combination of species affects farm productivity. In most of these studies/cases, it is not possible to single out effects of *A. suum*, e.g. a study in northeast India showed 22% better weight gains after repeated fenbendazole and mineral application in pigs infected with *A. suum*, *Oesophagostomum* spp., *Metastrongylus apri* and other parasites.[60] A recent meta-analysis of effects of intestinal helminths on performance of growing pigs indicated 5% reduced feed intake, 31% lower daily weight gain, and poorer (17%) feed conversion in infected animals.[61] The analysis was based on 18 studies of natural or experimental infections with *A. suum* (nine studies – some of them reported above), *T. suis*, *Oesophagostomum* spp., *S. ransomi* or/and *S. dentatus* including approximately 1500 pigs, but effects of single species were not evaluated. Interestingly the feed intake was increased in the early phase but reversed later on. Overall the reduced feed intake explained 59% of the reduction in weight gain. The crude protein intake was highly correlated with weight gain only in infected pigs, indicating the importance of protein in the adaptive responses.

There are also indications that slaughter yield and meatiness (percentage lean meat according to the EUROP scale) can be affected as they were decreased by 2.7 and 6.1%, respectively, by helminth infections (*A. suum* and *Oesophagostomum* spp.) when pigs are fattened in pens with slatted floors ($n = 60$).[62] Corresponding figures for pigs on shallow litter (5–20 cm of straw and fecal matter, removed frequently) were 1.7 and 4.2%. Infection status of animals was apparently determined by fecal egg counts twice during the fattening period but it was not clear when randomization took place. This study confirmed an earlier study in fatteners naturally infected with *A. suum* (post-mortem examination) showing significantly poorer carcass characteristics (4% reduction in relative ham size and increased fat-to-meat ratio) as compared to uninfected pigs.[17] However, the same study failed to show an effect on weight gains. Matthes et al.[63] reported increased weight gains, carcass weights and food intake after treatment with ivermectin at entry to the fattening unit as compared to untreated controls ($N = 3521$ pigs). However, the farm had a history of both *A. suum* and sarcoptic mange and it was not possible to separate the effects.

INDIRECT EFFECTS

Indirect effects of *A. suum* infections are related to co-transmission of other pathogens with the parasite eggs, enhanced pathogenicity associated with the migratory phase or the intestinal phase, and what can be termed as a "systemic immunomodulatory effect." With a few exceptions, these studies are based on few animals and are seldom repeated by other investigators. Nevertheless, infective eggs may potentially contain pathogenic *E. coli*[64] but whether they can in fact transmit other infectious diseases, as shown for the bird helminth *Heterakis gallinarum* and *Histomonas meleagridis*,[65] remains to be explored. Lung phase studies have focused on whether larvae with a contaminated cuticle may lead to lung infections, of which there is little evidence[66] or whether the associated lung pathology makes the infected pig more vulnerable to air-borne infections. Underdahl and Kelley[67] first reported a marked exacerbation of viral lung lesions by migrating ascarid larvae in experimentally infected pigs. A reduced lung clearance of bacteria was later found in a study reporting a sevenfold higher bacterial load in lungs of *A. suum*-infected pigs as compared to controls 2 hours after aerosol exposure to *E. coli*.[30] In addition, a higher pathogenic effect of an aerosol with *Pasteurella multocida* has been reported during the lung migration phase in *A. suum*-infected mice.[68] The migratory lung phase may perhaps in itself lead to lung lesions characterized by small spots of fibroplasia with surrounding hemorrhage as indicated by a higher occurrence of liver milk spots in pig with these lung lesions.[69]

In a study attempting to induce post-weaning diarrhoea by experimental *E. coli* infection in all piglets, those co-infected with *A. suum* and *T. suis* had markedly softer feces than controls not infected with helminths, although the induction of diarrhea was not successful.[70] Additionally, the helminth-infected pigs also had an increased *Campylobacter* spp. excretion level. Lastly, *A. suum* may negatively affect the post-vaccination immunity against *Mycoplasma hyopneumoniae*, resulting in increased pathology, which is probably related to interference with the immune response as this took place after the lung migratory phase.[4] *Ascaris suum* induces a classical Th2 response that may modulate or reduce the Th1 response needed for responding to mycoplasms. The impact of this finding in the context of commercial farming has not been assessed but will, besides other factors, depend on *A. suum* infection dynamics on each specific farm, e.g. the time of first acquisition of infection in relation to vaccination.

IMPACT OF *ASCARIS SUUM* IN OTHER LIVESTOCK

In cattle, several cases of clinical disease and even deaths due to *A. suum* have been reported, especially in young stock. Seven out of 17 heifers thus developed clinical signs of severe dyspnea, coughing and forced expiration after turnout on a *A. suum* contaminated pasture, and two animals eventually died.[71] In a total of three cases, acute respiratory distress, pneumonia or fatalities were registered 10 days after exposure either to pig manure in the feed or to pens previously used for pigs.[72–74] A sudden reduction in milk yield, respiratory symptoms, and eosinophilia were observed on two dairy farms after heifers had grazed pastures fertilized with pig slurry.[75] The animals were tested positive for *A. suum* antibodies. Only few incidents of mature *A. suum* in cattle have to our knowledge been described. Numerous worms were found in the bile duct of one calf on a farm whereas two other calves had 16 and 168 intestinal worms.[76] Based on morphology, *Toxocara vitulorum* was excluded.

Experimental infection studies support the above-mentioned clinical observations. *A. suum* larvae were thus recovered on day 10[77] p.i. and days 7–9 p.i. from the lungs of calves infected with embryonated eggs.[78] These findings coincide with the lung symptoms described in the above case reports. Greenway and McCraw[79] reported dyspnea, coughing, and increased respiration rates, consistent with atypical interstitial pneumonia, as most pronounced 10–13 days after a challenge infection in 7-week-old sensitized calves. Interestingly, the first sensitizing dose of 100,000 eggs was not enough to cause pronounced clinical signs. Similar lung pathology has also been described in yearling cattle that died after natural exposure to *A. suum.*[71,72] Pathology and clinical signs have been correlated with inoculation dose in surviving calves, but the link to mortality is less clear. Some animals have thus died after receiving "low doses" of 10,000–100,000 eggs whereas others have survived multiple doses adding up to 11 million eggs.[80,81]

A. suum has also been shown to cause varying levels of gross liver pathology in cattle. In one study, no milk spots or larvae were detected after inoculation of four 2–3-day-old calves with a single dose of 10,000 *A. suum* eggs[80] whereas milk spots were noted 3–5 days p.i. and larvae were isolated from the liver of 4-week-old calves infected with 2 million eggs.[78] Larval migration may have been underestimated due to suboptimal recovery methods. Alternatively, the difference in results may reflect the variation in age and/or dose level as it is likely that overall migration success is considerably lower in an inappropriate host. At slaughter, livers from cattle given 2000 or 5000 *A. suum* eggs were all condemned.[80]

Mature *A. suum* has been observed in the intestine or bile ducts of lambs at several abattoirs.[82–84] These infections were probably the result of exposure to pastures contaminated with pig feces[85] as worms from lambs have been genetically characterized to be of pig origin (Nejsum, unpublished data). Worms in the bile ducts have been shown to cause pathological changes[83] but clinical manifestations due to intestinal worms are to our knowledge unknown. As for cattle, transient lung symptoms days 5–14 p.i. have been reported for lambs and kids after experimental exposure to *A. suum.*[86,87] Symptoms included dry coughs, dyspnea and pneumonia, and an animal died. Larvae were isolated from the lungs in both studies.

Condemnation of 70% of the lamb livers from one farm was associated with *A. suum,*[88] and the pathological changes were similar to those described in an earlier *A. suum* suspected case in lambs.[83] Yellow–white plaques (15 mm in diameter) were observed on the liver from a lamb that also harbored *Ascaris* spp. in the bile ducts.[84] Histological examination of lamb livers after supposedly *Ascaris* spp. infection showed intensive eosinophil infiltration of the portal tracts, parasite tracks in the parenchyma and portal fibrosis.[83] However, other researchers have reported an apparent lack of macroscopic milk spots in lambs experimentally infected with *A. suum.*[87,89] The lack of interlobular connective tissue in lambs is considered the main reason why "white spots" do not develop in this host animal.[87] They are therefore critical to the reports suggesting that condemnation of liver at abattoirs should be related to white spots caused by *A. suum* but do not refute that pathological changes take place in the liver.

The importance of *A. suum* in other livestock species is not known. Pigs fed livers and lungs from *A. suum* inoculated chickens (clinically unaffected) became infected and it is possible that *A. suum* may cause illness in humans if raw chicken livers are consumed.[90]

DISCUSSION AND CONCLUSIONS

Losses due to ascariasis in pigs can be summarized as (1) farm economic losses due to clinical effects (although limited), reduced growth and feed conversion efficiency and costs of control (e.g. use of anthelmintics), (2) abattoir operator losses due to condemnation or downgrading of livers and lower product quality, and (3) potential interference with vaccinations and higher risk of co-infections. Apart from the cost of liver condemnations that can be extrapolated from available data, it is difficult to get an overall impression of relative importance and to assign values to these losses. An attempt to address some of the problems associated with economic evaluation of worm control has been reported,[91] indicating gross margin increases of 3–12 € per average present finisher per year after strategic de-worming during the fattening period. In the USA, the poorer feed conversion has been estimated to account for a total loss of US$60.1 million in 1999, based on a 5.6% higher feed intake to reach target weight in about 50% of the pigs slaughtered.[92] This loss was considerably larger than costs due to liver condemnations. Gross margin increases with 3 to 12 € per average present finisher per year, depending on the cyclic pig price conditions.

After reviewing the existing literature, it is most likely that pigs heavily exposed to *A. suum* will experience a reduction in feed utilization and perhaps also weight gain. The effect may be transient and perhaps more strongly associated with the intestinal phase than the early stages of infection. Further, it is striking how difficult it has been in more recent studies to demonstrate significant reduced productivity due to *A. suum* – both in experiments and on-farm.[42] As mentioned earlier, this may be related to the strong overdispersion of the *A. suum* infection – only a limited number of animals are infected with high numbers of worms at any one time – together with well-balanced diets rich in essential amino acids in most intensive systems. This can explain the often dramatic effects reported several decades ago and the pronounced effects of anthelmintic treatments in more extensive systems, e.g. in developing countries. *A. suum* and other intestinal helminths most likely constitute a serious problem of disease and productivity in poor farming communities worldwide. The impact of *A. suum* in these systems needs to be addressed in future studies. Another obvious explanation is, of course, that there is no measurable impact due to a high degree of co-adaptation between parasite and host. However, any researcher having necropsied *A. suum*-infected animals will find it hard to believe that the marked pathological changes do not impact upon function.

Lack of convincing and consistent findings on weight gain benefits in treatment trials in commercial herds create problems in motivating farmers and advisors, including decision making on when to intervene and how. From a farmer's perspective, the apparent paradox that better control and reduced transmission may in some cases lead to more livers condemned at slaughter due to later exposure and lower levels of immunity makes it even harder to understand. A later acquisition of *A. suum* (and *Oesophagostomum* spp.) by growing pigs in intensive systems may, however, still have a beneficial impact on the economics of production as numbers of both migrating larvae and adults may be negligible while the duration of the infections may be short. Therefore the overall impact of a relevant intervention may still be substantial.

A. suum can cause significant clinical manifestations and reduce carcass quality in cattle and sheep. However, in areas of industrialized farming systems we expect the clinical impact of *A. suum* to be limited since most farms are specialized for a single type of livestock, and pig slurry is seldom applied on ruminant grazing areas. In contrast, in more extensive livestock production systems with mixed species or in areas where livestock are roaming freely, as is the case in many developing countries, the impact of *A. suum* might be higher although not yet documented.

Lastly, the more subtle impact of *A. suum* on the course and impact of other infections, including the possible interference with vaccinations, should not be neglected. There is circumstantial evidence that *A. suum* together with other helminths result in more pronounced effects, bacterial and viral co-infections are potentiated, and Th1-dependent vaccinations may fail to protect as stipulated. There is an urgent need to assess the impact of *A. suum* in this context. This may well turn out to be just as important as any direct effect.

References

1. ILRI. In: *Investing in animal health research to alleviate poverty.* Kenya: ILRI-Nairobi; 2002. p. 140.
2. Nansen P, Roepstorff A. Parasitic helminths of the pig: factors influencing transmission and infection levels. *Int J Parasitol* 1999;**29**(6):877–91.
3. Roepstorff A, Mejer H, Nejsum P, Thamsborg SM. Helminth parasites in pigs: new challenges in pig production and current research highlights. *Vet Parasitol* 2011;**180**(1–2):72–81.
4. Steenhard NR, Jungersen G, Kokotovic B, Beshah E, Dawson HD, Urban JF, et al. *Ascaris suum* infection negatively affects the response to a *Mycoplasma hyopneumoniae* vaccination and subsequent challenge infection in pigs. *Vaccine* 2009;**27**(37): 5161–9.
5. Taylor MA, Coop RL, Wall RL. *Veterinary Parasitology.* 3rd ed. Blackwell Publishing; 2007. p. 874.
6. Hartwich G. [Type identity of the swine parasites *Globocephalus longemucronatus* and *G. urosubulatus* (Nematoda, Strongyloidea)]. *Angew Parasitol* 1986;**27**(4):207–14.

7. Roepstorff A, Nansen P. Epidemiology, diagnosis and control of helminth parasites of swine. In: *FAO Animal Health Manual*. 1st ed. Rome: FAO; 1998. p. 161.
8. Joachim A, Dulmer N, Daugschies A, Roepstorff A. Occurrence of helminths in pig fattening units with different management systems in Northern Germany. *Vet Parasitol* 2001;**96**(2):135–46.
9. Roepstorff A, Nilsson O, Oksanen A, Gjerde B, Richter SH, Ortenberg E, et al. Intestinal parasites in swine in the Nordic countries: prevalence and geographical distribution. *Vet Parasitol* 1998;**76**(4):305–19.
10. Haugegaard J. Prevalence of nematodes in Danish industrialized sow farms with loose housed sows in dynamic groups. *Vet Parasitol* 2010;**168**(1–2):156–9.
11. Wagner B, Polley L. *Ascaris suum* prevalence and intensity: an abattoir survey of market hogs in Saskatchewan. *Vet Parasitol* 1997;**73**(3–4):309–13.
12. Matsubayashi M, Kita T, Narushima T, Kimata I, Tani H, Sasai K, et al. Coprological survey of parasitic infections in pigs and cattle in slaughterhouse in Osaka. *Japan J Vet Med Sci* 2009;**71**(8):1079–83.
13. Nwanta JA, Shoyinka SVO, Chah KF, Onunkwo JI, Onyenwe IW, Eze JI, et al. Production characteristics, disease prevalence, and herd health management of pigs in Southeast Nigeria. *J Swine Health Prod* 2011;**19**(6):331–9.
14. Permin A, Yelifari L, Bloch P, Steenhard N, Hansen NP, Nansen P. Parasites in cross-bred pigs in the Upper East Region of Ghana. *Vet Parasitol* 1999;**87**:63–71.
15. Tamboura HH, Banga-Mboko H, Maes D, Youssao I, Traore A, Bayala B, et al. Prevalence of common gastrointestinal nematode parasites in scavenging pigs of different ages and sexes in eastern centre province, Burkina Faso. *Onderstepoort J Vet Res* 2006;**73**(1):53–60.
16. Nissen S, Poulsen I, Nejsum P, Olsen A, Roepstorff A, Rubaire-Akiiki C, et al. Prevalence of gastrointestinal nematodes in growing pigs in Kabale District in Uganda. *Trop Anim Health Pro* 2011;**43**(3):567–72.
17. Prosl H, Heimbucher J, Supperer R, Kläring WJ. Neue Gesichtspunkte hinsichtlich des Einflusses der Schweine-helminthen auf die Schalacht- und Mastleistung. *Wien Tierärztl Monat* 1980;**67**:14–9.
18. Boes J, Nansen P, Stephenson LS. False-positive *Ascaris suum* egg counts in pigs. *Int J Parasitol* 1997;**27**(7):833–8.
19. Nilsson O. Ascariasis in the pig. An epizootiological and clinical study. *Acta Vet Scand Sup* 1982;**79**:1–108.
20. Dangolla A, Bjorn H, Willeberg P, Roepstorff A, Nansen P. A questionnaire investigation on factors of importance for the development of anthelmintic resistance of nematodes in sow herds in Denmark. *Vet Parasitol* 1996;**63**(3–4):257–71.
21. Baumgartner J, Leeb T, Gruber T, Tiefenbacher R. Husbandry and animal health on organic pig farms in Austria. *Anim Welfare* 2003;**12**:631–5.
22. Gerwert S, Failing K, Bauer C. Husbandry management, worm control practices and gastro-intestinal parasite infections of sows in pig-breeding farms in Munsterland, Germany. *Deut Tierarztl Woch* 2004;**111**(10):398–403.
23. Eijck IA, Borgsteede FH. A survey of gastrointestinal pig parasites on free-range, organic and conventional pig farms in The Netherlands. *Vet Res Commun* 2005;**29**(5):407–14.
24. Bonde M, Hegelund L, Sørensen JT. Health in organic and conventional slaughter pigs assessed by meat inspection recordings and clinical evaluation. Danish Institute of Agricultural Sciences. *Internal Report* 2006;**1**:9–12.
25. Sanchez-Vazquez MJ, Smith RP, Kang S, Lewis F, Nielen M, Gunn GJ, et al. Identification of factors influencing the occurrence of milk spot livers in slaughtered pigs: A novel approach to understanding *Ascaris suum* epidemiology in British farmed pigs. *Vet Parasitol* 2010;**173**(3–4):271–9.

26. Roepstorff A. Ascaris suum *in pigs: population, biology and epidemiology. Doctoral Thesis.* Copenhagen: The Royal Veterinary and Agricultural University; 2003. p. 112.
27. Roepstorff A, Eriksen L, Slotved HC, Nansen P. Experimental *Ascaris suum* infection in the pig: worm population kinetics following single inoculations with three doses of infective eggs. *Parasitology* 1997;**115**(Pt 4):443–52.
28. Eriksen L. *Host parasite relations in* Ascaris suum *infections in pigs and mice. Doctoral Thesis.* Copenhagen: The Royal Veterinary and Agricultural University; 1981. p. 190.
29. Yoshihara S, Nakagawa M, Suda H, Ikeda K, Hanashiro K. White spots of the liver in pigs experimentally infected with *Ascaris suum. National Institute of Animal Health Quarterly* 1983;**23**(4):127–37.
30. Curtis SE, Tisch DA, Todd KS, Simon J. Pulmonary bacterial deposition and clearance during ascarid larval migration in weanling pigs. *Can J Vet Res* 1987;**51**:525–7.
31. Miskimins DW, Greve JH, Baker JR. The serious effects of ascarid larval migration on a group of market-weight swine. *Vet Med* 1994;**89**:247–53.
32. Spindler LA. The effect of experimental infections with ascarids on the growth of pigs. *P Helm Soc Wash* 1947;**14**:58–63.
33. Olsen LS, Kelley GW, Sen HG. Longevity and egg production of *Ascaris suum. Trans Am Microsc Soc* 1958;**77**:380–3.
34. Boes J, Medley GF, Eriksen L, Roepstorff A, Nansen P. Distribution of *Ascaris suum* in experimentally and naturally infected pigs and comparison with *Ascaris lumbricoides* infections in humans. *Parasitology* 1998;**117**:589–96.
35. Stephenson L, Pond WG, Nesheim MC, Krook LP, Crompton DWT. *Ascaris suum*: nutrient absorption, growth, and intestinal pathology in young pigs experimentally infected with 15-day-old larvae. *Exp Parasitol* 1980;**49**(1):15–25.
36. Dold C, Holland CV. *Ascaris* and ascariasis. *Microbes Infect* 2011;**13**:632–7.
37. Zimmerman DR, Spear ML, Switzer WP. Effect of *Mycoplasma hyopneumonia* infection, pyrantel treatment and protein nutrition on performance of pigs exposed to soil containing *Ascaris suum* ova. *J Anim Sci* 1973;**36**(5):894–7.
38. Forsum E, Nesheim MC, Crompton DWT. Nutritional aspects of *Ascaris* infection in young protein-deficient pigs. *Parasitology* 1981;**83**:497–512.
39. Hale OM, Stewart TB, Marti OG. Influence of an experimental infection of *Ascaris suum* on performance of pigs. *J Anim Sci* 1985;**60**(1):220–5.
40. Urban JF, Romanowski RD, Steele NC. Influence of helminth parasite exposure and strategic application of anthelmintics on the development of immunity and growth of swine. *J Anim Sci* 1989;**67**:1668–77.
41. Southern LL, Stewart TB, Bodak-Koszalka E, Leon DL, Hoyt PG, Bessette ME. Effect of fenbendazole and pyrantel tartrate on the induction of protective immunity in pigs naturally or experimentally infected with *Ascaris suum. J Anim Sci* 1989;**67**:628–34.
42. Bernardo TM, Dohoo IR, Donald A. Effect of ascariasis and respiratory diseases on growth rates in swine. *Can J Vet Res* 1990;**54**(2):278–84.
43. Stewart TB, Johnson JC, Hale OM. Effects of pyrantel HCl and dietary protein on growing pigs infected in different sequences with *Strongyloides ransomi, Ascaris suum* and *Oesophagostomum* spp. *J Anim Sci* 1972;**35**:561–8.
44. Stewart TB, Leon DL, Fox MC, Southern LL, Bodak-Koszalka E. Performance of pigs with mixed nematode infections before and after ivermectin treatment. *Vet Parasitol* 1991;**39**:253–66.
45. Wismer-Pedersen J, Juel Møller A, Eriksen L, Nansen P, Roepstorff A. Infection with round worms (*Ascaris suum*) in pig livers. *Dansk Veterinær Tidsskrift* 1990;**73**:126–32.
46. Bernardo TM, Dohoo IR, Ogilvie T. A critical assessment of abattoir surveillance as a screening test for swine ascariasis. *Can J Vet Res* 1990;**54**(2):274–7.
47. Urban Jr JF, Alizadeh H, Romanowski RD. *Ascaris suum*: development of intestinal immunity to infective second-stage larvae in swine. *Exp Parasitol* 1988;**66**(1):66–77.

48. Nejsum P, Thamsborg SM, Petersen HH, Kringel H, Fredholm M, Roepstorff A. Population dynamics of *Ascaris suum* in trickle-infected pigs. *J Parasitol* 2009;**95**(5): 1048–53.
49. Sanchez-Vazquez MJ, Nielen M, Gunn GJ, Lewis FI. National monitoring of *Ascaris suum* related liver pathologies in English abattoirs: a time-series analysis, 2005–2010. *Vet Parasitol* 2012;**184**(1):83–7.
50. Christensen G. The prevalence of pneumonia, pleuritis, pericarditis and milk spots in livers of Danish slaughter pigs (status of 1994). *Dansk Veterinærtidsskrift* 1995;**78**:554–61.
51. Christensen G, Enøe C. The prevalence of pathological changes in plucks of Danish slaughter pigs. Status of 1998 and developments since 1994. *Dansk Veterinærtidsskrift* 1999;**82**:1006–15.
52. Larsen MN, Roepstorff A. Seasonal variation in development and survival of *Ascaris suum* and *Trichuris suis* eggs on pastures. *Parasitology* 1999;**119**:209–20.
53. Goodall EA, McLoughlin EM, Menzies FD, McIlroy SG. Time series analysis of the prevalence of *Ascaris suum* in pigs using abattoir condemnation data. *Anim Prod* 1991;**53**:367–72.
54. Mkupasi E, Ngowi H, Nonga HE. Prevalence of extra-intestinal helminth infections and assessment of sanitary conditions of pig slaughter slabs in Dar es Salaam city, Tanzania. *Trop Anim Health Pro* 2011;**43**:417–23.
55. Vercruysse J, VanHoof D, DeBie S. Study on the prevalence of white spots of the liver in pigs in Belgium and its relationship to management practices and anthelmintic treatment. *Vlaams Diergen Tijds* 1997;**66**(1):28–30.
56. Pyz-Lukasik R, Prost EK. [Milk spots by *Ascaris suum* in pigs liver]. *Med Weter* 1999;**55**(6):351–420.
57. Clark P, Bilkei G. Production losses due to ascarid infestation of outdoor maintained pigs. *Tierarztl Umschau* 2003;**58**(8):425–31.
58. Kanora A. Effect on productivity of treating fattening pigs every 5 weeks with flubendazole in feed. *Vlaams Diergen Tijds* 2009;**78**(3):170–5.
59. Boes J, Kanora A, Havn KT, Christiansen S, Vestergaard-Nielsen K. Effect of *Ascaris suum* infection on performance of fattening pigs. *Vet Parasitol* 2010;**172**(3–4):269–76.
60. Kumaresan A, Bujarbaruah KM, Pathak KA, Anubrata D, Bardoloi RK. Integrated resource-driven pig production systems in a mountainous area of Northeast India: production practices and pig performance. *Trop Anim Health Pro* 2009;**41**:1187–96.
61. Kipper M, Andretta I, Monteiro SG, Lovatto PA, Lehnen CR. Meta-analysis of the effects of endoparasites on pig performance. *Vet Parasitol* 2011;**181**(2–4):316–20.
62. Knecht D, Jankowska A, Zalesny G. The impact of gastrointestinal parasites infection on slaughter efficiency in pigs. *Vet Parasitol* 2012;**184**(2–4):291–7.
63. Matthes W, Ilchmann G, Rehbock F. Economic impact of subclinical parasitic infestations in fattening pigs. *Prakt tierarzt* 1998;**79**(11):1067–71.
64. Lee YC, Liu CC. Isolation of *Escherichia coli, Alcaligenes faecalis* and *Pseudomonas aerugiunosa* from the eggs of *Ascaris suum*. *J Chin Soc Vet Sci* 1976;**2**:59–61.
65. Ruff MD, McDougald LR, Hansen MF. Isolation 313 of *Histomonas meleagridis* from 314 embryonated eggs of *Heterakis gallinarum*. *J Protozool* 1970;**17**:10–1.
66. Adedeji SO, Ogunba EO, Dipeolu OO. Synergistic effect of migrating *Ascaris* larvae and *Escherichia coli* in piglets. *J Helminthol* 1989;**63**:19–24.
67. Underdahl NR, Kelley GW. The enhancement of virus pneumonia of pigs by the migration of *Ascaris suum* larvae. *J Am Vet Med Assoc* 1957;**15**:173–6.
68. Tjørnehøj K, Eriksen L, Aalbæk B, Nansen P. Interaction between *Ascaris suum* and *Pasteurella multocida* in the lungs of mice. *Parasitol Res* 1992;**78**:525–8.
69. Liljegren CH, Aalbaek B, Nielsen OL, Jensen HE. Some new aspects of the pathology, pathogenesis, and aetiology of disseminated lung lesions in slaughter pigs. *APMIS* 2003;**111**(5):531–8.

70. Jensen AN, Mejer H, Mølbak L, Langkjær M, Jensen TK, Angen Ø, et al. The effect of a diet with fructan-rich chicory roots on intestinal helminths and microbiota with special focus on *Bifidobacteria* and *Campylobacter* in piglets around weaning. *Animal* 2011;**5**(6):851–60.
71. Morrow DA. Pneumonia in cattle due to migrating *Ascaris lumbricoides* larvae. *J AmVet Med Assoc* 1968;**153**(2):184–9.
72. Allen GW. Acute atypical bovine pneumonia caused by *Ascaris lumbricoides*. *Can Comp Med* 1961;**26**:241–3.
73. McCraw BM, Lautenslager JP. Pneumonia in calves associated with migrating *Ascaris suum* larvae. *Can Vet J* 1971;**12**(4):87–90.
74. McLennan MW, Humphries RB, Rac R. *Ascaris suum* pneumonia in cattle. *AustVet J* 1974;**50**(6):266–8.
75. Borgsteede FHM, Deleeuw WA, Dijkstra T, Alsma G, de Vries W. *Ascaris suum* infections causing clinical problems in cattle. *Tijdschr Diergeneeskd* 1992;**117**(10):296–8.
76. Roneus O, Christensson D. Mature *Ascaris suum* in naturally infected calves. *Vet Parasitol* 1977;**3**(4):371–5.
77. Henriksen SA, Krogh HV. Bovine infections of *Ascaris suum*. Preliminary observations. *Dansk Veterinær Tidsskrift* 1979;**73**:126–32.
78. McCraw BM. The development of *Ascaris suum* in calves. *Can Comp Med* 1975;**39**:354–7.
79. Greenway JA, McCraw BM. *Ascaris suum* infection in calves I. Clinical signs. *Can J Comp Med* 1970;**34**:227–37.
80. Kennedy PC. The migrations of the larvae of *Ascaris lumbricoides* in cattle and their relation to eosinophilic granulomas. *Cornell Vet* 1954;**44**(4):531–65.
81. McCraw BM, Greenway JA. *Ascaris suum* infection in calves III. Pathology. *Can Comp Med* 1970;**34**(3):247–55.
82. Johnston AA. Ascarids in sheep. *N Z Vet J* 1963;**11**:69–70.
83. Harcourt R, Costema P. Hepatic ascariasis in lambs. *Vet Rec* 1973;**92**:482–3.
84. Sauvageau R, Fréchette JL. Hepatic ascariasis in a lamb. *Can Vet J* 1980;**21**(2):66.
85. Pedersen K, Monrad J, Henriksen SA, Bindseil B, Nielsen JS, Jensen E, et al. Bovine infections of *Ascaris suum*. Preliminary observations. *Dansk Veterinærtidsskrift* 1992;**75**: 170–4.
86. Gaur SNS, Deo PG. Observations on the transmission of *Ascaris lumbricoides* (pig and human strains) in certain heterologous hosts. *Indian J Anim Sci* 1972;**42**(4):281–4.
87. Brown D, Hinton M, Wright AI. Parasitic liver damage in lambs with particular reference to the migrating larvae of *Ascaris suum*. *Vet Rec* 1984;**115**:300–3.
88. Mitchell GB, Linklater KA. Condemnation of sheep due to ascariasis. *Vet Rec* 1980;**107**:70.
89. Fitzgerald PR. The pathogenesis of *Ascaris lumbricoides var. suum* in lambs. *Am J Vet Res* 1962;**23**:731–6.
90. Permin A, Henningsen E, Murrell KD, Roepstorff A, Nansen P. Pigs become infected after ingestion of livers and lungs from chickens. *Int J Parasitol* 2000;**30**:867–8.
91. van Meensel J, Kanora A, Lauwers L, Jourquin J, Goosens L, van Huylenbroeck G. From research to farm: *ex ante* evaluation of strategic deworming in pig finishing. *Vet Med-Czech* 2010;**55**(10):483–93.
92. Stewart TB. Economics of endoparasitism of pigs. *Pigs News and Information* 2001;**22**(1):29N–30N.

CHAPTER

15

Approaches to Control of STHs including Ascariasis

Antonio Montresor, Albis Francesco Gabrielli, Lorenzo Savioli

World Health Organization, Geneva, Switzerland

OUTLINE

Ascaris: The Neglected Parasite

http://dx.doi.org/10.1016/B978-0-12-396978-1.00015-X

Copyright © 2013 Elsevier Inc. All rights reserved.

INTRODUCTION

Soil-transmitted helminthiasis (STH) represents a serious public health problem (see Chapter 13) in countries where sanitation and hygienic conditions are insufficient to respond to the needs of the population, and where effective drugs for their treatment and public health control are neither widely available nor affordable by those in need.

STH and socio-economic status are intimately linked. In countries where an improvement in sanitation as a natural component of the economic progress has taken place, a parallel progressive decline in the relevance of STH has invariably been observed. Where universal or targeted de-worming programs have accompanied such economic growth, results have been obtained in a much shorter time span and they have proven to be longer lasting. However, where chemotherapy has been implemented periodically, even in the absence of improvements in sanitation and economic growth, important achievements have also been obtained in terms of morbidity control.

STH control strategies should aim at controlling associated morbidity in the first place, and at reducing STH transmission in those cases where conditions permit a more comprehensive preventive effort. Different combinations of the aforementioned approaches have been implemented in endemic countries according to the local significance of the problem, and to the availability of resources. Results obtained from ongoing control programs in endemic areas are continuously monitored by the World Health Organization (WHO) to refine existing strategies for the control of STH.

THE EXPERIENCE FROM JAPAN AND KOREA

Japan has achieved successful and sustained control of STH and has led the way in this effort.

By 1949, the nationwide fecal examination survey reported an overall prevalence of 73.0% for intestinal nematodes: *Ascaris lumbricoides* (62.9%), *Trichuris trichiura* (50%), and hookworms (3.5%). Non-governmental organizations (NGOs) took the initiative, private laboratories were established, stool examinations were carried out, and treatment with anthelmintic drugs (initially thiabendazole, then mebendazole and albendazole) began. The intervention was organized as selective treatment in which school children regularly underwent mass stool examination and positive cases received treatment twice a year. In 1955, the Japan Association of Parasite Control was founded and the government passed the School Health Law in 1958 and issued guidance on control technologies. The cellophane thick smear method (now described as the Kato-Katz

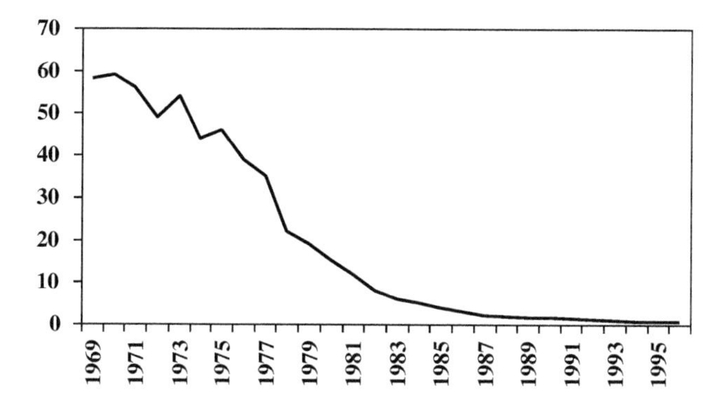

FIGURE 15.1 Decrease in the prevalence of *A. lumbricoides* in school children between 1969 and 1995 in the Republic of Korea.[2]

technique) was invented and was widely adopted for stool examinations. By 1990, the prevalence of *A. lumbricoides* had dropped to 0.9%, that of hookworms to 0.0%, and that of *T. trichiura* to 0.25%.[1]

A similar experience took place in Korea between 1969 and 1995.[2] In this case, too, the program focused on selective treatment of infected school children and the significant results obtained are presented in Figure 15.1.

The relevance of economic development (linked to the improvement of sanitation standards) to permanently solve the public health problem caused by STH is confirmed by the fact that in other countries significant prevalence reductions have been obtained without any specific disease control activity: for example, in Italy between 1965 and 1980 the prevalence of trichiuriasis dropped from 65% to less than 5% and the prevalence of ascariasis from 10% to 0%.[3]

When STH control measures are applied in a situation of economic development the results, in terms of decline in prevalence and health improvement, are rapid and definitive. In addition, the control of the morbidity due to STH can in itself contribute to the economic development of the country by boosting the capacity of school children to grow and learn better, and by increasing the physical fitness of adults and the health of adolescent girls and women of child-bearing age.[4]

THE "REALITY" IN DEVELOPING COUNTRIES

Unfortunately, this situation of rapid and general economic development does not apply to many tropical countries where even if progresses are observed, these are rarely equitable and most frequently apply to urban centers and not rural areas.[5]

In this context of limited resources, the population is more vulnerable to the damage caused by the STH and the need to protect them through periodical de-worming is greater. However, achieving control is more difficult and the results are less dramatic.

EPIDEMIOLOGICAL BASIS OF THE WHO STRATEGY

The comprehension of some epidemiological characteristics of STH is important to select the most appropriate control measures and to evaluate correctly their public health impact (see also Chapters 7 and 9).

1. **Children and women harbor peak worm burdens.**[6] In addition these population groups are characterized by intense metabolic and physical growth, resulting in increased nutritional needs. This explains why pre-school children, school-aged children and women of child bearing age are particularly vulnerable to the nutritional deficits related to the infections and are considered the population groups at greater risk of morbidity due to STH.
2. **Heavy-intensity infections are the major source of morbidity**: Morbidity is directly related to worm burden.[5] The greater the number of worms in the infected person, the greater the morbidity caused by these worms will be. For example, in the case of hookworms, the amount of blood lost in the feces (as an indicator of morbidity) is directly associated with hookworm egg count (as a measure of worm burden).[7]
3. **Until environmental and/or behavioral conditions have changed, the prevalence of infection will tend to return to original pre-treatment levels.** Reinfection occurs because the worm's infective stages will continue to contaminate the environment. However, repeated treatments will periodically decrease the worm burden of infected individuals, in spite of continuing reinfection episodes. Harboring fewer worms will thus significantly reduce the health damage caused by these parasites.[8]

Based on the points mentioned above, an appropriate and cost-effective control strategy can be designed, which would ensure as a priority the reduction of morbidity in high-risk population groups. This is done by reducing to minimal levels the proportion of heavily infected individuals and can be achieved with periodic distribution of de-worming drugs, an approach that is nowadays referred to as "preventive chemotherapy." At the same time, according to available resources, other complementary control measures such as health education, social mobilization, information, education and communication, as well as improvement of sanitation should be promoted in order to sustain the benefits of periodic treatment and to achieve long lasting control of transmission of infection.

PREVENTIVE CHEMOTHERAPY AS THE WHO STRATEGY FOR STH CONTROL

Preventive chemotherapy (PC) is "the use of anthelmintic drugs, either alone or in combination, as a public health tool against helminth infections"[9] and is the key public health strategy recommended by the WHO to reduce morbidity and transmission of STH.

Operationally, PC is characterized by population-level diagnosis, population-level treatment and by implementation at regular intervals.[10]

- **Population-based diagnosis.** Population-based diagnosis consists of assessing the significance of STH in a population through surveys applied to a sample of its individuals. Appropriate diagnostic tests, or standard questionnaires screening for pathognomonic symptoms or signs or for behaviors associated with risk of infection, can be used alternatively. Population-based diagnosis can also be carried out retrospectively by analyzing existing epidemiological data. Based on its results, an appropriate intervention is selected. Population-based diagnosis distinguishes PC from the clinical approach in which diagnosis is performed at the individual level prior to treatment.
- **Population-based treatment.** In PC, administration of anthelmintic drugs is not the outcome of a personalized, case management treatment approach performed by medical personnel on individuals reporting to health facilities. It rather entails actively targeting population groups at risk (pre-school and school-aged children and women of child-bearing age) with delivery of single-administration medicines by non-medical personnel (teachers, volunteers or community drug distributors).
- **Implementation at regular intervals.** PC is implemented at regular intervals of time; the retreatment interval is based on the epidemiological characteristics of the disease as measured by the population-based diagnosis; the intervention is repeated without the need for further diagnostic interventions,[8] although implementation of a monitoring system is recommended.

COLLATERAL BENEFITS OF PREVENTIVE CHEMOTHERAPY

Preventive chemotherapy for the control of neglected tropical diseases (NTDs) has the advantage of being effective against other parasitic infections that are not directly targeted by the intervention, for example the distribution of ivermectin and albendazole for the control of lymphatic filariasis and STH reduces the morbidity caused by strongyloidiasis, pinworm infections, and scabies; the praziquantel distributed for the control of schistosomiasis reduces the morbidity due to opisthorchiasis, clonorchiasis, paragonimiasis, and taeniasis.

It is therefore evident that an evaluation of the PC benefits based on the effects on the diseases targeted alone is frequently reductive because it does not provide the real extent of the benefits obtained. Furthermore, a complete evaluation of the benefits is frequently very difficult to conduct because of the lack of information on the baseline epidemiology of all the diseases prevented.

MONITORING AND EVALUATION

WHO sees monitoring and evaluation as an integral part of any STH control program, essential to ensure both efficient implementation and maximal benefit for the infected individuals, their families, and communities.

In a recently published manual,[11] in the monitoring and evaluation chapter, three classes of indicator are identified: process indicators, performance indicators, and impact indicators; and for each of these classes a number of indicators and their mode of calculation are presented.

One important aspect is that the cost of the monitoring process should not divert resources from the control activities. At the planning stage, it is recommended that approximately 5–10% of the budget be reserved for monitoring activities.

WHO is collecting annual coverage data from all the countries implementing preventive chemotherapy programs, in order to report on the achievement of the global target of 75% coverage for school-aged children.

SCALING UP STH CONTROL AND FUTURE TRENDS

The global goal endorsed in 2001 by the World Health Assembly (resolution WHA 54.19), i.e. attaining with regular de-worming at least 75% and up to 100% of all school-aged children at risk of morbidity by 2010, was not reached. By that date, only a third of all children in need of de-worming had received appropriate treatment.[12]

Anticipating this problem, in 2007, WHO convened the first Global Partners' Meeting on NTDs. Some 200 participants attended the meeting, including representatives of WHO member states, United Nations

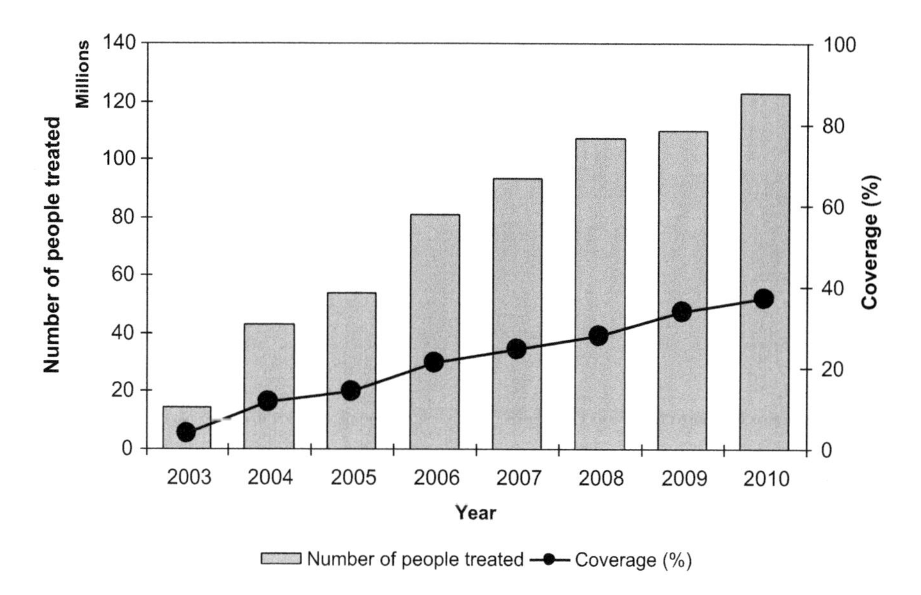

FIGURE 15.2 Number of pre-school-aged children treated with preventive chemotherapy for soil-transmitted helminthiasis, worldwide, 2003–2010.

agencies, philanthropic foundations, universities, pharmaceutical companies, international nongovernmental organizations, and other institutions dedicated to contributing with their time, efforts, and resources to tackle NTDs.[13] Since then, donors have made significant commitments, drug donation programs have been set up, and national governments in endemic countries have shown their engagement in implementing and scaling up activities to control and eliminate NTDs in general and STH in particular.

As a result of this global effort, the number of children receiving PC for STH has progressively increased to reach 328 million children in 2010,[14] corresponding to a global 30% coverage. Data show that while coverage of pre-school-aged children is steadily increasing (Figure 15.2), that of school-aged children is growing at a slower pace and seems to have reached a plateau (Figure 15.3).

A significant boost is, however, expected from 2011–2012 onwards when the effect of the recent and expanded drug donations will take place (from 2012, a quantity of 600 million tablets/year of albendazole 400 mg or mebendazole 500 mg will be available for the treatment of school-aged children in endemic countries). Coverage data for these two years are not yet available but the significant increase in requests for donated medicines submitted by endemic countries (+150 million tablets of albendazole or mebendazole in 2012 compared to 2011 data) suggests

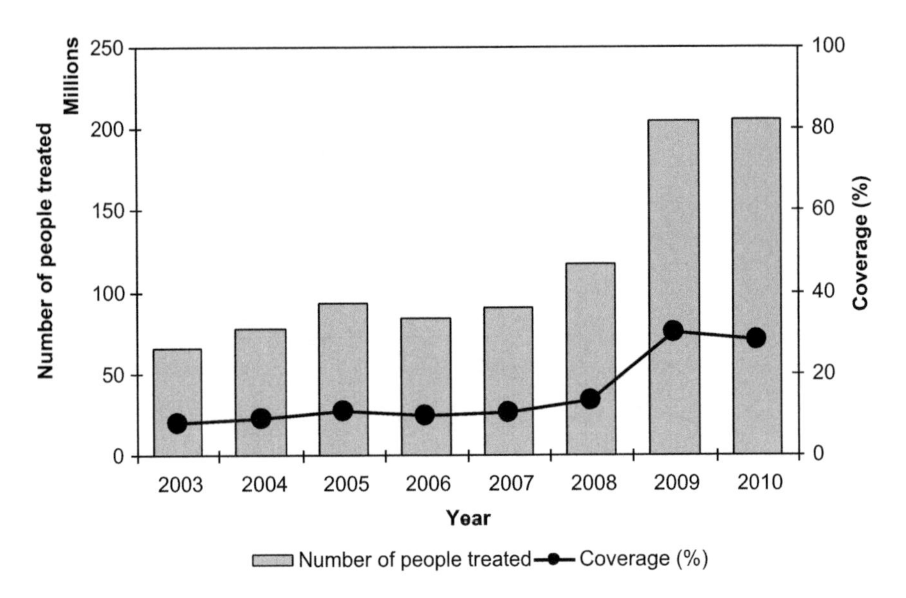

FIGURE 15.3 Number of school-aged children treated with preventive chemotherapy for soil-transmitted helminthiasis, worldwide, 2003–2010.

that the number of children treated will be substantially higher than before.

RISK OF REDUCED ANTHELMINTHIC EFFICACY

As a consequence of the considerable number of individuals treated for the control of STH, an increase of the risk of development of drug resistance has been evoked ([15,16] but see[17] for a recent review). However, the studies investigating this aspect are difficult to evaluate due to the lack of standardization and the absence of a reference threshold for drug efficacy.

WHO is seriously concerned by the possible reduction of drug efficacy and in 2007 established a working group that, since then, is meeting periodically. The working group developed guidelines on how to assess anthelmintic drug efficacy against STH and schistosomes that standardize the sample size, the interval between drug administration and evaluation of the drug efficacy, and point out egg reduction rate (ERR) as the only appropriate indicator for the evaluation of anthelmintic drug efficacy and provide a threshold for the acceptable level of drug efficacy for albendazole, mebendazole, and praziquantel.[18]

WHO hopes that this tool will facilitate the standardization of the study and an early identification of reduced drug efficacy should this develop.

INTEGRATED APPROACH

In all endemic countries, and particularly where only limited resources are available, strategies for the control of parasitic infections are being reconsidered in order to optimize human and financial resources and make the best use of personnel, expertise, surveillance and data collection, health infrastructure, and communication system. This approach of integrated control has enabled a broader range of health problems to be tackled more effectively and at affordable and sustainable costs. Integrated disease control is the merging of resources, services, and intervention at different levels and between sectors to improve health outcome.

Since 1997, countries have been developing programs based on integrated approaches to disease control. Coordination among activities previously implemented in a disease-specific fashion as well as their integration within national public health systems has been encouraged; multi-disease national plans of action are increasingly drawn up. WHO supports and endorses the development of such plans, which are then approved and adopted by the relevant governments.[18] This process has experienced a significant acceleration since the release of the guidelines on preventive chemotherapy, which promote coordination and integration among activities against neglected tropical diseases.[9]

STH are particularly suitable to integration as the approach to their control can be adapted and adopted to combat other parasitic diseases such as schistosomiasis and lymphatic filariasis. The Global Programme to Eliminate Lymphatic Filariasis, based on regular treatment of communities with single-administration drugs such as ivermectin and albendazole, also effective against STH, creates an excellent opportunity for integration. Indeed, control of STH can be the port of entry for the control of other endemic communicable and non-communicable diseases.[1] This is the approach that was successfully adopted by JOICFP (the Japanese Organization for International Cooperation in Family Planning) that utilized mass screening and treatment for intestinal nematodes to stimulate people's interest in family planning and in environmental and family hygiene.[20]

CONCLUSIONS

Preventive chemotherapy interventions to control STH can be adapted to the epidemiological characteristics of each endemic area, such as pattern of transmission and rate of reinfection, prevalence and intensity of infection, and prevalent parasite species. Although general guidelines have been recommended for targeting communities in endemic areas,[11] there is no prearranged package, and each country should tailor the approach to its particular eco-epidemiological conditions.

Socio-economic aspects such as availability of resources and health priorities are also important determinants to choose the most cost-effective approach to control STH. In a limited number of countries that are fast "developing," like the Seychelles, Iran and South Africa, it should be possible to replicate the experience of Japan and Korea (long-term elimination of the problem – with no need of further intervention). For slower "developing" countries, the objective is less ambitious (morbidity control in high-risk groups) but still necessary and relevant for the health status of those concerned.

Endemic countries should evaluate the need for integrated control of STH with the objective of improving effectiveness and reducing the cost of control programs. Priority areas for integration at the national levels, partners and opportunities for integrated STH control should be identified. As recommended by a workshop on integrated control of parasitic infections in East Mediterranean Countries,[21] where the health system allows, integration of parasitic and communicable diseases should be implemented at all levels: intersectoral (Health, Interior, Agriculture, Education), regional, district, and primary healthcare. Special efforts should be made to strengthen the intersectoral collaboration and coordination between ministries at central level, and the intrasectoral coordination within departments of the Ministry of Health.

The WHO strategy for control of STH is designed to meet the needs of endemic countries and to promote tools for diagnosis and disease control that are appropriate and sustainable. An essential component is the monitoring and evaluation that enables managers of helminth control programs and health planners to quantify the benefits of the intervention and to adapt the control strategy according to its outcome.[11]

References

1. World Health Organization. *Report of the WHO informal consultation on the use of chemotherapy for the control of morbidity due to soil-transmitted nematodes in humans*. Geneva 29 April to 1 May 1996. Division of Control of Tropical Diseases. WHO/CTD/SIP.96.2. Geneva: World Health Organization; 1996.
2. Ministry of Health and Social Affairs. *Korean Association for Parasite Eradication. Prevalence of intestinal parasitic infection in Korea, sixth report*. Seoul: Monographic series [in Korean] KAPE; 1996.
3. De Carneri I. Parasitologia generale ed umana [in Italian], Casa Editrice Ambrosiana Milano. 1989;44–45.
4. Hall A, Hewitt G, Tuffrey V, De Silva N. A review and meta-analysis of the impact of intestinal worms on child growth and nutrition. *Maternal and Child Nutrition* 2008;**4**(Suppl. 1):118–236.
5. McConnell CR, Brue SL. *Economics. The Economy of Developing Countries*. p. 757–90. McGraw-Hill; 2008.

6. Bundy DAP, Hall A, Medley GF, Savioli L. Evaluating measures to control intestinal parasitic infections. *World Health Stat Q* 1992;**45**:168–79.
7. Stoltzfus RJ, Albonico M, Chwaya HM, Savioli L, Tielsch J, Schulze K, et al. Hemoquant determination of hookworm-related blood loss and its role in iron deficiency in African children. *Am J Trop Med Hyg* 1996;**55**:399–404.
8. Guyatt HL, Bundy DA, Evans D. A population dynamic approach to the cost-effectiveness analysis of mass anthelminthic treatment: effects of treatment frequency on *Ascaris* infection. *Trans R Soc Trop Med Hyg* 1993;**87**:570–5.
9. World Health Organization. *Preventive chemotherapy in human helminthiasis. Coordinated use of anthelminthic drugs in control interventions: a manual for health professionals and programme managers.* Geneva: World Health Organization; 2006.
10. Gabrielli AF, Montresor A, Chitsulo L, Engels D, Savioli L. Preventive chemotherapy in human helminthiasis: theoretical and operational aspects. *Trans R Soc Trop Med Hyg* 2011;**105**(12):683–93.
11. World Health Organization. *Helminth control in school age children. Guide for managers of control programme.* 2nd ed. 2012.
12. World Health Organization. Soil-transmitted helminthiases: estimates of the number of children needing preventive chemotherapy and number treated, 2009. *Weekly Epidemiological Record* 2011;**25**(86):257–68.
13. World Health Organization. *A turning point. Report of the Global Partners' Meeting on Neglected Tropical Diseases.* Geneva: World Health Organization; 2007.
14. World Health Organization. Soil-transmitted helminthiases: number of children treated in 2010. *Weekly Epidemiological Record* 2012;**23**:225–32.
15. Geerts S, Coles GC, Gryseels B. Anthelmintic resistance in human helminths: learning from the problems with worm control in livestock. *Parasitol Today* 1997;**13**:149–51.
16. Sacko M, De Clercq D, Behnke JM, Gilbert FS, Dorny P, et al. Comparison of the efficacy of mebendazole, albendazole and pyrantel in treatment of human hookworm infections in the southern region of Mali, West Africa. *Trans R Soc Trop Med Hyg* 1999;**93**:195–203.
17. Vercruysse J, Levecke B. Pritchard, R. Human soil-transmitted helminthes: implications of mass drug administration. *Curr Opin* 2012;**25**(6):703–8.
18. World Health Organization Assessing the efficacy of anthelminthic drug against schistosomiasis and soil-transmitted helminthiases. Geneva: World Health Organization; 2013.
19. World Health Organization. *Integrating Disease Control: the challenge. Division of Control of Tropical Diseases.* WHO/CTD/98.7. Geneva: World Health Organization; 1998.
20. Yokogawa M. JOICFP'S experience in the control of ascariasis within an integrated programme. In: Crompton DWT, Nesheim MC, Pawloski ZS, editors. *Ascariasis and its Public Health Significance.* London and Philadelphia: Taylor and Francis; 1985. p. 265–77.
21. World Health Organization. Report of the WHO Regional Workshop on the integrated control of parasitic infections. Tunis 22–24 April 2001. Division of Control of Tropical Diseases. WHO-EM/CTD/2001. Alexandria; 2001.

CHAPTER

16

Diagnosis and Control of Ascariasis in Pigs

Johnny Vlaminck, Peter Geldhof

Ghent University, Belgium

OUTLINE

INTRODUCTION

Since their domestication, approximately 11,000 years ago, pigs have been raised as a source of meat for human consumption.[1] Today, the pig industry is the largest meat-producing industry in the world with a yearly production exceeding 100 million tonnes, making pork the meat of choice

Ascaris: The Neglected Parasite
http://dx.doi.org/10.1016/B978-0-12-396978-1.00016-1

Copyright © 2013 Elsevier Inc. All rights reserved.

worldwide (source: FAO). Since the early days of pig farming, farmers have seen the productivity of their animals compromised by the many parasitic infections affecting pigs. Fortunately, in the last couple of decades, the impact of these parasites on pigs and their productivity has been significantly reduced. The main reasons for this are the switch from smaller, open air, rural farming to more industrial-sized, high-intensity, indoor farming and the advent of pharmaceuticals directed against the parasites that affect pigs. Infections with *Oesophagostomum* spp., *Ascaris suum* and to a much lesser extent *Trichuris suis* have successfully survived the transition to conventional indoor conditions and are currently thriving in some farms.[2] Infections with *A. suum*, however, show the highest prevalence.[3–5]

In order to understand the problems associated with the methods of control and diagnosis discussed in the second and third part of this chapter, a proper understanding of the life-cycle of this parasite is indispensable. Pigs become infected with *A. suum* when they take up eggs from the environment in which an infective stage larva has developed. The larva emerging from the egg is not a second-stage larva (L2) as was previously presumed but rather a third-stage larva (L3) covered by a loosened second-stage cuticle.[6] The larvae of *Ascaris* complete two molts within the egg. After oral intake, hatching is induced by the altering chemical and physical factors of the new environment. The larvae inside the egg are stimulated to secrete proteinases and chitinases, which presumably help them degrade the different layers of the eggshell from the inside out.[7,8] When the larvae emerge from the egg, they start their hepatotracheal migration by penetrating the wall of the cecum and upper part of the colon.[9] The L3s are then transported by the mesenterial blood veins and can reach the liver as fast as 6 hours post-infection.[9,10] The larvae are stuck in the capillaries and destroy liver tissue in order to reach the efferent blood vessels.

The inflammatory reaction to this damage is manifested as the so-called "white spots" that are visible on the surface of the liver. These white spots appear as early as 3 days post-infection and start to resolve after about 2–3 weeks post-infection.[11,12] The bloodstream carries the larvae to the next capillary system, which is the lung, where they penetrate the alveoli, move up the respiratory tree, and are eventually swallowed again. From 8 days onwards, the L3s finally return to the small intestine where they start their first ecdysis inside the host, to reach the L4 stage. Between days 17 and 21 post-infection, most of the larvae are eliminated.[11,13] The mechanisms responsible for this "self-cure" reaction are currently still unknown.

The result of this elimination is a small, overdispersed population of adult worms in the intestine, characteristic of *Ascaris* infections.[11,14] After about 6 weeks, the worms have reached maturity and egg shedding by the female parasites starts. A study by Mejer and Roepstorff[15] showed that pigs that are naturally exposed to a paddock contaminated with *A. suum* show the highest egg excretion 17 weeks after being introduced onto the paddock. After this point, egg counts begin to drop. The lifespan of adult *A. suum* worms can be over 1 year. This is significantly longer than the average life of a fattening pig these days. Therefore, once adult worms are present in pigs, the egg shedding increases with the age of the pigs unless the worms are cleared from the intestine by anthelmintic therapy.

Ascariasis is present in most countries worldwide. However, few countries have up-to-date information on its prevalence. Many of the prevalence studies were performed decades ago, making the results unrepresentative for the current situation. Table 16.1 provides an overview of the studies in the literature investigating the prevalence of *A. suum*. Different studies applied different methods to determine the presence of *Ascaris* in pigs. Some used the percentage of rejected livers as a measure of parasite exposure; others checked the presence of worms by examining the intestinal tract at slaughter or by coprological examination of fecal samples to detect parasite eggs. The table shows the results of studies that were performed in different countries, in different times and settings. Because of this, it is not possible to draw conclusions concerning epidemiological trends of ascariasis on pig farms. However, two comparable studies from Denmark performed in 1989 and 2010, two decades apart, show that the situation has remained roughly the same with *A. suum* being present on 88% and 76% of the investigated farms, respectively.

There are several reasons why *Ascaris* is still so prevalent in current high intensity pig farms. First, because the ascarid has a direct life-cycle and is therefore not reliant on other organisms for its transmission to new hosts. Second, the female parasitic worms are highly fecund and are capable of producing hundreds of thousands of eggs per day that contaminate the surroundings instantaneously. Third, the eggs are extremely resistant to external environmental factors, ensuring their survival for up to several years in the appropriate conditions. Finally, and maybe most importantly, the lack of efficient diagnostic tools to shed light on the problem at a farm level, together with the seemingly unimportant health consequences of *A. suum* infections, have led to a certain negligent attitude towards roundworm infection on the part of the farmer. Yet, today, the economic pressure on the pig industry is of such extent that production efficiency is crucial and nobody can afford to neglect the treatment of a parasitic ailment like ascariasis, which has shown to significantly affect economic profitability.[16–20] However, in order to effectively control *A. suum* infections on a pig farm, it is necessary for the farmer to be informed of the evolution of the "*Ascaris*-status." In order to do so, good diagnostic tools are indispensable. These tools will allow for

TABLE 16.1 Results of epidemiological studies show that infections with *A. suum* in pigs remain highly prevalent throughout the world. Farms were considered infected when pigs on the farms were found positive for *A. suum* infection after examination

				% Infected pigs			
Year	**Country**	**Sample size and type**	**% Infected farms**	**Egg +**	**Worm +**	**Liver +**	**Reference**
1966	Sweden	200 sows			34%		a
1974	Canada	90 pigs			39%		b
1976	The Netherlands	653,540 pigs				16%	c
1980	Canada	2500 pigs			37%	46%	d
1980	Canada	2500 pigs				44%	
1980	England	468 pigs			16%		e
1989	Denmark	66 farms	88%	30%			f
1990	Nigeria	1000 pigs, 2 areas		53 & 10%			g
1990	Canada	15 herds		32%	35%	82%	41
1993	Zimbabwe	7128 pigs				17%	h
1997	Belgium	20,758 livers				36%	i
1997	Canada	2500 pigs				44%	25
1997	Canada	500 pigs			18%	50%	

1998	Denmark, Norway, & Sweden	516 herds		25–30%		67
1998	Iceland			13%		
1998	Finland			5%		
1999	Denmark	413 farms	56%			68
2000	China	100 outdoor pigs		37%		j
2001	Germany	13 farms		33%		3
2004	Tanzania	70 pigs			44%	k
2005	The Netherlands	9 conventional farms	11%			l
2005	The Netherlands	16 free range farms	50%			
2005	The Netherlands	11 organic farms	73%			
2005	China	3636 pigs		5%		m
2005	India	501 pigs			16%	n
2008	Kenia	230 pigs			29%	o
2010	USA	91 farms (finishing pigs)	39%			p

(*Continued*)

TABLE 16.1 Results of epidemiological studies show that infections with *A. suum* in pigs remain highly prevalent throughout the world. Farms were considered infected when pigs on the farms were found positive for *A. suum* infection after examination—cont'd

				% Infected pigs			
Year	**Country**	**Sample size and type**	**% Infected farms**	**Egg +**	**Worm +**	**Liver +**	**Reference**
2010	USA	40 farms (sows)	25%				q
2010	Switzerland	90 conventional farms	13%				r
2010	Switzerland	20 free range farms	35%				
2010	Sweden	2.4 million pigs				5%	s
2010	Denmark	79 farms, 1790 sows	76%	30%			5
2011	China	916 pigs		15%			65
2011	Uganda	106 pigs from 56 households,		40%	73%		t
2011	Tanzania	13,310 pigs				4%	u
2011	Tanzania	731 pigs			8%		v
2012	England	34,168 pigs				4%	86

[a] *Roneus O. Studies on the aetiology and pathogenesis of white spots in the liver of pigs. Acta Vet Scand. 1966;7:Suppl 16:1–112.*
[b] *Martin LJ, Gibbs HC, Pullin JW. Gastrointestinal parasites of swine in Quebec. I. An incidence survey. Can Vet J. 1974;15(3):72–6.*

[c] Tielen MJ, Truijen WT, Remmen JW. The incidence of diseases of the lung and liver in slaughtered pigs as a criterion in the detection of herds in which the disease is a recurrent problem (author's transl). Tijdschrift voor diergeneeskunde. 1976;101(17):962–71.
[d] Polley LR, Mostert PE. Ascaris suum in Saskatchewan pigs: an abattoir survey of prevalence and intensity of infection. Can Vet J. 1980;21(11):307–9.
[e] Pattison HD, Thomas RJ, Smith WC. A survey of gastrointestinal parasitism in pigs. Vet Rec. 1980;107(18):415–18.
[f] Roepstorff A, Jorsal SE. Prevalence of helminth infections in swine in Denmark. Vet Parasitol. 1989;33(3–4):231–9.
[g] Salifu DA, Manga TB, Onyali IO. A survey of gastrointestinal parasites in pigs of the Plateau and Rivers States, Nigeria. Revue d'elevage et de medecine veterinaire des pays tropicaux. 1990;43(2):193–6.
[h] Makinde MO, Majok AA, Hill FWG. The prevalence of subclinical diseases in abattoir pigs in Zimbabwe. Prev Vet Med. 1993;15(1):19–24.
[i] Vercruysse J, VanHoof D, DeBie S. Study on the prevalence of white spots of the liver in pigs in Belgium and its relationship to management practices and anthelmintic treatment. Vlaams Diergeneeskundig Tijdschrift. 1997;66(1):28–30.
[j] Boes J, Willingham AL, 3rd, Fuhui S, Xuguang H, Eriksen L, Nansen P, et al. Prevalence and distribution of pig helminths in the Dongting Lake Region (Hunan Province) of the People's Republic of China. J Helminthol. 2000;74(1):45–52.
[k] Ngowi HA, Kassuku AA, Maeda GE, Boa ME, Willingham AL. A slaughter slab survey for extra-intestinal porcine helminth infections in northern Tanzania. Trop Anim Health Prod. 2004;36(4):335–40.
[l] Eijck IA, Borgsteede FH. A survey of gastrointestinal pig parasites on free-range, organic and conventional pig farms in The Netherlands. Vet Res Comm. 2005;29(5):407–14.
[m] Weng YB, Hu YJ, Li Y, Li BS, Lin RQ, Xie DH, et al. Survey of intestinal parasites in pigs from intensive farms in Guangdong Province, People's Republic of China. Vet Parasitol. 2005;127(3–4):333–6.
[n] Gaurat RP, Gatne ML. Prevalence of helminth parasites in domestic pigs (Sus scrofa domestica) in Mumbai: an abbatoir survey. J Bombay Vet College. 2005;13(1/2):100–2.
[o] Nganga CJ, Karanja DN, Mutune MN. The prevalence of gastrointestinal helminth infections in pigs in Kenya. Trop Anim Health Prod. 2008;40(5):331–4.
[p] Pittman JS, Shepherd G, Thacker BJ, Myers GH, Fransisco CJ. Prevalence of internal parasites in a production system: Part II – Finishing pigs. 21st IPVS Congress; 2010 July 18–21, 2010; Vancouver, Canada; 2010. p. 807.
[q] Pittman JS, Shepherd G, Thacker BJ, Myers GH, Fransisco CJ. Prevalence of internal parasites in a production system. Part I – Sows. 21st IPVS Congress; 2010 July 18–21, 2010; Vancouver, Canada; 2010. p. 806.
[r] Eichhorn L, Zimmermann W, Gottstein B, Frey CF, Doherr MG, Zeeh F. Gastrointestinal nematodes in Swiss pig farms – an explorative overview. 21st IVPS Congress; 2010 July 18–21, 2010; Vancouver, Canada; 2010. p. 808.
[s] Lundenheim N, Holmgren N. Prevalence of lesions found at slaughter among Swedish fattening pigs. 21st IVPS Congress; 2010 July 18–21; Vancouver, Canada; 2010. p. 925.
[t] Nissen S, Poulsen IH, Nejsum P, Olsen A, Roepstorff A, Rubaire-Akiiki C, et al. Prevalence of gastrointestinal nematodes in growing pigs in Kabale District in Uganda. Trop Anim Health Prod. 2011;43(3):567–72.
[u] Mellau BL, Nonga HE, Karimuribo ED. Slaughter stock abattoir survey of carcasses and organ/offal condemnations in Arusha region, northern Tanzania. Trop Anim Health Prod. 2011;43(4):857–64.
[v] Mkupasi EM, Ngowi HA, Nonga HE. Prevalence of extra-intestinal porcine helminth infections and assessment of sanitary conditions of pig slaughter slabs in Dar es Salaam city, Tanzania. Trop Anim Health Prod. 2011;43(2):417–23.

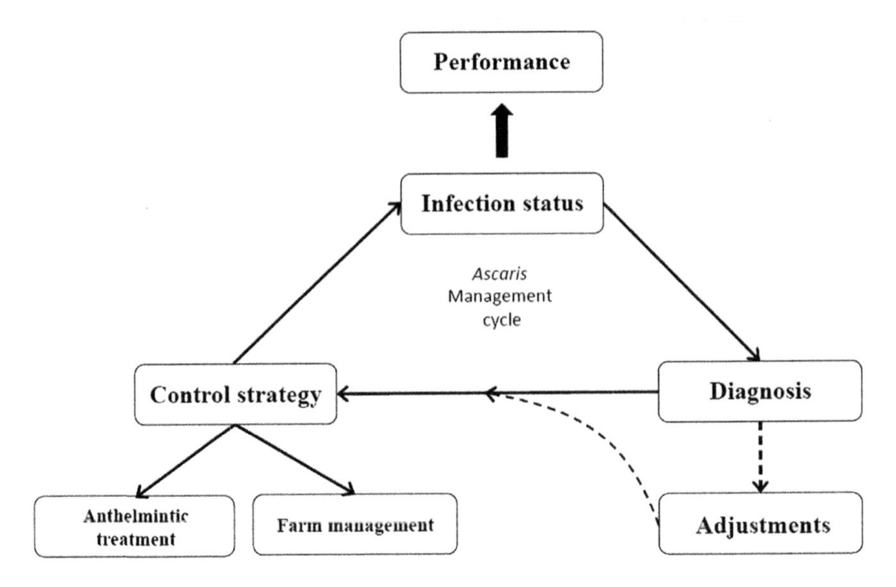

FIGURE 16.1 **The *Ascaris* control/management cycle.** This conceptual framework can be applied to continuously improve the worm control management on a pig farm. Parameters of pig performance are assessed as well as the worm status of the pig herd, using the diagnostic methods at hand. If necessary, adjustments to the anthelmintic treatment strategies and farm management can be implemented to increase performance and reduce *A. suum* presence on the farm. In time, these changes are evaluated and further adjustments could be applied during the next cycle.

the evaluation of the currently applied control strategies. If necessary, adjustments to the control program could be implemented and their effect on the infection intensity on the farm can be investigated in subsequent fattening rounds. The cyclic relationship between control practices and their evaluation by diagnostics is shown in Figure 16.1.

DIAGNOSIS OF PIG ASCARIASIS

Infections with *A. suum* very seldom cause clinical disease. Therefore, such infections typically remain unrecognized by farmers and their veterinarians. In a study by Theodoropoulos et al.[21] it was shown that most of the farmers considered the magnitude of worm infection in their own farm to be insignificant. Nevertheless, almost all farmers used anthelmintics to treat their stock prophylactically. The same story is applicable to Danish pig farmers,[22] and probably the majority of farmers worldwide. Very rarely do pig farmers investigate if the actions they undertake to control ascariasis on their farm are the most appropriate and whether these actions significantly improve the situation. To be able to evaluate the evolution of the infection status and therefore the efficacy of the applied control strategies, the use of correct diagnostic tools is crucial. Different possibilities are at hand to diagnose the presence of *A. suum* infections on pig farms.

Presence of Worms

Most farmers determine infection by observing expelled adult worms after anthelmintic treatment.[21] Although the expelled worms might not have conserved their exact shape and color, a farmer should be able to detect these spaghetti-like organisms mixed with the feces. When immature worms, which are notably smaller, are expelled, identification by the untrained eye might be harder. Clinical signs such as unthriftiness, difficulties breathing and wheezing, and indirect signs like decreased growth rate and overall lower food conversion efficiency could also indicate *A. suum* infection but are hardly ever associated with ascariasis by the farmer because of their non-specificity.

The presence of adult worms in the small intestine of pigs can be registered at slaughter. However, except in experimental settings, the detection of worms in the intestine is hardly ever done in practice. If performed, one must keep in mind the time between last anthelmintic treatment and slaughter, as the larval stages in the intestine are small and easily overlooked. This could lead to wrong conclusions when no worms could be detected. The detection of adult parasites is, however, only a qualitative measure, and not a quantitative one, as the number of adult worms is not representative of the amount of migrating larvae the pig has been exposed to. Regardless of the dose regimen, the number of worms that end up in the small intestine is generally inconsistent and independent of the intake of infective stages.[23] Furthermore, there seems to exist an inverse relationship between the number of adults found in the intestine and the amount of eggs given during a single experimental infection dose.[11,24] Additionally, *Ascaris* populations are strongly aggregated within the pig population with few pigs carrying the majority of the worms while most of the pigs carry few or even no adult worms at all.[11,25,26] The reason for this is currently still unknown (see Chapters 7 and 12). It seems that host genetics could be an important factor.[27]

Fecal Examination

Detecting the eggs excreted by adult female worms can prove the presence of adult worms in the intestine. For this, coprological techniques of sedimentation/flotation or McMaster can be applied.[28] Both techniques make use of the buoyancy of parasite eggs in a dense salt/sucrose solution to separate the eggs from the fecal debris. The sedimentation/flotation technique is a qualitative technique and is applied to investigate whether parasites are present in the host or not. To investigate and compare the quantity of eggs that are excreted, the McMaster test is used. *Ascaris* eggs are round or oval shaped with a thick brown irregular wall and are easily recognized. Although coprological examinations are easy to perform and require relatively cheap materials, these tests are time consuming and therefore not cost effective or optimal for the screening of large numbers of samples.

The amount of eggs excreted by a pig can give an impression of its intestinal worms burden because, in continuously exposed pigs, the quantity of eggs that are shed seems to be correlated with the number of adult worms in the intestine.[29,30] However, the aggregated distribution of adult worms within the infected population will be reflected in the number of pigs excreting eggs. When using coprological examinations for the detection of *A. suum* infections in pigs, there is a significant possibility of false positive and false negative egg counts. False positives are the result of coprophagia and/or geophagia and their prevalence and magnitude depend on different management and housing factors.[31] For diagnosis at farm level, however, these false positive samples are not as important as the false negatives. Boes et al.[31] has shown that as much as 23% of the investigated pigs that harbored worms in their intestine did not excrete any eggs.[31] These false negative results are possible when only immature worms are present or when only worms of a single sex are present. Evidently, these factors have an impact on the sensitivity of fecal examination for the diagnosis of *A. suum* infection. Moreover, as addressed above, the number of adult worms and therefore the number of excreted eggs are not representative of the level of parasite exposure.

Liver White Spots

Hepatic white spots are the most characteristic lesions caused by migrating *A. suum* larvae in pigs. After larvae migrated through the liver, the destroyed liver cells are replaced by interlobular depositions of fibrous tissue and cellular infiltrates, producing the typical white spots.[32,33] Milk spots have the property to heal, causing the livers to appear normal about 35 days after primary inoculation with *Ascaris* eggs.[34] In pigs that have primed immune responses due to vaccination or previous exposure to the parasite, white spots are larger, more distinct, and more persistent.[34–37] This would suggest that white spot formation is the result of secondary immunological reactions to antigenic material released by migratory or dying larvae, which seems to be in accordance with previous studies showing a significant relationship between the number of white spots and antibodies against *Ascaris* antigens.[12,36,38,39]

Although the number of liver lesions indicates the recent passing of migrating *A. suum* larvae, there exists no relationship whatsoever between the number of lesions and the number of adult parasites that eventually end up in the small intestine.[40] Presence of hepatic lesions predict the presence of intestinal stages with low (29%) positive and high (94%) negative predictive values.[25] In a study by Bernardo et al.,[41] the presence of white spots on the liver is more than twice as prevalent as worms in the small intestine at slaughter. The absence of worms in the intestine when white spots are visible on the liver can mean two things: (1) eggs have recently been ingested, but not enough time has elapsed to allow development of macroscopically detectable ascarids or (2) infection had been cleared by the host immune response but lesions had not yet resolved. When milk spots are present, it is highly unlikely that *A. suum* infection is absent. The presence of milk spots has a high sensitivity, very low specificity, and high negative predictive value as a screening test for ascariasis in individual hogs.[41] Under continuous exposure, the number of white spots on the liver increases until weeks 6–9, after which there is a gradual decline towards lower levels.[23] This decline could be attributed to the build-up of immunity as continuous exposure to *A. suum* infections stimulates the development of strong protective immunity. Both Eriksen et al.[35] and Urban et al.[42] suggested the existence of a prehepatic barrier in pigs repeatedly exposed to *A. suum* infections. This would impede freshly acquired larvae from reaching the liver and therefore prevent the formation of white spots. Hence, the number of liver white spots is a poor indicator of long-term *A. suum* exposure as it only reflects recent larval migration. Livers might therefore look normal or only mildly affected at slaughter even though pigs have been exposed to high numbers of infective eggs during the course of their life.

The visual assessment of livers is also very subjective. Especially when pigs with low numbers of white spots are slaughtered the decision on whether or not an abnormality on the liver is considered a true white spot largely depends on the perception of the person doing the assessment. In addition, this usually has to be performed within an extremely short timespan due to the high speed of slaughter. This supports the variability in the white spot counts and increases doubt on the reliability and uniformity of the data on liver lesions at slaughter. The quality of these data should be improved if they are to be of use in the future.

Serology

The use of serological tests is widespread and generally accepted in the pig industry. Tests are available for a number of important bacterial and viral diseases in pigs (e.g. salmonellosis, mycoplasma infection, Aujeszky's disease, PRRS virus, etc.). However, for *Ascaris*, a commercial serological test is currently not available. The use of a serological method, however, could in theory overcome the difficulties associated with the traditional methods (examination of feces or livers) of diagnosing ascariasis in pigs.

It can be assumed that all pigs within a unit are exposed to a similar amount of migrating larvae as they reside in the same infective environment. In contrast, only few pigs in the population will eventually harbor adult worms in their intestine.[11,25] Moreover, pig performance has shown to be negatively associated with the amount of larvae migrating through the body of the host.[18] Hence, serological diagnosis should be aimed at detecting larval exposure rather than the presence of adult worms.

Several studies have previously described the use of an Enzyme Linked ImmunoSorbent Assay (ELISA) for the detection of anti-*Ascaris* antibodies in the sera of swine. In this regard, the use of different adult and larval extracts or excretory/secretory products has been evaluated.[37,38,43–47] Although most of these tests have been shown to be effective in diagnosing *A. suum* infections, the sensitivity and specificity of these tests are unclear. The reason for this is that traditional methods, like the detection of parasite eggs in the stool or white spots on the liver at slaughter, were used to categorize the pigs used for the validation of the serodiagnostic test. Due to the lower sensitivity of these traditional methods (see previous sections) there is a significant chance to include false negative samples in the evaluation of the ELISA.

Recently, for the validation of an experimental ELISA, Vlaminck et al.[47] used sera from 190 pigs that were trickle-infected twice a week for 14 weeks, thereby ensuring that no false negative samples would be included for the validation of the ELISA test or the determination of its cut-off value. In the same study it was also advised to test serum samples from older fattening pigs, as the number of seropositive pigs increases with time.[47] This would reduce the chance of including false negatives when screening a farm for the presence of *Ascaris* infection using this serodiagnostic technique. This ELISA is currently being used to investigate whether or not the serological response to *A. suum* can be coupled to different production parameters of the investigated pig farm. If so, test results could appeal more to the farmer. It would help improve his awareness on the topic and convince him of the fact that a more effective helminth control program could significantly improve farm profitability. In this way, serological testing could serve as a cost-effective, supportive tool for the veterinarian to assess the worm status of a farm and to evaluate whether changes in worm-control management are necessary or not.

CONTROL

Once *A. suum* is present on a farm, it is rather challenging to completely eliminate its presence. Although the aim of control programs against this parasite should be to eradicate the parasite completely, a significant reduction in transmission intensity will readily result in a marked decrease of adverse effects on the health and productivity of the pig herd. To effectively control parasitic infections on farms, control-by-treatment alone will not suffice unless supported by good general farm management and increased hygienic standards.[40,48]

Chemotherapy

Before the advent of broad-spectrum anthelmintics in the 1960s, treatment of worm infections in pigs was hardly ever applied. Since then, the control of parasitic infections on pig farms has become increasingly reliant on mass treatment with anthelmintic drugs. When selecting a drug for the control of *Ascaris* infections, several things need to be taken into account. Namely, (1) the margin of safety of the used compound, (2) the efficacy of compound against the different stages of *A. suum*, (3) the spectrum of activity, (4) the mode of administration, and last but not least (5) the cost of the drug. All of the modern anthelmintics are found to be very safe. Treatments with doses exceeding the recommended dose did not show any clinical effect in pigs.[49] An overview of trials testing the effectiveness of different anthelmintic drugs used for treating pigs with *Ascaris* infections is shown in Table 16.2. For the effective control of helminth infections on a pig farm it is advantageous to select an anthelminticic that kills the larval stages and has a broad spectrum of activity. All compounds presented in Table 16.2 have proven to be highly effective against adult roundworms in pigs. Pyrantel, fenbendazole, flubendazole, and levamisole also have shown activity against migrating stages of *A. suum*.[50–53] Piperazine and pyrantel have restricted activity against the other common nematode infections (*Trichuris suis* and *Oesophagostomum* spp.)[49] and are therefore not the drugs of choice when co-infections with these parasites are present on a farm. Anthelmintics of the macrocyclic lactone family (e.g. ivermectin, doramectin) have a significantly longer residual effect in comparison with the other anthelmintics and are also highly efficient against ectoparasites like mange mites (*Sarcoptes scabiei* var *suis*) and hog lice (*Haematopinus suis*).[54–58]

Anthelmintics are available in a variety of formulations. Administration of the drug in either water or feed is the easiest and most economical way, but has a major disadvantage in variable dosages being delivered when numbers of pigs are being treated in groups from a single source of medication. Delivery by injection, on the other hand, is more accurate

TABLE 16.2 Efficacy of different anthelmintic drugs experimentally tested against adult and larval stages of *A. suum* in pigs

Drug	Formulation	Dose rate	Effectiveness to adult *A. suum.*	Effectiveness to larval stages	Reference
Dichlorvos	Orally	43 mg/kg, 1 d	100%		a
	In feed	17 mg/kg, 1 d	100%		b
Piperazine	In water	200 mg/kg, 1 d	100%		c
Fenbendazole	In feed	3 mg/kg, 3 d	100%		a
			100%		d
				99% liver and lung stage L3	50
			100%		e
			92.4%		b
		2.5 mg/kg, 3 d		100% to L4 stage	f
Flubendazole	In feed	30 ppm, 10 d	100%		g
		5 mg/kg, 1 d	100%		
		1.5 mg/kg, 5 d	100%	85% to L4 stage	h
Oxibendazole	Orally	15 mg/kg, 1 d	100%	92% to L4 stage	i
	In feed	40 ppm, 10 d	100%	100% to L4 stage	

Levamisole	Injection	7.5 mg/kg, 1 d	100%	92.3% to L4 stage and 63.5% to lung stage L3	53
	In feed	8 mg/kg, 1 d	97.9%		b
Pyrantel	In feed	96 g/ton feed, 24 d	100%	97–100% to lung stage L3	52
		106 mg/kg feed, cont.	100%	Reduced liver lesions	j
Ivermectin	In feed	2 ppm, 7 d	100%		k
		100 μg/kg, 7 d	97.7%	100% to L4 stage	l
		200 μg/kg, 7 d	100.0%	100% to L4 stage	
	Injection	300 μg/kg, 1 d	98.7%		a
			100 %		m
			97.5%		54
	In water	300 μg/kg, 1 d	100%	100% to L4 stage	m
Doramectin	Injection	300 μg/kg, 1 d	100%	100% to L4 stage	n
			100%		o
			100%		56
			100%		54

[a] *Marchiondo AA, Szanto J. Efficacy of dichlorvos, fenbendazole, and ivermectin in swine with induced intestinal nematcde infections. Am J Vet Res. 1987;48(8):1233–5.*
[b] *Marti OG, Stewart TB, Hale OM. Comparative efficacy of fenbendazole, dichlorvos, and levamisole HCI against gastrointestinal nematodes of pigs. J parasitol. 1978;64(6):1028–31.*

[c] *Steffan P, Olaechea F, Roepstorff A, Bjorn H, Nansen P. Efficacy of piperazine dihydrochloride against* Ascaris suum *and* Oesophagostomum *species in naturally infected pigs. Vet Rec. 1988;123(5):128–30.*

[d] *Stewart TB, Marti OG, Hale OM. Efficacy of fenbendazole against five genera of swine parasites. Am J Vet Res. 1981;42(7):1160–2.*

[e] *Corwin RM, Pratt SE, Muser RK. Evaluation of fenbendazole as an extended anthelmintic treatment regimen for swine. J Am Vet Med Ass. 1984;185(1):58–9.*

[f] *Stewart TB, Rowell TJ. Susceptibility of fourth-stage* Ascaris suum *larvae to fenbendazole and to host response in the pig. Am J Vet Res. 1986;47(8):1671–3.*

[g] *Vanparijs O, Hermans L, Marsboom R. Efficacy of flubendazole against gastrointestinal and lung nematodes in pigs. Vet Rec. 1988;123(13):337–9.*

[h] *Bradley RE, Guerrero J, Becker HN, Michael BF, Newcomb K. Flubendazole: dose range and efficacy studies against common internal parasites of swine. Am J Vet Res. 1983;44(7):1329–33.*

[i] *Pecheur M. Efficacy of Oxibendazole against* Oesophagostomum dentatum *and* Ascaris suum *of pigs. Ann Med Vet. 1983;127(3):203–8.*

[j] *Kennedy TJ, Conway DP, Bliss DH. Prophylactic medication with pyrantel to prevent liver condemnation in pigs naturally exposed to* Ascaris *infections. Am J Vet Res. 1980;41(12):2089–91.*

[k] *Primm ND, Hall WF, DiPietro JA, Bane DP. Efficacy of an in-feed preparation of ivermectin against endoparasites and scabies mites in swine. Am J Vet Res. 1992;53(4):508–12.*

[l] *Alva-Valdes R, Wallace DH, Foster AG, Ericsson GF, Wooden JW. Efficacy of an in-feed ivermectin formulation against gastrointestinal helminths, lungworms, and sarcoptic mites in swine. Am J Vet Res. 1989;50(8):1392–5.*

[m] *Schillhorn van Veen TW, Gibson CD. Anthelmintic activity of ivermectin in pigs naturally infected with* Ascaris *and* Trichuris. *Am J Vet Res. 1983;44(9):1732–3.*

[n] *Stewart TB, Fox MC, Wiles SE. Doramectin efficacy against gastrointestinal nematodes in pigs. Vet Parasitol. 1996;66(1–2):101–8.*

[o] *Mehlhorn H, Jones HL, Weatherley AJ, Schumacher B. Doramectin, a new avermectin highly efficacious against gastrointestinal nematodes and lungworms of cattle and pigs: two studies carried out under field conditions in Germany. Parasitol Res. 1993;79(7):603–7d = day.*

in dosage but is time and labor intensive, which adds to the overall cost of treatment. Injectable anthelmintics have a significantly longer withdrawal period in comparison with in-feed formulations[54] and are usually conserved for the treatment of the breeding stock. The overall cost of anthelmintic treatment will vary, depending on the choice of anthelmintic, number of treatments, and weight of the animals being treated. In general, for the choice of anthelmintic, pig farmers will base themselves on information received from their veterinarians or feed companies and, more importantly, on their previous experiences with a selected anthelmintic.[21]

To date, no studies have reported anthelmintic resistance in *A. suum* in pigs. Whether anthelmintic resistance would be easily induced in *A. suum* is unknown. However, the development of anthelmintic resistance in *Oesophagostomum* spp. and in the closely related horse parasite *Parascaris equorum*[59–62] suggests that continued malpractice in this area might eventually lead to anthelmintic resistance in *A. suum*.

Treatment Regime

Epizootiological knowledge on helminthic infections like ascariasis on pig farms is still rather limited and this will have definitely had its impact on the evolution of integrated parasite control programs. Routine anthelmintic treatments are commonly used these days in intensive pig rearing systems. Unfortunately, such treatments are not systematic and treatment schemes vary according to farm and animal category. Today, the breeding stock is usually treated at 3–4-month intervals or prior to farrowing. Piglets are typically treated upon entering the fattening units and maybe once more halfway through the fattening period. However, these treatment schemes seem to be unable to significantly decrease the infection intensity on a farm. Treatment frequency must be based on the prepatent period of the worm species. For *Ascaris* this is 42 days or 6 weeks. However, when anthelmintic treatment is ineffective in killing the migrating larval stages, it is possible that egg shedding occurs earlier as larvae remain in the lung up to 10 days post-infection.[11] Treatment of pigs every 5 weeks would therefore eradicate the worms before they are able to produce millions of eggs and contaminate the environment again. When a strict deworming scheme is applied, the infectivity of the environment will drop, as new hosts will take up eggs from the environment and eggs will be swept away by cleaning actions or eventually lose their viability. This will ultimately result in a stock of pigs with a reduced or absent parasitic burden. It is important to understand that persistence in the repeated application of the anthelmintic is necessary. If the environmental contamination is high, it will most likely take several fattening rounds to eventually reduce the infection pressure.[63,64]

In a Swedish study it appeared that the pens of sows and fatteners are the heaviest infected environments.[40] This was shown by the presence of eggs in old fecal deposits collected from the pens. These findings were supported by studies from China and Denmark where the highest prevalence of *A. suum* was also found in breeding sows and fatteners.[2,65,66] Although the prevalence in breeding boars is usually lower than in sows or fatteners,[65,67] farmers must not omit them from treatment. Infected boars could also be an important source of transmission of parasite eggs on the farm since they are often located in more traditional pens than the sows (e.g. with solid floor and/or bedding).[68] Moreover, boars are regularly used for contact stimulation of the sows, which allows them to spread the infection by defecating in areas other than their own pen.

Strong evidence exists that under indoor conditions, massive infection of piglets with *A. suum* usually occurs shortly after arrival in the highly infectious fattening units and not in the farrowing pens.[2,40,69] Several studies have shown that *A. suum* egg excretion was nearly absent in weaners despite the, sometimes substantial, presence of eggs in the farrowing pens of intensive herds.[2,3,65,66] Here, the increased hygienic standards in combination with the low humidity in the farrowing pens may significantly reduce the presence of favorable microenvironments for egg survival. On the other hand, in piglets raised under outdoor conditions, significant transmission is thought to occur soon after birth and pigs are infected before fattening.[2,14]

Anthelmintic treatment has limited to no residual activity, depending on which anthelmintic is used. As a consequence, pigs are immediately exposed to new infections and continually host migrating larvae in their bodies. This larval migration has a significant impact on pig performance. Increasing doses of infective eggs in experimentally infected pigs has shown to be correlated with decreasing productivity,[18] thereby questioning the traditional dogma that attributes the reduced productivity to the presence of adult worms in the intestine. Although the presence of adult worms in the intestine is important for the dispersal and epidemiology of the parasite, their effect on economical profitability of a herd does not seem to be as important as the migrating larval stages. This is underpinned by the fact that within an infected pig herd, the distribution of worms is highly aggregated, with the majority of adult worms being harbored in a minority of the host population.[23] Moreover, most of the pigs that do end up having patent infections only harbor a few worms[25,70] and it is hard to imagine that this would have a noticeable impact on the pig's production parameters. In contrast, we can assume that all pigs in the same pen are exposed to roughly an equal amount of infectious larvae as they forage in the same infected environment. This, again, emphasizes the importance of reducing the infection intensity on a farm.

Both the fact that infections usually seem to occur in the fattening units and the fact that anthelmintic treatment has limited persistent activity against recurring infectious larvae lead to the belief that the common practice of treating pigs prior to, or upon arrival in, the fattening units might be superfluous. In two experiments, it has been reported that a single anthelmintic treatment of fatteners at the start of the fattening period is not justified on economic grounds. It was shown to be ineffective in reducing parasite prevalence and it did not have a positive effect on performance results, nor did it influence the incidence of other lesions like white spots.[40,48,71] More importantly, Roepstorff[66] showed that, depending on the infection status, it might sometimes be more economically beneficial to switch from a routine anthelmintic treatment program to the conditional usage of anthelmintics, treating only when necessary. This should, however, be supported by routine diagnostic screening of pigs from different age categories and is probably only applicable to farms with moderate to highly intensive management.

Alternative Control Strategies

Over time, questions have been raised concerning the long-term impact of the massive application of these highly efficacious, broad-spectrum anthelmintic compounds on the environment. Due to the marginal absorbance rate of some of these drugs, high percentages of these substances are being excreted unchanged after oral or systemic administration.[72,73] Decomposing animal excrement/waste is spread all over our land for increased crop productivity, thereby also spreading the anthelmintic drugs and their residues. However, we remain oblivious to what the possible long-term ecological effects of these drugs and their residues are on pasture fauna and flora.[74] This concern, among others, is stimulating the need for alternative, more biological measures of parasite control.

In organic farming, the use of traditional synthetic drugs is not allowed and therefore farmers prefer a phytopharmaceutical approach for the control of parasitic infections on their farms.[75] Although many plants are suggested to have some nematocidal effects,[76] recent studies, e.g. by van Krimpen et al.,[75] could not show any significant reduction on the worm load of infected pigs after treatment with a herb mixture (*Thymus vulgaris*, *Melissa officinalis*, and *Echinacea purpurea*), *Papaya* fruits, *Bolbo* leaf or complete *Artemesia* plants in comparison with untreated controls. Nevertheless, continued efforts could be made to standardize the plant extracts with good anthelmintic activity and formulate the best alternative herbal preparations to replace or complement synthetic drugs, which are currently in use. In addition to phytopharmaceutical research, the possibility of using nematophageous fungi for the biological control of *A. suum* is currently being investigated. According to the study of Ferreira et al.,[77] mycelia of *Pochonia chlamydosporia*, which were added to the food of pigs, showed ovicidal activity against *A. suum* eggs after passage through the gastrointestinal tract.

Recently, it has been shown that a number of infection parameters like worm burden, total egg output, and raised antibody levels as well as natural resistance to *A. suum* infections are heritable traits representing the close genetic control of immunity against this parasite.[27,30] This, understandably, opens the door to selective breeding of pigs with increased resistance within the pig industry and could become relevant in free-range systems.[78]

Pigs build up natural resistance against recurring infections rather quickly. Significant levels of protection were detected in pigs following one or multiple inoculations with infective eggs or L3s of *A. suum*.[13,35] This provides some hope for future vaccination strategies. The development of vaccines against parasites is highly desirable. The search for a vaccine against *A. suum* has been ongoing for about half a century. Still, their development so far has been very limited and the information gathered on this topic remains scarce. An overview of the different vaccination trials published in the literature is given in Table 16.3. During the last decade, interest has shifted from native proteins towards recombinantly produced proteins. Recombinants were generally expressed in *E. coli*, though recently attempts have been made to produce larval antigens in rice plants.[79,80] Immunizing mice by feeding them transgenic rice elicited an antigen-specific antibody response, indicating that the transgenic rice was able to prime the immune response of mice. Orally immunized mice also showed a reduced amount of larvae recovered from the lungs after challenge infection in comparison with non-immunized controls.[79] In the future, this could provide an easy way of producing and administering parasite antigens. However, this technique should be further optimized and evaluated.

A vaccination protocol, targeting the necessary components of the host's immune system, could be efficient in providing protection to the host against newly acquired larvae. However, it is possible that this increased immunological response against migrating larvae invoked by the vaccination would manifest itself as increased numbers of lesions in the pig's liver. In a study by Vlaminck et al.,[36] significantly more white spots were counted on the livers of vaccinated pigs on day 14 (+86%) and day 28 (+118%) after infection compared with non-vaccinated controls. Furthermore, a vaccine against *Ascaris* will need to be highly protective. Female worms are extremely fecund and can produce hundreds of

TABLE 16.3 Overview of the vaccination experiments performed against A. *suum*

Year	Antigen	Immunization protocol	Host	Recovered stage	% Reduction	Reference
1978	UV-eggs	3 oral inoculations with 500 UV-eggs	Pigs	Adults	86%	a
1982	UV-eggs	3 oral inoculations with 10,000 UV- eggs	Pigs	Lung larvae	88%	b
1984	UV-eggs	2–5 oral inoculations with 100–10,000 UV-eggs	Pigs	Lung larvae	83–94%	c
1985	Hatching fluid L3 ES L3/4 ES UV-eggs	A number of immunization experiments using different antigens + alum w/wo oral inoculation with UV-eggs	Pigs	Lung larvae	Max 80%	37
1988	L3 larval body wall	2 oral immunizations with antigen in liposome complex, w/wo priming egg dose	Pigs	Lung larvae and L5/adults	N.S.	d
1992	Adult crude antigen	2 s.c. immunizations with the antigen in a liposome, w/wo immunomodulator	Mice	Lung and liver larvae together	89%	e
1994	L3 cuticle Ag adult cuticle Ag UV-eggs	1 i.m. immunizations with antigen + FIA and 2 i.p. immunizations with antigen + alum Or, 1 oral inoculation with 10,000 UV-eggs	Pigs	Lung larvae	49% 44% 89%	f

(Continued)

TABLE 16.3 Overview of the vaccination experiments performed against *A. suum*—cont'd

Year	Antigen	Immunization protocol	Host	Recovered stage	% Reduction	Reference
2001	As14 recombinant	3 i.n. immunizations with rAs14 coupled to CT-B	Mice	Lung larvae	64%	g
2002	As37 recombinant	3 immunizations with rAs37 in FCA	Mice	Lung larvae	N.S.	h
2001	14kDa PF fraction 42kDa PF fraction 97kDa BW fraction	6 s.c. immunizations of antigen in FIA	Pigs	Lung larvae	88% 77% 50%	i
2003	As16 recombinant	3 i.n. immunizations with rAs16 coupled to CT-B	Mice	Lung larvae	58%	j
2004	As16 recombinant	3 i.n. immunizations with rAs16 coupled to CT-B	Pigs	Lung larvae	58%	k
2005	As24 recombinant	3 s.c. immunizations with rAs24 in FCA	Mice	Lung larvae	58%	l
2008	As16 recombinant	As16 fused to CT-B expressed in rice and fed to mice	Mice	Lung larvae	Not mentioned but significant	79
2011	*A. suum* hemoglobin (AsHb)	3 i.m. immunizations with AsHb + QuilA	Pigs	L4, L5 and adults	N.S.	36
2012	Recombinant *A. suum* enolase	3 i.m. immunizations with pVAX-Enol DNA	Mice	Lung larvae	61%	m

PF= pseudocoelomic fluid, BW = adult body wall. UV-eggs = eggs attenuated by UV-radiation; N.S. = not significant; s.c. = subcutaneous; i.m. = intramuscular; i.n. = intranasal; w/wo = with or without; CT-B = Cholera Toxin subunit B; FCA = Freund's Complete Adjuvant; FIA = Freund's Incomplete Adjuvant, AsHb = *A. suum* haemoglobin.

[a] Tromba FG. *Immunization of pigs against experimental* Ascaris suum *infection by feeding ultraviolet-attenuated eggs. J Parasitol.* 1978;64(4):651–6.

[b] *Urban JF Jr, Tromba FG. Development of immune responsiveness to* Ascaris suum *antigens in pigs vaccinated with ultraviolet-attenuated eggs. Vet Immunol Immunopathol.* 1982;3(4):399–409.

[c] *Urban JF Jr, Tromba FG. An ultraviolet-attenuated egg vaccine for swine ascariasis: parameters affecting the development of protective immunity. Am J Vet Res.* 1984;45(10):2104–8.

[d] *Rhodes MB, Baker PK, Christensen DL, Anderson GA.* Ascaris suum *antigens incorporated into liposomes used to stimulate protection to migrating larvae. Vet Parasitol.* 1988;26(3–4):343–9.

[e] *Lukes S.* Ascaris suum *– vaccination of mice with liposome encapsulated antigen. Vet Parasitol.* 1992;43(1–2):105–13.

[f] *Hill DE, Fetterer RH, Romanowski RD, Urban JF, Jr. The effect of immunization of pigs with* Ascaris suum *cuticle components on the development of resistance to parenteral migration during a challenge infection. Vet Immunol Immunopathol.* 1994;42(2):161–9.

[g] *Tsuji N, Suzuki K, Kasuga-Aoki H, Matsumoto Y, Arakawa T, Ishiwata K, et al. Intranasal immunization with recombinant* Ascaris suum *14-kilodalton antigen coupled with cholera toxin B subunit induces protective immunity to* A. suum *infection in mice. Infect Immun.* 2001;69(12):7285–92.

[h] *Tsuji N, Kasuga-Aoki H, Isobe T, Arakawa T, Matsumoto Y. Cloning and characterisation of a highly immunoreactive 37 kDa antigen with multi-immunoglobulin domains from the swine roundworm* Ascaris suum. *Int J Parasitol.* 2002;32(14):1739–46.

[i] *Serrano FJ, Reina D, Frontera E, Roepstorff A, Navarrete I. Resistance against migrating* Ascaris suum *larvae in pigs immunized with infective eggs or adult worm antigens. Parasitology.* 2001;122(Pt 6):699–707.

[j] *Tsuji N, Suzuki K, Kasuga-Aoki H, Isobe T, Arakawa T, Matsumoto Y. Mice intranasally immunized with a recombinant 16-kilodalton antigen from roundworm* Ascaris *parasites are protected against larval migration of* Ascaris suum. *Infect Immun.* 2003;71(9):5314–23.

[k] *Tsuji N, Miyoshi T, Islam MK, Isobe T, Yoshihara S, Arakawa T, et al. Recombinant* Ascaris *16-kilodalton protein-induced protection against* Ascaris suum *larval migration after intranasal vaccination in pigs. J Infect Dis.* 2004;190(10):1812–20.

[l] *Islam MK, Miyoshi T, Tsuji N. Vaccination with recombinant* Ascaris suum *24-kilodalton antigen induces a Th1/Th2-mixed type immune response and confers high levels of protection against challenged* Ascaris suum *lung-stage infection in BALB/c mice. Int J Parasitol.* 2005;35(9):1023–30.

[m] *Chen N, Yuan ZG, Xu MJ, Zhou DH, Zhang XX, Zhang YZ, et al.* Ascaris suum *enolase is a potential vaccine candidate against ascariasis. Vaccine.* 2012;30(23):3478–82.

thousands of eggs per day. As a consequence, even the survival of a couple of worms will lead to an uncontrolled egg output, providing a highly contaminated environment. Moreover, the aim should be to develop a one-shot or two-shot vaccine and the price of vaccination will need to be comparable to the price of standard anthelmintic treatment in order to be profitable for the farmer. All of this, in combination with the "traditional" difficulties associated with vaccine development like producing a recombinant protein with equal effectiveness as its native counterpart, contributes to the fact that today only a few research groups are focusing on the development of new vaccines against this parasite and that the chances that a vaccine against *Ascaris* will be available soon is quite limited.

Impact of Management Practices

A routine application of anthelmintic drug seems to appeal to pig farmers for reasons of convenience.[62] However, merely resorting to anthelmintic treatment could result in little or no improvement due to continuous reinfection. The use of anthelmintics therefore needs to be complemented by higher quality farm hygiene and more effective farm management practices in order to reduce the roundworm infection pressure on the farm.

There are numerous aspects of farm management that can have a major impact on the epidemiology of *A. suum*. Undoubtedly, the type of production system that is in use will be a key factor regarding the infection intensity on a farm.[62] For obvious reasons, there is a higher diversity of parasite species and higher infection intensity on traditional farms and where pigs have access to outdoor facilities when compared to intensive indoor systems.[2,65]

In intensive indoor systems, the presence of a breeding stock on a farm seems to be associated with a lower chance of *A. suum* being present on the farm. More often than not, farmers are unaware of the parasite status of the pigs that are bought from external producers. Therefore, importing pigs from piglet producers evidently increases the risk of introducing new infections into the herd. Farmers with fattening herds are advised to buy piglets from larger piglet producers with good management and hygiene[3] and that were preferably treated prior to delivery. Whether weaners need to be treated at the beginning of fattening or not is open for discussion. However, if their worm-free status cannot be warranted, incoming pigs should be treated upon arrival, thereby reducing the risk of introducing the infection into their herd. Once *A. suum* is present on the farm, it is easily dispersed over the whole farm through moving animals and dirty materials and boots. Even flies have been shown to carry *A. suum* eggs.[81]

Other management practices like the use of the all-in-all-out system and early weaning are linked to lower prevalence of *A. suum*.[3,48,68,82] In the all-in-all-out production system, pigs are moved into and out of facilities in distinct groups with the hope that, by preventing the commingling of groups, the spread of disease is reduced. This practice also allows for a period of thorough cleaning and disinfecting between subsequent groups of animals. Early weaning of piglets (3–5 weeks) also seems to be associated with a reduced risk of *A. suum* infection.[68] However, it is likely that this can be attributed to the fact that both these parameters are associated with other factors that are more important for parasite survival, like, for instance, the use of more traditional rearing methods or poorer general pen hygiene.

Reducing contact of pigs with their own fecal deposits or from pigs from previous rounds is important. Consequentially, housing of pigs on slatted floors seems to reduce the chance of parasite infestation compared to solid or partially slatted floors.[48,82,83] Increased risk is associated with bedding being present in the pens. The use of bedding material may provide extra refugia that promote egg embryonation and hamper the effectiveness of repeated cleaning and disinfection measures. Evidently, the stocking density in the pig house will also be important,[82] as chances of pens containing pigs with patent infections will increase.

Positioning the water supply in the dung area instead of in the lying area or the feeding troughs seems to be associated with a decreased prevalence of *A. suum*.[68] Water spillage in areas where eggs are present will enhance the chances for survival of the eggs due to constant humidity. Despite the extremely rigid eggshell, eggs do not embryonate when relative humidity is low, for instance in the dry resting areas in pig pens, or when floor temperatures are lower than 15°C. Additionally, pig urine also seems to exhibit a strong inhibitory effect on the development of *A. suum* eggs.[40] Once in the environment, the eggs will embryonate until the infective larvae are present inside the egg. Only then is the egg infective for a new host. In a Danish study it was shown that *A. suum* eggs could embryonate on a pasture within 4–6 weeks during a normal Danish summer.[14] These results were supported by experiments in laboratory conditions, during which it was observed that although motile larvae were noticed from 4 weeks post-incubation, an incubation period of at least 6 weeks at room temperature was needed for eggs to become infective in mice.[84] When environmental factors like temperature, oxygen availability, and relative humidity are suboptimal, the time for larvae to become infective increases.[40] This is reflected by the fact that white spot levels seem to vary within farms depending on the season. Presumably, the rising temperatures in spring and summer allow simultaneous development of infectious stage larvae in the eggs that have spent the winter in the pig pens. This would increase the amount of infective eggs being present in the environment and subsequently cause higher liver condemnation rates in summer and early autumn.[40,65,85,86]

Frequent cleaning of the pens, preferably after each round, is indispensable to destroy or reduce the amount of eggs in the environment. Clearly, the presence of rough and uneven surfaces would provide a good microenvironment for egg development and have an impact on the efficiency of cleaning protocols. Joachim et al.[3] indicated that in older pens, the prevalence of *Ascaris* infections was significantly higher (63.0%) than in the new pens (27.9%). Moreover, Nilsson (1982)[40] found high numbers of eggs (up to 3,000 eggs per gram) in the crevices of the floor between the slatted dung area and the resting area of the pen. Approximately half of these eggs appeared to be embryonated. In a study by Beloeil et al.[69] 31% of the farms surveyed had residual dung in the fattening pens when restocking them with new fattening pigs, a practice favoring the transmission of *A. suum* eggs. Even though careful cleaning of the pens with high pressure water will remove most of the residual dung, it seems ineffective to completely remove all the infective eggs.[40] Although the use of disinfectants is effective against bacteria, most of them are rather ineffective against *A. suum* eggs.[87–89] Instead, steam cleaning[90] and drying of the pens is a highly recommended management practice for killing roundworm eggs. These are, however, hard to comply with in practice.

Looking to the future, the percentage of *Ascaris*-favorable production systems is expected to rise. Increased consumer awareness induces a shift from the industrial husbandry, with its high use of medication, feed additives, and questionable animal welfare, towards a more organic production system where pigs enjoy improved living comforts like bedding materials and outdoor runs and where prophylactic use of medication to prevent disease is not allowed. In addition, forthcoming implementation of new regulations for increased pig well-being will provide more parasite-friendly environments as well. In Denmark, for example, sprinkler systems should be placed in the pens of growing pigs and sows to accommodate a daily shower. Moreover, sows should not be tethered and should be housed in groups in enriched environments (bedding, wallowing) for part of the gestation.[78,91] As a result, the number of eggs surviving in the pens will increase and with it the prevalence of *Ascaris* and other pig parasites.[5] Good diagnostic assessments will be necessary to evaluate how these changes in farm management will affect parasite epidemiology.

CONCLUDING REMARKS

In summary, veterinarians that are assigned the task of developing an effective treatment plan for a specific facility are advised to first evaluate the production system, keeping in mind the different aspects important for the parasite dispersal before developing a personalized treatment program to reduce or control parasite transmission in that facility. The different diagnostic tools available to them should support them in this process and help with the evaluation of previously applied treatment strategies. In time, this should make it possible to attain a farm system that suffers least economical disadvantages by the presence of this parasite.

References

1. Larson G, Albarella U, Dobney K, Rowley-Conwy P, Schibler J, Tresset A, et al. Ancient DNA, pig domestication, and the spread of the Neolithic into Europe. *Proc Natl Acad Sci USA* 2007;**104**(39):15276–81.
2. Roepstorff A, Nansen P. Epidemiology and control of helminth infections in pigs under intensive and non-intensive production systems. *Vet Parasitol* 1994;**54**(1–3):69–85.
3. Joachim A, Dulmer N, Daugschies A, Roepstorff A. Occurrence of helminths in pig fattening units with different management systems in Northern Germany. *Vet Parasitol* 2001;**96**(2):135–46.
4. Eijck IA, Borgsteede FH. A survey of gastrointestinal pig parasites on free-range, organic and conventional pig farms in The Netherlands. *Vet Res Commun* 2005;**29**(5): 407–14.
5. Haugegaard J. Prevalence of nematodes in Danish industrialized sow farms with loose housed sows in dynamic groups. *Vet Parasitol* 2010;**168**(1–2):156–9.
6. Fagerholm HP, Nansen P, Roepstorff A, Frandsen F, Eriksen L. Differentiation of cuticular structures during the growth of the third-stage larva of *Ascaris suum (Nematoda, Ascaridoidea)* after emerging from the egg. *J Parasitol* 2000;**86**(3):421–7.
7. Geng J, Plenefisch J, Komuniecki PR, Komuniecki R. Secretion of a novel developmentally regulated chitinase (family 19 glycosyl hydrolase) into the perivitelline fluid of the parasitic nematode, *Ascaris suum*. *Mol Biochem Parasitol* 2002;**124**(1–2):11–21.
8. Hinck LW, Ivey MH. Proteinase activity in *Ascaris suum* eggs, hatching fluid, and excretions-secretions. *J Parasitol* 1976;**62**(5):771–4.
9. Murrell KD, Eriksen L, Nansen P, Slotved HC, Rasmussen T. *Ascaris suum*: a revision of its early migratory path and implications for human ascariasis. *J Parasitol* 1997;**83**(2): 255–60.
10. Douvres FW, Tromba FG, Malakatis GM. Morphogenesis and migration of *Ascaris suum* larvae developing to fourth stage in swine. *J Parasitol* 1969;**55**(4):689–712.
11. Roepstorff A, Eriksen L, Slotved HC, Nansen P. Experimental *Ascaris suum* infection in the pig: worm population kinetics following single inoculations with three doses of infective eggs. *Parasitology* 1997;**115**(Pt 4):443–52.
12. Eriksen L, Andersen S, Nielsen K, Pedersen A, Nielsen J. Experimental *Ascaris suum* infection in pigs. Serological response, eosinophilia in peripheral blood, occurrence of white spots in the liver and worm recovery from the intestine. *Nord Vet Med* 1980;**32**(6):233–42.
13. Helwigh AB, Nansen P. Establishment of *Ascaris suum* in the pig: development of immunity following a single primary infection. *Acta Vet Scand* 1999;**40**(2):121–32.
14. Roepstorff A, Murrell KD. Transmission dynamics of helminth parasites of pigs on continuous pasture: *Ascaris suum* and *Trichuris suis*. *Int J Parasitol* 1997;**27**(5):563–72.
15. Mejer H, Roepstorff A. *Ascaris suum* infections in pigs born and raised on contaminated paddocks. *Parasitology* 2006;**133**(Pt 3):305–12.

16. Stewart TB, Hale OM. Losses to internal parasites in swine production. *J Anim Sci* 1988;**66**(6):1548–54.
17. Knecht D, Jankowska A, Zalesny G. The impact of gastrointestinal parasites infection on slaughter efficiency in pigs. *Vet Parasitol* 2012;**184**(2–4):291–7.
18. Hale OM, Stewart TB, Marti OG. Influence of an experimental infection of *Ascaris suum* on performance of pigs. *J Anim Sci* 1985;**60**(1):220–5.
19. Kipper M, Andretta I, Monteiro SG, Lovatto PA, Lehnen CR. Meta-analysis of the effects of endoparasites on pig performance. *Vet Parasitol* 2011;**181**(2–4):316–20.
20. Bernardo TM, Dohoo IR, Donald A. Effect of ascariasis and respiratory diseases on growth rates in swine. *Can J Vet Res* 1990;**54**(2):278–84.
21. Theodoropoulos G, Theodoropoulou E, Melissaropoulou G. Worm control practices of pig farmers in Greece. *Vet Parasitol* 2001;**97**(4):285–93.
22. Dangolla A, Bjorn H, Willeberg P, Roepstorff A, Nansen P. A questionnaire investigation on factors of importance for the development of anthelmintic resistance of nematodes in sow herds in Denmark. *Vet Parasitol* 1996;**63**(3–4):257–71.
23. Eriksen L, Nansen P, Roepstorff A, Lind P, Nilsson O. Response to repeated inoculations with *Ascaris suum* eggs in pigs during the fattening period. I. Studies on worm population kinetics. *Parasitol Res* 1992;**78**(3):241–6.
24. Andersen S, Jorgensen RJ, Nansen P, Nielsen K. Experimental *Ascaris suum* infection in piglets. Inverse relationship between the numbers of inoculated eggs and the numbers of worms established in the intestine. *Acta Pathol Micr Sc* 1973;**81**(6):650–6.
25. Wagner B, Polley L. *Ascaris suum* prevalence and intensity: an abattoir survey of market hogs in Saskatchewan. *Vet Parasitol* 1997;**73**(3–4):309–13.
26. Boes J, Medley GF, Eriksen L, Roepstorff A, Nansen P. Distribution of *Ascaris suum* in experimentally and naturally infected pigs and comparison with *Ascaris lumbricoides* infections in humans. *Parasitology* 1998;**117**(6):589–96.
27. Skallerup P, Nejsum P, Jorgensen CB, Goring HH, Karlskov-Mortensen P, Archibald AL, et al. Detection of a quantitative trait locus associated with resistance to *Ascaris suum* infection in pigs. *Int J Parasitol* 2012;**42**(4):383–91.
28. Thienpont D, Rochette F, Vanparijs OFJ. *Diagnosing Helminthiasis through Coprological Examination. Beerse, Belgium: Janssen Research Foundation*; 1979.
29. Bernardo TM, Dohoo IR, Donald A, Ogilvie T, Cawthorn R. Ascariasis, respiratory diseases and production indices in selected Prince Edward Island swine herds. *Can J Vet Res* 1990;**54**(2):267–73.
30. Nejsum P, Roepstorff A, Jorgensen CB, Fredholm M, Goring HH, Anderson TJ, et al. High heritability for *Ascaris* and *Trichuris* infection levels in pigs. *Heredity* 2009;**102**(4):357–64.
31. Boes J, Nansen P, Stephenson LS. False-positive *Ascaris suum* egg counts in pigs. *Int J Parasitol* 1997;**27**(7):833–8.
32. Nakagawa M, Yoshihara S, Suda H, Ikeda K. Pathological studies on white spots of the liver in fattening pigs. *Natl Inst Anim Health Q (Tokyo)* 1983;**23**(4):138–49.
33. Perez J, Garcia PM, Mozos E, Bautista MJ, Carrasco L. Immunohistochemical characterization of hepatic lesions associated with migrating larvae of *Ascaris suum* in pigs. *J Comp Pathol* 2001;**124**(2–3):200–6.
34. Copeman DB, Gaafar SM. Sequential development of hepatic lesions of ascaridosis in colostrum-deprived pigs. *Aust Vet J* 1972;**48**(5):263–8.
35. Eriksen L. Experimentally induced resistance to *Ascaris suum* in pigs. *Nord Vet Med* 1982;**34**(6):177–87.
36. Vlaminck J, Martinez-Valladares M, Dewilde S, Moens L, Tilleman K, Deforce D, et al. Immunizing pigs with *Ascaris suum* haemoglobin increases the inflammatory response in the liver but fails to induce a protective immunity. *Parasite Immunol* 2011;**33**(4): 250–4.

37. Urban Jr JF, Romanowski RD. *Ascaris suum*: protective immunity in pigs immunized with products from eggs and larvae. *Exp Parasitol* 1985;**60**(2):245–54.
38. Lind P, Eriksen L, Nansen P, Nilsson O, Roepstorff A. Response to repeated inoculations with *Ascaris suum* eggs in pigs during the fattening period. II. Specific IgA, IgG, and IgM antibodies determined by enzyme-linked immunosorbent assay. *Parasitol Res* 1993;**79**(3):240–4.
39. Frontera E, Carron A, Serrano FJ, Roepstorff A, Reina D, Navarrete I. Specific systemic IgG1, IgG2 and IgM responses in pigs immunized with infective eggs or selected antigens of *Ascaris suum*. *Parasitology* 2003;**127**:291–8.
40. Nilsson O. Ascariasis in the pig. An epizootiological and clinical study. *Acta Vet Scand Sup* 1982;**79**:1–108.
41. Bernardo TM, Dohoo IR, Ogilvie T. A critical assessment of abattoir surveillance as a screening test for swine ascariasis. *Can J Vet Res* 1990;**54**(2):274–7.
42. Urban Jr JF, Alizadeh H, Romanowski RD. *Ascaris suum*: development of intestinal immunity to infective second-stage larvae in swine. *Exp Parasitol* 1988;**66**(1):66–77.
43. Frontera E, Serrano F, Reina D, Alcaide M, Sanchez-Lopez J, Navarrete I. Serological responses to *Ascaris suum* adult worm antigens in Iberian finisher pigs. *J Helminthol* 2003;**77**(2):167–72.
44. Yoshihara S, Oya T, Furuya T, Goto N. Use of body fluid of adult female *Ascaris suum* as an antigen in the enzyme-linked immunosorbent assay (ELISA) for diagnosis of swine ascariosis. *J Helminthol* 1993;**67**(4):279–86.
45. Roepstorff A. Natural *Ascaris suum* infections in swine diagnosed by coprological and serological (ELISA) methods. *Parasitol Res* 1998;**84**(7):537–43.
46. Bogh HO, Eriksen L, Lawson LG, Lind P. Evaluation of an enzyme-linked-immunosorbent-assay and a histamine-release test system for the detection of pigs naturally infected with *Ascaris suum*. *Prev Vet Med* 1994;**21**(3):201–14.
47. Vlaminck J, Nejsum P, Vangroenweghe F, Thamsborg SM, Vercruysse J, Geldhof P. Evaluation of a serodiagnostic test using *Ascaris suum* haemoglobin for the detection of roundworm infections in pig populations. *Vet Parasitol* 2012;**189**(2–4):267–73.
48. Roepstorff A, Jorsal SE. Relationship of the prevalence of swine helminths to management practices and anthelmintic treatment in Danish sow herds. *Vet Parasitol* 1990;**36**(3–4):245–57.
49. Rochette F. Chemotherapy of gastrointestinal nematodiasis in pigs. In: Vanden Bossche H, Thienpont D, Janssens PG, editors. *Chemotherapy of Gastrointestinal Helminths*. Springer-Verlag; 1985. p. 463–86.
50. Stewart TB, Bidner TD, Southern LL, Simmons LA. Efficacy of fenbendazole against migrating *Ascaris suum* larvae in pigs. *Am J Vet Res* 1984;**45**(5):984–6.
51. Thienpont D, Vanparijs O, Niemegeers C, Marsboom R. Biological and pharmacological properties of flubendazole. *Arznei-Forschung* 1978;**28**(4):605–12.
52. Kennedy T, Lucas MJ, Froe DL. Comparative efficacy of pyrantel tartrate, ivermectin and fenbendazole against experimentally induced immature *Ascaris suum* in pigs. *Agri-Practice* 1987;**8**(2):19–21.
53. Oakley GA. Activity of levamisole hydrochloride administered subcutaneously against *A. suum* infections in pigs. *Vet Rec* 1974;**95**(9):190–3.
54. Lichtensteiger CA, DiPietro JA, Paul AJ, Neumann EJ, Thompson L. Persistent activity of doramectin and ivermectin against *Ascaris suum* in experimentally infected pigs. *Vet Parasitol* 1999;**82**(3):235–41.
55. Lee RP, Dooge DJ, Preston JM. Efficacy of ivermectin against *Sarcoptes scabiei* in pigs. *Vet Rec* 1980;**107**(22):503–5.
56. Yazwinski TA, Tucker C, Featherston H, Johnson Z, Wood-Huels N. Endectocidal efficacies of doramectin in naturally parasitized pigs. *Vet Parasitol* 1997;**70**(1–3):123–8.

57. Stewart TB, Marti OG, Hale OM. Efficacy of ivermectin against 5 genera of swine nematodes and the hog louse, *Haematopinus suis*. *Am J Vet Res* 1981;**42**(8):1425–6.
58. Logan NB, Weatherley AJ, Jones RM. Activity of doramectin against nematode and arthropod parasites of swine. *Vet Parasitol* 1996;**66**(1–2):87–94.
59. Boersema JH, Eysker M, Nas JWM. Apparent resistance of *Parascaris equorum* to macrocylic lactones. *Vet Rec* 2002;**150**(9):279–81.
60. Nareaho A, Vainio K, Oksanen A. Impaired efficacy of ivermectin against *Parascaris equorum*, and both ivermectin and pyrantel against strongyle infections in trotter foals in Finland. *Vet Parasitol* 2011;**182**(2–4):372–7.
61. Laugier C, Sevin C, Menard S, Maillard K. Prevalence of *Parascaris equorum* infection in foals on French stud farms and first report of ivermectin-resistant *P. equorum* populations in France. *Vet Parasitol* 2012;**188**(1–2):185–9.
62. Nansen P, Roepstorff A. Parasitic helminths of the pig: factors influencing transmission and infection levels. *Int J Parasitol* 1999;**29**(6):877–91.
63. Jourquin J. Strategic de-worming to boost performance. *Int Pig Topics* 2007;**22**(6):7–9.
64. Bakker J. Long-term effects of a deworming program using flubendazole and levamisole on the percentage of condemned livers in slaughtering pigs. *Tijdschr Diergeneesk* 1984;**109**(20):815–9.
65. Lai M, Zhou RQ, Huang HC, Hu SJ. Prevalence and risk factors associated with intestinal parasites in pigs in Chongqing, China. *Res Vet Sci* 2011;**91**(3):121–4.
66. Roepstorff A. Helminth surveillance as a prerequisite for anthelmintic treatment in intensive sow herds. *Vet Parasitol* 1997;**73**(1–2):139–51.
67. Roepstorff A, Nilsson O, Oksanen A, Gjerde B, Richter SH, Ortenberg E, et al. Intestinal parasites in swine in the Nordic countries: prevalence and geographical distribution. *Vet Parasitol* 1998;**76**(4):305–19.
68. Roepstorff A, Nilsson O, O'Callaghan CJ, Oksanen A, Gjerde B, Richter SH, et al. Intestinal parasites in swine in the Nordic countries: multilevel modelling of *Ascaris suum* infections in relation to production factors. *Parasitology* 1999;**119**(Pt 5): 521–34.
69. Beloeil PA, Chauvin C, Fablet C, Jolly JP, Eveno E, Madec F, et al. Helminth control practices and infections in growing pigs in France. *Livest Prod Sci* 2003;**81**(1):99–104.
70. Nejsum P, Thamsborg SM, Petersen HH, Kringel H, Fredholm M, Roepstorff A. Population dynamics of *Ascaris suum* in trickle-infected pigs. *J Parasitol* 2009;**95**(5):1048–53.
71. Boes J, Kanora A, Havn KT, Christiansen S, Vestergaard-Nielsen K, Jacobs J, et al. Effect of *Ascaris suum* infection on performance of fattening pigs. *Vet Parasitol* 2010;**172**(3–4): 269–76.
72. Plumb DC. *Plumb's Veterinary Drug Handbook.* 6th ed. Blackwell Publishing; 2008.
73. Farkas R, Gyurcso A, Borzsonyi L. Fly larvicidal activity in the faeces of cattle and pigs treated with endectocide products. *Med Vet Entomol* 2003;**17**(3):301–6.
74. Spratt DM. Endoparasite control strategies: implications for biodiversity of native fauna. *Int J Parasitol* 1997;**27**(2):173–80.
75. van Krimpen MM, Binnendijk GP, Borgsteede FH, Gaasenbeek CP. Anthelmintic effects of phytogenic feed additives in *Ascaris suum* inoculated pigs. *Vet Parasitol* 2010; **168**(3–4):269–77.
76. Chitwood DJ. Phytochemical based strategies for nematode control. *Annu Rev Phytopathol* 2002;**40**:221–49.
77. Ferreira SR, de Araujo JV, Braga FR, Araujo JM, Frassy LN, Ferreira AS. Biological control of *Ascaris suum* eggs by *Pochonia chlamydosporia* fungus. *Vet Res Commun* 2011; **35**(8):553–8.
78. Thamsborg SM, Roepstorff A, Nejsum P, Mejer H. Alternative approaches to control of parasites in livestock: Nordic and Baltic perspectives. *Acta Vet Scand* 2010;**52**(Suppl. 1):S27.

79. Matsumoto Y, Suzuki S, Nozoye T, Yamakawa T, Takashima Y, Arakawa T, et al. Oral immunogenicity and protective efficacy in mice of transgenic rice plants producing a vaccine candidate antigen (As16) of *Ascaris suum* fused with cholera toxin B subunit. *Transgenic Res* 2009;**18**(2):185–92.
80. Nozoye T, Takaiwa F, Tsuji N, Yamakawa T, Arakawa T, Hayashi Y, et al. Production of *Ascaris suum* As14 protein and its fusion protein with cholera toxin B subunit in rice seeds. *J Vet Med Sci* 2009;**71**(7):995–1000.
81. Forster M, Klimpel S, Sievert K. The house fly (*Musca domestica*) as a potential vector of metazoan parasites caught in a pig-pen in Germany. *Vet Parasitol* 2009;**160**(1–2):163–7.
82. Tielen MJ, Truijen WT. v d Groes CA, Verstegen MA, de Bruin JJ, Conbey RA. Conditions of management and the construction of piggeries on pig-fattening farms as factors in the incidence of diseases of the lung and liver in slaughtered pigs. *Tijdschr Diergeneesk* 1978;**103**(21):1155–65.
83. Sanchez-Vazquez MJ, Smith RP, Kang S, Lewis F, Nielen M, Gunn GJ, et al. Identification of factors influencing the occurrence of milk spot livers in slaughtered pigs: a novel approach to understanding *Ascaris suum* epidemiology in British farmed pigs. *Vet Parasitol* 2010;**173**(3–4):271–9.
84. Geenen PL, Bresciani J, Boes J, Pedersen A, Eriksen L, Fagerholm HP, et al. The morphogenesis of *Ascaris suum* to the infective third-stage larvae within the egg. *J Parasitol* 1999;**85**(4):616–22.
85. Menzies FD, Goodall EA, Taylor SM. The epidemiology of *Ascaris suum* infections in pigs in Northern Ireland, 1969–1991. *Br Vet J* 1994;**150**(2):165–72.
86. Sanchez-Vazquez MJ, Nielen M, Gunn GJ, Lewis FI. National monitoring of *Ascaris suum* related liver pathologies in English abattoirs: a time-series analysis, 2005–2010. *Vet Parasitol* 2012;**184**(1):83–7.
87. van den Burg WP, Borgsteede FH. Effects of various disinfectants on the development and survival possibilities of the pre-parasitic stages of *Ostertagia ostertagi, Cooperia oncophora* and *Ascaris suum*. *Tijdschr Diergeneesk* 1987;**112**(13):769–78.
88. Plachy P, Juris P, Placha I, Venglovsky J. Use of hydrated lime for disinfection of sewage sludge containing *Salmonella typhimurium* and *Ascaris suum* as model pathogens. *Vet Med (Praha)* 1996;**41**(8):255–9.
89. Massara CL, Ferreira RS, de Andrade LD, Guerra HL, Carvalho Odos S. Effects of detergents and disinfectants on the development of *Ascaris lumbricoides* eggs. *Cad Saude Publica* 2003;**19**(1):335–40.
90. Haas A, Platz S, Eichhorn W, Kaaden OR, Unshelm J. Effect of steam application based on microbiological and parasitologic test procedures. *Zentralbl Hyg Umweltmed* 1998;**201**(4–5):337–47.
91. Roepstorff A, Mejer H, Nejsum P, Thamsborg SM. Helminth parasites in pigs: new challenges in pig production and current research highlights. *Vet Parasitol* 2011; **180**(1–2):72–81.

Index

Note: Page numbers with "f" denote figures; "t" tables; "b" boxes.

D

E

F

G

L

M

N

O

P

Printed and bound by CPI Group (UK) Ltd, Croydon, CR0 4YY
08/05/2025
01864979-0001